Hazardous Materials Technician

Second Edition

Libby Snyder and Leslie Miller
Editors and Project Managers

Lindsey Dugan and David Schaap
Editors and Lead Instructional Developers

Validated by the International Fire Service Training Association
Published by Fire Protection Publications • Oklahoma State University

RECYCLABLE

The International Fire Service Training Association (IFSTA) was established in 1934 as a *nonprofit educational association of fire fighting personnel who are dedicated to upgrading fire fighting techniques and safety through training.* To carry out the mission of IFSTA, Fire Protection Publications was established as an entity of Oklahoma State University. Fire Protection Publications' primary function is to publish and distribute training materials as proposed, developed, and validated by IFSTA. As a secondary function, Fire Protection Publications researches, acquires, produces, and markets high-quality learning and teaching aids consistent with IFSTA's mission.

IFSTA holds two meetings each year: the Winter Meeting in January and the Annual Validation Conference in July. During these meetings, committees of technical experts review draft materials and ensure that the professional qualifications of the National Fire Protection Association standards are met. These conferences bring together individuals from several related and allied fields, such as:

- Key fire department executives, training officers, and personnel
- Educators from colleges and universities
- Representatives from governmental agencies
- Delegates of firefighter associations and industrial organizations

Committee members are not paid nor are they reimbursed for their expenses by IFSTA or Fire Protection Publications. They participate because of a commitment to the fire service and its future through training. Being on a committee is prestigious in the fire service community, and committee members are acknowledged leaders in their fields. This unique feature provides a close relationship between IFSTA and the fire service community.

IFSTA manuals have been adopted as the official teaching texts of many states and provinces of North America as well as numerous U.S. and Canadian government agencies. Besides the NFPA requirements, IFSTA manuals are also written to meet the Fire and Emergency Services Higher Education (FESHE) course requirements. A number of the manuals have been translated into other languages to provide training for fire and emergency service personnel in Canada, Mexico, and outside of North America.

Copyright © 2017 by the Board of Regents, Oklahoma State University

All rights reserved. No part of this publication may be reproduced in any form without prior written permission from the publisher.

ISBN 978-0-87939-626-8   Library of Congress Control Number: 2017940850

Fifth Edition, First Printing, June 2017   *Printed in the United States of America*

10   9   8   7   6   5   4   3   2   1

If you need additional information concerning the International Fire Service Training Association (IFSTA) or Fire Protection Publications, contact:

Customer Service, Fire Protection Publications, Oklahoma State University
930 North Willis, Stillwater, OK 74078-8045
800-654-4055   Fax: 405-744-8204

For assistance with training materials, to recommend material for inclusion in an IFSTA manual, or to ask questions or comment on manual content, contact:
Editorial Department, Fire Protection Publications, Oklahoma State University
930 North Willis, Stillwater, OK 74078-8045
405-744-4111   Fax: 405-744-4112   E-mail: editors@osufpp.org

Oklahoma State University in compliance with Title VI of the Civil Rights Act of 1964 and Title IX of the Educational Amendments of 1972 (Higher Education Act) does not discriminate on the basis of race, color, national origin or sex in any of its policies, practices or procedures. This provision includes but is not limited to admissions, employment, financial aid and educational services.

Chapter Summary

Chapters

1. Introduction to the Hazmat Technician 8
2. Analyzing the Incident: Understanding how Matter Behaves 26
3. Analyzing the Incident: Understanding Atomic Structure and the Chemistry of Hazardous Materials 46
4. Analyzing the Incident: Understanding Common Families and Special Hazards of Hazardous Materials/WMDs 80
5. Analyzing the Incident: Detection, Monitoring, and Sampling Procedures 132
6. Analyzing the Incident: Detecting, Monitoring, and Sampling Hazardous Materials .. 158
7. Analyzing the Incident: Collecting and Interpreting Hazard and Response Information 214
8. Analyzing the Incident: Assessing Container Condition, Predicting Behavior, and Estimating Outcomes 246
9. Planning the Response: Developing Response Objectives 320
10. Implementing and Evaluating the Action Plan: Incident Management 350
11. Implementing the Action Plan: Personal Protective Equipment 372
12. Implementing the Action Plan: Decontamination 436
13. Implementing the Action Plan: Product Control 478
14. Implementing the Action Plan: Incident Demobilization and Termination 546

Appendices

A. Chapter and Page Correlation to NFPA 1072 Requirements 560
B. Protective Clothing Materials and Manufacturing Processes 561

Glossary 563
Index 581

Table of Contents

Acknowledgements

Introduction 1
Purpose and Scope 1
Book Organization 1
Terminology 2
Key Information 3
Metric Conversions 4

1 Introduction to the Hazmat Technician 8
APIE Process 11
Review of Awareness and Operations Level Qualifications 12
    Awareness Level Personnel 12
        Analyzing the Incident 12
        Planning the Response 13
        Implementing the Response 13
        Evaluating Progress 14
    Operations Level Responders 14
        Analyzing the Incident 14
        Planning the Response 15
        Implementing the Response 15
        Evaluating Progress 15
    Operations Mission Specific Level 15
Hazmat Technician 16
    Technician Training 17
        U.S. OSHA Standards 17
        NFPA Standards 17
        Training Sources 18
    Analyzing the Incident 18
        Collecting and Interpreting Hazard and Response Information 19
        Detecting, Monitoring, and Sampling Hazardous Materials 19
        Estimating Container Damage 19
        Predicting Behavior 20
    Planning the Initial Response 20
        Developing Response Objectives and Options ..21
        Selecting Appropriate Personal Protective Equipment 21
        Selecting Decontamination Procedures 21
        Developing Action Plans 22
    Implementing the Planned Response 22
        Performing Assigned ICS/IMS Duties 22
        Using Personal Protective Equipment 23
        Performing Control Functions 23
        Performing Decontamination 23
        Evaluating Progress 23
        Terminating the Incident 24
Chapter Notes 25
Chapter Review 25

2 Analyzing the Incident: Understanding How Matter Behaves 26
Physical States of Matter 29
    Solids 31
    Liquids 31
    Gases 31
        Compressed Gases 33
        Liquefied Gases 33
        Cryogenic Liquids 33
Physical Properties of a Material 34
    Density 35
    Specific Gravity 35
    Vapor Density 36
        Viscosity 37
    Odor 37
    Appearance 38
Temperature and Pressure 38
    Temperature 38
        Flash Points and Fire Points 38
        Flammable Range 38
    Pressure 40
        Vapor Pressure 40
        Vapor Expansion 41
    Molecular Weight and Polarity 41
Phase Changes and Related Properties 41
    Temperature Changes 42
        Melting and Freezing 42
        Boiling and Condensation 42
        Sublimation 43
        Critical Points 44
    Relationships of Physical Properties 44
Chapter Review 45

3 Analyzing the Incident: Understanding Atomic Structure and the Chemistry of Hazardous Materials 46
Atomic Theory 49
    Atoms 49
    Elements 50
        Metals 52
        Nonmetals 52
    Compounds 53

　　　　Salts..53
　　　　Nonsalts...54
　　Mixtures...55
　　　　Solutions..55
　　　　Slurries...56
　　　　Alloys..56
The Periodic Table of Elements..............................56
　　Atomic Number..58
　　Material Types..59
Four Significant Family Groups............................60
　　Group I – The Alkali Metals.................................61
　　Group II – The Alkaline Earths............................61
　　Group VII – The Halogens...................................62
　　Group VIII – The Noble Gases............................64
Bonding...65
　　Stability of Elements and Compounds...............65
　　Diatomic Molecules (Compounds)......................66
　　Covalent Bonds..66
　　Ionic Bonds..67
　　Resonant Bonds..68
　　Bond Energy..70
　　Bonding in Hazmat Response.............................70
Reactions...70
　　Exothermic and Endothermic Reactions............72
　　Oxidation and Combustion..................................72
　　Polymerization..74
　　　　Catalysts..74
　　　　Inhibitors..75
　　Decomposition..75
　　Synergistic Reactions..75
　　The Fundamentals of a Reaction........................75
Mixing Materials...77
　　Concentration...77
　　Solubility..78
　　Miscibility..79
　　Polarity...79
Chapter Review...79

4　Analyzing the Incident: Understanding Common Families and Special Hazards of Hazardous Materials/WMDs..............80

Common Families of Hazardous Materials...........83
　　Inorganic Compounds...83
　　Organic Compounds..83
　　　　Hydrocarbons...84
　　　　Hydrocarbon Derivatives..............................86
　　Oxidizing Agents..87
　　　　Inorganic Peroxides.......................................87
　　　　Organic Peroxides..87
　　　　Chlorates and Perchlorates...........................88
　　Reactive Materials..88
　　　　Air-Reactive..88
　　　　Water-Reactive...89
　　　　Shock- and Friction-Sensitive......................92
　　　　Light-Sensitive...93
　　　　Temperature-Sensitive..................................93
　　Corrosives..94
　　Radiation...95
　　　　Ionizing and Nonionizing Radiation...........96
　　　　Radioactive Decay...98
　　　　Half-Life..98
　　　　Activity..99
Special Hazards of Chemicals and Weapons of Mass Destruction..99
　　Chemical Warfare Agents..................................100
　　　　Nerve Agents..100
　　　　Blister Agents/Vesicants.............................103
　　　　Blood Agents..104
　　　　Choking Agents...104
　　　　Riot Control Agents/Irritants....................105
　　Biological Agents and Toxins............................106
　　　　Dose-Response Relationships...................108
　　　　Measuring and Expressing Dose and
　　　　　　Concentration..108
　　　　Toxicity..109
　　　　Pesticides and Agricultural Chemicals.....110
　　Radioactive and Nuclear...................................111
　　　　Radioactive Material Exposure and
　　　　　　Contamination......................................111
　　　　Ionizing Radiation for Technician Response..112
　　　　Radiation Health Hazards.........................114
　　　　Protection from Radiation........................114
　　Explosives and Incendiaries.............................116
　　　　Understanding the Danger.......................116
　　　　Danger at Any Division.............................117
　　　　Types of Explosives....................................117
　　Homemade Explosives......................................118
　　　　Peroxide-Based Explosives........................120
　　　　Chlorate-Based Explosives........................121
　　　　Nitrate-Based Explosives...........................121
　　Improvised Explosive Devices.........................121
　　　　Identification of IEDs................................122
　　　　IED Types Categorized by Containers....123
　　　　Mail, Package, or Letter Bombs................125
　　　　Person-Borne Improvised Devices (PBIED)...126
　　　　Vehicle-Borne Improvised Explosive Devices
　　　　　　(VBIEDs)..128
　　　　Response to Explosive/IED Events..........129
Chapter Notes...130
Chapter Review...130

5 Analyzing the Incident: Detection, Monitoring, and Sampling Procedures 132

Detection and Monitoring Operations 135
- Fundamentals of Monitoring 135
  - *Technology* *136*
  - *Instrument Response Time* *138*
- Calibration 139
  - *Bump Test* *140*
  - *Relative Response/Correction Factors* *140*
  - *Cross-Sensitivities and Interference* *142*
- Action Levels 143
- Individual Monitoring and Area Monitoring ... 144
  - *Individual Monitoring* *144*
  - *Area Monitoring* *145*
- Confined Spaces 146
- Cargo Containers 146

Sampling Techniques 146
- Sample Collection for Hazard Identification ... 147
- Field Screening Samples 147
  - *Sampling Plans* *147*
  - *Chain of Custody for Sampling* *150*

Evidence Collection Techniques 150
- Evidence Collection 151
- Chain of Custody for Evidence 152
- Evidence Collection and Preservation 152
- Decontaminating Samples and Evidence 152

Maintenance of Monitoring, Detection, and Sampling Equipment 152

Chapter Review 153

Skill Sheets 154

6 Analyzing the Incident: Detecting, Monitoring, and Sampling Hazardous Materials 158

Exposure 161
- Routes of Entry 161
- Contamination versus Exposure 162
- Acute versus Chronic Exposure 163

Radiological and Biological Exposures 164
- Exposure Limits 164
  - *Immediately Dangerous to Life and Health (IDLH)* *164*
  - *Other Exposure Limits* *164*
- Radiological Exposures 168
- Biological Exposures 168

Sensor-Based Instruments and Other Devices .. 170
- Oxygen Indicators 170
- Combustible Gas Indicators (CGIs) 172
- Conversion Factors 173
- Electrochemical Cells 175
- Metal Oxide Sensors 175
- pH Meters 176
- Multi Sensor Instruments 176
- Colorimetric Methods 176
  - *Colorimetric Tubes/Chips* *177*
  - *pH Paper/Strips* *179*
  - *Reagent Papers* *181*
  - *M8, M9, and M256 Chemical Agent Detection Papers* *182*
- Radiation Detection Instruments 183
  - *Geiger-Mueller* *183*
  - *Scintillation Detectors* *184*
  - *Dosimeters and Badges* *184*
- Photoionization Detectors (PIDs) 185
- Biological Immunoassay Indicators 187
- Thermal Imagers 187
- Infrared Thermometers 188

Other Detection Devices 189
- Halogenated Hydrocarbon Meters 189
- Flame Ionization Detectors 190
- Gas Chromatography 190
- Mass Spectroscopy 191
- Ion Mobility Spectrometry 192
- Surface Acoustic Wave 192
- Gamma-Ray Spectrometer 193
- Fourier Transform IR 194
- Infrared Spectrophotometers 194
- Raman Spectroscopy IR 195
- Mercury Detection 197
- Wet Chemistry 198
- DNA Fluoroscopy 199
- Polymerase Chain Reaction Devices 199

Chapter Notes 200

Chapter Review 200

Skill Sheets 201

7 Analyzing the Incident: Collecting and Interpreting Hazard and Response Information 214

Introduction to the Technical Research Job Function 217
- Product Information and Behavior Prediction ... 217
- Data Interpretation for PPE Selection 218
- Technical Information and Research Documentation 220

Research 221
- Hazardous Materials Profiling 221
- Chemical Assessment 224

Written Technical References 225
- Reference Manuals 225
  - *Emergency Response Guidebook (ERG)* *226*

Emergency Handling of Hazardous Materials in Surface Transportation 227
Chemical Hazards Response Information System (CHRIS) 227
NIOSH Pocket Guide to Chemical Hazards (NPG) 228
NFPA Fire Protection Guide to Hazardous Materials 228
Hawley's Condensed Chemical Dictionary (CCD) 229
Jane's CBRN Response Handbook 229
Merck Index 230
Sax's Dangerous Properties of Industrial Materials 230
Symbol Seeker: Hazard Identification Manual 231
The Threshold Limit Values and Biological Exposure Indices (TLVs®& BEIs®) 231
Cryogenic Heat Transfer 231
Safety Data Sheet (SDS) 231
Shipping Papers 232
Facility Documents 233
Electronic Technical Resources **234**
Computer-Aided Management of Emergency Operations (CAMEO).................. 234
Wireless Information System for Emergency Responders (WISER) 236
Palmtop Emergency Action for Chemicals (PEAC®-WMD) 237
E-Plan for First Responders............................. 237
Chemical Hazards Emergency Medical Management (CHEMM) 237
Mobile Applications (Apps) 237
National Pipeline Mapping System (NPMS) 238
Weather Data 239
Plume Modeling 239
Other Internet Resources 240
Technical Information Centers and Specialists **240**
Emergency Response Centers 240
Poison Control Centers 242
Chlorine Emergency Plan (CHLOREP) 242
Interagency Radiological Assistance Plan (IRAP) 243
U.S. WMD-Civil Support Teams 243
EPA Emergency Response Team (ERT) 243
Explosive Ordnance Disposal (EOD) 243
FBI WMD Coordinator 244
Other Federal Resources 244
Chapter Notes **244**
Chapter Review **244**
Skill Sheets **245**

8 Analyzing the Incident: Assessing Container Condition, Predicting Behavior, and Estimating Outcomes **246**
Non-Bulk Containers **249**
Bags 250
Carboys and Bottles 251
Drums and Pails 252
Cylinders 253
Basic Identification 253
Construction Features 254
Cryogenic Cylinders 254
Y Cylinders 256
Other Common Small Containers 256
Intermediate Bulk Containers (IBCs) **256**
Ton Containers (Pressure Drums) **258**
Highway Cargo Containers **260**
Tank Markings 261
Specification Plates 261
Nonpressure Cargo Tanks 263
Basic Identification 265
Construction Features 266
Low-Pressure Cargo Tanks 267
Basic Identification 267
Construction Features 270
Corrosive Liquid Tanks 270
Basic Identification 270
Construction Features 273
High-Pressure Cargo Tanks 273
Basic Identification 274
Construction Features 274
Cryogenic Tanks 276
Basic Identification 277
Construction Features 277
Tube Trailers 279
Dry Bulk Carriers 281
Railway Tank Cars **282**
Tank Car Markings 282
Tank Car Structure 285
Safety Features of Railway Tank Cars 287
Head Shields 287
Insulation 288
Thermal Protection 288
Heating Coils 288
Top and Bottom Shelf Couplers 288
Skid Protection 289
Tank Car Fittings 289
Manways 289
Valves and Venting Devices 289
Safety Relief Devices 290
Other Fittings 290

vii

General Service (Nonpressure) Railway Tank Cars 291
   Basic Identification *292*
   Construction Features *292*
Pressure Railway Tank Cars 293
   Basic Identification *294*
   Construction Features *294*
Cryogenic Railway Tank Cars 295
   Basic Identification *295*
   Construction Features *295*
Specialized Tank Cars 296
   Pneumatically Unloaded Hopper Cars *297*
   Refrigerated Cars *297*
Intermodal Containers 297
   Intermodal Tanks 299
   Intermodal Tanks for Liquids and Solid Hazardous Materials 300
   Intermodal Tanks for Non-Refrigerated Liquefied Compressed Gases 300
   Intermodal Tanks for Refrigerated Liquefied Gases 301
Marine Tank Vessels 302
   Tankers 302
      Petroleum Carriers *302*
      Chemical Carriers *303*
      Liquefied Flammable Gas Carriers *303*
   Cargo Vessels 304
   Barges 305
Air Freight Cargo 305
Pipelines 306
   Principles of Pipeline Operation 306
   Basic Identification 306
   Construction Features 307
Fixed Facility Containers 309
   Nonpressure Tanks 309
   Pressure Tanks 312
   Cryogenic Tanks 313
Other Storage Facility Considerations 313
   Laboratories 313
   Batch Plants 314
   Non-Regulated and Illicit Container Use ... 314
Radioactive Materials Packaging 314
   Excepted 314
   Industrial 314
   Type A Packaging 315
   Type B Packaging 315
   Type C Packaging 316
   Descriptions and Types of Radioactive Labels 317
Chapter Notes 317
Chapter Review 318

Skill Sheets 319

9 Planning the Response: Developing Response Objectives 320

Risk-Based Response (RBR) 323
Risk versus Harm 326
Response Models 326
   APIE 326
   GEDAPER 327
   Eight Step Process© 327
   HazMatIQ© 327
   D.E.C.I.D.E. 327
   Emergency Response Guidebook 328
Developing an Incident Action Plan 329
   Preincident Surveys 330
   Incident Priorities 330
   Response Modes 331
      Nonintervention Operations *332*
      Defensive Operations *333*
      Offensive Operations *334*
   Strategic Goals and Tactical Objectives 335
      Isolation *335*
      Identification *336*
      Notification *336*
      Protection *337*
      Rescue *338*
      Fire Control *338*
      Confinement *338*
      Containment *339*
      Recovery and Termination *339*
   PPE Selection and Decontamination Selection 341
Developing a Site Safety Plan 341
   Medical Surveillance Plan 343
   Pre-Entry Evaluation 344
   Post-Entry Evaluation 344
   Backup and Rapid Intervention 345
   Emergency Procedures 345
Chapter Notes 345
Chapter Review 346
Skill Sheets 347

10 Implementing and Evaluating the Action Plan: Incident Management 350

Incident Management Systems 353
   Incident Command 353
      General Responsibilities *354*
      Specific Responsibilities *355*
   Unified Command 355
Section Leaders and Supervisors in the Hazmat Branch 356

Hazmat Branch Director/Group Supervisor.... 356	
Entry Team Leader.. 357	
Decontamination Leader 358	
Site Access Control Leader 358	
Hazmat Safety Officer 358	
Hazmat Logistics Officer 359	
Safe Refuge Area Manager 359	
Medical Officer 360	
Information and Research Officer 360	

Hazardous Materials Incident Levels 361
    Level I 361
    Level II 361
    Level III 362
    NIMS Incident Typing System 362
    Evaluating Effectiveness 363

Forms and Logs 364
    Activity Logs 364
    Exposure Records 365
    Entry/Exit Logs 365

Medical Monitoring and Exposure Reporting ... 366
    Initial and Annual Hazmat Team Physical 366
    Immediate Care 367
    Post-Exposure Monitoring 367

Chapter Review 368
Skill Sheets 369

11 Implementing the Action Plan: Personal Protective Equipment 372

Respiratory Protection 375
    Standards for Respiratory Protection at Hazmat/
        WMD Incidents 377
    Self-Contained Breathing Apparatus (SCBA) .. 378
    Supplied Air Respirators 380
    Air-Purifying Respirators 380
        Particulate-Removing Filters 382
        Vapor-and-Gas-Removing Filters 383
        Powered Air-Purifying Respirators (PAPR) 383
        Combined Respirators 383
        Supplied-Air Hoods 384
        Respiratory Equipment Limitations 384

Protective Clothing Overview 385
    Standards for Protective Clothing and Equipment
        at Hazmat/WMD Incidents 386
    Structural Firefighters' Protective Clothing 388
    High Temperature-Protective Clothing 389
    Flame-Resistant Protective Clothing 390
    Chemical-Protective Clothing (CPC) 390
        Liquid Splash-Protective Clothing 391
        Vapor-Protective Clothing 393
        Mission Specific Operations Requiring Use of
            Chemical-Protective Clothing 394
        Written Management Programs 394
        Service Life 395

PPE Ensembles and Classification 395
    Levels of Protection 396
        Level A 297
        Level B 400
        Level C 401
        Level D 402
    Typical Ensembles of Response Personnel 402
        Fire Service Ensembles 403
        Law Enforcement Ensembles 404
        EMS Ensembles 404

PPE Selection Factors 404
    Selecting PPE for Unknown Environments 407
    Selecting Thermal Protective Clothing 408
    Selecting for Chemical Resistance/
        Compatibility 408
        Permeation 409
        Degradation 410
        Penetration 410

PPE-Related Stresses 410
    Heat Emergencies 411
    Heat-Exposure Prevention 411
    Cold Emergencies 413
    Psychological Issues 414

PPE Use 414
    Pre-Entry Inspection 415
    Safety and Emergency Procedures 416
        Safety Briefing 416
        Air Management 417
        Contamination Avoidance 417
        Communication 418
        Buddy System 419
        Suit Integrity 420
        Mission Work Duration 420
    Donning and Doffing of PPE 423
        Donning of PPE 423
        Doffing of PPE 424

**PPE Inspection, Testing, Maintenance, Storage,
and Documentation** 425
    Inspection 425
    Testing 426
    Maintenance and Storage 426
    Documentation and Written PPE Program 427

Chapter Review 428
Skill Sheets 429

12 Implementing the Action Plan: Decontamination 436

ix

Introduction to Decontamination 439
Decontamination Methods 440
   Mass Decontamination 443
      Ambulatory Victims *445*
      Nonambulatory Victims *445*
   Technical Decontamination 446
   Absorption 448
   Adsorption 449
   Chemical/Biological Degradation 449
   Dilution 450
   Disinfection 451
   Evaporation 451
   Isolation and Disposal 452
   Neutralization 452
   Solidification 453
   Sterilization 453
   Vacuuming 454
   Washing 455
Assessing Decontamination Effectiveness 455
Decontamination Implementation 457
   Site Selection 457
   Factors Affecting Decontamination 458
   Decontamination Corridor Layout 459
      Segregation of the Decontamination Line *461*
      Containment *461*
      Personnel *461*
      Equipment *461*
      Procedures *463*
   Other Implementation Considerations .. 463
   Cold Weather Decontamination 464
   Medical Monitoring 465
   Evidence Collection and Decontamination 465
   Termination 466
Special Considerations 466
   Radiation 467
   Pesticides 467
   Infectious Agents 467
   Decontamination during a Terrorist Event 468
Chapter Review 468
Skill Sheets 469

13 Implementing the Action Plan: Product Control 478
Introduction to Product Control 481
   Nonintervention 483
   Defensive Operations 483
   Offensive Operations 485
Damage Assessment and Predicting Behavior .. 486
   Early Size-Up 487
   Container Information 487
      Aluminum Containers *487*
      Steel Containers *487*
      High Strength Low Alloy Containers *488*
      Austenitic Stainless Containers *488*
   Potential Container Stress 488
   Types of Container Damage 488
      Cracks *490*
      Dents *490*
      Scores and Gouges *490*
      Heat-Affected Areas: Welds, Rail Burn, Wheel Burn, and Road Burn *491*
   Predicting Likely Behavior 492
Product Containment: Plugging and Patching .. 494
   Plugging 495
   Patching 496
      Box Patch *497*
      Hook Bolts and Patch *497*
      Pipe Patches *498*
   Specialized Plugging Equipment 499
      A-Kit *499*
      B-Kit *499*
      C-Kit *500*
      Propane A- and B-Kits *500*
      Midland Kit *500*
      Recovery Vessels *501*
Cargo Tanks 501
   Methodology of Leak Control 502
      Dome Cover Leaks *502*
      Addressing Breaches in Cargo Tanks *503*
   Methods and Precautions for Fire Control 504
   Product Removal and Transfer Considerations 505
      Safety Considerations *506*
      Container Stability *506*
      Air Monitoring *506*
      Bonding and Grounding *506*
Pressurized Containers 507
   Understanding Fittings on Pressure Containers 508
   Working with Fittings on Pressure Containers 509
      Fusible Plug *509*
      Fusible Plug Threads *511*
      Cylinder Wall *511*
      Valve Blowout *511*
      Valve Gland *512*
      Valve Inlet Thread *512*
      Valve Seat *512*
      Valve Stem Blowout *512*
Drums 513

Bung Leaks	513
Chime Leaks	513
Sidewall Punctures	513

Overpacking **514**
    Tasks 515
    Lab Packs 516

Other Basic Product Control Techniques **517**
    Vapor Suppression and Dispersion 517
    Blanketing/Covering 517

Specialized Product Control Techniques **518**
    Hot and Cold Tapping 518
    Product Transfer 519
    Flaring 519
    Venting 520
    Applying Large Plugging and Patching
      Devices 520
    Vent and Burn 520

Chapter Review **521**
Skill Sheets **522**

14 Implementing the Action Plan: Incident Demobilization and Termination 546

Demobilization **549**
Termination **550**
    Debriefing 550
      Concepts of an Effective Debriefing *551*
      When to Conduct a Debriefing *552*
    Postincident Critique 552
      Components of a Post-Incident Critique *553*
      Postincident Critique Participation *553*
    After Action Report 554
    Documentation 555
      National Fire Incident Reporting System
        (NFIRS) Reports *555*
      Debriefing Records *555*
      Critique Records *555*
      Reimbursement Logs *555*
      Records Management *555*

Chapter Review **556**
Skill Sheets **557**

Appendix A
Chapter and Page Correlation to NFPA 1072
    Requirements 560

Appendix B
Protective Clothing Materials and Manufacturing
    Processes 561

Glossary 563
Index 581

List of Tables

Table 2.1	Flammable Ranges for Selected Materials	40
Table 3.1	Diatomic Molecules	51
Table 3.2	Special Hazards of Binary Salts	53
Table 3.3	Hazards of Salts	54
Table 3.4	Hazards of Nonsalts	55
Table 3.5	Four Family Groups Summary	60
Table 3.6	Covalent Compounds	67
Table 3.7	Examples of Common Salts and Their Uses	68
Table 3.8	Reactions Caused by Mixing Incompatible Chemicals	76
Table 4.1	Aromatic Hydrocarbons and their Hazards	85
Table 4.2	Hydrocarbon Derivatives and their Hazards	86
Table 4.3	Nerve Agents and their Characteristics	101
Table 4.4	Riot Control Agents and their Characteristics	105
Table 4.5	Types of Pesticides and their Target Organisms	110
Table 4.6	Pesticide Categories and Functions	110
Table 5.1	Comparison of Actual LEL and Gas Concentrations with Typical Instrument Readings	140
Table 5.2	Sample Conversion Factors	141
Table 6.1	Exposure Limits Technology	165
Table 6.2	Sample PID Correction Factors	174
Table 6.3	Ionization Potential and Lamp Strength	186
Table 8.1	Nameplate Information	262
Table 8.2	Specification Plate Information	263
Table 8.3	Nonpressure Liquid Tank	265
Table 8.4	Low-Pressure Chemical Liquid Tank	269
Table 8.5	Corrosive Liquid Tank	272
Table 8.6	High-Pressure Cargo Tank	275
Table 8.7	Cryogenic Liquid Tank	278
Table 8.8	Compressed Gas/ Tube Trailer	280
Table 8.9	Dry Bulk Cargo Trailer	281
Table 8.10	Tank Car Class Numbers and Approving Authority	282
Table 8.11	Stenciled Commodity Names	286
Table 8.12	Portable Tank Instruction Codes and Contents	299
Table 8.13	Intermodal Tank Container Descriptions	299
Table 8.14	Atmospheric/Nonpressure Storage Tanks	310
Table 8.15	Low-Pressure Storage Tanks and Pressure Vessels	312
Table 8.16	Cryogenic Liquid Storage Tank	313

Table 9.1	Health and Safety Plan and NIMS/ICS Local Forms	342
Table 10.1	Examples of Level I/ II/ III Hazmat Incidents (US National Response Team System)	362
Table 10.2	Characteristics of Level 5/ 4/ 3/ 2/ 1 Hazmat Incidents (NIMS Incident Typing System)	363
Table 10.3	NIMS-ICS Form Numbers and Titles	364
Table 11.1	Sample Compatibility Chart	409
Table 11.2	NOAA's Weather Service Heat Index	413
Table 11.3	Wind Chill Chart	414
Table 11.4	Hand Signals	418
Table 11.5	Breathing Air Cylinder Capabilities	421
Table 11.6	Air Cylinder Safety Buffers	421
Table 11.7	Energy Expenditures during Common Sports and Fire Fighting Activities	422
Table 12.1	Advantages and Disadvantages of Technical Decon Methods	446
Table 13.1	Limiting Score Depths for 340W Tanks	491
Table 13.2	Limiting Score Depths for 400W Tanks	491

Dedication

This manual is dedicated to the men and women who hold devotion to duty above personal risk, who count on sincerity of service above personal comfort and convenience, who strive unceasingly to find better and safer ways of protecting lives, homes, and property of their fellow citizens from the ravages of fire, medical emergencies, and other disasters

...The Firefighters of All Nations.

*This manual is also dedicated to **Kristina Kreutzer**. Without Krissy's technical assistance, the first and second editions of this manual would not be as technically accurate as they are. We are indebted to her for her contributions to this project, and we laud her for sharing her experience, knowledge, and expertise with us and the hazmat community at large.*

Acknowledgements

The second edition of the **Hazardous Materials Technician** is designed to meet the requirements of NFPA 1072, *Standard for Hazardous Materials/Weapons of Mass Destruction Emergency Response Personnel Professional Qualifications* (2017).

Acknowledgement and special thanks are extended to the members of the IFSTA validating committee who contributed their time, wisdom, and knowledge to the development of this manual.

IFSTA Hazardous Materials Technician Second Edition Validation Committee

Chair
Scott D. Kerwood
Fire Chief
Hutto Fire Rescue
Hutto, TX

Vice Chair and Secretary
Rich Mahaney
Manager/Instructor
Mahaney Loss Prevention Services
Gobles, MI

Committee Members

Tyler Bones
Chief
Fairbanks North Star Borough Hazardous Materials Response Team
Fairbanks, AK

Dennis Clinton
Missouri HAZ-MAT Consultants, LLC
Ozark, Missouri

David Coates
Hazardous Materials Coordinator
South Carolina Fire Academy
Columbia, South Carolina

Brent Cowx
Program Coordinator, Hazardous Materials, Technical Rescue
Justice Institute of British Columbia
(Retd. Vancouver Fire Rescue Services)

Bryn Crandell
Training Developer
DoD Fire Academy
San Angelo, TX

Michael Fortini
Los Angeles Fire Department
Camarillo, CA

Doug Goodings
Continuing Education Coordinator
Blue River College
Missouri

CJ Haberkorn
Assistant Chief, Shift Commander
Denver Fire Department
Denver, CO

Butch Hayes
Firefighter/Hazmat Technician
Houston Fire Department
Conroe, TX

Steve Hergenreter
IAFF Haz Mat Training
Fort Dodge, IA

Robert Kronenberger
City of Middletown, CT Fire Department
Middletown, CT

Barry Lindley
Specialized Professional Services, Inc
Charleston, WV

Thomas Miller
Instructor III
Sissonville Volunteer Fire Department, National Volunteer Fire Council
Charleston, WV

Carlos Rodriguez
Fire Captain
City of Wichita Fire Department
Wichita, KS

IFSTA Hazardous Materials Technician
Second Edition Validation Committee
Committee Members (cont.)

Walter G. M. Schneider III, Ph.D., P.E., CBO, MCP
Agency Director
Centre Region Code Administration
State College, Pennsylvania
Department Chief
Bellefonte Fire Department
Bellefonte, Pennsylvania

Fred Terryn
Fire Program Manager
U.S. Air Force Fire Emergency Services
Tyndall AFB, FL

Brian D. White
US Department of Justice
Washington DC

Much appreciation is given to the following individuals and organizations for contributing information, photographs, and technical assistance instrumental in the development of this manual:

- Barry Lindley
- Bill Hand, Houston Fire Department (ret)
- Brent Cowx
- Brent Cowx and Jonathan Gormick, Vancouver Fire & Rescue Services
- David Lewis
- Joan Hepler
- John Demyan
- Rich Mahaney
- Scott Kerwood
- Steve George
- Steven Baker, New South Wales Fire Brigades
- Tom Clawson
- William D. Stewart
- DOD, photo by SrA Christopher J. Wiant
- FEMA News Photos, photo by Jocelyn Augustino
- MSA
- National Nuclear Security Administration
- National Nuclear Security Administration, Nevada Site Office
- U.S. Agency for Toxic Substances and Disease Registry (ATSDR)
- U.S. Air Force, photo by Taylor Marr
- U.S. Army, photo by SSG Fredrick P. Varney, 133rd Mobile Public Affairs Detachment
- U.S. Bureau of Alcohol, Tobacco, Firearms, and Explosives, and the Oklahoma Highway Patrol
- U.S. Coast Guard, photo by PA1 Charles C. Reinhart
- U.S. E.P.A.
- U.S. Marine Corps, photo by Warren Peace
- United States Air Force
- USAF, photo by Chiakia Iramina
- USMC, photo by Cpl Brian A. Tuthill

Special thanks go to the following people who contributed to this manual in exceptional ways:
- Rich Mahaney, for providing so many photos used throughout the manual, including in tables and skill sheets. Rich's photos have enriched IFSTA's hazmat manuals for several editions.
- Barry Lindley, for fixing the little details, providing pictures, and giving the editors valuable chemistry lessons. He also answered innumerable questions about product and container behavior.
- Carlos Rodriguez, for sharing his technical expertise, providing exceptional feedback, and troubleshooting. As well, Carlos kept us on task with reminders to make this text usable for the target audience.
- The entire NFPA 472/1072 technical committee for answering questions about the standards' intent, as well as random technical questions.

Thanks also go to the agencies and organizations producing various resources used throughout this manual:

- Health Canada
- Los Alamos National Laboratory
- Sandia National Laboratories
- Transport Canada
- Union Pacific Railroad
- Oklahoma Highway Patrol Bomb Squad
- U.S. Air Force
- U.S. Army
- U.S. Centers for Disease Control and Prevention
- U.S. Coast Guard
- U.S. Department of Defense
- U.S. Department of Energy
- U.S. Department of Homeland Security
- U.S. Department of Justice
- U.S. Department of Transportation; Pipeline, Hazardous Materials, and Safety Administration
- U.S. Drug Enforcement Agency
- U.S. Environmental Protection Agency
- U.S. Federal Bureau of Investigation
- U.S. Federal Emergency Management Agency
- U.S. Fire Administration
- U.S. Marines
- U.S. National Institute for Occupational Safety and Health
- U.S. National Nuclear Security Administration
- U.S. Navy
- U.S. Nuclear Regulatory Commission
- U.S. Occupational Safety and Health Administration

Last, but certainly not least, gratitude is extended to the following members of the Fire Protection Publications staff whose contributions made the final publication of this manual possible.

Hazardous Materials Technician, Second Edition, Project Team

Lead Senior Editors
Libby Snyder, Senior Editor
Leslie A. Miller, Senior Editor

Lead Instructional Developers
Lindsey Dugan, Curriculum Developer
David Schaap, Curriculum Developer

Lead Graphic Designer
Errick Braggs, Senior Graphic Designer

Director of Fire Protection Publications
Craig Hannan

Curriculum Manager
Lori Raborg
Leslie A. Miller
Colby Cagle

Editorial Manager
Clint Clausing

Production Manager
Ann Moffat

Editors
Alex Abrams, Senior Editor
Brad McLelland, Senior Editor
Cindy Brakhage, Senior Editor
Jeff Fortney, Senior Editor
Rikka Strong, Senior Editor
Tony Peters, Senior Editor

Illustrators and Layout Designers
Errick Braggs, Senior Graphic Designer
Clint Parker, Senior Graphic Designer

Curriculum Development
Lindsey Dugan, Curriculum Developer
Tara Moore, Curriculum Developer
David Schaap, Curriculum Developer
Simone Rowe, Curriculum Developer
Alyssa Williams, Curriculum Developer

Photographer(s)
Jeff Fortney, Senior Editor
Leslie A. Miller, Senior Editor
Alex Abrams, Senior Editor

Editorial Staff
Tara Gladden, Editorial Assistant

Indexer
Nancy Kopper

The IFSTA Executive Board at the time of validation of the **Hazardous Materials Technician, Second Edition** was as follows:

IFSTA Executive Board

Executive Board Chair
Steve Ashbrock
Fire Chief
Madeira & Indian Hill Fire Department
Cincinnati, OH

Vice Chair
Bradd Clark
Fire Chief
Ocala Fire Department
Ocala, FL

IFSTA Executive Director
Mike Wieder
Fire Protection Publications
Stillwater, OK

Board Members

Steve Austin
Project Manager
Cumberland Valley Volunteer Firemen's Association
Newark, DE

Mary Cameli
Assistant Chief
City of Mesa Fire Department
Mesa, AZ

Dr. Larry Collins
Associate Dean
Eastern Kentucky University
Safety, Security, & Emergency Department
Richmond, KY

Chief Dennis Compton
Mesa & Phoenix, Arizona
Chairman of the National Fallen Firefighters
Foundation Board of Directors

John Hoglund
Director Emeritus
Maryland Fire & Rescue Institute
New Carrollton, MD

Tonya Hoover
State Fire Marshal
CA Department of Forestry & Fire Protection
Sacramento, CA

Dr. Scott Kerwood
Fire Chief
Hutto Fire Rescue
Hutto, TX

Wes Kitchel
Assistant Chief
Sonoma County Fire & Emergency Services Dept.
Santa Rosa, CA

Brett Lacey
Fire Marshal
Colorado Springs Fire Department
Colorado Springs, CO

Robert Moore
Division Director
Texas A&M Engineering Extension Services
College Station, TX

Dr. Lori Moore-Merrell
Assistant to the General President
International Association of Fire Fighters
Washington, DC

Jeff Morrissette
State Fire Administrator
State of Connecticut
Commission on Fire Prevention and Control
Windsor Locks, CT

Josh Stefancic
Division Chief
Largo Fire Rescue
Largo, FL

Paul Valentine
Senior Engineer
Nexus Engineering
Oakbrook Terrace, IL

Steven Westermann
Fire Chief
Central Jackson County Fire Protection District
Blue Springs, MO

Introduction

Introduction Contents

Introduction...................................1	Terminology 2
Purpose and Scope 1	Key Information 3
Book Organization 1	Metric Conversions 4

Introduction

Hazardous materials are found in every jurisdiction, community, workplace, and modern household. These substances possess a wide variety of harmful characteristics. Some hazardous materials can be quite deadly or destructive, and terrorists and other criminals may use them to deliberately cause harm. Hazardous materials technicians must be able to direct first responders (personnel who are likely to arrive first at an incident scene) and advise the Incident Commander about the safe and effective resolution of an incident. Technicians also research substances and the hazards they can present under various conditions, plus those substances' detection and mitigation strategies.

Purpose and Scope

The purpose of this manual is to prepare emergency responders who conduct advanced, technical, offensive operations at hazardous materials incidents to meet the Technician Level certification requirements of NFPA 1072, *Standard for Hazardous Materials/Weapons of Mass Destruction Emergency Response Personnel Professional Qualifications, 2016 Edition*. This manual builds on the principles and techniques presented in IFSTA's **Hazardous Materials for First Responders, Fifth Edition.**

The scope of this manual includes the job performance requirements for the Technician level as described in NFPA 1072. This manual also addresses the U.S. Occupational Safety and Health Administration (OSHA) training requirements found in Title 29 *Code of Federal Regulations (CFR) 1910.120 Hazardous Waste Operations and Emergency Response* (HAZWOPER), paragraph (q), for emergency responders at the Technician Level. In addition, NFPA 472, *Standard for Competence of Responders to Hazardous Materials/Weapons of Mass Destruction Incidents, 2013 Edition* is referenced in this manual where appropriate.

Book Organization

To meet the competencies of NFPA 1072 Chapter 7, this book is organized in order of APIE:

Analyzing

Chapter 1 – Introduction to the Hazmat Technician

Chapter 2 – Analyzing the Incident: Understanding How Matter Behaves

Chapter 3 – Analyzing the Incident: Understanding Atomic Structure and the Chemistry of Hazardous Materials

Chapter 4 – Analyzing the Incident: Understanding Common Families and Special Hazards of Hazardous Materials/WMDs

Chapter 5 – Analyzing the Incident: Detection, Monitoring, and Sampling Procedures

Chapter 6 – Analyzing the Incident: Detecting, Monitoring, and Sampling Hazardous Materials

Chapter 7 – Analyzing the Incident: Collecting and Interpreting Hazard and Response Information

Chapter 8 – Analyzing the Incident: Assessing Container Condition, Predicting Behavior, and Estimating Outcomes

Planning

Chapter 9 – Planning the Response: Developing Response Objectives

Implementing and Evaluating

Chapter 10 – Implementing and Evaluating the Action Plan: Incident Management

Chapter 11 – Implementing the Action Plan: Personal Protective Equipment

Chapter 12 – Implementing the Action Plan: Decontamination

Chapter 13 – Implementing the Action Plan: Product Control

Chapter 14 – Implementing the Action Plan: Incident Demobilization and Termination

Terminology

This manual is written with a global, international audience in mind. For this reason, it often uses general descriptive language in place of regional- or agency-specific terminology (often referred to as *jargon*). Additionally, in order to keep sentences uncluttered and easy to read, the word *state* is used to represent both state and provincial level governments (or their equivalent). This usage is applied to this manual for the purposes of brevity and is not intended to address or show preference for only one nation's method of identifying regional governments within its borders.

The glossary at the end of the manual will assist the reader in understanding words that may not have their roots in the fire and emergency services. The sources for the definitions of fire-and-emergency-services-related terms will be the *NFPA Dictionary of Terms* and the IFSTA **Fire Service Orientation and Terminology** manual. Additionally, when reading this text, remember the following points:

1. The terms *emergency*, *incident*, and *hazmat incident* are often used interchangeably, with the understanding that the types of incidents addressed by this book are emergencies.

2. NFPA and OSHA have different terms for persons trained to the Awareness Level. NFPA 1072 refers to these individuals as *personnel* whereas OSHA's 29 *CFR* 1910.120 uses the term *responders*. When the term *first responder* is used in this manual, it generally refers to both Awareness- and Operations-Level responders as defined by OSHA. The authority having jurisdiction

(AHJ) is responsible for defining the actions allowed by persons trained to the Awareness Level, depending on the standard to which they are trained.

3. There are many different ways to refer to hazardous materials. You may see *hazmat*, *haz mat*, *dangerous goods*, or *hazardous materials*. This manual will use the abbreviation *hazmat*.

4. Weapons of mass destruction (WMDs) as addressed in this manual are considered to be hazardous materials. When the term *hazmat* is used throughout this manual in a general sense, it should be understood that WMDs are included.

Key Information

Various types of information in this book are given in shaded boxes marked by symbols or icons. See the following definitions:

Case Study

A case study analyzes an event. It can describe its development, action taken, investigation results, and lessons learned.

Safety Alert

Safety alert boxes are used to highlight information that is important for safety reasons. (In the text, the title of safety alerts will change to reflect the content.)

Information

Information boxes give facts that are complete in themselves but belong with the text discussion. It is information that needs more emphasis or separation. (In the text, the title of information boxes will change to reflect the content.)

What This Means To You

These boxes take information presented in the text and synthesize it into an example of how the information is relevant to (or will be applied by) the intended audience, essentially answering the question, "What does this mean to you?"

A **key term** is designed to emphasize key concepts, technical terms, or ideas that first responders need to know. They are listed at the beginning of each chapter, highlighted in bold **red** font, and the definition is placed in the margin for easy reference.

Self-Reading Dosimeter (SRD) — Detection device that displays the cumulative reading without requiring additional processing. *Also known as* Direct-Reading Dosimeters (DRDs) *and* Pencil Dosimeters.

Introduction 3

Three key signal words are found in the book: **WARNING**, **CAUTION**, and **NOTE**. Definitions and examples of each are as follows:

- **WARNING** indicates information that could result in death or serious injury to first responders. See the following example:

- **CAUTION** indicates important information or data that first responders need to be aware of in order to perform their duties safely. See the following example:

WARNING
SCBA must be worn during emergency operations at terrorist/hazmat incidents until air monitoring and sampling determines other options are acceptable.

CAUTION
Water may react violently with some materials.

- **NOTE** indicates important operational information that helps explain why a particular recommendation is given or describes optional methods for certain procedures. See the following example:

  NOTE: Hydrogen, which is located in the upper-left corner above the thick dividing line, is also a nonmetal.

Metric Conversions

Throughout this manual, U.S. units of measure are converted to metric units for the convenience of our international readers. Be advised that we use the Canadian metric system. It is very similar to the Standard International system, but may have some variation.

We adhere to the following guidelines for metric conversions in this manual:

- Metric conversions are approximated unless the number is used in mathematical equations.

- Centimeters are not used because they are not part of the Canadian metric standard.

- Exact conversions are used when an exact number is necessary such as in construction measurements or hydraulic calculations.

- Set values such as hose diameter, ladder length, and nozzle size use their Canadian counterpart naming conventions and are not mathematically calculated. For example, 1½ inch hose is referred to as 38 mm hose.

The following two tables provide detailed information on IFSTA's conversion conventions. The first table includes examples of our conversion factors for a number of measurements used in the fire service. The second shows examples of exact conversions beside the approximated measurements you will see in this manual.

U.S. to Canadian Measurement Conversion

Measurements	Customary (U.S.)	Metric (Canada)	Conversion Factor
Length/Distance	Inch (in) Foot (ft) [3 or less feet] Foot (ft) [3 or more feet] Mile (mi)	Millimeter (mm) Millimeter (mm) Meter (m) Kilometer (km)	1 in = 25 mm 1 ft = 300 mm 1 ft = 0.3 m 1 mi = 1.6 km
Area	Square Foot (ft^2) Square Mile (mi^2)	Square Meter (m^2) Square Kilometer (km^2)	1 ft^2 = 0.09 m^2 1 mi^2 = 2.6 km^2
Mass/Weight	Dry Ounce (oz) Pound (lb) Ton (T)	gram Kilogram (kg) Ton (T)	1 oz = 28 g 1 lb = 0.5 kg 1 T = 0.9 T
Volume	Cubic Foot (ft^3) Fluid Ounce (fl oz) Quart (qt) Gallon (gal)	Cubic Meter (m^3) Milliliter (mL) Liter (L) Liter (L)	1 ft^3 = 0.03 m^3 1 fl oz = 30 mL 1 qt = 1 L 1 gal = 4 L
Flow	Gallons per Minute (gpm) Cubic Foot per Minute (ft^3/min)	Liters per Minute (L/min) Cubic Meter per Minute (m^3/min)	1 gpm = 4 L/min 1 ft^3/min = 0.03 m^3/min
Flow per Area	Gallons per Minute per Square Foot (gpm/ft^2)	Liters per Square Meters Minute (L/(m^2.min))	1 gpm/ft^2 = 40 L/(m^2.min)
Pressure	Pounds per Square Inch (psi) Pounds per Square Foot (psf) Inches of Mercury (in Hg)	Kilopascal (kPa) Kilopascal (kPa) Kilopascal (kPa)	1 psi = 7 kPa 1 psf = .05 kPa 1 in Hg = 3.4 kPa
Speed/Velocity	Miles per Hour (mph) Feet per Second (ft/sec)	Kilometers per Hour (km/h) Meter per Second (m/s)	1 mph = 1.6 km/h 1 ft/sec = 0.3 m/s
Heat	British Thermal Unit (Btu)	Kilojoule (kJ)	1 Btu = 1 kJ
Heat Flow	British Thermal Unit per Minute (BTU/min)	watt (W)	1 Btu/min = 18 W
Density	Pound per Cubic Foot (lb/ft^3)	Kilogram per Cubic Meter (kg/m^3)	1 lb/ft^3 = 16 kg/m^3
Force	Pound-Force (lbf)	Newton (N)	1 lbf = 0.5 N
Torque	Pound-Force Foot (lbf ft)	Newton Meter (N.m)	1 lbf ft = 1.4 N.m
Dynamic Viscosity	Pound per Foot-Second (lb/ft.s)	Pascal Second (Pa.s)	1 lb/ft.s = 1.5 Pa.s
Surface Tension	Pound per Foot (lb/ft)	Newton per Meter (N/m)	1 lb/ft = 15 N/m

Conversion and Approximation Examples

Measurement	U.S. Unit	Conversion Factor	Exact S.I. Unit	Rounded S.I. Unit
Length/Distance	10 in	1 in = 25 mm	250 mm	250 mm
	25 in	1 in = 25 mm	625 mm	625 mm
	2 ft	1 in = 25 mm	600 mm	600 mm
	17 ft	1 ft = 0.3 m	5.1 m	5 m
	3 mi	1 mi = 1.6 km	4.8 km	5 km
	10 mi	1 mi = 1.6 km	16 km	16 km
Area	36 ft²	1 ft² = 0.09 m²	3.24 m²	3 m²
	300 ft²	1 ft² = 0.09 m²	27 m²	30 m²
	5 mi²	1 mi² = 2.6 km²	13 km²	13 km²
	14 mi²	1 mi² = 2.6 km²	36.4 km²	35 km²
Mass/Weight	16 oz	1 oz = 28 g	448 g	450 g
	20 oz	1 oz = 28 g	560 g	560 g
	3.75 lb	1 lb = 0.5 kg	1.875 kg	2 kg
	2,000 lb	1 lb = 0.5 kg	1 000 kg	1 000 kg
	1 T	1 T = 0.9 T	900 kg	900 kg
	2.5 T	1 T = 0.9 T	2.25 T	2 T
Volume	55 ft³	1 ft³ = 0.03 m³	1.65 m³	1.5 m³
	2,000 ft³	1 ft³ = 0.03 m³	60 m³	60 m³
	8 fl oz	1 fl oz = 30 mL	240 mL	240 mL
	20 fl oz	1 fl oz = 30 mL	600 mL	600 mL
	10 qt	1 qt = 1 L	10 L	10 L
	22 gal	1 gal = 4 L	88 L	90 L
	500 gal	1 gal = 4 L	2 000 L	2 000 L
Flow	100 gpm	1 gpm = 4 L/min	400 L/min	400 L/min
	500 gpm	1 gpm = 4 L/min	2 000 L/min	2 000 L/min
	16 ft³/min	1 ft³/min = 0.03 m³/min	0.48 m³/min	0.5 m³/min
	200 ft³/min	1 ft³/min = 0.03 m³/min	6 m³/min	6 m³/min
Flow per Area	50 gpm/ft²	1 gpm/ft² = 40 L/(m².min)	2 000 L/(m².min)	2 000 L/(m².min)
	326 gpm/ft²	1 gpm/ft² = 40 L/(m².min)	13 040 L/(m².min)	13 000 L/(m².min)
Pressure	100 psi	1 psi = 7 kPa	700 kPa	700 kPa
	175 psi	1 psi = 7 kPa	1225 kPa	1 200 kPa
	526 psf	1 psf = 0.05 kPa	26.3 kPa	25 kPa
	12,000 psf	1 psf = 0.05 kPa	600 kPa	600 kPa
	5 psi in Hg	1 psi = 3.4 kPa	17 kPa	17 kPa
	20 psi in Hg	1 psi = 3.4 kPa	68 kPa	70 kPa
Speed/Velocity	20 mph	1 mph = 1.6 km/h	32 km/h	30 km/h
	35 mph	1 mph = 1.6 km/h	56 km/h	55 km/h
	10 ft/sec	1 ft/sec = 0.3 m/s	3 m/s	3 m/s
	50 ft/sec	1 ft/sec = 0.3 m/s	15 m/s	15 m/s
Heat	1200 Btu	1 Btu = 1 kJ	1 200 kJ	1 200 kJ
Heat Flow	5 BTU/min	1 Btu/min = 18 W	90 W	90 W
	400 BTU/min	1 Btu/min = 18 W	7 200 W	7 200 W
Density	5 lb/ft³	1 lb/ft³ = 16 kg/m³	80 kg/m³	80 kg/m³
	48 lb/ft³	1 lb/ft³ = 16 kg/m³	768 kg/m³	770 kg/m³
Force	10 lbf	1 lbf = 0.5 N	5 N	5 N
	1,500 lbf	1 lbf = 0.5 N	750 N	750 N
Torque	100	1 lbf ft = 1.4 N.m	140 N.m	140 N.m
	500	1 lbf ft = 1.4 N.m	700 N.m	700 N.m
Dynamic Viscosity	20 lb/ft.s	1 lb/ft.s = 1.5 Pa.s	30 Pa.s	30 Pa.s
	35 lb/ft.s	1 lb/ft.s = 1.5 Pa.s	52.5 Pa.s	50 Pa.s
Surface Tension	6.5 lb/ft	1 lb/ft = 15 N/m	97.5 N/m	100 N/m
	10 lb/ft	1 lb/ft = 15 N/m	150 N/m	150 N/m

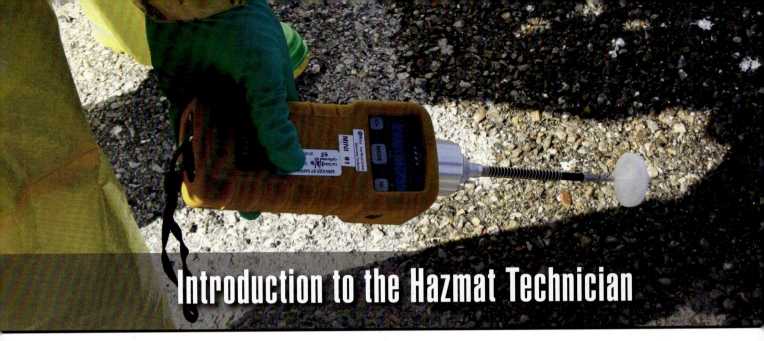

Introduction to the Hazmat Technician

Chapter Contents

APIE Process 11	Analyzing the Incident 18
Review of Awareness and Operations Level Qualifications 12	Planning the Initial Response 20
Awareness Level Personnel 12	Implementing the Planned Response 22
Operations Level Responders 14	Evaluating Progress 23
Operations Mission Specific Level 15	Terminating the Incident 24
Hazmat Technician 16	**Chapter Notes** 25
Technician Training 17	**Chapter Review** 25

chapter 1

Key Terms

Hazardous Materials Technician 16
Hazardous Waste Operations and
Emergency Response (HAZWOPER) 17

Overpack 23
Response Objective 15
Response Option 15

Introduction to the Hazmat Technician

Learning Objectives

After reading this chapter, students will be able to:

1. Explain the APIE response model used to mitigate hazmat incidents.
2. Describe Awareness and Operations Level qualifications as they apply to the APIE response model.
3. Identify Technician Level qualifications as they apply to the APIE response model.

1. Analyze
   Plan
   Implement
   Evaluate

2. awareness + ops: if you know something tell the correct people
   analyze: what is the problem? do you know what the chemical is?
   ~~isolate?~~ evacuate.
   plan: what were you told to do if an accident happened
   implement: do what you were told to do
   evaluate: did you get out safely

3. tech level
   A: what is it? (gather info on PPE, tools, personnel needed.) how will the environment be affected.
   P: ↓ what will their specific jobs be
   I: put plan into action
   E: make necessary changes.

Chapter 1
Introduction to the Hazmat Technician

Information presented in this manual will reflect best practices in the field and will focus on safe operations based on NFPA 1072, *Standard for Hazardous Materials/Weapons of Mass Destruction Emergency Response Personnel Professional Qualifications* (2016). This chapter is an introduction to elements necessary for that goal:

- APIE process
- Review of awareness and operations level qualifications
- Hazmat technician

APIE Process

APIE is a simple four-step response model that can guide responders' actions at hazardous materials incidents. These steps form a consistent problem-solving process that can be used at any incident, regardless of size or complexity:

Step 1: Analyze the incident — During this phase of the problem-solving process, personnel and responders attempt to understand the current situation. For example, first responders attempt to identify the hazmat involved, what kind of containers are present, the quantity of materials released, the number of exposures, potential hazards, and other relevant information needed to plan a safe and effective response.

Step 2: Plan the initial response — During this phase, responders use the information gathered during the analysis phase to determine what actions need to be taken to mitigate the incident. For example, the Incident Commander (IC) will develop the incident action plan (IAP) and assign tasks to first responders.

Step 3: Implement the response — During this phase, responders perform the tasks determined in the planning stage. When implementing the response, responders direct actions to mitigate the incident.

Step 4: Evaluate progress — During this phase – which continues throughout the incident until termination – responders monitor progress to see whether the response plan is working. For example, first responders should report if their actions are completed successfully or if they notice changing conditions.

Responders with different levels of training have different responsibilities in each of these steps. Because Awareness Level personnel and Operations Level responders have limited responsibilities at hazmat incidents, not all aspects of *APIE* are addressed in the Awareness and Operations Levels. As responsibilities increase, so do the components of APIE, and hazardous materials technicians have responsibilities in all four steps. These responsibilities will be addressed throughout this manual, often in the context of APIE.

Review of Awareness and Operations Level Qualifications

First responders must be trained to the Awareness and Operations Levels before training to become a technician. While some first responders may have received training to Mission Specific levels, these are not prerequisite training for technicians.

Awareness Level Personnel

Personnel trained and certified to the Awareness Level may, in the course of their normal duties, be the first to arrive at or witness a hazardous material/weapon of mass destruction incident. Awareness Level personnel serve an important role at hazmat incidents and their initial actions can affect the course of the incident for better or worse.

Analyzing the Incident

Awareness personnel must be able to recognize that hazardous materials may be involved in an incident. There are seven clues that Awareness and Operations Level responders can use to detect the presence of hazardous materials. Some clues may be easily identified from a distance while others require responders to be much closer. The closer responders need to be in order to identify the material, the greater their chances of being in an area where they could be exposed to its harmful effects.

The seven clues to the presence of hazardous materials are as follows **(Figure 1.1)**:

1. **Preincident surveys, occupancy types, and locations** — Preincident surveys can help first responders identify businesses, industries, and other occupancies that store, manufacture, use, and/or ship hazardous materials before an incident occurs. Hazmat incidents are more likely to occur at certain locations such as ports, docks or piers, railroad sidings, airplane hangars, truck terminals, and other places of material transfer (such as trucking warehouses).

2. **Container shapes** — A wide variety of storage and shipping containers may contain solid, liquid, gaseous, and cryogenic hazardous materials.

3. **Transportation placards, labels, and markings** — Canada, the U.S., and Mexico require placards, labels, and markings based on the UN hazard class system to identify transported hazardous materials (OSHA).

4. **Other markings and colors (non-transportation)** — Other hazmat markings and colors include the NFPA 704 system, common hazardous communication labels, ISO safety symbols, pipeline markings, and others.

5. **Written resources** — Fixed facilities should have safety data sheets, inventory records, and other facility documents; at transportation incidents, the current *Emergency Response Guidebook* and shipping papers can provide written information.

6. **Senses** — Vision is the safest sense to use to detect hazardous materials, but other senses such as smell, hearing, touch, and taste can provide input, as well. Incident witnesses and victims can provide useful first-hand information about what they experienced.

    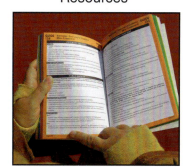 

Figure 1.1 Indicators of a hazardous materials incident include container shape and labels on the container.

7. **Monitoring and detection devices** — These devices require specialized skills and training beyond Awareness Level.

Planning the Response

Awareness Level personnel are not responsible for planning the response to a hazmat incident. However, standard operating procedures (SOPs) and/or predetermined procedures may provide Awareness Level personnel with the initial actions they should take in the event of an emergency. For example, Awareness Level personnel may be authorized to use the *Emergency Response Guidebook* to determine isolation distances **(Figure 1.2)**.

Implementing the Response

Individuals trained to the Awareness Level are expected to assume the following responsibilities at an incident involving hazardous materials:

- Recognize the presence or potential presence of a hazardous material
- Initiate protective actions for themselves and others
- Transmit information to an appropriate authority and call for appropriate assistance
- Establish scene control by isolating the hazardous area and denying entry

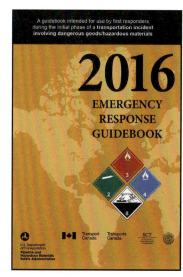

Figure 1.2 The *Emergency Response Guidebook (ERG)* is an Awareness Level document that may offer general guidance to a hazmat technician.

Evaluating Progress

As with planning the response, Awareness Level personnel are not expected to evaluate the incident response. However, if personnel have pertinent information about the incident or its status, that information should be relayed to an appropriate authority.

Operations Level Responders

Responders who are trained and certified to the Operations Level are individuals who respond to hazmat releases (or potential releases) as part of their normal duties. Operations Level responders are expected to protect individuals, the environment, and property from the effects of the release in a primarily defensive manner (see Information Box for exceptions).

> **Offensive Tasks Allowed by U.S. OSHA and Canada**
> U.S. Occupational Safety and Health Administration (OSHA) and the Canadian government recognize that first responders at the Operations Level who have appropriate training (including demonstration of competencies and certification by employers), appropriate protective clothing, and adequate/appropriate resources can perform offensive operations involving flammable liquid and gas fire control of the following materials:
>
> - Gasoline
> - Diesel fuel
> - Natural gas
> - Liquefied petroleum gas (LPG)

Responsibilities of the first responder at the Operations Level include the Awareness Level responsibilities. Additionally, first responders at the Operations Level must be able to perform the following actions:

- Identify the potential hazards involved in an incident if possible
- Identify response options
- Implement the planned response to mitigate or control a release from a safe distance by performing assigned tasks to lessen the harmful incident and keep it from spreading
- Evaluate the progress of the actions taken to ensure that response objectives are safely met

Analyzing the Incident

At the Operations Level, responders are expected to identify potential hazards at incidents, including (but not limited to) the following:

- Type of container involved
- Hazardous material involved
- Hazards presented by the material
- Potential behavior of the material

Operations Level responders are expected to analyze the surrounding conditions and determine the location and amount of any release, if possible. Once they have an understanding of the material involved and the hazards present, they can begin to plan an appropriate response based on their training, SOPs, and/or predetermined procedures.

Planning the Response

Operations Level responders must be able to identify **response options** for hazmat incidents. While they may not be responsible for planning the actual response, they must understand the tasks they may be asked to perform and why. In order to protect themselves, they must be aware of safety precautions, the suitability of the personal protective equipment available, and emergency decontamination needs. Response options are the tactics used to meet these objectives, for example, absorbing a gasoline spill with an absorbent material or using foam to extinguish a flammable liquid fire.

Some standard **response objectives** of hazmat incidents include:

- Isolation
- Notification
- Identification
- Protection (life safety)
- Rescue
- Spill/leak control and confinement
- Crime scene and evidence preservation
- Fire control
- Recovery/termination

> **Response Option** — Specific operations performed in a specific order to accomplish the goals of the response objective.
>
> **Response Objective** — Statement based on realistic expectations of what can be accomplished when all allocated resources have been effectively deployed that provide guidance and direction for selecting appropriate strategies and the tactical direction of resources.

Implementing the Response

Operations Level responders are expected to perform response options within the framework of the incident management system. They are expected to establish scene control, implement protective actions such as evacuation, and perform emergency decontamination when necessary. Additionally, they must be able to:

- Follow safety procedures
- Use personal protective equipment in the proper manner
- Avoid hazards and complete their assignments
- Identify and preserve potential evidence if the incident is a suspected crime

Evaluating Progress

At the Operations Level, responders are expected to evaluate the progress of their assigned tasks. Responders must report this information to their supervisor or other appropriate authority so the action plan can be adjusted if necessary.

Operations Mission Specific Level

First responders do not need to be trained to any Operations Mission Specific levels prior to their technician training. Mission Specific competencies are optional proficiencies that allow a community and the authority having jurisdiction (AHJ) to match a task with the requirements to perform those tasks. These specialized competencies are over and above the Operations-Core competencies. For the Operations Level responder, it is understood that those personnel working in a Mission Specific competency will be doing so under the watchful eye of a hazardous materials technician.

The Mission Specific competencies are:

- Personal protective equipment
- Mass decontamination
- Technical decontamination
- Evidence preservation and sampling
- Product control
- Air monitoring and sampling
- Victim rescue and recovery
- Response to illicit laboratory incidents

Hazmat Technician

Hazardous Materials Technician — Individual trained to use specialized protective clothing and control equipment to control the release of a hazardous material.

Technician Level competencies are gained by building on the knowledge and skills that have been acquired at the prerequisite Awareness and Operations Levels. Because the **hazardous materials technician** will be expected to respond and work in an offensive mode, these competencies must be more advanced. This manual will concentrate on the core competencies outlined in the current edition of NFPA 1072, *Standard for Hazardous Materials/Weapons of Mass Destruction Emergency Response Personnel Professional Qualifications (2016)*.

For personnel functioning at the Technician Level, there currently are no Mission Specific competencies. Instead, there are specialties for the Technician Level responder in radioactive materials and other disciplines.

Hazardous Materials Technicians Terminology
Several reference abbreviations apply to Hazardous Materials Technicians, including:

- Hazmat Technicians
- Hazmat Techs
- Technicians
- Techs
- HMTs

The following sections will outline and briefly describe the technician's responsibilities and requirements in regards to the following:

- Training
- Analyzing the incident

- Planning the initial response
- Implementing the response
- Evaluating progress

Technician Training

Hazmat technician training must come in phases. You must be competent at the Awareness and Operations Levels before you can work at the Technician Level. As you gain the knowledge and skills for each level of response, you can then move on to the next level.

Responders in the U.S. must be trained to U.S. OSHA standards. NFPA standards may also apply. Always, the requirements of the AHJ take priority because they are the final authority and may exceed minimum standards established by law.

U.S. OSHA Standards

The Occupational Safety and Health Administration (OSHA), which falls under the Department of Labor, outlines the training requirements for employees who work with chemicals. OSHA establishes standards such as 29 *CFR* 1910.120, the **Hazardous Waste Operations and Emergency Response (HAZWOPER)** standard and has the ability to legally enforce these standards.

OSHA's HAZWOPER standard is broken down into seventeen paragraphs and addresses subjects such as medical monitoring and health and safety plans. Paragraph q of the standard refers to emergency response. This paragraph has a direct influence on emergency response and consists of the following components:

- Development of an emergency response plan
- Procedures for handling emergency response
- Skilled support personnel
- Specialist employees
- Training
- Trainers
- Refresher training
- Medical surveillance
- Chemical protective clothing
- Postemergency response operations

> **Hazardous Waste Operations and Emergency Response (HAZWOPER)** — U.S. regulations in Title 29 (Labor) CFR 1910.120 for cleanup operations involving hazardous substances and emergency response operations for releases of hazardous substances.

NFPA Standards

The NFPA's hazardous materials requirements are detailed in the following standards **(Figure 1.3)**:

- NFPA 1072, *Standard for Hazardous Materials/Weapons of Mass Destruction Emergency Response Personnel Professional Qualifications (2016)*
- NFPA 472, *Standard for Professional Competence of Responders to Hazardous Materials/Weapons of Mass Destruction Incidents (2008)*

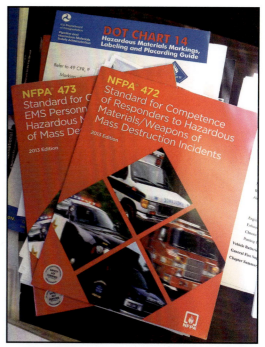

Figure 1.3 Documents that hazmat technicians should consult include national standards that contain professional qualifications for response at specific types of incidents.

- NFPA 473, *Standard for Competencies for EMS Personnel Responding to Hazardous Materials/Weapons of Mass Destruction Incidents (2008)*

NFPA 1072, *Standard for Hazardous Materials/Weapons of Mass Destruction Emergency Response Personnel Professional Qualifications (2016)* outlines the professional qualifications for response at the Awareness, Operations, Technician, and Incident Commander Levels. NFPA 472 also provides competencies for the following Technician specialties:

- Cargo Tank Specialty
- Intermodal Tank Specialty
- Marine Tank Vessel Specialty
- Tank Car Specialty

Training Sources

The hazardous materials technician should not be satisfied with simply meeting minimum training standards. Numerous training opportunities are available that can help you increase your skills and competencies. Federal agencies such as the Centers for Disease Control and Prevention (CDC) offer training for the hazardous materials responder. CDC training topics include:

- Emergency response
- Disaster management
- Terrorism response
- Personal Protective Equipment (PPE)
- Chemical hazards

Other public agencies can be of assistance in training. State and local health departments can offer a variety of training options. These agencies can be a great resource for training and information in chemical and biological topics. State and local emergency management departments along with the Department of Homeland Security (DHS) and the Federal Emergency Management Agency's (FEMA) Emergency Management Institute and National Fire Academy can offer assistance in both disaster management and training in the National Incident Management System (NIMS).

Finally, local jurisdictions may require specialized training for hazards common to the community or area. As the AHJ, local emergency response leaders may dictate requirements for specialized training in various areas and at all levels of response to address a critical need in the community.

Analyzing the Incident

The hazardous materials technician must consider many variables when analyzing an incident and is expected to have in-depth knowledge of containers. The technician must also have a general understanding of chemistry along with chemical and physical properties. In addition, the technician must be able to use metering devices and interpret relevant data in order to make intelligent, informed decisions **(Figure 1.4)**. These different components are all part of the process of analyzing and mitigating a hazardous materials incident.

Figure 1.4 Monitoring and detection devices are common resources used at a hazmat incident.

Collecting and Interpreting Hazard and Response Information

One of the most critical aspects of a hazmat response is product identification. Knowing exactly what product or products are involved is important for safe mitigation. Once a product is identified, the next step is to identify the properties of that chemical. This information allows the hazmat technician to make informed decisions on everything from response strategies and tactics to PPE selection.

When specific information is not available, the technician may be required to research the product to gain the information needed. A variety of references and resource materials are available to assist in chemical research. Chapter 7 of this manual will discuss hazardous materials research along with hazard and response information. It will address the principles of chemical research, and the varieties of different research and reference materials available.

Detecting, Monitoring, and Sampling Hazardous Materials

In order to identify, quantify, and classify the materials involved in an incident, hazmat technicians may need to use basic detection, monitoring, and sampling tools and equipment. In addition to using devices according to manufacturers' specifications, technicians must correctly interpret the data provided and apply this information to ensure a safe response. Finally, technicians must maintain and service tools and equipment to keep them working properly.

Estimating Container Damage

Many containers are in use today. Before a hazmat technician can estimate the amount of damage to a container, the technician must have a general knowledge of containers, their potential contents, and the associated hazards. Based on need, a container may be transported or located at a fixed facility. The container may hold a solid, liquid, or gas and may contain a great amount of pressure. The hazmat technician should be able to identify a container, its construction features, and its typical contents. Furthermore, the technician should be able to estimate the container's capacity and any common leak points associated with the container. Containers are discussed further in Chapter 8.

Predicting Behavior

To assist in predicting the likely behavior of a product, the hazmat technician must understand the chemical and physical properties of substances **(Figure 1.5)**. This will help the technician not only analyze a situation, but safely mitigate an incident.

Figure 1.5 In addition to knowing what types of materials are shipped in what types of containers, a hazmat technician will need to know how those materials react under certain circumstances. *Courtesy of Steve George.*

Once the likely behavior of a product is predicted, responders must use their knowledge and skill to estimate the scope of the incident. Understanding the chemical and physical properties of a product can assist in determining this information. Chemical reference manuals and databases not only assist in estimating the scope but also provide pertinent information, such as plume modeling that assists in establishing the product's path of travel and impact.

The hazmat technician should also have a general knowledge of chemistry and the associated properties of substances. Understanding chemistry helps a technician predict likely behavior of a product and can be helpful in both product control and decontamination.

Planning the Initial Response

Once the research is underway and the hazards and products involved are understood in a basic way, the hazmat technician must use that information to plan the response. Thinking through the situation and developing a solid strategy with realistic response objectives should deliver a successful and safe outcome. Before initiating a response, technicians must answer all of the following questions:

- What response option will be used?
- Will the mitigation be done offensively or defensively?
- What equipment and PPE will be needed to support the chosen option?
- What decontamination process will be used?

Developing Response Objectives and Options

Response to a hazardous materials incident is different from other responses that emergency personnel see in the course of their careers. Incidents such as structure fires and emergency medical responses require quick decisions and quick actions. These types of responses are time sensitive, and quick intervention will usually help those in need.

For a hazardous materials response, a quick response and a quick intervention may not only be counterproductive, but may have deadly consequences. Hazardous materials response at any level must be organized and methodical. Many factors will affect the outcome of a hazardous materials incident. Some of these factors are not in the responder's control, but those that are must be taken into consideration and evaluated appropriately.

The use of a risk-based response model must be considered to analyze the problems at hand and used to make appropriate decisions. A risk-based approach should be taken with regards to considering protective clothing options, control measures, and decontamination options. There are a variety of risk-based models to choose from. A few of those models will be discussed in later chapters.

Selecting Appropriate Personal Protective Equipment

When responding at the Operations Level, responders should know what PPE is available and understand the limits of that PPE. At the Technician Level, PPE selection becomes critical because the responder will be working in an offensive capacity. The hazmat technician must be familiar with PPE options from both a respiratory and chemical protective clothing (CPC) standpoint. Chapter 11 of this manual includes much more information on personal protective equipment, including chemical protective equipment.

Selecting Decontamination Procedures

Another duty of the Technician Level responder is to select appropriate decontamination procedures. This activity may include Operations Level response personnel, and the technician must be proficient in establishing and implementing mass and technical decon operations for ambulatory and nonambulatory victims and responders **(Figure 1.6)**. Decontamination operations are discussed in Chapter 12 of this manual.

Figure 1.6 Decontamination options must be selected for their suitability to the intended outcome.

Developing Action Plans

As stated earlier in this chapter, an IAP must be developed before a response plan can be implemented. The IAP may need to be written, depending on the scope of the incident. All responders must understand what issues are going to be dealt with, what products are involved, and what actions will be taken to safely mitigate the incident.

Some IAP components are common to emergency responses; however, the majority of the components must be developed with the unique incident and hazards in mind. Today, emergency responders have the assistance of the NIMS to help guide them through the development of an IAP. Incident action plans are discussed further in Chapter 9 of this manual.

Implementing the Planned Response

As stated earlier, an organized and controlled response methodology is needed to safely mitigate a hazardous materials incident. To successfully implement these actions, the duties of a hazmat technician must be examined. In the Awareness and Operations response Levels, the responder keeps his or her distance from the actual product. With the exception of Mission Specific competencies, response at these lower levels will always be in a defensive mode. Responders trained to the Technician Level are expected to move forward and take an offensive posture and be "hands-on" in their work **(Figure 1.7)**. This requires these individuals to have an in-depth background in all facets of the Technician Level response.

Figure 1.7 Hazmat technicians must be ready to take offensive actions when they can do so safely.

Performing Assigned ICS/IMS Duties

The duties of the hazmat technician may be diverse based on the chosen response model and the area of responsibility the technician is assigned. Primarily, the technician must work within the structure of the Incident Command System and Incident Management System (ICS/IMS). Whether the technician is in a command role or has been assigned to function within a hazardous materials group or branch, personnel responding at this level must have an understanding of these roles and responsibilities. Incident management is discussed further in Chapter 10 of this manual.

Using Personal Protective Equipment

Because the technician is operating in an offensive mode, response personnel at this level must be proficient in using respiratory protection and protective clothing. The technician must be fully aware of all safety procedures when working in chemical protective clothing. When wearing vapor-protective clothing, these safety procedures escalate even farther. The technician also must be aware of safety procedures that include:

- Actions necessary when there is a loss of air supply
- Actions necessary when there is a loss of suit integrity
- Actions necessary when there may be a responder down in the hot zone

Performing Control Functions

The primary objective of a hazardous materials technician is to control a release of product from a variety of containers, including tanks or vessels of varying sizes. The hazmat technician must be able to select and utilize the proper tools, equipment, and materials to successfully control the release of a hazardous material from nonbulk containers. They must also be able to **overpack** a variety of nonbulk containers including such things as drums and radioactive materials packaging.

> **Overpack** — (1) To enclose or secure a container by placing it in a larger container. (2) An outer container designed to enclose or secure an inner container.

Performing Decontamination

In any given hazardous materials incident, the hazmat technician will be called upon to enter a contaminated area and perform work that may put the responder in close proximity to a dangerous chemical or hazardous substance. As a result, decontamination is needed to clean the responder, victims, and equipment when leaving the hazard area. Based on the contaminant and information gathered from multiple references, the hazmat technician will choose the correct level and method of decon process.

Evaluating Progress

The last phase of the APIE process is evaluation. The plan must be evaluated to ensure that progress is being made and that the desired outcome will be achieved. If findings reveal that the incident is not improving or the situation is escalating, then the plan must be reevaluated. In some instances it may be necessary to transition from offensive to defensive or nonintervention activities if continuing operations become unsafe for responders.

If during the evaluation phase of this response model it is determined that the expected outcome may not be realized, the Incident Commander (IC) must refer back to the IAP and begin APIE again. It will be necessary to reanalyze the incident to determine if responders achieve their expected outcome. If new information is gathered, it will be necessary to perform another planning session to determine a proper course of action. This cyclical process will continue until the incident is considered stable.

When evaluating the incident, the hazmat technician must also determine the effectiveness of the decon process. Because cross contamination is a significant concern, the decon process must be adequate in removing all contaminants.

Terminating the Incident

An incident does not truly conclude until all necessary paperwork has been completed. Whether reporting and documenting a structure fire or a hazardous materials incident, there are requirements that must be met to formally terminate an incident. While the documentation may be viewed as an administrative requirement, there is much to be learned from each and every incident.

Written documentation is not the only requirement in the termination process. Debriefing is a critical component of this phase of the incident. The Command team must provide valuable information to responders. Some critical aspects of the incident must be relayed to all response personnel. Some of the components include relevant observations, actions taken, and a timeline of the incident.

In addition, one very important step in this process is to provide information to personnel concerning the signs and symptoms of overexposure to the hazardous material or materials involved in the incident. This is called the *hazard communication briefing* and is required by OSHA in the United States. This debriefing process must be thoroughly documented. Each person attending must receive and understand the instructions and sign a document stating those facts. The information provided to responders before they leave the scene includes the following:

- Identity of material involved
- Potential adverse effects of exposure to the material
- Actions to be taken for further decontamination
- Signs and symptoms of an exposure
- Mechanism by which a responder can obtain medical evaluation and treatment
- Exposure documentation procedures

Documentation will serve as a timeline of the actions taken and a permanent record of the events that transpired at the incident. Some documents and reports that may be required include:

- Incident reports
- Debriefing records
- Critique records
- Reimbursement logs

Finally, a postincident critique must be performed. OSHA Title 29 *CFR* 1910.120 requires that incidents be critiqued for the purpose of identifying operational deficiencies and learning from mistakes. As with all critiques performed by the fire and emergency services, hazmat incident critiques need to occur as soon as possible after the incident and involve all responders, including law enforcement, public works, and emergency medical services

(EMS) responders. As with other administrative and emergency response functions, the critique is documented to identify those in attendance as well as any operational deficiencies that were identified.

Chapter Notes

Occupational Safety and Health Administration (OSHA). "Foundation of Workplace Chemical Safety Programs." https://www.osha.gov/dsg/hazcom/global.html

Chapter Review

1. What are the four steps of the APIE process?
2. What are the seven clues that Awareness and Operations level responders can use to detect the presence of hazardous materials?
3. How do the responsibilities of the Operations Core level differ from the Operations Mission Specific level?
4. What are the Hazardous Materials Technician's responsibilities in analyzing the incident?
5. What are the Hazardous Materials Technician's responsibilities in planning the initial response?
6. What are the Hazardous Materials Technician's responsibilities in implementing the planned response?
7. What are the Hazardous Materials Technician's responsibilities in evaluating progress and terminating?

Analyzing the Incident: Understanding How Matter Behaves

Chapter Contents

Physical States of Matter **29**
    Solids 31
    Liquids 31
    Gases 31

Physical Properties of a Material **34**
    Density 35
    Specific Gravity 35
    Vapor Density 36
    Viscosity 37
    Odor 37

    Appearance 38

Temperature and Pressure **38**
    Temperature 38
    Pressure 40
    Molecular Weight and Polarity 41

Phase Changes and Related Properties ... **41**
    Temperature Changes 42
    Relationships of Physical Properties 44

Chapter Review **45**

chapter 2

Key Terms

Anhydrous 34	International System of Units (SI) 32
Bar 32	Liquefied Gas 33
Boiling Point 42	Liquid 29
Celsius Scale 38	Lower Flammable (Explosive) Limit (LFL) 39
Combustible Liquid 38	
Condensation 43	Melting Point 42
Critical Point 44	Millimeters of Mercury (mmHg) 32
Density 35	Molecular Weight 37
Evaporation 43	Oxidizer 39
Evaporation Rate 43	Pascals (Pa) 32
Expansion Ratio 32	Phase 41
Fahrenheit Scale 38	Safety Data Sheet (SDS) 38
Flammability 39	Solid 29
Flammable Liquid 38	Specific Gravity 35
Flammable Range 39	Sublimation 43
Flash Point 38	Torr 32
Freezing Point 42	Upper Flammable Limit (UFL) 39
Gas 29	Vapor Density 37
Hydrophilic 34	Viscosity 37
Hydrophobic 34	Volatility 43

JPRs addressed in this chapter

7.2.1, 7.2.4

Analyzing the Incident: Understanding How Matter Behaves

Learning Objectives

After reading this chapter, students will be able to:

1. Describe characteristics of matter in solid, liquid, and gaseous forms. [NFPA 1072, 7.2.1, 7.2.4]
2. Explain ways that a material's physical properties can influence its behavior. [NFPA 1072, 7.2.1, 7.2.4]
3. Describe how temperature and pressure relate to each other and to physical states of a material. [NFPA 1072, 7.2.1, 7.2.4]
4. Explain phase changes and related properties of materials. [NFPA 1072, 7.2.1, 7.2.4]

---

1. solid: defined shape
   liquid: flows
   gas: hardest to control

2. vapor pressure: evaporate? how quickly?
   vapor density: will it rise or sink in the air
   specific gravity
   viscosity: how thick   is it a polymer?
   LEL and UEL   volatylity: evaporates at a low temperature.
   will it mix with H2O   sublimation = solid to a gas

3. as temp goes up vapor pressure goes up

4. water/liquid can go to a gaseous state if it reaches its boiling point in a chemical reaction. Will expand.

NFPA 1072

Chapter 2
Analyzing the Incident: Understanding How Matter Behaves

At a hazmat incident, it is important to identify the physical state of a material as early as possible. The material's state of matter will indicate how mobile that material may become and can help determine if there will be far-reaching hazardous properties. This chapter will discuss the following topics:

- Physical States of Matter
- Physical Properties of a Material
- Temperature and Pressure
- Phase Changes and Related Properties

Physical States of Matter

Everything as we know it exists as some form of matter. Matter exists in three primary forms (**Figure 2.1**):

- **Solid**
- **Liquid**
- **Gas**

Solid — Substance that has a definite shape and size; the molecules of a solid generally have very little mobility.

Liquid — Incompressible substance with a constant volume that assumes the shape of its container; molecules flow freely, but substantial cohesion prevents them from expanding as a gas would.

Gas — Compressible substance, with no specific volume, that tends to assume the shape of a container. Molecules move about most rapidly in this state.

Figure 2.1 The three primary states of matter, in increasing order by mobility, are solids, liquids, and gases.

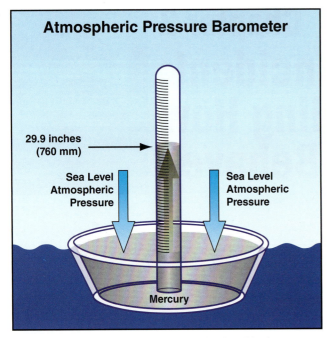

Figure 2.2 Atmospheric pressure varies by altitude.

Generally speaking, solids are the least mobile and gases have the greatest mobility. Liquids are less mobile than gases and will flow from their container and follow the law of gravity. Responders equipped with this knowledge can determine control zones and evacuation distances.

The physical state of a material is determined at normal temperatures and pressures. For the purpose of this chapter, assume temperatures between 68°F and 77°F (20°C and 25°C) and the pressure to be atmospheric pressure (14.7 psi) **(Figure 2.2)**.

Some matter can exist in different states. For example, water is liquid in its natural state and can become a solid (ice) below 32°F (0°C) or a gas (steam) above 212°F (100°C). Temperature is the determining factor, as is the case with many materials that change state. Pressure is sometimes used to cause a change in state, as is the case with liquefied compressed gases. Both temperature and pressure are used to transform gases into cryogenic liquids.

Matter can exist in multiple states within the same container at a hazardous materials incident, such as when a liquefied gas vaporizes as it escapes a breached container. In these instances, the product exists as a liquid inside the container and as a gas outside. Even within a container, there may be liquid space at the bottom and a vapor space at the top—a distinction that affects the incident. For example, fire impinging on the vapor space stresses a container far more so than fire impinging on the liquid space because the liquid helps absorb the heat whereas vapors do not.

The following items act like a gas and can be very mobile **(Figure 2.3)**:

Figure 2.3 Releases that act like gases can take several forms and have unique characteristics.

- **Vapor** — Gaseous state of a material that may normally be a solid or a liquid.
- **Aerosol** — Microscopic particles that may be a solid or a liquid.
- **Dust** — Airborne solid particles that may be 0.1-50 microns in diameter. Dust particles less than 50 microns cannot be seen without a microscope.
- **Mist** — Aerosol of liquid particles suspended in air.
- **Fog** — Visible aerosol of a liquid formed by condensation.

The state of matter is largely dependent on the temperature of a substance. The state may change if the temperature changes. For example, a solid may change to a liquid if the temperature increases. If a material is located outdoors, temperature and weather can have a profound effect on a material's state of matter.

NOTE: Weather factors such as atmospheric pressure, relative humidity, dew points, and temperature can affect the behavior and chemical and physical properties of materials.

Physical properties of a state of matter are a factor of intermolecular attraction. For example, there is a weak interaction between gas molecules and strong interaction between the molecules of a solid. Attractions and bonding will be discussed in Chapter 3.

Solids

A solid will have specific mass and volume, and it will take on a specific shape. Solids are the least mobile of the three states of matter. Based on its mass and surface area, a solid may have the ability to become airborne and travel with air movement and currents. For example, in its natural state, and based on its dimensions, a piece of wood may be able to move only in strong winds. However, reducing the piece of wood to sawdust changes the surface area, and a light breeze or even natural air currents may cause the material to become airborne. Dust is even more prone to movement by natural air currents.

Liquids

Liquids often present unique challenges. From a response standpoint, a liquid may be more difficult to contain than a solid, and its vapor has the characteristics of a gas. This conversion can increase its mobility and, consequently, the challenges that technicians will face when dealing with the material. The molecules of a liquid may flow freely. Like a solid, a liquid will have specific mass and volume, but because of the loose molecular attraction, the liquid will not have a definitive shape and will take on the shape of its container.

Gases

Hazardous materials in gaseous form are typically the most challenging to control. Like solids and liquids, gases will have a specific mass but will not have a specific volume or shape. A gas is a substance that will expand to fill any given volume. Gas molecules move with a great deal of energy. Gases are the least dense of the three states of matter and have the ability to compress. The cooling of gases can sometimes cause them to take liquid form.

Emergency responders will typically be at the greatest risk when dealing with a gas. Many hazmat related contaminations and exposures are due to

Expansion Ratio — 1) Volume of a substance in liquid form compared to the volume of the same number of molecules of that substance in gaseous form. 2) Ratio of the finished foam volume to the volume of the original foam solution. *Also known as* Expansion.

the inhalation of vapors or gases. A gaseous material will have many variables and hazards. These materials:

- May or may not have an odor
- May be colorless and/or tasteless
- May be toxic, corrosive, or flammable
- May be under extremely high pressure in excess of 15,000 psi (103 500 kPa)
- May be extremely cold upon release and may have a large **expansion ratio** if liquefied **(Figure 2.4)**

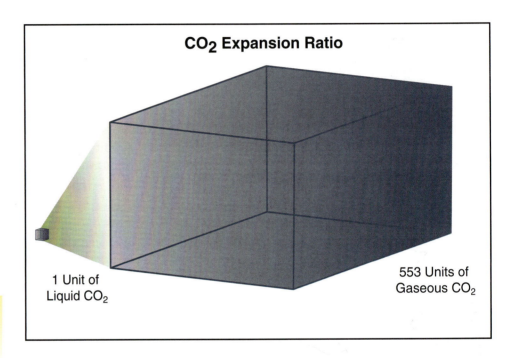

Figure 2.4 When liquefied carbon dioxide (CO_2) returns to a gaseous form, it will expand to over 500 times its liquid volume.

International System of Units (SI) — Modern form of the metric system of measurement that standardizes mathematical quantification.

Pascals (Pa) — SI unit of measure used to indicate internal pressure and stress on a container.

Millimeters of Mercury (mmHg) — Unit of pressure measurement; not part of the SI. Currently defined as a rate rounded to 133 Pascals. Rough equivalent to 1 torr.

Bar — Unit of pressure measurement; not part of the SI. Equals 100 000 Pa.

Torr — Unit of pressure measurement; not part of the SI. Measured as 1/760 of a standard atmosphere.

Gases are the only state of matter that can be compressed tightly or expanded to fill a large space. Hazmat incidents involving gases typically include products contained under pressure. Therefore, it is important to understand how gases behave when under pressure.

Atmospheres are normally used as a standard unit of pressure in the United States. In the **International System of Units (SI)**, units of pressure are typically measured in **Pascals (Pa)**. One atmosphere (atm) is equal to 101.325 Pa. One atm is defined as 760 **millimeters of mercury (mmHg)** and is the pressure normally found at sea level. Less commonly used units of pressure are **bar** and **torr**.

Gases have a vapor pressure greater than 760 mm Hg at 68°F. Common equivalent measures for pressure include:

- 1 atm = 760 mm Hg = 407 in H2O = 14.7 psi = 101 kPa = 760 torr = 29.92 inches Hg
- 1 bar = 14.5 psi
- 1 ft H2O = 0.43 psi

Compressed Gases

Hazardous materials technicians may encounter gases in different forms or states. Compressed gases are compounds and mixtures in a gaseous form that are shipped in pressurized cylinders at ambient temperatures **(Figure 2.5)**. Pressures in these cylinders may range from a few psi (kPa) to 15,000 psi (103 500 kPa).

Figure 2.5 Compressed hydrogen gas is stored at high pressures and ambient temperatures. *Courtesy of Rich Mahaney.*

Figure 2.6 Liquefied petroleum gas boils below ambient temperatures, so it is often stored in pressurized steel vessels.

Liquefied Gases

Liquefied gases are defined as gases kept liquefied under pressure **(Figure 2.6)**. The critical temperature and pressure of a gas refer to the temperature and pressure needed to liquefy the gas. A gas cannot be liquefied by pressure alone if it is above its critical temperature. For example, carbon dioxide (CO_2) has a critical temperature of 88°F (31°C). Above this critical temperature, CO_2 cannot be liquefied regardless of the amount of pressure applied to the product.

> **Liquefied Gas** — Confined gas that at normal temperatures exists in both liquid and gaseous states.

Cryogenic Liquids

Cryogenic liquids are liquefied gases at very low temperatures **(Figure 2.7)**. The U.S. DOT defines a cryogenic liquid as a refrigerated liquid having a boiling point lower than -130°F (-90°C) at 14.7 psi (103 kPa). The distinction between cryogenic liquids and liquefied gases is that cryogens, such as liquid nitrogen or liquid oxygen, remain liquid at these low temperatures at atmospheric pressure. Cryogens also have high expansion ratios.

Cryogenic liquids, liquefied gases, and compressed gases have their own unique hazards in addition to those associated with the chemical reactions the materials might undergo or the toxicity of the materials. These hazards include:

Figure 2.7 Some gases must be refrigerated for storage as a liquid.

- Vaporizing liquid or the gas released from a pressurized cylinder may displace oxygen, causing asphyxiation.
- The extreme cold of cryogenic liquids and liquefied gases may cause severe frostbite and the destruction of exposed body tissue.
- The extreme cold of cryogenic liquids may cause metals to become brittle, leading to failure of the container, supporting structures, or other exposed metal parts.
- Rapid expansion of liquefied gases or cryogenic liquids can lead to the buildup of extremely high pressure and the violent failure of pressurized equipment. This explosive release of gas is similar in effect to a boiling liquid expanding vapor explosion (BLEVE).
- Some pressurized gases may generate large static-electric charges when they are released, leading to the potential to ignite flammable atmospheres.
- Some cryogenic liquids and liquefied gases possess unique reactive hazards beyond the hazards associated with the same materials at ambient temperatures. For example, liquid oxygen can react with asphalt to form a powerful and sensitive contact explosive which can be detonated by the pressure of a single footstep **(Figure 2.8)**. Similarly, liquid chlorine reacts violently with almost any organic material, including wood and cloth, leading to extreme fire danger in the vicinity of a spill.

Figure 2.8 Liquid oxygen reacts with some materials differently than oxygen in its gaseous form, and must be stored under specific conditions.

Physical Properties of a Material

Materials can be characterized by their water content and reactions to water:

- **Anhydrous** is a term that means "dry" or without water. A product can be a highly impure mixture but still be considered anhydrous. For example, Class D breathing air is not a single or "pure" element, but contains oxygen, hydrogen, and nitrogen along with other elements. If the moisture content in the air cylinder has a dew point of less than -50°F (-45°C), it can be considered anhydrous.
- **Hydrophobic** elements will repel water. For example, car wax repels water, causing it to bead and run off.
- **Hydrophilic** elements and materials absorb water. For example, most common fabrics wick up water.

Materials can also be characterized by their physical properties:

- Density
- Specific gravity
- Vapor density

Anhydrous — Material containing no water.

Hydrophobic — Material that is incapable of mixing with water.

Hydrophilic — Material that is attracted to water. This material may also dissolve or mix in water.

- Viscosity
- Odor
- Appearance

Density

Density is a measure of how heavy a unit volume of a substance is, or the mass of a known volume. The density of a substance changes with temperature. Usually, as matter gets colder, it gets denser. As matter gets warmer, it gets less dense. The comparison of densities tells you whether an item will sink or float in a type of liquid. For example, a wax candle will float in water, but sinks in rubbing alcohol because wax is less dense than water but more dense than the rubbing alcohol **(Figure 2.9)**.

Density — Mass per unit of volume of a substance; obtained by dividing the mass by the volume.

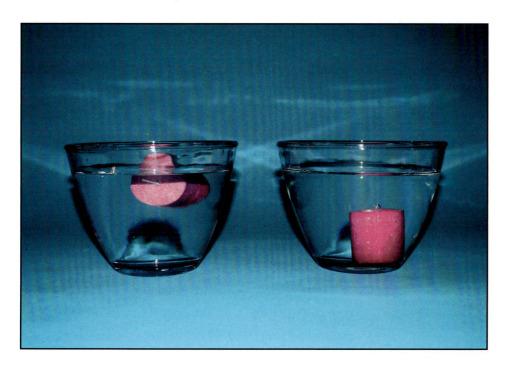

Figure 2.9 Relative density of a material will help you predict whether a material will float or sink in another material.

Density is useful to determine how heavy the material is. When you look up density in a reference source, it is usually given in units of grams per milliliter (g/ml). For hazardous materials response, a more useful measurement is pounds per grams (lb/g). To convert g/ml to lb/gal, multiply g/ml by 8.34.

NOTE: Grams per milliliter (g/ml) is the same measurement as grams per cubic centimeter (g/cc).

Specific Gravity

Every solid and liquid in the environment occupies a specific volume of space and has a certain weight at standard temperature and pressure. For example, water weighs 62.4 pounds per cubic foot (lb/ft^3) or one gram per cubic centimeter (g/cm^3). Different materials will have different weights per volume.

Another way of expressing the weight of a solid or a liquid is **specific gravity**. Specific gravity can be determined by dividing the density of a substance by

Specific Gravity — Mass (weight) of a substance compared to the weight of an equal volume of water at a given temperature. A specific gravity less than one indicates a substance lighter than water; a specific gravity greater than one indicates a substance heavier than water.

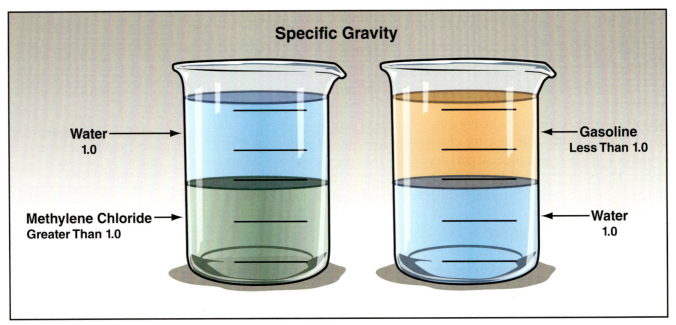

Figure 2.10 Materials heavier than water, including most chlorinated solvents, will sink. Materials lighter than water, including most hydrocarbons, will float.

the density of water **(Figure 2.10)**. Water has a specific gravity equal to 1. If a material has a specific gravity less than 1, it is lighter than water and therefore would float on water. If a material has a specific gravity greater than 1, it will sink in water.

Reference manuals usually express specific gravity at a certain temperature because specific gravity will vary depending on the temperature of the material. If the temperature of the material increases, the material will expand and the density will decrease. Therefore, the weight by volume will decrease.

Vapor Density

Similar to specific gravity, **vapor density** is the ratio of the density of pure gas or vapor to the density of air **(Figure 2.11)**. The density of air is equal to 1. If a material has a vapor density less than 1, it is determined to be lighter than air and that material will rise. If a material has a vapor density greater than 1, then it is determined to be heavier than air and that material will sink. The Compressed Gas Association (CGA) uses the specific gravity of the gas for vapor density. While this may appear confusing, the numbers are the same.

The NIOSH RgasD (relative gas density) measurement is identical to vapor density. NIOSH defines RgasD as the relative density of gases referenced (compared) to air, where air = 1. Like vapor density, this value will indicate how many times a gas is heavier than air at the same temperature.

If a reference manual does not list the vapor density of a material, it is still possible to estimate the measurement. Simply divide the **molecular weight (MW)** of the

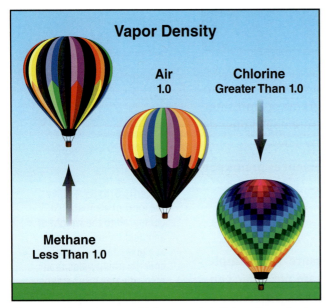

Figure 2.11 As a general rule, materials lighter than air at the ambient temperature will rise and materials heavier than air will sink.

36 Chapter 2 • Analyzing the Incident: Understanding How Matter Behaves material by the average molecular weight of air (29) to obtain the estimated vapor density. For example, chlorine has a molecular weight of 71. Dividing 71 (chlorine's molecular weight) by 29 (the molecular weight of air) yields an estimated vapor density of 2.45. Reference materials list the vapor density of chlorine at 2.49. While not exact, this estimate is adequate for response. A product with a molecular weight greater than 29 will sink; a molecular weight less than 29 will rise.

NOTE: Hydrogen fluoride (HF) is the exception to this rule: HF vapor density is 3, but its molecular weight is 20. An introduction to basic molecules is included in Chapter 3.

When discussing vapor density, the temperature of the vapor must also be considered. Hot vapors will rise, but unless dispersed they will sink when cooled. Cold vapors are very dense and will tend to rise when they warm. Most materials are heavier than air.

Viscosity

Viscosity is the measure of the internal friction of a liquid at a given temperature. Viscosity determines the ease with which a product will flow **(Figure 2.12)**. Usually, the hotter a liquid, the thinner or more fluid it becomes. Likewise, the cooler a liquid, the thicker or less fluid it becomes.

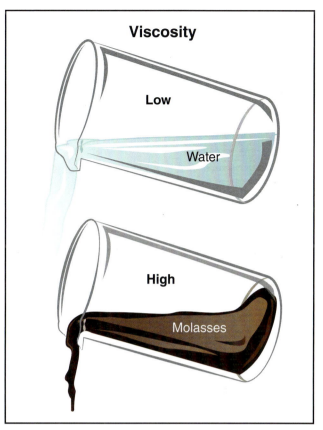

Figure 2.12 Viscosity is the internal friction, or flowability, of a liquid.

Odor

Odor is an expression of how people perceive chemical scents. Some chemicals have little or no odor, while others have a strong characteristic odor. An unexpected odor may be a warning that a substance has escaped from its container. Callers/victims may provide odor characteristics when they alert authorities. Responders can use these reports for initial information, but they must use detection and monitoring to evaluate an atmosphere.

Vapor Density — Weight of pure vapor or gas compared to the weight of an equal volume of dry air at the same temperature and pressure. A vapor density less than one indicates a vapor lighter than air; a vapor density greater than one indicates a vapor heavier than air.

WARNING!
Inhalation of materials can cause injury or death.

WARNING!
Never attempt to use odors to define safe areas.

Molecular Weight (MW) — Average mass of one molecule. This can be calculated as the sum of the atomic masses of the component atoms.

Odor Detection and Recognition Thresholds

The use of odor detection and recognition thresholds is a dangerous metric because not everyone will detect materials in the same way and at the same levels. Using the strength, prevalence, or location of a smell as a metric for detection is not measurable or testable, and may place victims and responders in unwarranted jeopardy.

Viscosity — Measure of a liquid's internal friction at a given temperature. This concept is informally expressed as thickness, stickiness, and ability to flow.

Safety Data Sheet (SDS) — Reference material that provides information on chemicals that are used, produced, or stored at a facility. Form is provided by chemical manufacturers and blenders; contains information about chemical composition, physical and chemical properties, health and safety hazards, emergency response procedures, and waste disposal procedures. *Also known as* Material Safety Data Sheet (MSDS) *or* Product Safety Data Sheet (PSDS).

Celsius Scale — International temperature scale on which the freezing point is 0°C (32°F) and the boiling point is 100°C (212°F) at normal atmospheric pressure at sea level. *Also known as* Centigrade Scale.

Fahrenheit Scale — Temperature scale on which the freezing point is 32°F (0°C) and the boiling point at sea level is 212°F (100°C) at normal atmospheric pressure.

Flash Point — Minimum temperature at which a liquid gives off enough vapors to form an ignitable mixture with air near the surface of the liquid.

Flammable Liquid — Any liquid having a flash point below 100°F (37.8°C) and a vapor pressure not exceeding 40 psi absolute (276 kPa) {2.76 bar}, per NFPA.

Combustible Liquid — Liquid having a flash point at or above 100°F (37.8°C) and below 200°F (93.3°C), per NFPA.

Appearance

A material's appearance is important. Elements, compounds, and mixtures have a characteristic appearance consistent with their composition. A description of the appearance of a material (physical state, color, etc.) will normally appear on a **safety data sheet (SDS)**. For example, a clear liquid may be described as "water white."

For many industrial products, the color listed on the SDS may represent an "average" and the product shipped may vary significantly in color and still be the same product. In other instances, a significant difference in color may also show contamination or high levels of impurities that may have their own hazards.

Temperature and Pressure

Many variables can affect the physical properties of a product. In many cases, the effects of the variables are interrelated.

Temperature

Common temperature measurement scales are **Celsius (°C)** and **Fahrenheit (°F)**; some measurements use other scientific scales such as Kelvin (K) or Rankine (°Ra). Increasing heat in an object will increase the movement of molecules in the object.

Flash Points and Fire Points

Flash point is the lowest temperature at which a liquid can form an ignitable mixture in air near the surface of the liquid and will not support continuous burning. The NFPA defines **flammable liquids** as products that have a flash point below 100°F (38°C) and **combustible liquids** as products that have a flash point above 100°F (38°C). The U.S. DOT defines a flammable liquid as having a flash point below 141°F (60°C).

Flash point is expressed in terms of temperature of the liquid being tested: the lower the flash point of a product, the higher the **flammability** of the product. Remember that liquids do not burn. Instead, the vapors produced by the liquids actually burn.

Molecular weight is a factor that can affect the flash point of a product. In general, the lower the molecular weight, the lower the flash point. Also, if the product is a liquid, the vapors are typically heavier than air. Conversely, if a material has a high molecular weight, it will typically have a high flash point (**Figure 2.13**). While this is not a scientific certainty, it is good information to remember in the field.

The fire point is the minimum temperature to which a substance must be heated so it produces enough vapor to support sustained combustion. The fire point is generally a few degrees above the flash point.

Flammable Range

The **flammable range** (also known as explosive, or combustible range), is the percentage of the gas or vapor concentration in air that will burn or explode if ignited (**Figure 2.14**). The **lower flammable limit (LFL)**, or lower explosive limit (LEL), of a vapor or gas is the lowest concentration of a material in air

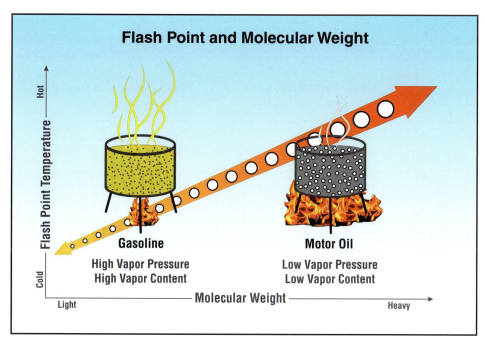

Figure 2.13 Gasoline has both a low flash point and a low molecular weight. Motor oil has both a high flash point and a high molecular weight.

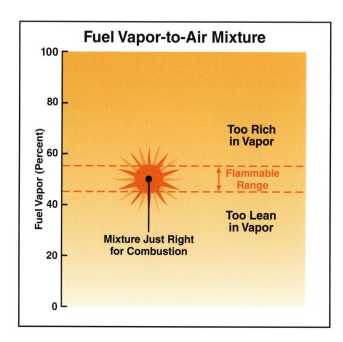

Figure 2.14 Materials within the flammable range will readily ignite.

Flammability — Fuel's susceptibility to ignition.

Flammable Range — Range between the upper flammable limit and lower flammable limit in which a substance can be ignited. *Also known as* Explosive Range.

Lower Flammable (Explosive) Limit (LFL) — Lower limit at which a flammable gas or vapor will ignite and support combustion; below this limit the gas or vapor is too *lean* or *thin* to burn (too much oxygen and not enough gas, so lacks the proper quantity of fuel). *Also known as* Lower Explosive Limit (LEL).

Upper Flammable Limit (UFL) — Upper limit at which a flammable gas or vapor will ignite; above this limit the gas or vapor is too rich to burn (lacks the proper quantity of oxygen). *Also known as* Upper Explosive Limit (UEL).

Oxidizer — Any material that readily yields oxygen or other oxidizing gas, or that readily reacts to promote or initiate combustion of combustible materials. (Reproduced with permission from NFPA 400-2010, *Hazardous Materials Code*, Copyright©2010, National Fire Protection Association)

that will ignite when an ignition source is present. At concentrations lower than the LEL, the mixture is too lean to burn.

The **upper flammable limit (UFL)**, or upper explosive limit (UEL), of a vapor or gas is the highest concentration of a material in air that will ignite when an ignition source is present. At higher concentrations, the mixture is too rich to burn. Within the upper and lower limits, the gas or vapor concentration will burn rapidly if ignited. Atmospheres within the flammable range are particularly dangerous **(Table 2.1, pg. 40)**. The oxygen concentration may also be a factor, when **oxidizers** are not present. Oxidizers are discussed in Chapter 4 of this manual.

Chapter 2 • Analyzing the Incident: Understanding How Matter Behaves **39**

Table 2.1
Flammable Ranges for Selected Materials

Material	Lower Flammable Limit (LFL) (percent by volume)	Upper Flammable Limit (UFL) (percent by volume)
Acetylene	2.5	100.0
Carbon Monoxide	12.5	74.0
Ethyl Alcohol	3.3	19.0
Fuel Oil No. 1	0.7	5.0
Gasoline	1.4	7.6
Methane	5.0	15.0
Propane	2.1	9.5

Source: *NIOSH Pocket Guide to Chemical Hazards*

Combustible Metal Reaction, July 13, 2010

Close to midnight, first responders arrived at the scene of a large commercial structure with heavy fire. Reinforcement resources were requested. Because the conditions at the scene were unstable and deteriorating, a defensive mode attack was implemented after a short while.

A large explosion created new hazards: fires on the north and south of the structure, injuries to seven firefighters, and damage to apparatus. The IC realized that the structure contained combustible materials. New tactics used more unmanned ladder pipes and an indirect approach to the areas that were likely to include burning metals. A second explosion occurred some time later, when water contacted burning combustible metals, but no firefighters were injured. The structure was a recycling collection station for combustible metals.

Source: NIOSH Report #F2010-30

Pressure

Pressure can affect the physical state and the state change of a material. Atmospheric pressure can control different variables such as the boiling point of a substance. As pressure decreases, the amount of vaporization will increase at a given temperature. As pressure increases, the amount of vaporization will decrease at that given temperature.

Vapor Pressure

Vapor pressure is the measurement of the ability of a material to evaporate or change from a liquid to a vapor or gaseous state. Substances with high vapor pressure are volatile and give off significant amounts of vapor **(Figure 2.15)**. Materials with a vapor pressure over 760 mmHg will be gases under normal conditions. Materials with low boiling points have high vapor pressure.

Vapor pressure increases with temperature. In most cases, if the product within a container is cooled, the vapor pressure can be reduced. At a hazmat incident this may be done by pumping a colder material through the cooling coils (if equipped), placing the container in dry ice, or spraying water on it. Vapor pressure can also be reduced by venting the container. This action also provides a cooling effect but should only be done in extreme circumstances, because an environmental problem could result. This effect is known as *autorefrigeration*.

Temperature affects vapor pressure. A liquid can be heated above its boiling point in a closed container. If the container is large enough, only vapor may exist. If the container is not large enough to contain the vapor, tremendous pressure will be created. This is extremely serious because if the container fails, superheated fluid will expand at tremendous velocities in a BLEVE. BLEVEs will be discussed in greater detail in Chapter 8.

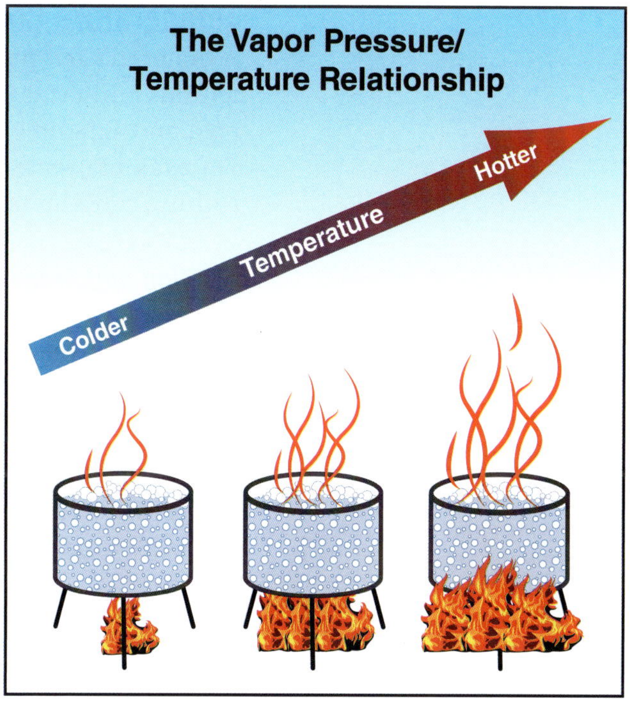

Figure 2.15 Vapor pressure increases with temperature.

Vapor Expansion

When a material volatilizes or evaporates, the vapors occupy a specific volume. The ratio of how much vapor is created in relation to the volume of the original liquid or solid is called the vapor expansion ratio. Some compounds that are shipped as liquids under pressure or as cryogenic liquids will expand at several hundred times their original volume. For example, chlorine has a vapor expansion of 460:1. This means that for every gallon of liquid chlorine that evaporates, an equivalent of 460 gallons of vapor is produced.

Molecular Weight and Polarity

Other variables that can affect the state of a product include molecular weight and polarity. Hydrocarbons and derivatives with lower molecular weight tend to be gases or nonviscous liquids. Hydrocarbons and derivatives with higher molecular weight tend to change from liquids to viscous liquids and eventually to solids.

Polarity is another variable that can affect the physical properties of a material. Because of the intermolecular attraction caused by their polarity, molecules and atoms have a decreased distance between each other. Polarity will be discussed in greater detail later in this chapter. Bonding will be discussed in greater detail in Chapter 4.

Phase Changes and Related Properties

Given the right conditions, matter can change its state from a solid to a liquid, from a liquid to a gas, and back again. Yet its chemical composition is unchanged. When a substance changes from one state of matter to another, we say that it has undergone a "change of state" or a "change of **phase**." These changes of phase always occur with a change of heat. Heat, which is energy, either comes into the material during a change of phase or heat goes out of the material during this change.

> **Phase** — Distinguishable part in a course, development, or cycle; aspect or part under consideration. In chemistry, a change of phase is marked by a shift in the physical state of a substance caused by a change in heat.

Temperature Changes

A material's temperature will influence both the range of hazards and potential countermeasures. As a normal rule, materials contract when they get cold and expand when they get hot; however, significant exceptions apply. For example, water expands when it freezes, and has sufficient pressure to rupture containers.

Melting and Freezing

The temperature at which a solid turns into a liquid is called its **melting point**. The melting point is a critical property to analyze for emergency response. A solid material may change to a liquid if heated sufficiently **(Figure 2.16)**. It is typically easier to control a solid than a liquid.

The **freezing point** is the temperature at which a liquid will solidify or crystallize. The freezing point is typically the same as or similar to the temperature of the melting point; however, there are some exceptions. Some materials are known to "supercool." In other words, they do not crystallize or freeze at the same temperature they melted. The liquid may need to be 5 or 10 degrees cooler before it starts to crystallize or solidify.

Melting Point — Temperature at which a solid substance changes to a liquid state at normal atmospheric pressure.

Freezing Point — Temperature at which a liquid becomes a solid at normal atmospheric pressure.

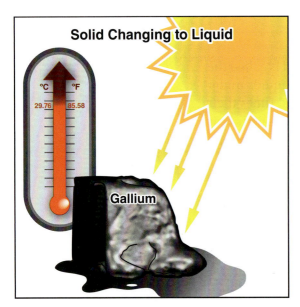

Figure 2.16 The melting point of a solid may vary; for example, gallium will melt at the relatively low temperature of 85 degrees Fahrenheit (30 degrees Celcius).

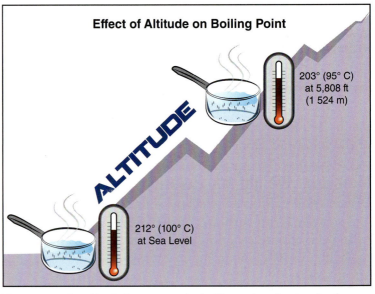

Figure 2.17 As altitude increases, the boiling point of water decreases.

Boiling and Condensation

Boiling is a process in which molecules anywhere in the liquid escape, resulting in the formation of vapor bubbles within the liquid. The **boiling point** is the temperature at which the vapor pressure of a liquid is equal to the atmospheric pressure around the liquid **(Figure 2.17)**. The standard boiling point is defined as the temperature at which boiling occurs under a pressure of 1 bar (1 atmosphere).

Water boils at different temperatures depending on the altitude. For example, water boils at 212°F (100°C) at sea level, or at one atmosphere of pressure. However at 5,000 feet (1 500 m) above sea level, water boils at 203°F (95°C)

Boiling Point — Temperature of a substance when the vapor pressure equals atmospheric pressure. At this temperature, the rate of evaporation exceeds the rate of condensation. At this point, more liquid is turning into gas than gas is turning back into a liquid.

due to the reduced atmospheric pressure. Similarly, increasing pressure (as in a pressure cooker) raises the temperature of the contents above the open air boiling point. Adding a water soluble substance, such as salt or sugar, also increases the boiling point. This is called *boiling-point elevation*.

Liquids may change to vapor at temperatures below their boiling points through the process of **evaporation**. Evaporation is a surface phenomenon in which molecules located near the liquid's edge escape into the atmosphere as vapor. The **evaporation rate** is the speed at which some material changes from a liquid to a vapor. Temperatures will affect evaporation rates — the hotter the temperature, the faster the evaporation rate; the colder the temperature, the slower the evaporation rate.

Some materials change to a vapor more easily than others. Materials that change readily to gases are termed volatile. **Volatility** is usually reported as a comparative value versus another compound.

Once the temperature of a liquid drops below its boiling point, the vapors that are produced will start to cool and form condensation. In other words, the vapors will start to condense back to a liquid. **Condensation** is the conversion of a vapor into a liquid as it is cooled down to, or below, the liquid's boiling point.

> **Evaporation** — Process of a solid or liquid turning into gas.
>
> **Evaporation Rate** — Speed at which some material changes from a liquid to a vapor. Materials that change readily to gases are considered volatile.
>
> **Volatility** — Ability of a substance to vaporize easily at a relatively low temperature.
>
> **Condensation** — Process of a gas turning into a liquid state.
>
> **Sublimation** — Vaporization of a material from the solid to vapor state without passing through the liquid state.

Sublimation

Sublimation is the transformation of a material from a solid into a vapor without passing through a liquid state. For example, dry ice is a solid form of carbon dioxide that is used as a cooling agent in many commercial applications **(Figure 2.18)**. When vaporization occurs, the material goes from a solid to a gas without becoming a liquid.

Figure 2.18 Carbon dioxide (CO_2) in its solid form will sublimate directly into a gas.

Another product that will sublimate easily is iodine. First responders may encounter iodine in areas where illicit methamphetamine is produced. Iodine has the ability to sublimate and travel until it finds a cool spot where it can condense, causing movement of this hazardous chemical to other locations within a production area.

> **Critical Point** — The end point of an equilibrium curve. In liquid and vapor response, the conditions under which liquid and its vapor can coexist.

Critical Points

Critical points are behaviors that can harm people. When a material reaches any critical point, the chemical reaction initiated by that threshold cannot be stopped until the reaction is completed.

The following are several critical points with regards to the temperature and pressure of hazardous materials:

- **Critical temperature** — The minimum temperature above which a gas cannot be liquefied no matter how much pressure is applied.
- **Critical pressure** — The pressure necessary to liquefy a gas at its critical temperature.
- **Autoignition temperature** — The lowest temperature at which a substance will ignite in air when there is no ignition source.
- **Self-accelerating decomposition temperature (SADT)** — The temperature above which the decomposition of an unstable substance continues unimpeded, regardless of the ambient or external temperature.
- **Maximum safe storage temperature (MSST)** — The maximum safe temperature at which a product can be stored. This temperature is well below the SADT.
- **Polymerization** — A special chemical reaction in which small-molecule compounds called monomers react with themselves to form larger molecules called polymers.

Relationships of Physical Properties

Temperature and pressure are directly proportional to one another: as temperature increases and decreases, pressure does the same. As the temperature of matter increases, so does the speed of its molecules **(Figure 2.19)**. Pressure in turn rises due to the molecules' force on the container.

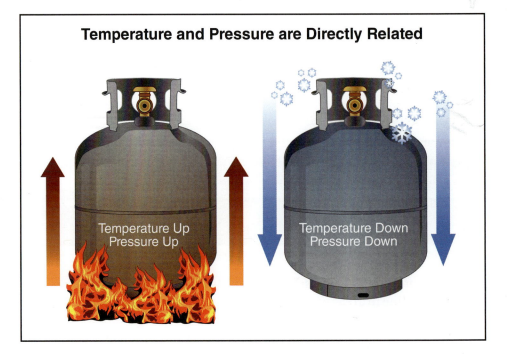

Figure 2.19 Pressure and temperature will increase and decrease in response to each other.

Understanding how the physical properties of a chemical will interrelate can help hazmat technicians understand the product and also the situation around them. If a material has a low boiling point, one can presume that the flash point will also be low. A product with a low boiling point and flash point will produce a large quantity of vapor — thus the product will have a high vapor pressure and vapor content. Conversely, if a product has a high boiling point and flash point, it can be presumed to have a low vapor pressure and vapor content **(Figure 2.20)**.

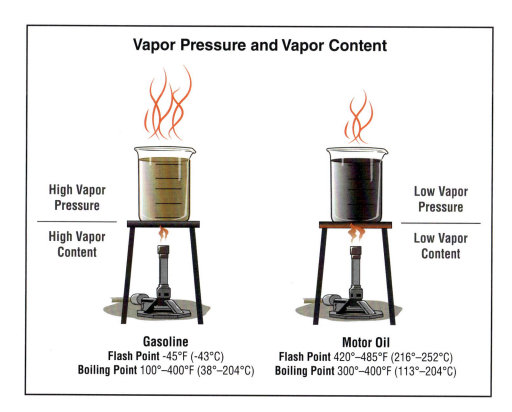

Figure 2.20 Boiling points and flash points are directly related.

Chapter Review

1. In what ways do temperature and pressure influence the physical states of matter?
2. In what ways can understanding the physical properties of a material help responders predict its behavior?
3. How are temperature and pressure influenced by each other?
4. What are the critical points and what changes to physical properties do they describe?

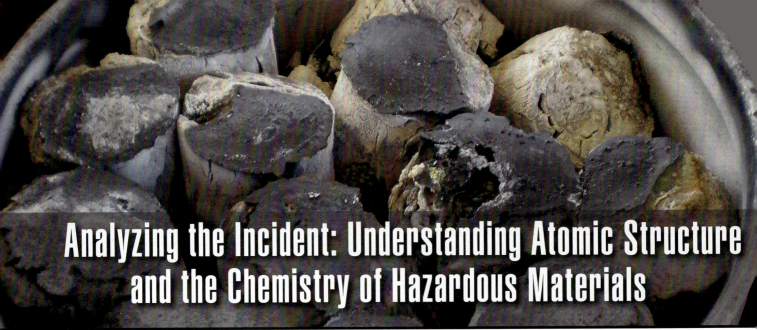

Analyzing the Incident: Understanding Atomic Structure and the Chemistry of Hazardous Materials

Photo courtesy of Barry Lindley.

Chapter Contents

Atomic Theory **49**	Ionic Bonds 67
Atoms 49	Resonant Bonds 68
Elements 50	Bond Energy 70
Compounds 53	Bonding in Hazmat Response 70
Mixtures 55	**Reactions** **70**
The Periodic Table of Elements **56**	Exothermic and Endothermic Reactions ... 72
Atomic Number 58	Oxidation and Combustion 72
Material Types 59	Polymerization 74
Four Significant Family Groups **60**	Decomposition 75
Group I – The Alkali Metals 61	Synergistic Reactions 75
Group II – The Alkaline Earths 61	The Fundamentals of a Reaction 75
Group VII – The Halogens 62	**Mixing Materials** **77**
Group VIII – The Noble Gases 64	Concentration 77
Bonding **65**	Solubility 78
Stability of Elements and Compounds ... 65	Miscibility 79
Diatomic Molecules (Compounds) 66	Polarity 79
Covalent Bonds 66	**Chapter Review** **79**

Chapter 3

Key Terms

Air-Reactive Material 72	Monomer 74
Alloy 56	Neutron 50
Anions 68	Nucleus 49
Atom 49	Octet Rule 65
Atomic Number 50	Oxidation 73
Atomic Stability 65	Oxidation Number 61
Atomic Weight 50	Oxidation-Reduction (Redox) Reaction 72
Bond Energy 70	Oxidizing Agent 72
Catalyst 74	Periodic Table of Elements 56
Cation 68	Polar Solvent 78
Compound 50	Polarity 79
Concentration 77	Polymer 74
Covalent Bond 66	Polymerization 74
Decomposition 75	Proton 49
Diatomic Molecules 66	Reactive Material 53
Duet Rule 65	Reducing Agent 72
Electron 50	Resonant Bond 68
Endothermic Reaction 71	Shell 50
Exothermic Reaction 71	Slurry 56
Hypergolic 72	Solubility 78
Immiscible 79	Soluble 78
Inert Gas 61	Solution 55
Inhibitor 75	Synergistic Effect 75
Ion 68	Unstable Material 66
Ionic Bond 67	Water Solubility 79
Miscibility 79	Water-Reactive Material 72
Miscible 78	
Mixture 55	

Analyzing the Incident: Understanding Atomic Structure and the Chemistry of Hazardous Materials

JPRs Addressed In This Chapter

7.2.4

Learning Objectives

After reading this chapter, students will be able to:

1. Use concepts of atomic structure to describe how hazardous materials behave. [NFPA 1072, 7.2.4]
2. Explain how the Periodic Table of Elements can be used to understand hazardous materials. [NFPA 1072, 7.2.4]
3. Describe characteristics of the four significant family groups within the Periodic Table of Elements. [NFPA 1072, 7.2.4]
4. Explain the significance of chemical bonding in predicting the behavior of hazardous materials. [NFPA 1072, 7.2.4]
5. Describe ways in which chemical reactions alter and create hazardous materials. [NFPA 1072, 7.2.4]
6. Explain ways in which hazardous materials may be mixed together. [NFPA 1072, 7.2.4]

Chapter 3
Analyzing the Incident: Understanding Atomic Structure and the Chemistry of Hazardous Materials

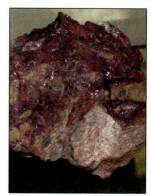

This chapter provides a brief and basic introduction to chemistry — enough to address terminology and concepts that will make it easier to understand the behavior of hazardous materials. As much as possible, these concepts are presented in terms of what it means to the hazardous materials technician in the field. This chapter explores the following topics:

- Atoms and molecules
- The Periodic Table of Elements
- Four significant family groups
- Matter
- Bonding
- Reactions
- Common families of hazardous materials
- Radiation

> **Atom** — The smallest complete building block of ordinary matter in any state.
>
> **Nucleus** — The positively charged central part of an atom, consisting of protons and neutrons.
>
> **Proton** — Subatomic particle with a physical mass and a positive electric charge.

Atomic Theory

Some knowledge of chemistry can help technicians comprehend how and why chemicals behave the way they do. This skill-set is especially important when reference sources are not immediately available. Examples provided in this chapter illustrate concepts and are not meant to be all-inclusive.

Atoms

The basic building block of any substance is the **atom**. Each atom consists of several different parts. At the center of the atom is a **nucleus** or core consisting of **protons**, which are positively charged, and **neutrons** that have no charge. **Electrons** orbit the nucleus and have a negative charge. These electrons reside in **shells**, sometimes called orbits, orbitals, or rings **(Figure 3.1)**. How elements interact with one another is primarily a function of their electrons. A pure atom of an element has an equal number of protons and electrons, and the atom is electrically neutral.

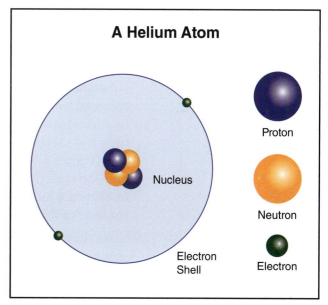

Figure 3.1 In an atom, protons and neutrons are at the core and electrons orbit the core.

Neutron — Component of the nucleus of an atom that has a neutral electrical charge yet produces highly penetrating radiation; ultrahigh energy particle that has a physical mass but no electrical charge.

Electron — Subatomic particle with a physical mass and a negative electric charge.

Shell — Layer of electrons that orbit the nucleus of an atom. The innermost shell can hold up to two electrons, and each subsequent shell can hold eight. *Also known as* Orbit, Orbital, *and* Ring.

Atomic Number — Number of protons in an atom.

Atomic Weight — Physical characteristic relating to the mass of molecules and atoms. A relative scale for atomic weights has been adopted, in which the atomic weight of carbon has been set at 12, although its true atomic weight is 12.01115.

Compound — Substance consisting of two or more elements that have been united chemically.

Elements differ from one another by the number of protons in the nucleus. For example, hydrogen always has one proton in its nucleus, oxygen always has eight protons, and sulfur always has sixteen protons.

Some elements have an equal number of protons and neutrons, but it is not unusual for elements to have more neutrons than protons. For example, sodium (atomic number 11) has eleven protons and twelve neutrons. Hydrogen, the smallest of all elements, is an oddity; its dominant form has one proton but no neutrons **(Figure 3.2)**.

The **atomic number** of an element is the number of its protons. The number of protons usually equals the number of electrons in an atom. The **atomic weight**, the mass of an atom, is determined by the number of neutrons and protons present in the nucleus.

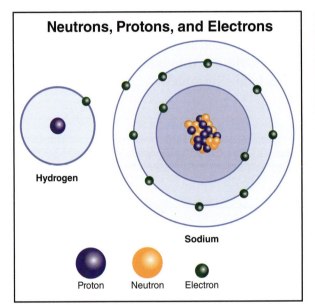

Figure 3.2 Hydrogen and sodium are examples of elements that do not have the same number of protons and neutrons.

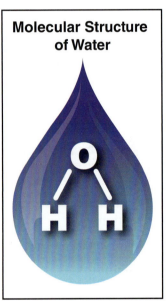

Figure 3.3 Elements may be understood as a formula.

Elements

Elements cannot be broken down into anything simpler without getting into subatomic structure (protons, neutrons, and electrons). Elements are represented by a single capital letter (e.g., H or O) or by a capital letter followed by a lowercase letter (e.g., Na or Cl). When the formula contains numbers, it indicates that more than one atom of the same element is present, for example, H_2O **(Figure 3.3)**. However, some elements such as oxygen (O_2), Chlorine (Cl_2), and Nitrogen (N_2) can exist as diatomic molecules **(Table 3.1)**. The behavior of elements may differ significantly from **compounds**, as explained in later sections in this chapter **(Figure 3.4)**.

As shown in the following sections, elements can be divided into metals, nonmetals, and metalloids. Metalloids are only briefly discussed in this chapter because technicians cannot easily identify/quantify them during response.

Table 3.1
Diatomic Molecules

Element	Elemental State (77° F [25° C])	Color	Molecule
Hydrogen	Gas	Colorless	H_2
Nitrogen	Gas	Colorless	N_2
Oxygen	Gas	Pale Blue	O_2
Fluorine	Gas	Pale Yellow	F_2
Chlorine	Gas	Pale Green	Cl_2
Bromine	Liquid	Reddish Brown	Br_2
Iodine	Solid	Dark Purple, Lustrous	I_2

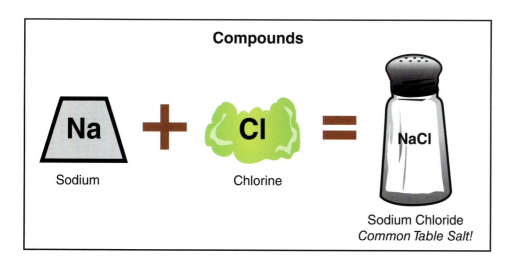

Figure 3.4 When elements combine, they can become a type of matter with different properties than their component elements' pure forms.

Metalloids

While tables displaying the elements show a distinction between metals and nonmetals, some elements, called metalloids, do not fit the classic profile of either **(Figure 3.5)**. Though the technicalities are beyond the scope of this book, technicians must recognize that any comparison of metals and nonmetals is a generalization because exceptions may apply.

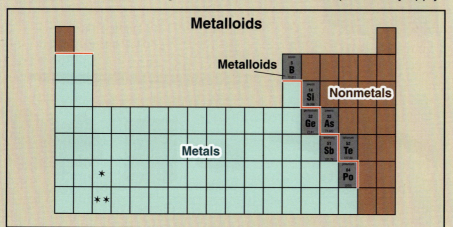

Figure 3.5 Metalloids are difficult to identify in the field at a hazmat incident.

Metals

Metals generally have a characteristic lustrous appearance and are good conductors of heat and electricity. They can also be hammered to form sheets or drawn into wires. Most metals are solids in their natural states **(Figure 3.6)**.

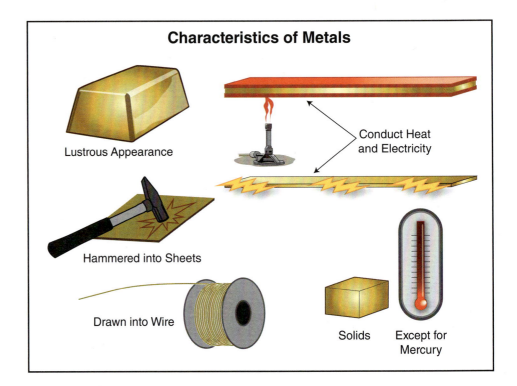

Figure 3.6 Metals typically have several characteristics in common.

Nonmetals

Nonmetals typically do not have a lustrous appearance and are not good conductors of heat and electricity. They cannot be shaped into sheets or wire. All nonmetals are either solids or gases in their elemental form, with the exception of bromine (Br), which is a liquid **(Figure 3.7)**.

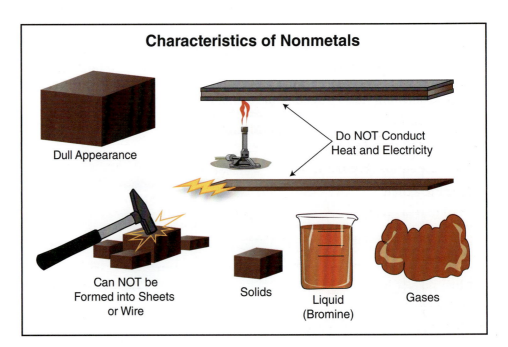

Figure 3.7 Nonmetals may have some characteristics in common with other nonmetals, but they will be different from metals.

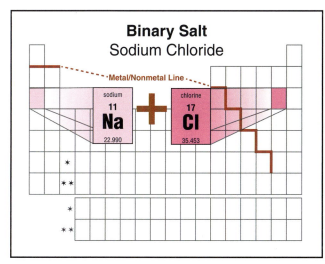

Figure 3.8 Binary salts are made of a metal and non-oxygen nonmetal.

Table 3.2 Special Hazards of Binary Salts

Binary Salt	Special Hazards
Nitrides	• Release irritating ammonia gas when in contact with moisture
Carbides	• Release unstable and flammable gas when in contact with moisture • Calcium carbide - generates acetylene • Aluminum carbide - generates methane
Hydrides	• Release heat and hydrogen gas when in contact with moisture
Phosphides	• Release poisonous phosphine gas when in contact with moisture

Compounds

Compounds are substances made up of two or more elements. Compounds are the result of chemical bonding. For example, water is two parts hydrogen (H_2) and one part oxygen (O).

Salts

Compounds are divided into salts and nonsalts, both of which are described in more detail later in this chapter in the chemical bonding section. In general, salt compounds are comprised of a metal element bonded to one or more nonmetal elements.

Binary salts consist of a metal and a nonmetal that is not oxygen (**Figure 3.8**). When a binary salt is named, the metal will always be named first with the nonmetal following, and the nonmetal name will be altered to end in a different suffix ("-ide"). For example, a salt made of sodium (Na) and chlorine (Cl) becomes sodium chloride (NaCl).

Binary salts can have wide-ranging properties. Some binary salts such as table salt (sodium chloride) can be food-safe, while others such as sodium fluoride may be toxic (**Table 3.2**). Binary salts may also be **reactive** with water (nitrides, carbides, hydrides, and phosphides). Other salts are described in **Table 3.3, p. 54**.

Common traits of salts include:

- Most are solids (**Figure 3.9**).
- Many are toxic.
- Metals give up electrons.
- Nonmetals accept electrons.
- Many can stimulate combustion.
- Most do not burn with a flame, though some may undergo smoldering combustion.
- Most dissolve in water or react with water.

Reactive Material — Substance capable of chemically reacting with other substances; for example, material that reacts violently when combined with air or water.

Figure 3.9 Cinnabar (HgS) is an example of a solid binary salt.

Table 3.3
Hazards of Salts

Salt	Make-Up	Example (Name)	Formula	Hazard
Cyanide Salts	Metal/cyanide	Sodium cyanide	NaCN	• Toxic by every route • Characteristic of cyanide poisoning
Metal Oxides	Metal/oxygen	Sodium oxide	Na_2O	• May produce an intense reaction • Produce caustic liquid and intense heat when mixed with water
Hydride Salts	Metal/hydride radical	Lithium aluminum hydride	$LiAlH_4$	• Produce a caustic solution, hydrogen gas, and intense heat when mixed with water • May be pyrophoric
Peroxide Salts	Metal/peroxide anion	Sodium peroxide	Na_2O_2	• Will form corrosive liquids when dissolved in water along with a liberal amount of oxygen and heat
Oxy-Salts	Metal/oxy-anion	Sodium perchlorate	$NaClO_4$	• Potent oxidizer
Ammonium Salts	Ammonium ion	Ammonium chlorate	NH_4ClO_3	• Act like a metal in chemical reactions • May be strong oxidizer

- Most will conduct electricity when dissolved in water.
- Salts form through ionic bonding (covered later in this chapter).

Nonsalts

Nonsalt compounds can be divided into two categories: organic and inorganic. Organic nonsalts are comprised of hydrocarbon and hydrocarbon derivatives. These will be discussed later in the chapter.

Most inorganic nonsalts are liquids and gases with very few being solid. They primarily tend to be toxic and poisonous and many have corrosive and/or flammable properties. For naming purposes, the name of one nonmetal will fall first with the suffix "-ide." Some examples of inorganic nonsalts are phosphorous and chlorine creating phosphorous chloride, or fluorine and chlorine creating fluorine trichloride.

Similar to the binary salts, the hazards of the nonsalts can be far ranging. Nonsalt compounds include nonmetal oxides, hydrogen compounds, binary acids, and oxy acids **(Table 3.4)**.

**Table 3.4
Hazards of Nonsalts**

Nonsalt	Make-Up	Example (Name)	Formula	Primary Hazards
Nonmetal Oxides	Nonmetal/oxygen	Sulfur dioxide	SO_2	• Toxic • Corrosive
Hydrogen Compounds	Hydrogen/nonmetal	Hydrogen chloride	HCl	• Toxic • Corrosive
Binary Acids	Hydrogen halide gas dissolved in water	Hydrofluoric acid	HF	• Corrosive • May migrate through tissue
Oxy Acids	Nonmetal oxide that is ionized and dissolved in water	Nitric acid	HNO_3	• Corrosive • Oxidizer

Some hazards of the nonsalts may include:

- Highly water reactive
- Highly toxic
- Flammable
- Oxidizing capabilities
- Corrosive properties

Mixtures

A **mixture** consists of two or more substances that are physically mixed, but not chemically bonded to one another. For instance, sodium chloride and water mix to form salt water, but neither the water nor salt are chemically altered within the mixture. Gasoline, kerosene, and fuel oil are common examples of other mixtures. Unlike pure substances, mixtures can vary from one sample to another. For example, gasoline purchased from two different service stations may contain ingredients and additives in different percentages. The following sections identify different types of mixtures.

Mixture — Substance containing two or more materials not chemically united.

Solutions

A **solution** is a mixture in which all the ingredients are completely dissolved. Solutions are often assumed to be liquids because many of them are. However, solids and gases can also be solutions, or they can be components of a solution in liquid form. Salt water is composed of a solid dissolved into a liquid. Carbonated beverages are comprised of a gas (carbon dioxide) dissolved into a liquid. Air is a solution made up primarily of two gases (nitrogen and oxygen). Alloys, such as steel, brass, and bronze, are solutions composed of two or more solids. With any solution, the result is uniform blend of all ingredients.

Solution — Uniform mixture composed of two or more substances.

Slurry — Suspension formed by a quantity of granulated or powdered solid material that is not completely soluble mixed into a liquid.

Alloy — Substance or mixture composed of two or more metals (or a metal and nonmetallic elements) fused together and dissolved into each other to enhance the properties or usefulness of the base metal.

Some solutions are easily separated into their component parts. When the water in salt water evaporates, it leaves the salt behind.

Heat normally increases the solubility of substances. For example, it is easier to dissolve sugar in hot tea than in iced tea. However, the opposite applies with gases dissolved into liquids and some solids. Raising the temperature decreases the solubility. Because carbon dioxide is a gas dissolved into a liquid, warm carbonated beverages lose their carbonation faster than cold beverages do. However, most solutions do not separate so easily.

To put this in the context of a hazardous materials incident, one can take the example of a release of ammonia gas at a refrigerant plant. One of the current tactics is to use a fine fog stream of cold water to pull a significant amount of the ammonia out of the air, because ammonia is soluble in cold water. Doing so creates a weak caustic solution of ammonium hydroxide (NH_4OH) but significantly decreases the risk of ignition. This tactic will not remove the problem; it will change the hazard profile and may allow responders to evacuate an area to establish another type of response. Hazard profiling is discussed in more detail in Chapter 6 of this manual.

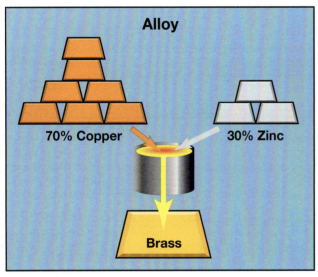

Figure 3.10 Alloys are customized to their purpose.

Slurries

When a material cannot completely dissolve in water, the end result may be **slurry**. Slurries are a suspension of insoluble particles, usually in water. Slurry is a pourable mixture of a solid and a liquid (although not necessarily one that remains pourable). Examples of common slurries include cement and plaster of Paris, which are both created by adding solids to water.

Alloys

Alloys (or metal alloys) are generally described as a mixture of two or more metals, the purpose of which is to create a material with improved chemical and/or mechanical properties for a particular application **(Figure 3.10)**.

Alloys often contain a small percentage of nonmetal elements. For example, steel is comprised primarily of iron (Fe) and carbon (C). Stainless steel is iron (Fe), chromium (Cr), nickel (Ni), and carbon (C).

The processes to make alloys are dangerous ones. The details are beyond the scope of this manual, but the processes often include reactive materials, steam, high-voltage electricity, and high-temperature metals. Some alloy-producing foundries may disable sprinkler systems to prevent accidental water discharge **(Figure 3.11)**.

Periodic Table of Elements — Organizational chart showing chemical elements arranged in order by atomic number, electron configuration, and chemical properties.

The Periodic Table of Elements

The **Periodic Table of Elements** is a tool for organizing and displaying the elements in a way that provides some basic information about their characteristics **(Figure 3.12)**. The first 94 elements are naturally occurring, though a few were synthesized in laboratories before being found in nature. The remainders are

Figure 3.11 Lithium reacts with nitrogen, a common inerting agent. Some materials are similarly reactive with water. *Courtesy of Barry Lindley.*

Figure 3.12 A commonly used Periodic Table shows basic information about elements.

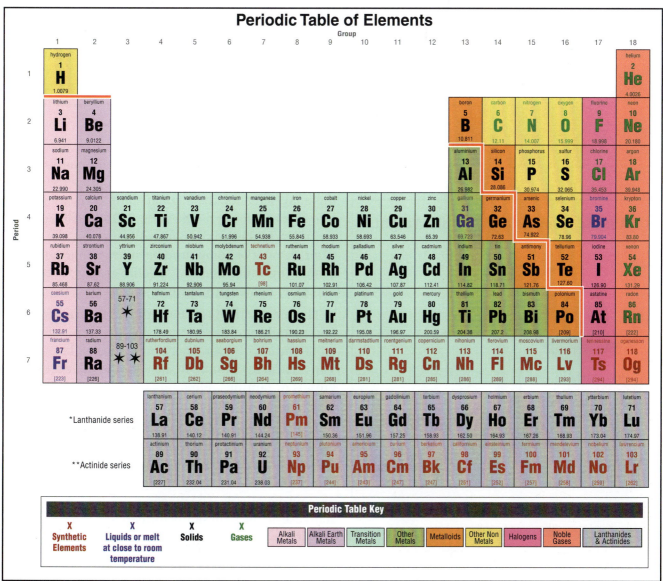

human-made. As such, the Periodic Table continues to evolve as new elements are created and named. Tables may show different numbers of elements, but those differences are not significant to the hazmat technician because they involve elements not likely to be encountered. Each element in the Periodic Table is identified by name, symbol, atomic number, and atomic weight **(Figure 3.13)**. Two pieces of information that a technician will especially need to consider are the atomic number and material type.

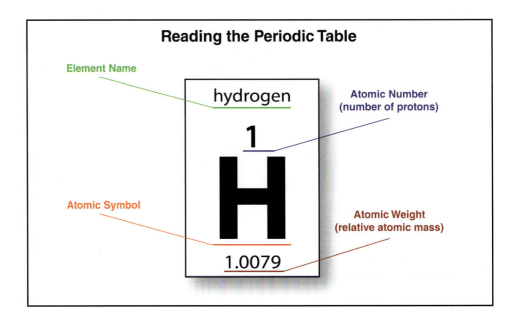

Figure 3.13 Each element on the table includes identifying information.

Atomic Number

For the most part, as the atomic number increases, so does the atomic weight **(Figure 3.14)**. Therefore, when considering the hazardous properties of individual elements, the hazmat technician can predict that elements lower on the Periodic Table will be solids and those higher on the table will be gases. The weight of an element or compound becomes important when evaluating vapor density, specific gravity, detection, and other related properties.

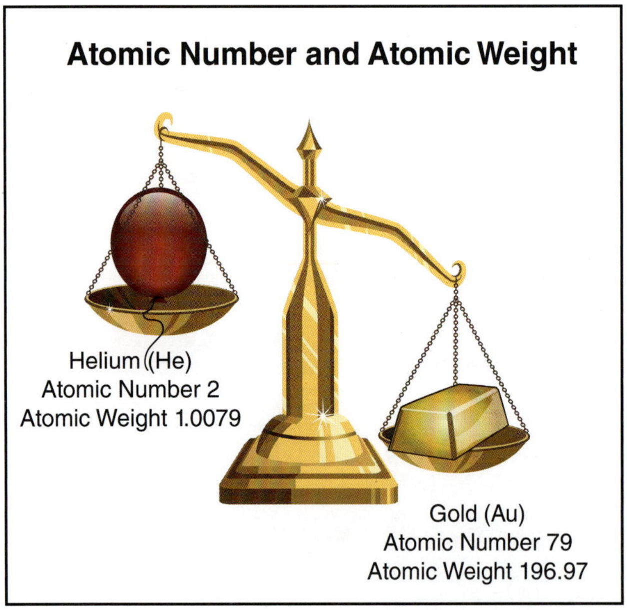

The elements of the Periodic Table are arranged horizontally by order of increasing atomic number. The table consists of several horizontal rows called periods. The vertical rows found in the Periodic Table are called groups or families. Elements within a specific group have similar chemical properties.

The similar chemical properties of elements within a group occur because these elements have the same number of electrons in their outer shells. These groups can be divided into three categories: metals, nonmetals, and metalloids. The diagonal line (that resembles a staircase) in the right side of the Periodic Table separates the metals from the nonmetals.

Figure 3.14 In some elements, the atomic number and atomic weight may differ significantly.

Material Types

A key feature of the Periodic Table is the dividing line between metals and nonmetals **(Figure 3.15)**. Metals are located on the left side of the Periodic Table; nonmetals are located on the right. Metalloids and semiconductors straddle that line. Elements 113 through 118 are artificial or synthetic elements created in laboratories.

NOTE: Hydrogen, which is located in the upper-left corner above the thick dividing line, is also a nonmetal.

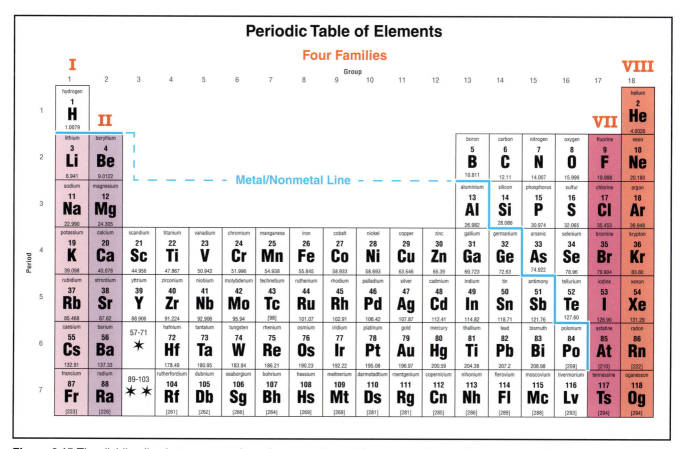

Figure 3.15 The dividing line between metals and nonmetals provides some indicator of the behavior of different elements.

The three major regions or areas on the Periodic Table are metals, nonmetals, and metalloids. The four family groups explained in the next section are the most important to the study of hazardous materials. As a category, they are more chemically active and encountered more often than the transition metals (metalloids) and the rare earth elements.

What This Means To You
Generalities from Material Types
When determining an effective response, you can quantify the type and anticipated reactions of a material based on the properties you can observe using tools and reports. For example, a solid will probably be a metal, and a non-solid will probably be a non-metal. The behavior of a material is the bedrock of risk-based response.

Four Significant Family Groups

Elements are arranged vertically in family groups based on their chemical structure. All members of the same family have similar chemical characteristics, though each has unique properties. Four family groups are significant because they are the most predictable:

- **Group I** — Alkali metals
- **Group II** — Alkaline earths
- **Group VII** — Halogens
- **Group VIII** — Noble gases

Family groups are important to hazmat technicians because they allow generalizations that can then be researched further **(Table 3.5)**. The following sections briefly describe each family with one or two specific examples.

Table 3.5
Four Family Groups Summary

Group	Name	Elements	Characteristics
I	Alkali Metals	Lithium Sodium Potassium Rubidium Caesium Francium	• Highly reactive • In contact with water, produce: – Hydrogen gas – Strong caustic runoff – Excessive heat
II	Alkaline Earths	Beryllium Magnesium Calcium Strontium Barium Radium	• Often water reactive • Burn intensely
VII	Halogens	Fluorine Chlorine Bromine Iodine Astatine Tennessine	• Highly reactive • Highly toxic • Nonflammable, but powerful oxidizers
VIII	Noble Gases	Helium Neon Argon Krypton Xenon Radon Oganesson	• Inert gases • Nonreactive but simple asphyxiants • Often stored/transported as cryogenic liquids

Group I – The Alkali Metals

Group I elements are highly reactive. Reactivity increases with the elements lower down in the column within the Periodic Table. When these elements come in contact with water, they produce flammable hydrogen gas, a strong caustic runoff, and excessive heat. This heat may be sufficient to ignite the hydrogen.

Group I includes the alkali metals **(Figure 3.16)**:

- Lithium
- Sodium
- Potassium
- Rubidium
- Caesium
- Francium

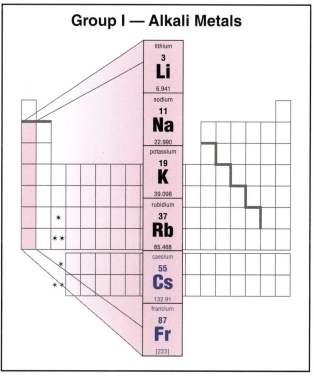

Figure 3.16 Group I alkali metals are highly reactive.

Sodium and potassium are the most common of the alkali metals. Both are highly water-reactive, enough to ignite the hydrogen spontaneously if exposed to moist air **(Figure 3.17)**. Small (pea size) pieces of sodium will move quickly on top of the water, while larger pieces will generally explode. These metals are usually stored submerged in oil, in blankets of **inert gas** or nitrogen, or in vacuum-packed containers to prevent exposure to air or water.

Contact between these metals and water separates or disassociates the water molecules, allowing other compounds to form from the individual elements. Sodium and potassium combine with free oxygen and hydrogen atoms to form sodium hydroxide (NaOH) and potassium hydroxide (KOH) solutions, respectively. Both are corrosive. Another by-product of the reaction in both cases is hydrogen gas (H_2) which is highly flammable. The reaction with water is so violent that it generates enough heat to melt the metal, ignite the hydrogen gas, and ignite nearby combustibles. Responders must use special extinguishing agents on fires involving alkali metals to avoid intensifying the fire.

Group II – The Alkaline Earths

Group II elements are often water-reactive, though less so than the alkali metals (Group I). For example, magnesium powder can explode on contact with water depending on the **oxidation number** (oxidation level) of the metal.

Group II includes the alkaline earths (metals) **(Figure 3.18, p. 62)**:

- Beryllium
- Magnesium
- Calcium
- Strontium
- Barium
- Radium

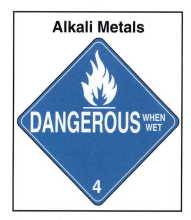

Figure 3.17 Alkali metals react violently when exposed to water.

Inert Gas — Gas that *does not* normally react chemically with another substance or material; any one of six gases: helium, neon, argon, krypton, xenon, and radon.

Oxidation Number — A theoretical number assigned to individual atoms and ions to track whether an oxidation-reduction reaction has taken place. *Also known as* Oxidation Level.

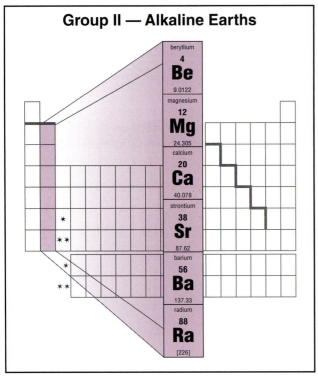

Figure 3.18 Group II alkaline earths are dangerous when wet but less reactive than alkali metals.

Magnesium, a metal often found in machine shops, burns intensely with such a bright white flame that prolonged unprotected viewing of it can cause retinal damage. With an ignition temperature of 1202°F (650°C), magnesium does not ignite easily. However, the risk of ignition increases in proportion to the amount of surface area exposed to air. In other words, magnesium powder, dust, chips, and shavings are far more dangerous than a solid block of magnesium.

Magnesium is water-reactive when burning **(Figure 3.19)**. Water used as an extinguishing agent intensifies the fire because it has been broken down to an oxidizer (oxygen) and a fuel (hydrogen). Water molecules that are not separated can be instantly converted to steam by the heat of the fire, resulting in a steam explosion. Both of these instances are dangerous for emergency responders. Flooding with large volumes of water can cool fires involving solid blocks of magnesium, but most magnesium fires require the application of special extinguishing agents.

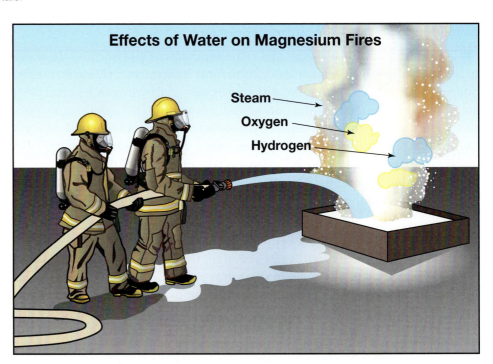

Figure 3.19 Magnesium is water-reactive when it burns.

Group VII – The Halogens

Group VII elements are highly reactive and toxic. They are nonflammable but are powerful oxidizers that support combustion.

Group VII includes halogens such as **(Figure 3.20)**:

- Fluorine
- Chlorine

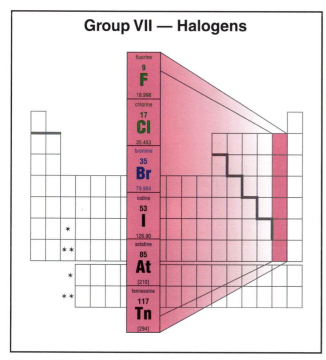

Figure 3.20 Group VII halogens are toxic and support flaming combustion.

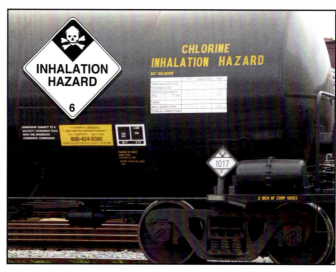

Figure 3.21 Halogens, like chlorine, are oxidizers that support combustion. *Courtesy of Rich Mahaney.*

- Bromine
- Iodine

Chlorine is the most common of the halogens **(Figure 3.21)**. Its characteristic suffocating odor is easy to recognize. What makes chlorine so toxic is that the gas reacts with moisture in the respiratory system to form corrosive hydrochloric acid (HCl) and hypochlorous acid (HClO). Chlorine reacts the same way with moisture on the skin, which is why chemical protective clothing is more appropriate than structural fire fighting clothing for handling chlorine emergencies in the hot zone. Fluorine, another halogen, reacts with water to form hydrofluoric acid (HF) and hypoflourous acid (HOF) which are even more destructive to human bone by leaching out calcium and unbalancing electrolytes.

While chlorine and fluorine do not burn, they are such powerful oxidizers that fires can burn intensely in the presence of chlorine or fluorine even when no oxygen is present. For example, fluorine is efficient enough as an oxidizer that asbestos will ignite and burn in a fluorine atmosphere.

WARNING!
Exposure to strong acids or weak acids alike can cause irreversible health damage.

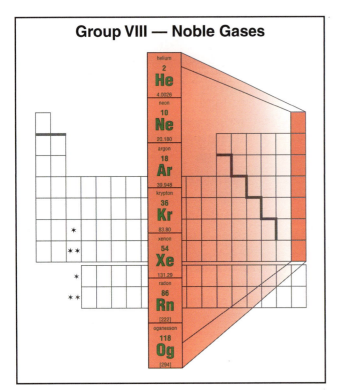

Figure 3.22 Group VIII noble gases displace oxygen in an environment.

Figure 3.23 Noble gases like helium are nonreactive and nontoxic.

Group VIII – The Noble Gases

Group VIII elements are inert gases, nonreactive, but simple asphyxiants.

Group VIII noble gases are (**Figure 3.22**):

- Helium
- Neon
- Argon
- Krypton
- Xenon
- Radon

Helium is familiar to responders and the public alike (**Figure 3.23**). It is not toxic or flammable and will not react with other materials. However, an uncontrolled release of helium, like any other noble gas, can still cause significant problems. When gas expands to fill an enclosed space, such as a small room, it will displace oxygen in the environment. Like all noble gases, helium can asphyxiate anyone who is unable to get an adequate supply of fresh air.

Helium, like the other noble gases, is often transported as a cryogenic liquid. The U.S. DOT establishes cryogenic liquids as those colder than -130°F (-90°C). These products can cause immediate and severe damage to human tissue and other materials such as steel, rubber, and plastic.

Like all cryogenic liquids, the noble gases in cryogenic form have high expansion ratios. This information is important because, in the event of a leak, there is a significant risk of displacing oxygen in the atmosphere. There is also a risk of catastrophic container failure in the event of a fire or other scenario where the product temperature is elevated beyond the capacity of pressure-relief devices.

Bonding

The study of hazardous materials is really about how substances react – how they react with their environment and how they react with other materials. At the core of most of those reactions is chemical bonding, with bonds being formed and/or broken between elements, compounds, or some combination thereof. The behavior of the material is affected by the molecules' internal bonding. The following sections will explore what is happening at the molecular level when materials react with one another.

Stability of Elements and Compounds

As indicated earlier, all atoms must have an equal number of protons and electrons to be electrically neutral. **Atomic stability** is achieved when elements possess filled outer shells **(Figure 3.24)**. To be stable, an atom must have a completely filled outer shell. For example, hydrogen, with only one shell, needs one more electron to obtain a total of two. This is called the **Duet Rule**. Elements with a second shell will attempt to achieve eight electrons on their outer shells, in a reaction called the **Octet Rule (Figure 3.25)**.

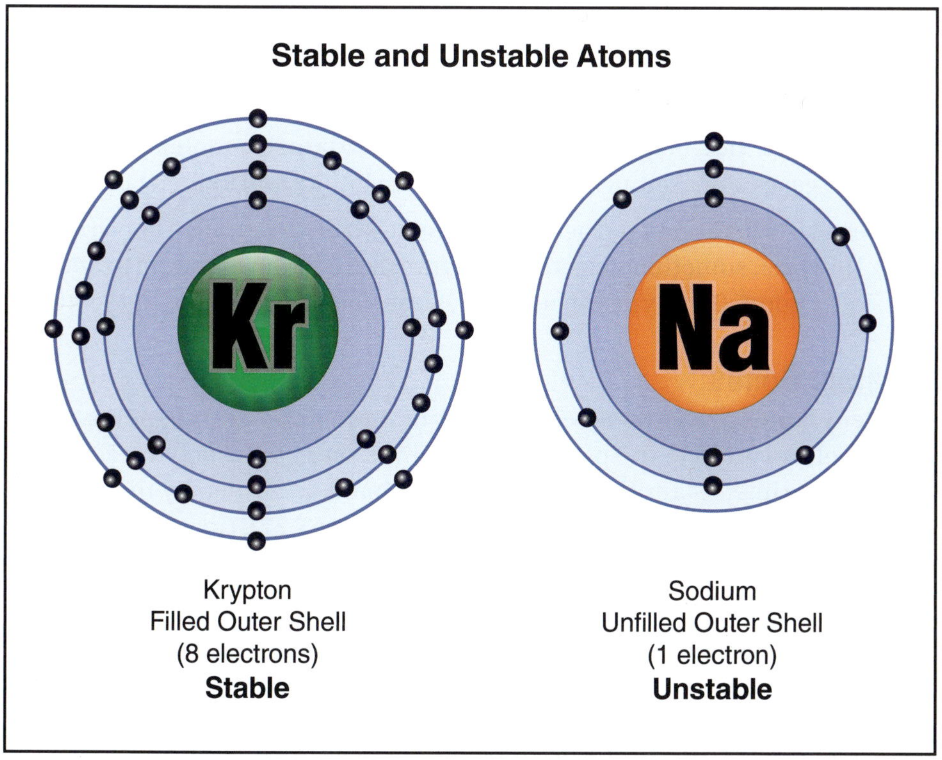

> **Atomic Stability** — Condition where an atom has a filled outer shell and is not seeking electrons. Stable atoms also have the same number of protons and electrons.
>
> **Duet Rule** — Atoms with only one shell will attempt to maintain two electrons to fill the outer shell at all times, whether by gaining or losing electrons. A complete outer electron shell makes elements very stable.
>
> **Octet Rule** — Atoms with two or more shells will attempt to maintain eight electrons to fill the outermost shell at all times, whether by gaining or losing electrons. A complete outer electron shell makes elements very stable.

Stable and Unstable Atoms

Krypton — Filled Outer Shell (8 electrons) — **Stable**

Sodium — Unfilled Outer Shell (1 electron) — **Unstable**

Figure 3.24 Filled outer shells are more important to an atom's stability than the overall number of electrons.

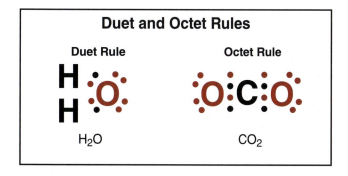

Duet and Octet Rules — Duet Rule: H_2O — Octet Rule: CO_2

Figure 3.25 The innermost shell only holds two electrons; every shell outside of that holds eight.

Unstable Material — Materials that are capable of undergoing chemical changes or that can violently decompose with little or no outside stimulus.

Diatomic Molecules — Molecules composed of only two atoms that may or may not be the same element.

Covalent Bond — Chemical bond formed between two or more nonmetals. This chemical bond results in a nonsalt.

Most of the time, the discussion of stability and instability refers to the tendency of some chemical compounds to break down into their component parts with little provocation. (The term *instability* is often used interchangeably with the term *reactivity*.) **Unstable materials** (compounds) are sensitive to heat, shock, friction, or contamination and often decompose violently. For example, contaminated nitroglycerin decomposes explosively if heated or shocked. Unstable compounds do not need to mix with other chemicals or be contaminated by other chemicals to react. For example, organic peroxides will decompose rapidly if simply overheated. Once the reaction starts, it will run its course to completion.

This is the basis for chemical reactivity: the need to fill the outer shell causes elements to react (bond) with one another chemically. Elements in Group VIII, the noble gases, are the exception to the chemical bonding process. Noble gases already have filled outer shells, so they are inert and do not react with other elements. When it comes to chemical bonding, these elements are not normally available to bond. Other elements are available to bond or share their electrons because their outer shells are not filled to the maximum. These elements must bond or share with others to create a filled outer shell.

Diatomic Molecules (Compounds)

Gases that exist as a compound of two identical atoms are called *diatomic gases*. Oxygen (O_2) is the most common diatomic gas. Others are hydrogen (H_2), nitrogen (N_2), fluorine (F_2), and chlorine (Cl_2). **Diatomic molecules** are not limited to gases. For example, bromine (Br_2) is a liquid, and iodine (I_2) is a solid.

When electrons are shared between atoms, it is called a covalent bond (**Figure 3.26**). However, sometimes electrons are transferred from one atom to another in an ionic bond. Both types of bonds are described in the following sections.

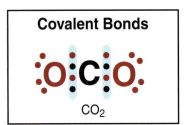

Figure 3.26 Covalent bonds are formed when electrons are shared between atoms.

Covalent Bonds

Covalent bonds are formed between two or more nonmetal elements. Their outer shells overlap to the point that electrons seem to belong to each atom at the same time. A familiar compound, methane, is formed by the union of one carbon atom and four hydrogen atoms (**Figure 3.27**). Covalently bonded molecules are referred to as nonsalts, in contrast to ionically bonded salts, explained in the next section.

Many covalent compounds are only composed of nonmetal elements. However, some *organometallic* compounds are covalently bonded. The majority of covalent compounds that hazardous materials technicians encounter consist solely of carbon, hydrogen, and oxygen in various combinations. **Table 3.6** lists examples of covalent compounds containing some of these elements.

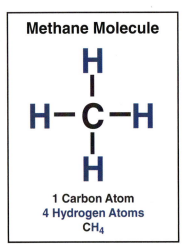

Figure 3.27 Methane is a covalently bonded molecule.

Of the compounds that contain other elements, most are limited to:

- Nitrogen
- Phosphorus
- Sulfur
- Silicon
- The halogens (chlorine, fluorine, bromine, and iodine)

Table 3.6
Covalent Compounds

Compound Name	Formula	Hazards
Carbon Monoxide	CO	• Toxic to humans and animals in high concentrations • Highly flammable
Nitrogen Trioxide (also known as Dinitrogen Trioxide)	N_2O_3	• Toxic • Water and air reactive • Produces toxic gas when heated
Hydrogen Iodide, anhydrous	HI	• Toxic • Water and air reactive • Dissolves exothermically in water
Phosphine (also known as Phosphorus Trihydride)	PH_3	• Highly flammable • Strong reducing agent • Air reactive • Can explode with powerful oxidizers • Highly toxic gas

Ionic Bonds

Ionic bonds are formed by the transfer of electrons from a metal element to a nonmetal element **(Figure 3.28)**. Compounds comprised of a metal element bonded with one or more nonmetal elements all belong in the category called salts. For example, sodium chloride forms when sodium "donates" the one electron on its outer shell, essentially shedding its outer shell and exposing a filled shell beneath it. Chlorine, which has seven electrons on its outer shell, "accepts" the donated electron from the sodium atom, also ending up with a filled outer shell.

Ionic Bond – Chemical bond formed by the transfer of electrons from a metal element to a nonmetal element. This chemical bond results in two oppositely charged ions.

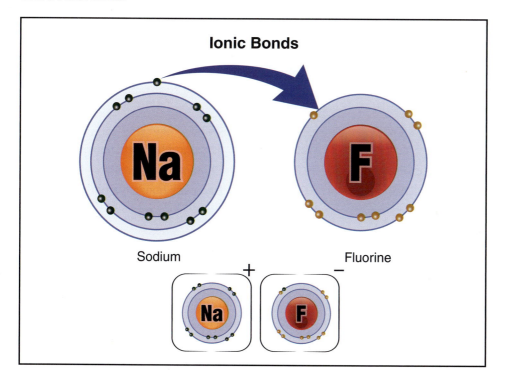

Figure 3.28 Ionic bonds form charged elements.

Ion — Atom that has lost or gained an electron, thus giving it a positive or negative charge.

Cation — Atom or group of atoms carrying a positive charge.

Anion — Atom or group of atoms carrying a negative charge.

While the two elements are now stable, they are no longer electrically balanced. Sodium has one more proton than electrons; chlorine has one electron more than protons. Atoms that have gained or lost electrons are no longer referred to as atoms; rather, they are called **ions**. An ion with more protons than electrons (the sodium ion in this example) has a net positive charge, whereas one with more electrons than protons (the chloride ion) has a net negative charge. Positively and negatively charged ions are referred to as **cations** and **anions**, respectively **(Figure 3.29)**.

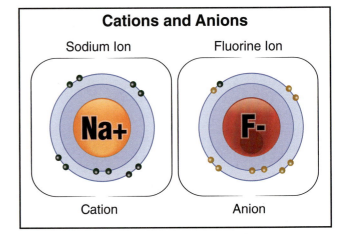

Figure 3.29 Anions and cations are stable and drawn to each other via electrical charge.

Oppositely charged ions attract each other. In essence, although there has been a transfer of electrons, the elements involved cannot stray far apart because ions cannot exist by themselves.

Table 3.7 shows examples of common salts. Sodium chloride (NaCl) is often used as an example because it is familiar to most people. Its close cousin, sodium fluoride (NaF), is very hazardous. The characteristics and hazards of salts are explored later in this chapter.

Table 3.7 Examples of Common Salts

Salt	Formula	Use
Sodium Chloride	NaCl	Seasoning of food
Sodium Bisulfate	$NaHSO_4$	Photographic purposes
Potassium Dichromate	$K_2Cr_2O_7$	Photographic purposes
Calcium Chloride	$CaCl_2$	Removes humidity from packaging

Resonant Bonds

Resonant Bond — Type of chemical bond in which electrons move freely between the compound atoms. *Also known as* Delocalized Bond.

Resonant bonds resemble covalent bonds in that electrons are shared between elements. However, when compounds contain resonant bonds, as do the aromatic hydrocarbons, the electrons actually rotate or alternate rapidly between the carbon atoms. Benzene (C_6H_6) is a classic example of a resonant bond because it includes all the standard features of resonant bonds without additional features. Benzene's ring structure can be represented either by elements connected with dashes or by a ring within a hexagon **(Figure 3.30)**.

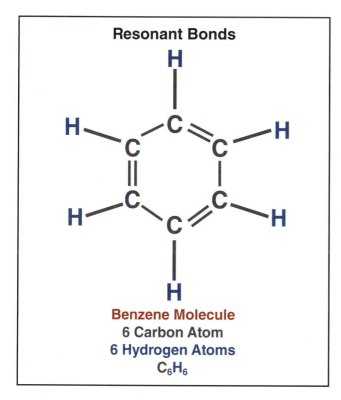

Figure 3.30 Resonant bonds are similar to covalent bonds, but in resonant bonds, the shared electrons move between multiple carbon atoms.

Many other types of aromatic hydrocarbons (toluene, xylene, styrene, and cumene) build from the benzene ring **(Figure 3.31)**.

NOTE: Refer to Chapter 4 for more information on aromatic hydrocarbons and resonant bonds.

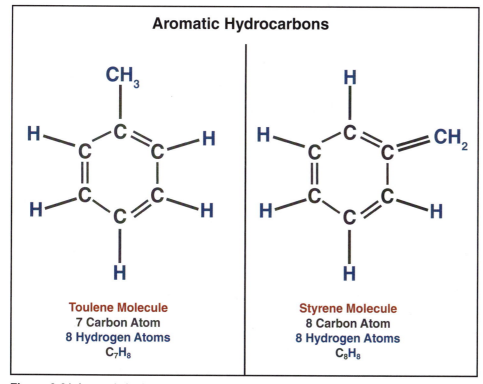

Figure 3.31 Aromatic hydrocarbons build from the benzene ring structure.

WARNING!
Aromatic hydrocarbons are flammable, volatile, toxic, and carcinogenic.

Bond Energy

Bond Energy — The amount of energy needed to break covalent bonds.

Every bond stores some amount of energy that is the measure of the strength of the connection(s) between atoms in a chemical bond. This energy is called **bond energy**, and it is released when the bond is broken. The amount of energy released will depend on the number and type of elements that comprise the bond and on what causes the bond to be broken. Some compounds can release a tremendous amount of energy in the form of light and heat when their bonds are broken.

Bonding in Hazmat Response

Knowing the difference between ionic and covalent bonds can help the responder predict the behavior of the chemical and the detection capability of some instruments.

Characteristics of ionically bonded compounds:

- Are solids
- Generate limited or no vapor
- Do not have flash points (FP)
- Do not have lower explosive limits (LEL)
- Do not polymerize
- May form caustic or basic solutions
- May be air or water-reactive
- Are generally not detectable via infrared (IR) spectroscopy and/or PID (photoionizing detectors) / FID (flame ionizing detectors)
- Some can be seen by Raman spectroscopy

Covalently bonded compounds:

- Are solids, liquids, and gases
- Produce vapors
- Have flash points (FP)
- Have lower explosive limits (LEL)
- Can polymerize
- May be seen by infrared (IR) spectroscopy and/or PID (photoionizing detectors) / FID (flame ionizing detectors)
- Can be detected via Raman spectroscopy and FTIR (Fourier transform infrared) spectroscopy

Reactions

Hazardous materials emergency responders are always concerned when two or more chemicals mix during any incident or when chemicals are exposed to heat from a fire or other source. This is especially true when the incident is complex such as a major industrial fire, a motor vehicle accident involving a transport container, or a train derailment.

All chemical reactions require changes in energy. The majority of reactions liberate energy as heat and sometimes heat, light, and sound. These reactions are **exothermic**. Some exothermic reactions are strong enough to ignite a fire. In a minority of reactions, the products absorb energy from the atmosphere. When this occurs the reaction is **endothermic** (Figure 3.32).

> **Exothermic Reaction** — Chemical reaction between two or more materials that changes the materials and produces heat.
>
> **Endothermic Reaction** — Chemical reaction in which a substance absorbs heat energy.

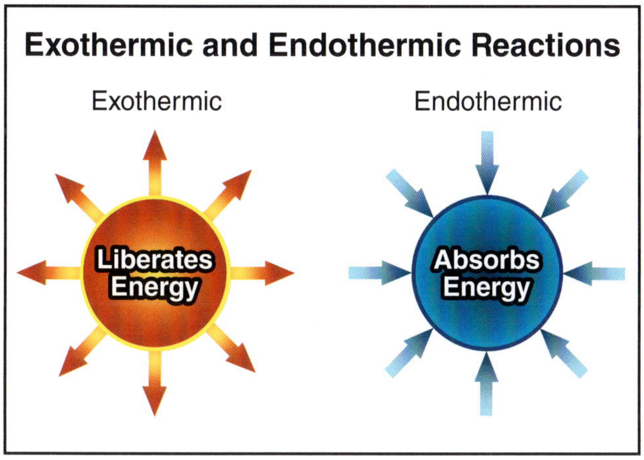

Figure 3.32 Exothermic and endothermic reactions are characterized by whether they release or absorb energy.

The questions that normally arise in such situations include:

- Will the mixing or heating of these chemicals cause the release of large amounts of energy as heat or with explosive force?
- Will the mixing or heating of these chemicals cause a fire or will it generate a greater flammability hazard than that found with the same chemicals without a chemical reaction?
- Will the mixing or heating of these chemicals increase the toxicity hazards at the scene compared to the toxicity of the unreacted chemicals?
- Will the mixing or heating of these chemicals form products with more vapors and/or fumes than the original chemicals?
- Will the products formed by the mixing or heating of these chemicals create additional environmental hazards or problems beyond those from the unreacted chemicals?

The answers to these questions will determine every aspect of the emergency response including the following:

- Size of the control zones, especially the hot zone
- Need for a shelter-in-place or civilian evacuation
- Choice of PPE needed for a safe response
- Scope and type of decontamination used
- Choice of response option to yield a positive effect on the situation
- Type of remediation needed to limit both long- and short-term environmental damage

Hypergolic — Substance that ignites when exposed to another substance.

Air-Reactive Material — Substance that reacts or ignites when exposed to air at normal temperatures. *Also known as* Pyrophoric.

Water-Reactive Material — Substance, generally a flammable solid, that reacts when mixed with water or exposed to humid air.

A reactive material can undergo a chemical reaction under specific conditions. Reactive materials can undergo a violent or abnormal reaction in the presence of water or air, or under normal ambient conditions. Examples:

- Reactive materials can be shock or friction sensitive.
- **Hypergolic** materials ignite on contact with other materials.
- **Air-reactive materials** primarily react with dry air, but they may also react with moist air.
- **Water-reactive materials** react on contact with water.

A water-reactive material is defined by the gases it gives off – toxic, corrosive, or flammable. The reaction can be mild to severe as seen with the alkali metals; the reaction may occur slowly as with lithium or violently as with potassium. The results of these reactions can be explosive and may result in a corrosive solution. More information on reactive materials will be covered in this chapter.

Exothermic and Endothermic Reactions

Removing the heat from an exothermic reaction slows and eventually stops the reaction. Exothermic reactions can be the most dangerous because of the potential for fire or explosion.

Adding heat to an endothermic reaction slows the reaction. A common example of an endothermic reaction is a cold pack used in first aid. When the pack is broken, the reaction of the two chemicals inside rapidly absorbs heat.

Oxidation and Combustion

When ionic bonds were described earlier in this chapter, it was explained how sodium loses one electron and chlorine gains one to form table salt. This change in the individual atoms' charges is an example of an **oxidation-reduction reaction** in which electrons transfer from one atom, compound, or molecule to another.

When sodium gives up an electron (which carries a negative charge of 1), it has a net positive charge (+1). Conversely, when chlorine gains the electron, it has a net negative charge (-1). However, the combined sodium chloride molecule is electrically neutral. Every oxidation-reduction reaction involves the simultaneous loss and gain of an equal number of electrons. Neither reaction (oxidation or reduction) can happen without the other.

The substance that loses electrons is called a **reducing agent**. The one that gains electrons is called an **oxidizing agent** (**Figure 3.33**). Substances that either give up or gain electrons easily are said to be strong reducing agents or oxidizing agents, respectively.

An element that only needs one electron to have a filled outer shell will be highly reactive. Earlier in the chapter, the elements called halogens (particularly fluorine and chlorine) were described as strong oxidizers. With seven electrons on their outer shells, fluorine and chlorine are "anxious" to gain an electron to fill the outer shell. Oxygen has six electrons on its outer shell. It needs two more to satisfy the Octet Rule, and is less reactive than elements that need only one.

Oxidation-Reduction (Redox) Reaction — Chemical reaction that results in a molecule, ion, or atom gaining or losing an electron. *Also known as* Redox Reaction.

Reducing Agent — Fuel that is being oxidized or burned during combustion. *Also known as* Reducer.

Oxidizing Agent — Substance that oxidizes another substance; can cause other materials to combust more readily or make fires burn more strongly. *Also known as* Oxidizer.

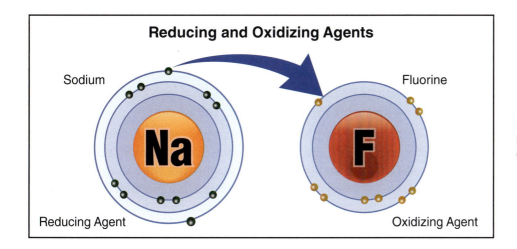

Figure 3.33 Reducing agents lose electrons; oxidizing agents gain electrons.

What can be confusing are the terms used to describe what happens to oxidizing and reducing agents after the transfer of electrons takes place:

- The oxidizing agent is said to have been *reduced*.
- The reducing agent is said to have been *oxidized*.

A more practical approach is to consider how the oxidation-reduction reaction leads to a hazardous materials incident. **Oxidation** is the chemical combination of oxygen (or another oxidizer) with another substance. The chemical reaction can vary in speed. A slow reaction may be illustrated as oxygen combining with iron to create rust. A fast reaction will manifest as a fire or explosion. In fact, fire (combustion) is nothing more than a complex oxidation-reduction reaction involving fuel (the reducing agent) and an oxidizing agent (usually oxygen).

Methane (CH_4) burns in an atmosphere of oxygen, and releases carbon dioxide and water vapor **(Figure 3.34)**. Electrical charges and electrons are not shown in this illustration. This reaction includes a transfer of elements, not just electrons, creating new substances and generating heat in the process.

Oxidation — Chemical process that occurs when a substance combines with an oxidizer such as oxygen in the air; a common example is the formation of rust on metal.

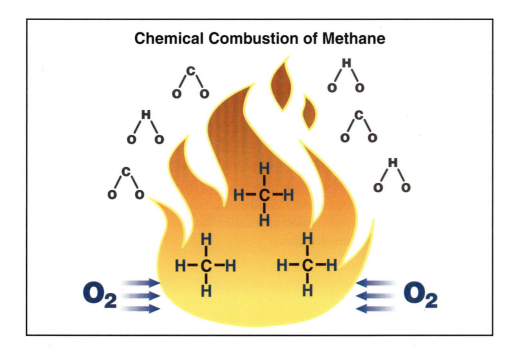

Figure 3.34 Methane combustion produces light, heat, carbon dioxide (CO_2), and water vapor.

Firefighters are familiar with the concept of complete combustion versus incomplete combustion. These reactions can be explained as oxidation-reduction reactions. Complete combustion includes a complete bond between carbon and oxygen resulting in a noncombustible byproduct (carbon dioxide). Incomplete combustion includes an incomplete or weak bond with oxygen resulting in a combustible byproduct (carbon monoxide).

A material's propensity to give up oxygen atoms, making them available to react with other substances, is a measure of the material's oxidation potential. The more readily a material yields oxygen, the greater the hazard it presents. The more readily a material yields oxygen or any other element that acts as an oxidizer, the more powerful the entire compound.

Polymerization

Polymerization is a special chemical reaction in which small compounds called **monomers** react with themselves to form larger molecules called **polymers**. These polymers are "repeating units" that resemble the original molecule. For example, ethylene (C_2H_4), the most common of all the monomers, transforms into polyethylene, a repeating chain of C_2H_4 units **(Figure 3.35)**. Polyethylene is a common manufactured plastic.

> **Polymerization** — Chemical reactions in which two or more molecules chemically combine to form larger molecules; this reaction can often be violent.
>
> **Monomer** — A molecule that may bind chemically to other molecules to form a polymer.
>
> **Polymer** — Large molecule composed of repeating structural units (monomers).
>
> **Catalyst** — Substance that modifies (usually increases) the rate of a chemical reaction without being consumed in the process.

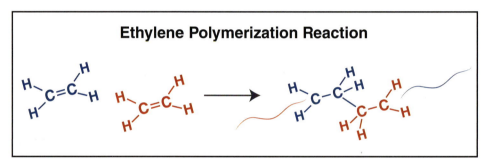

Figure 3.35 Under the right conditions, the monomer ethylene can react with itself to form the polymer polyethelene.

Ethylene in its natural state has a double bond between its two carbon atoms. Double bonds between carbon atoms are not as stable as single bonds; therefore, ethylene is prone to polymerization. To create polyethylene, manufacturers initiate a chemical reaction through the careful application of heat, pressure, and a catalyst. This reaction breaks one of the bonds between carbon atoms, leaving a single bond in place of the double bond and an open bond on either side. These incomplete molecules combine with other incomplete molecules to create more stable compounds (polymers). Different polymers can be created by varying the available types of monomers and the rate of the reaction.

Polymers are not as dense as monomers, and therefore they take up more space. The reaction of monomers combining into polymers also generates heat. Heat and overpressurization can cause catastrophic container failure from uncontrolled polymerization.

Catalysts

Polymerization is initiated by the use of a **catalyst**. Catalysts are substances added to other products either to initiate or to speed up chemical reactions. Catalysts themselves are not used up in the reaction; they may sometimes be recovered and reused. If catalysts are used improperly, they can increase the speed of a reaction beyond the point where the container can withstand the buildup of pressure and heat.

Inhibitors

Inhibitors, sometimes referred to as *stabilizers* or *negative catalysts*, are added to other products to stop or slow a reaction such as to prevent uncontrolled polymerization. Inhibitors have a limited shelf life, meaning their capacity to stop or slow a reaction changes over time. If an inhibitor is allowed to degrade, escape, or dilute below the necessary concentration, the resulting polymerization may cause catastrophic container failure. The concentration of an inhibitor may vary across storage tanks and shipments of the same product.

Decomposition

A **decomposition** reaction, sometimes referred to as an analysis or breakdown reaction, is one of the most common types of chemical reactions. Simply stated, when a decomposition reaction occurs, a compound breaks down into smaller components. Chemical decomposition is often an undesired chemical reaction that can range from extremely violent or not noticeable, depending on the chemical.

The electrolysis of water is a good example of a decomposition reaction. When an electric current is passed through water (H_2O), the molecule will break into oxygen (O_2) and hydrogen gas (H_2).

Synergistic Reactions

A *synergistic reaction* can be described as an interaction between two or more individual compounds that produce an effect that is different from the original starting materials. A synergistic reaction will make the end product greater as a whole than the individual compounds. A **synergistic effect** is usually thought of as two or more items working together in tandem, but with a synergistic reaction, the effects may not always be desired.

An example of a synergistic reaction is the mixing of household bleach and household ammonia to generate a compound called *chloramine*. Chloramine (NH_2Cl) can be used in low concentrations as a disinfectant in municipal water supplies and is starting to replace chlorine (free chlorine) in water treatment use. Chloramine tends to be more stable than chlorine and will not dissipate in the water before reaching consumers.

> **Inhibitor** — Material that is added to products that easily polymerize in order to control or prevent an undesired reaction. *Also known as* Stabilizer.

> **Decomposition** — Chemical change in which a substance breaks down into two or more simpler substances. Result of oxygen acting on a material that results in a change in the material's composition; oxidation occurs slowly, sometimes resulting in the rusting of metals.

> **Synergistic Effect** — Phenomenon in which the combined properties of substances have an effect greater than their simple arithmetical sum of effects.

WARNING!
Chloramine is ten times more toxic than hydrogen cyanide (HCN).

The Fundamentals of a Reaction

Many hazardous situations at a hazmat incident involve the inadvertent mixing of chemicals. Knowing the compatibility of the mixed materials is important. If the materials are incompatible, the results of the mixing could range from the formation of an innocuous gas or liquid to a violent explosion to one involving toxic reaction products. Per risk-based response, technicians must identify the risks associated with materials matching the behavior profile of the current incident before determining compatibilities. Determining the compatibility of more than two reactants can be difficult and may require additional research beyond the information provided in safety data sheets (SDS). **Table 3.8** provides examples of what could happen when incompatible materials combine.

Table 3.8 Reactions Caused by Mixing Incompatible Chemicals	
Incompatible Chemicals	**Reactions**
Acids with cyanide salts	Produces highly toxic hydrogen cyanide gas
Acids with sulfide salts	Produces highly toxic hydrogen sulfide gas
Acids with bleach	Produces highly toxic chlorine gas
Oxidizing acids with alcohols	May result in fire
Silver salts with ammonia and a strong base	May produce an explosively unstable solid
Unsaturated compounds (containing carbonyls or double bonds) and an acid or a base	May polymerize violently
Hydrogen peroxide and acetic acid mixtures	May explode upon heating
Hydrogen peroxide and sulfuric acid mixtures	May detonate spontaneously

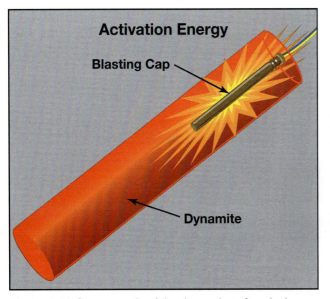

Figure 3.36 One example of the detonation of explosive material uses the ignition of a blasting cap to initiate the chemical reaction.

For a chemical reaction to occur, a certain amount of energy must be overcome. Activation energy is the minimum amount of energy needed to start a chemical reaction **(Figure 3.36)**. If you can prevent a chemical reaction from reaching its activation energy, the reaction may either stop or slow down. Inhibitors increase the activation energy by slowing the polymerization process. Catalysts decrease the activation energy, thus speeding up the reaction. A chemical reaction may be altered by changing the temperatures of the chemicals involved. If cooling occurs, the reaction will likely slow down. If you increase the temperature of a chemical compound, the reaction rate will increase as well. Pressure can use the same philosophy: an increase in pressure will increase the reaction.

Particulate size can also affect a chemical reaction. When discussing particle size, the smaller the particle, the faster the reaction and vice versa, based on the total surface of the particles. For example, when comparing sawdust versus a log in a fireplace, the sawdust in quantity may explode based on the overall surface area whereas in the case of the log, the reaction will not be an explosion, but rather a fire.

The state of matter can also have an effect on a reaction as well. In most cases, a solid will have a much slower reaction rate than a liquid. Similarly, gases may have faster reaction rates than liquids, depending on density and concentration.

Mixing Materials

The mixing of chemicals can be a dangerous event if not done correctly. Even household grade chemicals can be dangerous if mixed. For example, toxic vapors created by mixing bleach and ammonia-based cleaning products can prove fatal. With industrial grade chemicals, these types of incidents can be catastrophic.

Working with mixed materials takes a great understanding of not only the primary products, but also the end product. To fully understand mixed materials, it is important to have a good working knowledge of the terminology associated with these materials.

Concentration

The amount of each component in a mixture can be measured. Measures of how much of each compound is present in the mixture are expressed as **concentrations**. Concentration is usually expressed in percentages: parts per million (ppm), parts per billion (ppb), or milligrams per cubic meter (mg/m^3). For example, 5g of salt in 100 ml of water is a 5 percent solution of salt water **(Figure 3.37)**.

> **Concentration** — (1) Percentage (mass or volume) of a material dissolved in water (or other solvent). (2) Quantity of a chemical material inhaled for purposes of measuring toxicity.

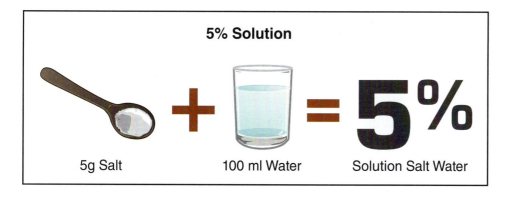

Figure 3.37 Salt water is a common example of a solution's concentration.

When dealing with hazardous materials, the concentration of the mixture will usually have a significant effect on the hazard. For example, 1 percent caustic (sodium hydroxide) will be a skin irritant and can cause skin burns over a long period of time if not washed off. However, 50 percent caustic will cause serious burns immediately on contact. Strength is sometimes used to describe concentration. However, this is used more in the description of acid or base strength or the amount of medicine in a prescription than the concentration of a liquid solution.

> **WARNING!**
> Some materials are harmful at low concentrations.

Solubility

Soluble materials will dissolve in a liquid. **Solubility** refers to the polarity of the solute in relation to the polarity of the solvent. Tables listing solubility in water and in common organic solvents are routinely available and these data can usually be found on an SDS. The values listed are normally the maximum amounts of material that will dissolve in the listed solvent at room temperature.

NOTE: The information on an SDS about solubility and solutes may not be easily usable to a technician because the values presented are listed in terms of a predetermined temperature which may not correspond to conditions at an incident response.

Solubility information can be useful in determining spill cleanup methods and extinguishing agents. For example, when a nonwater-soluble liquid such as a hydrocarbon (gasoline, diesel fuel, pentane) combines with water, the two liquids remain separate. When a water-soluble liquid such as a **polar solvent** combines with water, the two liquids mix easily.

Heat generally increases solubility for solids dissolved in liquids. However, heat will generally decrease the solubility of gases in solution. Solubility is an important concept in decontamination and product control, and will be discussed in greater detail in Chapters 9 and 11 of this manual.

Any liquid that is soluble but not completely **miscible** has a solubility limit. Solubility limits are usually expressed as percentages. For example, toluene is only 0.07 percent soluble in water. The degree of solubility is an indication of the solubility and/or miscibility of the material. Knowing whether a material is soluble in water may suggest, or rule out, an effective method of controlling the hazard.

Degrees of solubility that may be encountered in product research materials are as follows:

- Negligible (insoluble) — Less than 0.1 percent dissolved in water
- Slight (slightly soluble) — From 0.1 to 1 percent dissolved in water
- Moderate (moderately soluble) — From 1 to 10 percent dissolved in water
- Appreciable (partly soluble) — More than 10 to 25 percent dissolved in water
- Soluble — From 25 to 100 percent (or greater) dissolved in all proportions in water

NOTE: Some materials may be soluble at percentages higher than 100 percent. For example, NIOSH lists materials that are over 300 percent soluble.

> **Soluble** — Capable of being dissolved in a liquid (usually water).

> **Solubility** — Degree to which a solid, liquid, or gas dissolves in a solvent (usually water).

> **Polar Solvents** — 1) A material in which the positive and negative charges are permanently separated, resulting in their ability to ionize in solution and create electrical conductivity. Examples include water, alcohol, esters, ketones, amines, and sulfuric acid. 2) Flammable liquids with an attraction for water.

> **Miscible** — Materials that are capable of being mixed in all proportions.

> **CAUTION**
> All materials that are soluble at any percentage will react with water and other chemicals.

Miscibility

Miscibility is a special term for when a liquid or gas completely dissolves in another liquid or gas. This implies that the materials have the ability to mix in any combination or concentration to form a uniform blend with each other. For example, ethyl alcohol is miscible with water. Materials that are **immiscible** will not mix or dissolve in each other. For example, oil and water are immiscible.

Polarity

Polarity is a reflection of how strongly one atom attracts the electrons of another atom in a covalent bond. Polar materials have an affinity for other polar materials, which explains **water solubility** to an extent. Water is highly polar so other polar materials, like alcohols, dissolve or are soluble in water. Hydrocarbons are nonpolar so they do not dissolve in water to any appreciable amount.

Some molecules, such as soaps, have both polar and nonpolar areas. Though soaps dissolve in water (a polar substance), they are also able to remove grease (a nonpolar substance) during washing. Soap molecules have small polar ends attached to large nonpolar sections. The grease particles attract the nonpolar sections of the soap molecules, leaving the polar ends protruding from the grease particles to dissolve in water. The grease particles are coated with a water soluble layer of soap molecules. When decontaminating clothing and equipment that are contaminated with oily substances, soapy water may be the only decontamination solution needed.

The same principles that apply to the solubility of liquids apply to solids. Solids with polar molecules will be soluble in polar solvents. However, unlike polar liquids, polar solids are not miscible with water. They have a limited capacity to dissolve in water. For example, table salt is water soluble, but only to a certain extent. If water and salt mix, a certain amount of the salt will dissolve. After enough salt has been added, the water will be "saturated" and unable to dissolve any more salt.

> **Miscibility** — Two or more liquids' capability to mix together.
>
> **Immiscible** — Incapable of being mixed or blended with another substance.
>
> **Polarity** — Property of some molecules to have discrete areas with negative and positive charges.
>
> **Water Solubility** — Ability of a liquid or solid to mix with or dissolve in water.

Chapter Review

1. What is the difference between compounds and mixtures?
2. Describe two important pieces of information that the Periodic Table of Elements provides to Hazmat Technicians.
3. What is one distinguishing characteristic of each of the four significant family groups?
4. What is the difference between ionic and covalent bonds?
5. Give a brief description or example of each type of chemical reaction.
6. What is the difference between a material's solubility, miscibility, and polarity?

Analyzing the Incident: Understanding Common Families and Special Hazards of Hazardous Materials/WMD

Chapter Contents

Common Families of Hazardous Materials 83
- Inorganic Compounds 83
- Organic Compounds 83
- Oxidizing Agents 87
- Reactive Materials 88
- Corrosives 94
- Radiation 95

Special Hazards of Chemicals and Weapons of Mass Destruction 99

- Chemical Warfare Agents 100
- Biological Agents and Toxins 106
- Radioactive and Nuclear 111
- Explosives and Incendiaries 116
- Homemade Explosives 118
- Improvised Explosive Devices 121

Chapter Notes 130
Chapter Review 130

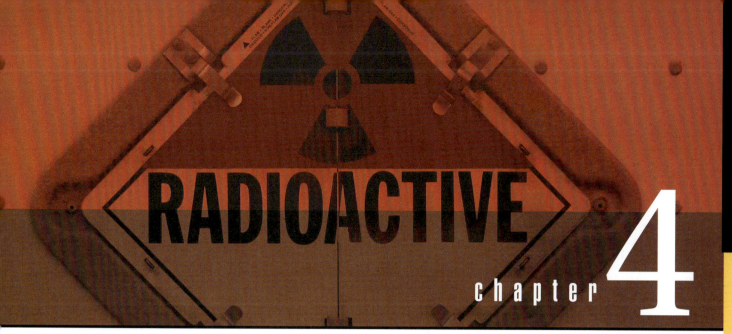

RADIOACTIVE

chapter 4

Key Terms

- Acetone Peroxide (TATP) 120
- Acid 94
- Activity 99
- Acute 114
- Alkane 84
- Alkene 84
- Alkyne 84
- Aromatic Hydrocarbon 85
- Bacteria......................................106
- Base..94
- Becquerel (Bq) 99
- Bioassay 109
- Biological Agent......................106
- Blister Agent............................103
- CBRNE 99
- Chemical Asphyxiant.............104
- Chemical Burn...........................95
- Choking Agent.........................104
- Chronic....................................... 114
- Class D Fire 90
- Code of Federal Regulations (CFR) 117
- Contaminant 111
- Contamination......................... 111
- Curie (Ci) 99
- Deflagrate 117
- Detonate.................................... 118
- Dissociation (Chemical) 94
- Division Number 117
- Dose 107
- Dose-Response Relationship 108
- Dry Powder 91
- Explosive Ordnance Disposal (EOD).. 116
- Exposure.................................... 111
- G-Series Agents......................102
- Half-Life..98
- Halogenated Agent 86
- Hexamethylene Triperoxide Diamine (HMTD) 120
- High Explosive 118
- Homemade Explosive (HME)... 118
- Hydrocarbon...............................83
- Improvised Explosive Device (IED)...121
- Inverse Square Law 115
- Ionize...97
- Ionizing Radiation 96
- Low Explosive 117
- Maximum Safe Storage Temperature (MSST)..............93
- Median Lethal Concentration, 50 Percent Kill (LC$_{50}$) 109
- Median Lethal Dose, 50 Percent Kill (LD$_{50}$)................................109
- Mobile Data Terminal (MDT)...129
- Munitions 122
- Nerve Agent.............................100
- Nonionizing Radiation 96
- Organophosphate Pesticides 110
- Person-Borne Improvised Explosive Device (PBIED).....126
- Photoionization Detector (PID)..94
- Photon 113
- Primary Explosive.................. 118
- Radiation Absorbed Dose (rad) 112
- Radioactive Decay 98
- Radioactive Material (RAM) 95
- Radioisotope 98
- Reactivity.....................................88
- Rickettsia 106
- Riot Control Agent 105
- Scintillator................................ 113
- Secondary Explosive............. 118
- Self-Accelerating Decomposition Temperature (SADT)..............93
- Thermal Burn..............................95
- Toxicity......................................108
- Toxicology 107
- Toxin 106
- Transmutation 98
- Vehicle-Borne Improvised Explosive Device (VBIED).....128
- Virus 106

Chapter 4 • Analyzing the Incident: Understanding Common Families and Special Hazards of Hazardous Materials/WMD   **81**

Analyzing the Incident: Understanding Common Families and Special Hazards of Hazardous Materials/WMD

JPRs Addressed In This Chapter

7.2.1, 7.2.4

Learning Objectives

After reading this chapter, students will be able to:

1. Describe the common families of hazardous materials. [NFPA 1072, 7.2.1, 7.2.4]
2. Recognize special hazards of chemicals and weapons of mass destruction. [NFPA 1072, 7.2.1, 7.2.4]

Chapter 4
Analyzing the Incident: Understanding Common Families and Special Hazards of Hazardous Materials/WMD

This chapter will address the following topics:
- Common families of hazardous materials
- Special hazards of chemicals and weapons of mass destruction

Common Families of Hazardous Materials

Hazmat technicians need to be familiar with the properties and behavior of the common families (categories) of hazardous materials, including:

- Inorganic compounds
- Organic compounds
- Oxidizing agents
- Reactive materials
- Corrosives

> **Hydrocarbon** — Organic compound containing only hydrogen and carbon and found primarily in petroleum products and coal.

Inorganic Compounds

Inorganic chemistry is a branch of chemistry concerned with the properties and behavior of inorganic compounds. This field can cover all chemical compounds except organic compounds. Many inorganic compounds can be considered ionic compounds that are joined by ionic bonding.

Organic Compounds

There are predominantly two types of organic compounds: **hydrocarbons** and hydrocarbon derivatives. Hydrocarbons contain only two elements, carbon and hydrogen, in any quantity **(Figure 4.1)**. Hydrocarbon derivatives contain carbon, hydrogen, and one or more additional elements.

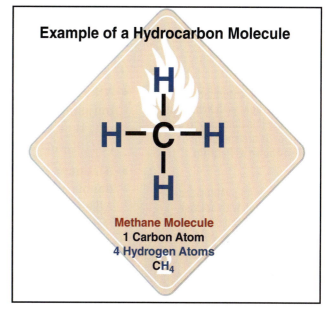

Figure 4.1 As implied by their name, hydrocarbons consist entirely of hydrogen and carbon atoms.

Hydrocarbons

Hydrocarbons have common characteristics that can allow hazardous materials technicians to predict their behavior and hazards even before checking reference sources:

- All hydrocarbons will burn.
- All hydrocarbons are toxic to some degree.
- All hydrocarbons are insoluble in water.
- Most hydrocarbons will float on the surface of water.

Many of these properties are directly related to the size of the molecule. Comparisons between individual hydrocarbons and their derivatives are most accurate when evaluating similar compounds.

The types of bonds between carbon atoms also have a significant effect on chemical behavior **(Figure 4.2)**. The four basic types of hydrocarbons are:

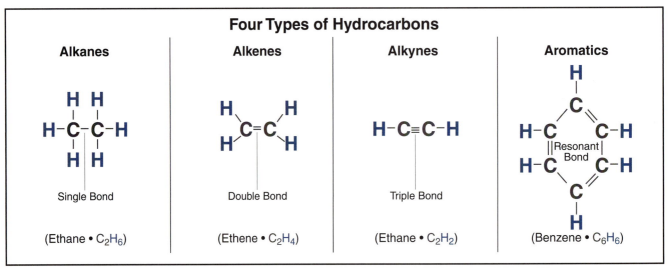

Figure 4.2 Four types of hydrocarbons are defined by the types of bonds between their carbon atoms.

Alkane — A saturated hydrocarbon, with hydrogen in every possible location. All bonds are single bonds. *Also known as* Paraffin.

Alkene — An unsaturated hydrocarbon with at least one double bond between carbon atoms. *Also known as* Olefin.

Alkyne — An unsaturated hydrocarbon with at least one triple bond. *Also known as* Acetylene.

- **Alkanes** — Most alkanes (single bonds) are relatively stable. They have only single bonds between carbon atoms. Alkanes have a general formula of C_nH_{2n+2}. Examples of alkanes are methane (CH_4), ethane (C_2H_6), propane (C_3H_8), and butane (C_4H_{10}).
- **Alkenes** — Alkenes are prone to polymerization. They have at least one double bond between carbon atoms, making them less stable than alkanes. Alkenes have a general formula of C_nH_{2n}. Examples of alkenes are ethylene (ethene [C_2H_4]), propylene (propene [C_3H_6]), and butylene (butene [C_4H_8]).
- **Alkynes** — Alkynes have explosive potential. They have at least one triple bond between carbon atoms, and are highly unstable. Double and triple bonds are reactive with triple bonds, more so than double bonds. Alkynes have a general formula of C_nH_{2n-2}. An example of an alkyne is acetylene (ethyne [C_2H_2]).
- **Aromatics** — Most aromatics are stable even though they appear to have three double bonds because the electrons resonate between the carbon

atoms in the resonant bonds. Aromatics have a ring (cyclic) structure with resonant bonds. Some are carcinogenic. As a group, they are also fairly toxic **(Table 4.1)**. The aromatics typically burn with sooty smoke that often gives the appearance of a spider web. **Aromatic hydrocarbons** have a general formula of C_nH_n. Examples of aromatic hydrocarbons are benzene (C_6H_6), toluene (C_7H_8), xylene (C_8H_{10}), styrene (C_8H_8), and cumene (C_9H_{12}).

> **Aromatic Hydrocarbon** — A hydrocarbon with bonds that form rings. *Also known as* Aromatics, *or* Arene.

Table 4.1
Aromatic Hydrocarbons and Their Hazards

Aromatic Hydrocarbons	Formula	Hazards
Benzene	C_6H_6	• Highly flammable liquid • Can form explosive mixtures with air • Fire can produce irritating, corrosive, or toxic gases • Can cause toxic effects if inhaled or absorbed through the skin • Can irritate or burn the skin and eyes
Biphenyl	$C_{12}H_{10}$	• Combustible • Emits toxic fumes under fire conditions • Irritating to eyes, nose, throat, and skin • Incompatible with oxidizers
Durene	$C_{10}H_{14}$	• Flammable/combustible • Emits irritating or toxic gases under fire conditions • Vigorous reactions to include explosions when in contact with strong oxidizers
Toluene	C_7H_8	• Highly flammable liquid • Insoluble in water • Vapors irritate eyes and upper respiratory tract • Vigorous reactions when in contact with alkyl chloride or other alkyl hallides

The descriptions of the categories may not strictly apply to all compounds within each category, but there are enough common characteristics that technicians can use these generalizations for a quick assessment.

For example, while most aromatic hydrocarbons are stable, styrene (C_8H_8) contains a double bond between carbon atoms outside the resonant structure. Styrene will polymerize at relatively low temperatures if not mixed with an inhibitor.

> **WARNING!**
> Response Guides may not include all of the features that a responder may want during a response. Responders may need to consult specialized resources depending on the incident.

Hydrocarbon Derivatives

Some hydrocarbon derivatives and their associated hazards are included in **Table 4.2**. The range of hydrocarbon derivatives is beyond the scope of this text.

Table 4.2
Hydrocarbon Derivatives and Their Hazards

Hydrocarbon Derivatives	Formula	Hazards
alpha-Naphthylamine	$C_{10}H_7NH_2$	• Combustible material • May form explosive mixtures with air when heated • Toxic if inhaled, ingested, or through skin contact
Ethyl Ether	$(C_2H_5)_2O$	• Extremely flammable liquid • Susceptible to peroxide formation that can form explosive mixtures
Ethyl Hexaldehyde	$C_8H_{16}O$	• Highly flammable liquid • Can form explosive mixtures with air • Fire can produce irritating, corrosive, or toxic gases • Can irritate the skin and eyes on contact • Can irritate the nose, throat, and lungs causing coughing and/or shortness of breath
Formaldehyde	CH_2O	• Flammable/combustible material • May form explosive mixtures with air • Can be toxic if inhaled/ingested • Can cause burns to the skin and eyes • Fire can produce irritating, corrosive, or toxic gases
Methyl Ethyl Ketone (also called Butanone)	$CH_3C(O)CH_2CH_3$	• Highly flammable material • Containers may explode when heated • Can be toxic if inhaled/ingested • May irritate or burn the skin and eyes • Fire can produce irritating, corrosive, or toxic gases

> **Halogenated Agent** — Chemical compounds (halogenated hydrocarbons) that contain carbon plus one or more elements from the halogen series. Halon 1301 and Halon 1211 are most commonly used as extinguishing agents for Class B and Class C fires. *Also known as* Halogenated Hydrocarbons.

Some hydrocarbon derivatives are relatively common. For example, the chemical difference between the hydrocarbon methane (CH_4) and hydrocarbon derivative methanol (CH_3OH) is that the original methane compound loses one hydrogen atom (H) and gains a hydroxyl group (OH) in its place **(Figure 4.3)**. The "-ol" suffix on these derivative names indicates that the material is an alcohol.

Halogenated agents contain a halogen (chlorine, bromine, iodine, fluorine). They are typically more toxic than the parent materials. Many of these materials do not have flash points because the halogen suppresses the flashing of these materials, but these same materials may have explosive ranges. When they combust they release toxic acid gases and other toxic materials. Chlorinated hydrocarbons can release extremely toxic materials such as phosgene ($COCl_2$) when burned.

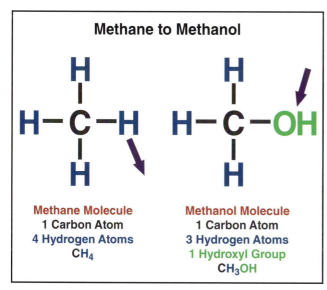

Figure 4.3 The difference between methane and methanol is the replacement of one hydrogen atom with a hydroxyl group.

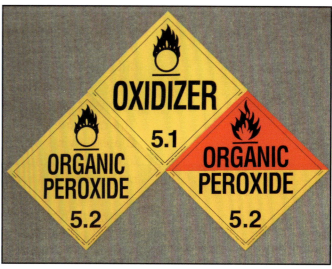

Figure 4.4 Oxidizers can increase the hazard potential of fuels.

Oxidizing Agents

Some oxidizing agents are relatively stable while some can be very unstable and even explosive. An oxidizer does not burn, but it can make a fire burn much hotter and faster **(Figure 4.4)**. For example, oxygen is not a flammable element but it can significantly accelerate combustion. Oxidizers can be categorized by their reactions. Oxidizing agents were introduced in Chapter 3 of this manual.

Inorganic Peroxides

Inorganic peroxides can act as both an oxidizer and a corrosive. Hydrogen peroxide is an example of inorganic peroxide. In contrast to grocery store hydrogen peroxides, which are roughly 3 percent product with the remainder being water, commercial grade hydrogen peroxide can range from 30-70 percent product. Hydrogen peroxide in this concentration can spontaneously combust if mixed with Class A fuels (organic materials). Peroxides may have the ability to combust and even explode in the absence of air.

Organic Peroxides

Organic peroxides are a unique element and typically contain both an O-O which supplies oxygen and a part that can act as a fuel in its molecular structure. For these materials, just a small amount of heat may be needed to create a fire or explosion.

Chapter 4 • Analyzing the Incident: Understanding Common Families and Special Hazards of Hazardous Materials/WMD

Organic peroxides are commonly used as catalysts and/or initiators for a polymerization reaction. Organic peroxides may be liquids or solids. They may be dissolved into solvents that may also be flammable. Organic peroxides are often shipped refrigerated and stored in cold storage cabinets.

Chlorates and Perchlorates

Chlorates and perchlorates are unstable molecules containing excess oxygen. These elements are not flammable, but they decompose rapidly when subjected to heat. When these elements decompose, they release oxygen and support rapid burning in any nearby combustible material.

Reactive Materials

Reactivity describes a material's propensity to release energy or undergo change either on its own or in contact with other materials. Reactivity is the tendency of a material or combination of materials to undergo chemical change under the right conditions. Although dangerous, reactive materials can benefit some processes. Chemical reactivity can be a highly desirable trait that permits a wide variety of useful materials to be synthesized. It may also allow products to be made under moderate conditions of pressure and temperature, thus saving energy and reducing the risks of working with high pressure and temperature equipment.

> **Reactivity** — Ability of a substance to chemically react with other materials, and the speed with which that reaction takes place.

Reactive Materials

Reactive materials, by design, rely on a chemical reaction for a specific purpose. These materials improve quality of life for people using them in controlled environments, but are highly dangerous for first responders and technicians in uncontrolled conditions. Reactives present an invisible hazard that may not be associated with their related chemical names. These materials may be reactive with water/moisture, oxygen, or other common materials. The reaction can be as quick as a detonation.

Air-Reactive

Air-reactive materials ignite, decompose, or otherwise release energy when exposed to air. These reactions can be violent, with the added potential of container failure due to overpressurization. For example, reaction between aluminum phosphide and the humidity in air creates a toxic material, phosphine. Air-reactive materials were introduced in Chapter 3 of this manual.

Technically, pyrophoric materials react with the oxygen in dry air, whereas materials that react in moist air might be considered water-reactive. Hazmat technicians should assume that pyrophoric materials are also water-reactive until they have determined otherwise by checking at least three reference sources. Water-reactive materials were introduced in Chapter 3 of this manual.

CAUTION
Air-reactive materials may also react to water, even if reference documents do not list that risk.

Materials that are air-reactive but not water-reactive may be stored under water to prevent contact with air. Examples of air-reactive materials that may be stored under water include white and yellow phosphorus.

Air-reactive materials that will react with moisture in the air must be stored under some other substance, such as an inert gas, mineral oil, or kerosene. Examples of air-reactive materials that cannot be stored under water include sodium, potassium, and other alkali (Group I) metals.

When pyrophoric materials burn, any extinguishing agent that excludes air should be effective, which means that water can be used to fight the fire. However, because some pyrophoric materials are also water-reactive, it is essential that hazmat technicians check available reference sources to determine appropriate extinguishing agents for the substance(s) involved.

Water-Reactive

Water reactivity is the tendency of a material to react, or chemically change, upon contact with water. Reactions involving water-reactive materials can range from mild to severe, with some materials reacting explosively. The reactions and their by-products will vary depending on the materials involved. The following are some of the more common consequences of allowing these materials to come in contact with water **(Figure 4.5)**:

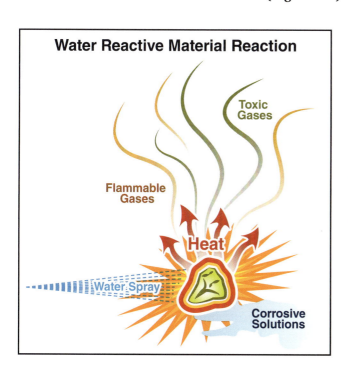

Figure 4.5 Water reactive materials may react violently on contact with water, even if the water is only in the air.

- **Generation of flammable gases** — Hydrogen gas (H_2) is often created when hydrogen atoms that are liberated from water molecules separate via the reaction. However, hydrogen is not the only flammable gas possible. For example, when calcium carbide (CaC_2) reacts with water, carbon and hydrogen combine to form highly flammable acetylene (C_2H_2) gas. Some of the other compounds that contain carbon, such as aluminum carbide (Al_4C_3) and beryllium carbide (Be_2C), evolve into methane gas (CH_4) upon contact with water.

- **Heat** — Water reactions are always exothermic, some very much so. A reaction that generates its own heat and flammable gases while liberating extra oxygen upon contact with water contains all four parts of the fire tetrahedron. Fires should be expected. There is also often enough heat to ignite nearby combustibles. Because fires involving combustible metals can be significantly hotter than other fires, a steam explosion may occur if hazmat technicians attempt to use water as an extinguishing agent.

- **Corrosive solutions** — Many of the water-reactive materials are either metals or salts (compounds comprised of a metal element and one or more nonmetal elements). When they react with water, metals frequently combine with oxygen and hydrogen from the water molecules to form highly caustic solutions, often recognized by the word "hydroxide" in the name. Examples include sodium hydroxide (NaOH) and potassium hydroxide (KOH). Some of the materials that are comprised of a metal element and chlorine, such as aluminum chloride (AlCl3) and stannic chloride ($SnCl_4$), react with water to form hydrogen chloride (HCl), either in the form of an acidic gas (if only small amounts of water are involved) or the acidic gas in solution (if enough water is present).

- **Toxic gases** — Many of the reactions also produce gases that are irritating or toxic. For example, compounds made with nitrogen, such as lithium nitride (Li_3N), may generate ammonia gas (NH_3), while those made with phosphorus, such as aluminum phosphide (AlP), may generate toxic phosphine gas (PH_3).

When water-reactive materials are present, hazardous materials technicians must consider all the different ways that these substances may come in contact with moisture, including moisture in the air and skin, eyes, or respiratory system. To overlook any of these less obvious scenarios could put responders at risk for serious injury. The best course of action for large fires involving water-reactive materials is typically to isolate the area, protect exposures, and allow the fire to burn until it consumes all of the fuel.

When facilities are equipped with automatic flooding systems that fill the area with an inert gas such as argon, the exclusion of oxygen can effectively extinguish the fire. If a system is already in place but has not been discharged, hazmat technicians may want to consider activating the system as the safest means to mitigate the incident.

Using water may be an acceptable option in some situations. For example, flooding amounts of water can extinguish burning magnesium parts that are solid or large such as castings and fabricated structures, because copious amounts of water can eventually cool the fire enough to overcome the exothermic reaction. Water, in any amount, should not be used for **Class D** fires involving magnesium powder, dust, chips, and shavings.

Class D Fire — Fires of combustible metals such as magnesium, sodium, and titanium.

> **WARNING!**
> All metal fires are water-reactive. Water should not be used to extinguish these fires unless sufficient quantities are accessible and a large defensible space is available.

Hazmat technicians must not use water to fight fires involving water-reactive materials unless they know what they are dealing with, have thoroughly assessed all the risks, and are properly trained in extinguishing these fires. The dangers are too great.

> **WARNING!**
> The use of water or water-based agents on Class D fires will cause the fire to react violently and emit bits of molten metal.

If the fire is still small, hazmat technicians may be able to control it with specially formulated **dry powders**, such as Pyrene G-1 or Met-L-X. Sand can sometimes be used to extinguish a small fire involving combustible metals, but the sand must be completely dry. If any moisture is present, the burning metal will react with the water in the sand and intensify the problem.

Dry Powder — Extinguishing agent suitable for use on combustible metal fires.

> **CAUTION**
> Do NOT confuse dry powder extinguishers with dry chemical units used on Class A, B, and C fires.

Class D dry powder agents and extinguishers are designed for these types of fires. Use only extinguishers that are rated for Class D fires, whether liquid or powder, to extinguish metal fires **(Figure 4.6)**.

> **CAUTION**
> Use only Class D rated fire extinguishers to extinguish metal fires.

Class D Fires and Extinguishers
Class D fires involve combustible metals and alloys such as:

- Titanium
- Magnesium
- Lithium
- Potassium
- Sodium

These metals can react violently with water, especially in powders, chips, and turnings (fine machine waste) rather than larger pieces. Other combus-

Figure 4.6 Always choose a fire extinguisher that is compatible with the material on fire.

Chapter 4 • Analyzing the Incident: Understanding Common Families and Special Hazards of Hazardous Materials/WMD   **91**

tible metals may have similar characteristics but are less likely to explode on contact with incompatible materials. Many of these materials will melt rather than burn in common situations.

Common uses of lithium include small (personal) and large (electric car) batteries, as well as illicit labs such as one-pot methamphetamine production.

Magnesium fires can be identified by the bright white light emissions during the combustion process. Common locations for magnesium include:

- Cameras
- Laptops
- Luggage
- Metal box springs for beds
- Vehicle wheels
- Transmissions
- Other vehicle components

Titanium, potassium, and sodium are less common in industry and transportation applications. Information about fires involving these metals can be found during preincident surveys, and should be addressed per AHJ.

Test fires for establishing Class D ratings vary with the type of combustible metal being tested. The following factors are considered during each test:

- Reactions between the metal and the agent
- Toxicity of the agent
- Toxicity of the fumes produced and the products of combustion
- Time to allow metal to burn completely without fire suppression compared to the time to extinguish the fire using the extinguisher

CAUTION
Do NOT look directly at a fire burning with a white light. Choose eye protection for these incidents based on AHJ protocols.

Shock- and Friction-Sensitive

Shock- and friction-sensitive reactive materials are potentially explosive and can be extremely sensitive to heat or shock. The chemicals may decompose violently if struck or heated **(Figure 4.7)**. The most common example of a shock- or friction-sensitive material is trinitrophenol, commonly known as picric acid. Other examples of shock- or friction-sensitive materials include:

- Ammonium perchlorate
- Calcium nitrate
- Nitroglycerin
- Organic peroxides

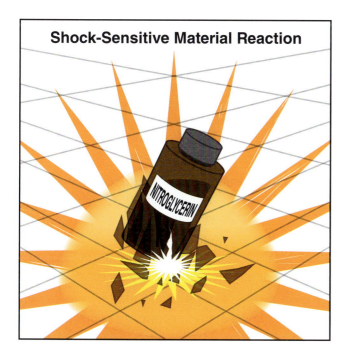

Figure 4.7 Materials reactive to shock or friction should be handled accordingly.

Light-Sensitive

Light-sensitive chemicals have the ability to change their composition if exposed to natural light. Often, these changes will lead to less stable substances that may be more dangerous than their original compound. In addition, light catalyzes many monomers to promote polymerization, such as the methacrylates used as plastic fillings.

Some examples of light-sensitive chemicals are:

- Ammonium dichromate
- Hydrogen peroxide
- Silver salts
- Mercuric salts

Temperature-Sensitive

Some chemicals may break down and decompose based on their temperature. While some temperature reactions may be mild, others may be violent or even explosive. Many peroxides tend to be temperature-sensitive materials. Some chemicals may be sensitive to high temperatures and some to low temperatures. Of course, the actual temperature that may affect a chemical is relative to the chemical itself. This critical information may be found in many chemical reference manuals.

Temperature change in the product/container, as identified via an infrared thermometer (temperature gun), is an important clue indicating a reaction. Organic peroxides should be stored below the **maximum safe storage temperature (MSST)** for routine purposes. Should organic peroxides reach the **self-accelerating decomposition temperature (SADT)**, they undergo a chemical change and may violently release from their packaging. The length of time before this release depends upon how much the SADT is exceeded, which can greatly accelerate the decomposition.

> **Maximum Safe Storage Temperature (MSST)** — Temperature below which the product can be stored safely. This is usually 20-30 degrees cooler than the SADT temperature, but may be more depending on the material.

> **Self-Accelerating Decomposition Temperature (SADT)** — Lowest temperature at which product in a typical package will undergo a self-accelerating decomposition. The reaction can be violent, usually rupturing the package, dispersing original material, liquid and/or gaseous decomposition products considerable distances.

> **WARNING!**
> Immediately evacuate the area if the SADT is reached. If decomposition occurs, observe it from a safe distance and take only those measures necessary to preserve life and nearby property.

During a SADT event, hazmat technicians should note that the heat generated might autoignite flammable vapors. In addition, the decomposition products may be much more toxic or corrosive than the original compound.

After the material reaches its SADT threshold, there is generally a period of time before its decomposition becomes violent. The length of time depends upon how much the SADT is exceeded, which can greatly accelerate the decomposition.

The SADT of a material is usually lower at higher concentrations. Dilution with a compatible, high boiling point diluent will usually increase the SADT since the material is dilute and the diluent can absorb much of the heat, minimizing the temperature increase. Also, larger containers generally support a lower SADT because of the poorer heat transfer due to lower surface area to volume ratio.

Most organic peroxides react to some extent with their decomposition products during thermal decomposition. This reaction often increases the rate since the decomposition proceeds more rapidly as the decomposition products are generated. Monitoring equipment, such as a **photoionization detector (PID)** and lower explosive limit (LEL) indicator, will indicate when decomposition occurs.

In addition to organic peroxides, many polymerization initiators or reactive chemicals have SADTs. The technician must recognize the behaviors of these materials. Resources that may aid that identification include safety data sheets and other reference sources. Many times the SADT is written into the SDS as *decomposition temperature*.

Corrosives

Corrosives are classified as either acids or bases based on their chemical behavior when in contact with water. **Acids** are compounds that release hydronium ions [H_3O]+1 when dissolved in water, a process sometimes referred to as **dissociation**. **Bases** (also called *caustics* or *alkalis*) release hydroxide ions [OH]-1 when dissolved in water. Hazmat technicians cannot see the release of ions, but can detect them and distinguish between acids and bases by measuring the strength (pH) of a corrosive. Corrosives were introduced in Chapter 3 of this manual.

All corrosives, whether acid or base, present a variety of hazards. The profiles provided in this section give some good examples. Acids and bases can both cause **chemical burns**. However, one of the most significant differences between acids and bases is the way in which they damage human tissue. An acid in contact with the skin will cause the tissue to harden even as it eats away at that tissue, thereby limiting the damage to some degree. However,

---

Photoionization Detector (PID) — Gas detector that measures volatile compounds in concentrations of parts per million and parts per billion.

Acid — Compound containing hydrogen that reacts with water to produce hydronium ions; a proton donor; a liquid compound with a pH less than 7. Acidic chemicals are corrosive.

Dissociation (Chemical) — Process of splitting a molecule or ionic compounds into smaller particles, especially if the process is reversible. *Opposite of* Recombination.

Base — Any alkaline or caustic substance; corrosive water-soluble compound or substance containing group-forming hydroxide ions in water solution that reacts with an acid to form a salt.

a base will soften and dissolve the tissue, creating far more penetrating and severe injuries. In both cases, the damage will continue until the corrosive is thoroughly flushed from the body. The extent of the injury is often not immediately obvious, as is also the case with **thermal burns**.

Corrosives often produce immediate irritation, but it is not uncommon for pain to be delayed. This is particularly true with corrosives in solid form, which essentially do not start eating away at the tissue until they react with the moisture on the skin. Such a corrosive on dry skin might not be noticed until a person starts sweating or takes a shower later in the day.

NOTE: Both acids and bases are corrosives. The Detection and Monitoring chapter in this manual will discuss corrosives in more depth.

> **Chemical Burn** — Injury caused by contact with acids, lye, and vesicants such as tear gas, mustard gas, and phosphorus.

> **Thermal Burn** — 1) Injury caused by contact with flames, hot objects, and hot fluids; examples include scalds and steam burns. 2) Any injury to living tissue from contact with extreme hot or cold materials.

CAUTION
Hazmat technicians should never disregard the hazard potential of weak corrosive materials at an incident.

CAUTION
Atypical results from detection resources should be evaluated; they may indicate unexpected, higher level hazards.

Radiation

Incidents involving **radioactive materials** are hyped in media as potentially catastrophic occurrences that will affect the general public. In practice, these types of incidents are uncommon because of the strict requirements governing their use and transportation. Examples of radioactive materials include cobalt, uranium hexafluoride, and medical isotopes such as barium.

In contrast, radioactive materials are becoming more commonplace in society as we find beneficial uses of their unique properties. These uses can include:

- Residential smoke detectors
- Medical diagnostic imaging
- Cancer treatment
- Soil density measurement
- Industrial food preservation

As a result, the hazmat technician may come in contact with radioactive materials and radiation almost anywhere, and needs to understand the basic protection strategies if radioactive materials or radiation is present at an incident and what resources may be used to detect the presence of radiation. This section is an overview of the topic, and will be revisited later in this chapter in terms of weapons of mass destruction.

> **Radioactive Material (RAM)** — Material with an atomic nucleus that spontaneously decays or disintegrates, emitting radiation as particles or electromagnetic waves at a rate of greater than 0.002 microcuries per gram (Ci/g).

Ionizing and Nonionizing Radiation

Atoms perpetually try to achieve a state of electrical equilibrium. As some types of atoms seek a strong, stable balance of energy, they may release excess atomic energy as radiation. This radiation takes the form of rays or high-speed particles. All radioactive materials give off at least one type of radiation. Many give off two or three **(Figure 4.8)**.

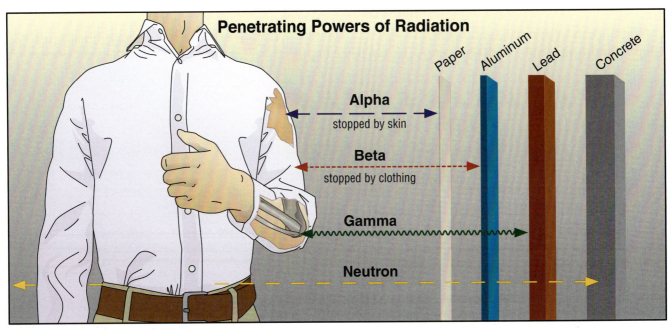

Figure 4.8 Radioactive materials may emit multiple types of radiation.

Ionizing Radiation — Radiation that causes a chemical change in atoms by removing their electrons.

Nonionizing Radiation — Series of energy waves composed of oscillating electric and magnetic fields traveling at the speed of light. Examples include ultraviolet radiation, visible light, infrared radiation, microwaves, radio waves, and extremely low frequency radiation.

Types of radiation vary in energy levels. The most energetic and hazardous form is **ionizing radiation**. This type of radiation is of greatest concern to first responders. The least energetic form is **nonionizing radiation** such as visible light and radio waves.

Types of ionizing radiation particles are:

- **Alpha (α) particles** — Relatively large particles that can travel only a few inches in air. Alpha particles cannot penetrate intact skin and are an internal hazard only. They may enter the body through inhalation, ingestion, or contamination of an open wound. Alpha particles are stopped by shielding as thin as a single sheet of paper.

- **Beta (β) particles** — Much smaller than alpha particles, and can travel several yards in the air. Beta particles can penetrate intact skin, damaging the skin and possibly internal organs. Beta particles can be inhaled or ingested like alpha particles. Shielding necessary to stop beta particles includes wood or metal, including aluminum foil.

- **Neutron particles** — Considerably larger than beta particles but smaller than alpha particles. Neutron radiation is normally associated with nuclear power plants but is becoming more common in medical treatment. It has strong penetrating power and presents health hazards similar to other forms of radiation.

Types of electromagnetic waves that carry ionizing radiation with a high frequency and low wavelength include (**Figure 4.9**):

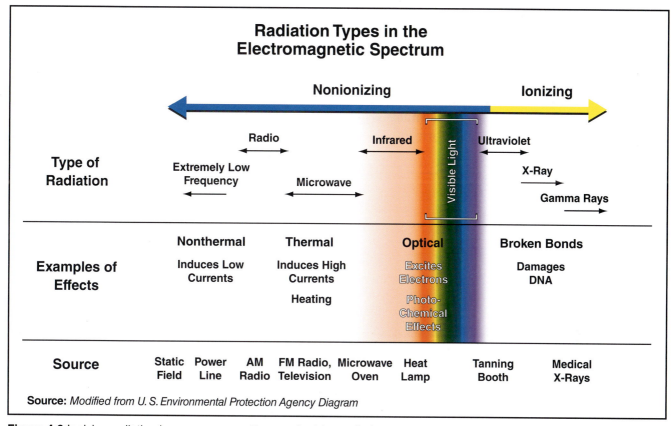

Figure 4.9 Ionizing radiation has more energy than nonionizing radiation.

- **Gamma (γ) rays** — Electromagnetic waves of high energy and short wavelength that travel at the speed of light. Gamma radiation has strong penetrating power, able to travel considerable distances and through heavy objects. It also can penetrate intact skin, causing skin burns and severe internal damage. Dense shielding necessary to stop gamma rays includes lead, concrete, or several feet of water.

- **X-rays** — Electromagnetic waves that are similar to gamma rays, though not as hazardous. X-rays are primarily produced by machines. They are rarely a by-product of natural radioactive decay. The chances of encountering them at a hazardous materials incident are remote, so they are considered much less of a threat than alpha, beta, and gamma radiation.

In contrast to ionizing radiation, nonionizing radiation does not carry enough energy to **ionize** an atom or a molecule. Visible light, microwaves, and radio waves are some examples of nonionizing radiation. The light from the sun that reaches the earth is primarily composed of nonionizing radiation, but most of the ionizing radiation from the sun is filtered out by the earth's atmosphere. Nonionizing radiation will not completely remove an electron from an atom. There is only sufficient energy to excite the electron into a higher energy state.

Ionize — Process in which an atom or molecule gains a negative or positive charge by gaining or losing electrons.

Ionizing radioactive materials have some types of reactions in common:
- Radioactive decay
- Half-life
- Activity

Radioactive Decay

Radioactive decay is the spontaneous breakdown of an atomic nucleus resulting in the release of energy and matter from the nucleus. During this process, the unstable nucleus of a **radioisotope** will release energy and matter and often transform into a new element in a process called **transmutation** (Figure 4.10).

Radioactive Decay — Process in which an unstable radioactive atom loses energy by emitting ionizing radiation and conversion electrons.

Radioisotope — Unstable atom that releases nuclear energy.

Transmutation — Conversion of one element or isotope into another form or state.

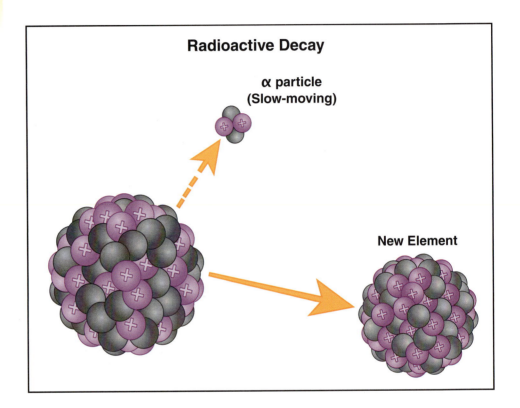

Figure 4.10 Transmutation is the process of radioactive decay where new molecules are formed by the shedding of energy and matter.

Half-Life

Half-life is the measure of time it takes for one half of a given amount of radioactive material to decay or change to a less hazardous form. For example, as uranium 238 decays, it changes first to thorium 230, which becomes radium 226, which decays to radon 218, which changes to bismuth 214, which finally becomes the stable element lead 206. This change occurs because as uranium 238 decays, it "throws off" both neutrons and protons from the nucleus of the atom. The half-life of U238 is 4,468 million years. As a rule of thumb, a radioactive isotope decreases to less than 1 percent of its original value after seven half-lives. The majority of radioactive materials have extremely long half-lives, from several thousand years to several million years.

Knowing a material's half-life can help hazmat technicians determine how long an area affected by radiation must be sealed off and whether the incident can be allowed to self-mitigate. For example, a scene involving radioactive material with a half-life of a few days may be secured for a couple of weeks

Half-Life — The time required for a radioactive material to reduce to half of its initial value.

until the hazard expires. An incident involving a material with an extended half-life may require the assistance of an appropriate cleanup company to mitigate the hazard.

Activity

The energy of radiation gives it the ability to penetrate matter. Higher energy will be able to penetrate a higher volume and denser matter than lower level radiation. The strength of a radioactive source is called its **activity**. More accurately, the activity of a radioactive source can be defined as the rate at which a number of atoms will decay and emit radiation in one second **(Figure 4.11)**.

> **Activity** — Rate of decay of the isotope in terms of decaying atoms per second. Measured in becquerels (Bq) for small quantities of radiation, and curies (Ci) for large quantities of radiation.

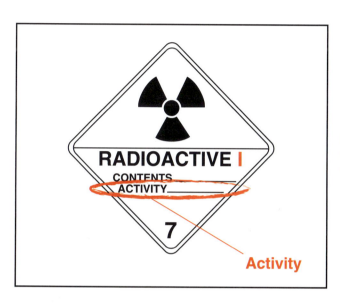

Figure 4.11 Activity refers to the strength of a radioactive source.

The International System (SI) unit for activity is the **becqueral (Bq)**, which is the quantity of radioactive material in which one atom transforms per second. The becqueral tends to be a small unit. The **curie (Ci)** is also used as the unit for activity of a particular source material. The curie is a quantity of radioactive material in which 3.7×10^{10} atoms disintegrate per second.

> **Becquerel (Bq)** — International System unit of measurement for radioactivity, indicating the number of nuclear decays/disintegrations a radioactive material undergoes in a certain period of time.

Special Hazards of Chemicals and Weapons of Mass Destruction

Up to this point, this manual has concentrated on the chemistry of various chemical families and compounds. If a hazmat technician is called to intervene in an emergency situation, the technician should understand the makeup of the product and know how it will behave in this situation. Now that the technician has a background of the chemistry of a compound or a product, it is important to describe the special hazards of various chemical compounds. The following sections, arranged in order of **CBRNE** categories, will explain the special hazards of chemicals to which the responder may be exposed during an emergency incident.

Experts have not reached consensus on which types of WMDs first responders are most likely to encounter. However, given the availability of parts, relative ease of production, and ease of deployment, the following is a probable WMD threat spectrum from most likely to least likely:

> **Curie (Ci)** — English System unit of measurement for radioactivity, indicating the number of nuclear decays/disintegrations a radioactive material undergoes in a certain period of time.

> **CBRNE** — Abbreviation for Chemical, Biological, Radiological, Nuclear, and Explosive. These categories are often used to describe WMDs and other hazardous materials characteristics.

1. Explosives
2. Biological toxins
3. Industrial chemicals
4. Biological pathogens
5. Radiological materials
6. Military-grade chemical weapons
7. Nuclear weapons

NOTE: Conventional attacks such as hijackings, sniper attacks, and/or shootings are also highly likely, but not considered a WMD threat for purposes of this list.

Other Acronyms for WMDs
While this chapter discusses types of attacks based on the CBRNE acronym, other organizations may use other acronyms to indicate essentially the same thing. These terms include COBRA (chemical, ordinance, biological, radiological agents); B-NICE (biological, nuclear, incendiary, chemical, explosive); and NBC (nuclear, biological, chemical) in addition to others. This manual may refer to CBR materials when describing chemical, biological, and radiological attacks.

Ranking of WMD Threats
There is no way to predict which types of terrorist or WMD attacks will occur with what frequency. Several organizations have developed a framework to guide how they teach these topics, but no single framework is necessarily more accurate than any other.

Chemical Warfare Agents
While many chemicals can be used as weapons, five categories are most likely to be used in this way. The following sections will discuss these agents in greater detail.

Nerve Agents
Nerve agents are similar to organophosphate pesticides, or carbamate, in their chemical makeup and in the way they attack the nervous system and cause uncontrolled muscular contractions. However, nerve agents are typically significantly more potent than the pesticide formulated materials. Exposure to even minute quantities can kill quickly. Nerve agents are considered the most dangerous of the chemical warfare agents. The common nerve agents include **(Table 4.3)**:

- Tabun (GA)
- Sarin (GB)
- Soman (GD)
- V agent (VX)

Nerve Agent — A class of toxic chemical that works by disrupting the way nerves transfer messages to organs.

Table 4.3
Nerve Agents and Their Characteristics

Nerve Agent (Symbol)	Descriptions	Symptoms (All Listed Agents)
Tabun (GA)	• Clear, colorless, and tasteless liquid • May have a slight fruit odor, but this feature cannot be relied upon to provide sufficient warning against toxic exposure • **Probable Dispersion Method:** Aerosolized liquid	***Low or moderate dose by inhalation, ingestion (swallowing), or skin absorption:*** Persons may experience some or all of the following symptoms within seconds to hours of exposure: • Runny nose • Diarrhea • Watery eyes • Increased urination • Small, pinpoint pupils • Confusion • Eye pain • Drowsiness • Blurred vision • Weakness • Drooling and excessive sweating • Headache • Cough • Nausea, vomiting, and/or abdominal pain • Chest tightness • Slow or fast heart rate • Rapid breathing • Abnormally low or high blood pressure ***Skin contact:*** Even a tiny drop of nerve agent on the skin can cause sweating and muscle twitching where the agent touched the skin ***Large dose by any route:*** These additional health effects may result: • Loss of consciousness • Convulsions • Paralysis • Respiratory failure possibly leading to death ***Recovery Expectations:*** • Mild or moderately exposed people usually recover completely • Severely exposed people are not likely to survive • Unlike some organophosphate pesticides, nerve agents have *not* been associated with neurological problems lasting more than 1 to 2 weeks after the exposure
Sarin (GB)	• Clear, colorless, tasteless, and odorless liquid in pure form • **Probable Dispersion Method:** Aerosolized liquid	
Soman (GD)	• Pure liquid is clear, colorless, and tasteless; discolors with aging to dark brown • May have a slight fruity or camphor odor, but this feature cannot be relied upon to provide sufficient warning against toxic exposure • **Probable Dispersion Method:** Aerosolized liquid	
Cyclohexyl sarin (GF)	• Clear, colorless, tasteless, and odorless liquid in pure form • Only slightly soluble in water • **Probable Dispersion Method:** Aerosolized liquid	
V Agent (VX)	• Clear, amber-colored odorless, oily liquid • Miscible with water and dissolves in all solvents • Least volatile nerve agent • Very slow to evaporate (about as slowly as motor oil) • Primarily a liquid exposure hazard, but if heated to very high temperatures, it can turn into small amounts of vapor (gas) • **Probable Dispersion Method:** Aerosolized liquid	

Source: Information on symptoms provided by the Centers for Disease Control and Prevention (CDC).

Although nerve agents are generally clear and colorless, colors and odors can vary with impurities. Impure G-series agents may have a slight fruity odor. VX is odorless. Although people sometimes use the term *nerve gas*, the term is a misnomer. Nerve agents are liquids at ambient temperatures and dispersed as an aerosolized liquid (vapor, not gas).

WARNING!
Odor is not a safe indicator of a hazard.

G-Series Agents — Nonpersistent nerve agents initially synthesized by German scientists.

G-series agents are described as volatile and nonpersistent. They are less volatile than water. These agents are sometimes made more persistent by adding various thickeners. VX is a persistent, oily liquid that evaporates at about the same rate as motor oil. The vapors of all four agents are heavier than air.

Considering the low vapor pressures, nerve agent vapors will not travel far under normal conditions. Therefore, the size of the endangered area may be relatively small. However, the vapor hazard can significantly increase if the liquid is exposed to high temperatures, spread over a large area, or aerosolized.

Nerve agents usually enter the body through inhalation of the vapors. These agents are also toxic by eye absorption, skin absorption, or ingestion. Onset of signs and symptoms following inhalation and eye exposure can happen within minutes, sometimes seconds. Onset can take several minutes to several hours with skin absorption, although nerve agents do penetrate the skin rapidly and effectively. Speed of onset also varies with amount of exposure.

Because nerve agents are similar to organophosphate pesticides, attacking the nervous system by inhibiting acetylcholinesterase, the signs and symptoms of exposure will be much like those of organophosphate poisoning, though usually more severe. The most significant signs of nerve agent poisoning are:

- Rapid onset of pinpoint pupils
- Muscular twitching
- Seizures
- Excess secretion of body fluids (runny nose, tearing, sweating, salivation, vomiting, urination, and defecation)

NOTE: Following inhalation exposure, these signs can appear almost immediately.

SLUDGEM or DUMBELS

Some people teach the symptoms of exposure to chemical warfare agents using one of two common acronyms, SLUDGEM or DUMBELS. SLUDGEM indicators are:

- Salivation (drooling)
- Lacrimation (tearing)
- Urination
- Defecation
- Gastrointestinal upset/aggravation (cramping)
- Emesis (vomiting)
- Miosis (pinpointed pupils) or Muscular twitching/spasms

DUMBELS indicators are:

- Defecation
- Urination
- Miosis or Muscular twitching
- Bronchospasm (wheezing)

- Emesis
- Lachrimation
- Salivation

Blister Agents/Vesicants

Blister agents are extremely toxic chemicals that produce characteristic blisters on exposed skin. Common types of blister agents include mustard (H), distilled mustard (HD), nitrogen mustard (HN), lewisite (L), and phosgene oxime (CX).

Mustards and lewisite are oily liquids often described as anything from colorless to pale yellow to dark brown, depending on purity. Also depending on purity and concentration, mustards may be odorless or may smell like mustard, onion, or garlic. Lewisite reportedly smells like geraniums.

Pure phosgene oxime is a colorless crystalline solid, but phosgene oxime can also be found as a yellowish-brown liquid. It has an intense, irritating odor.

All of the blister agents are relatively persistent, which means that it can take several days or weeks for them to evaporate. The oily consistency also means that these agents will be more difficult to remove during decontamination than less viscous products.

Blister agents are extremely toxic and many are carcinogenic. The primary routes of entry are inhalation and absorption. Blister agents easily penetrate clothing and are quickly absorbed through the skin. Only a few drops on the skin can cause severe injury and death.

Common signs and symptoms are irritation and burns to the skin, eyes, and respiratory tract; difficulty breathing; and blisters. If a large area of the skin is involved, significant amounts of the agent can be absorbed into the bloodstream, causing severe systemic poisoning.

A few notable differences between the effects produced by the three types of blister agents can help responders distinguish which one may be involved:

- Mustard causes tissue damage almost immediately after exposure. The clinical (noticeable) effects are delayed from two to twenty-four hours, with four to eight hours being the most common. Mustard is the only chemical warfare agent that does not produce symptoms within minutes of exposure.

- Phosgene oxime causes immediate pain or irritation to the skin, eyes, and lungs. Another sign that distinguishes phosgene oxime from other blister agents is that the skin lesions look more like gray wheals (hives) than blisters. Phosgene oxime causes more severe tissue damage than do other blister agents.

- Lewisite's effects are felt within seconds to minutes after exposure. With lewisite and phosgene oxime, the irritation to eyes, skin, and mucous membranes may initially resemble the effects of riot control agents. However, the pain is more severe and will not decrease upon moving the patient to fresh air, as would be the case with riot control agents.

> **Blister Agent** — Chemical warfare agent that burns and blisters the skin or any other part of the body it contacts. *Also known as* Vesicant *and* Mustard Agent.

Blood Agents

> **Chemical Asphyxiant** — Substance that reacts to prevent the body from being able to use oxygen. *Also known as* Blood Agent.

Blood agents are **chemical asphyxiants**. They interfere with the body's ability to use oxygen either by preventing red blood cells from carrying oxygen to other cells in the body or by inhibiting the ability of cells to use oxygen for producing the energy required for metabolism. The most common blood agents are hydrogen cyanide (AC) and cyanogen chloride (CK).

These substances are volatile, nonpersistent, colorless liquids under pressure, although at higher temperatures, hydrogen cyanide is a colorless gas. Hydrogen cyanide has a reported odor of bitter almonds or peach kernels, but at least 40 percent of the population is genetically unable to smell the odor. Hydrogen cyanide (AC) is also flammable. Cyanogen chloride (CK) has an extremely irritating odor.

The primary route of entry is inhalation. However, cyanide can also be absorbed through skin or eyes. Exposed patients experience rapid onset of respiratory difficulty, ranging from initial gasping and rapid breathing to subsequent respiratory arrest if exposed to higher concentrations. Other signs and symptoms often include:

- Dizziness
- Nausea
- Vomiting
- Headache
- Irritation of eyes, nose, and mucous membranes

Like nerve agents, blood agents can produce seizures. However, cyanide does not produce pinpoint pupils and excessive secretions, common with nerve agents. Another distinguishing characteristic sometimes seen with exposure to cyanide is abnormally red skin.

Specific treatment for hydrogen cyanide and cyanogen chloride includes amyl nitrite and sodium nitrite to draw the cyanide out of cells in the body and sodium thiosulfate to detoxify the cyanide and help remove it from the body. The three compounds (amyl nitrite, sodium nitrite, and sodium thiosulfate) are packaged together in cyanide antidote kits.

Choking Agents

> **Choking Agent** — Chemical warfare agent that attacks the lungs, causing tissue damage.

Choking agents primarily attack and damage the respiratory tract. While a number of common industrial chemicals can act as choking agents, the two most often cited in terrorism training are chlorine and phosgene. Each agent will cause the following:

- Irritation to the eyes, nose, and throat
- Coughing and choking
- Respiratory distress
- Nausea and vomiting
- Headache
- Tightness in the chest
- Potentially fatal pulmonary edema at small or low-level exposures

Riot Control Agents/Irritants

Riot control agents (sometimes called *tear gas* or *irritating agents*) cause temporary incapacitation by irritating the eyes and respiratory system (**Table 4.4**). Common agents include: pepper spray (OC), tear gas (CS or CR), mace (CN), and adamsite (DM).

> **Riot Control Agent** — Chemical compound that temporarily makes people unable to function, by causing immediate irritation to the eyes, mouth, throat, lungs, and skin.

Table 4.4
Riot Control Agents and Their Characteristics

Riot Control Agent (Symbol)	Descriptions	Symptoms (All Listed Agents)
Chlorobenzylidene malononitrile (CS)	• White crystalline solid • Pepper-like smell	***Immediately after exposure:*** People exposed may experience some or all of the following symptoms: • ***Eyes:*** Excessive tearing, burning, blurred vision, and redness • ***Nose:*** Runny nose, burning, and swelling • ***Mouth:*** Burning, irritation, difficulty swallowing, and drooling • ***Lungs:*** Chest tightness, coughing, choking sensation, noisy breathing (wheezing), and shortness of breath • ***Skin:*** Burns and rash • ***Other:*** Nausea and vomiting Long-lasting exposure or exposure to a large dose, especially in a closed setting, may cause severe effects such as the following: • Blindness • Glaucoma (serious eye condition that can lead to blindness) • Immediate death due to severe chemical burns to the throat and lungs • Respiratory failure possibly resulting in death Prolonged exposure, especially in an enclosed area, may lead to long-term effects such as the following: • Eye problems including scarring, glaucoma, and cataracts • May possibly cause breathing problems such as asthma ***Recovery Expectations:*** If symptoms go away soon after a person is removed from exposure, long-term health effects are unlikely to occur.
Chloroacetophenone (CN, mace)	• Clear yellowish brown solid • Poorly soluble in water, but dissolves in organic solvents • White smoke smells like apple blossoms	
Oleoresin Capsicum (OC, pepper spray)	• Oily liquid, typically sold as a spray mist • ***Probable Dispersion Method:*** Aerosol	
Dibenzoxazepine (CR)	• Pale yellow crystalline solid • Pepper-like odor • ***Probable Dispersion Method:*** Propelled	
Chloropicrin (PS)	• Oily, colorless liquid • Intense odor • Violent decomposition when exposed to heat	

Source: Information on symptoms provided by the Centers for Disease Control and Prevention (CDC).

Riot control agents are solids, dispersed either as a fine powder or as an aerosol (a powder suspended in a liquid.) Some are sold in small containers as personal defense devices containing either a single agent or a mixture. Some devices also contain a dye to visually mark a sprayed assailant.

Riot control agents quickly incapacitate a person with the symptoms such as:

- Severe irritation to the eyes, nose, and respiratory tract
- Difficulty keeping eyes open
- Sneezing

- Coughing
- Runny nose
- Shortness of breath

Adamsite will cause nausea and vomiting. Exposure to riot control agents can also trigger asthma attacks or other secondary problems in some individuals.

The primary route of entry is inhalation, but these agents can also be absorbed through the skin and eyes. Onset of signs and symptoms usually occurs within seconds. The effects seldom persist more than a few minutes once victims are removed to fresh air, but they can last several hours, depending on the dose and duration of exposure. Decontamination can often be limited to flushing the eyes and washing the face or other parts of the body directly affected.

Biological Agents and Toxins

The **biological agents** most likely to be used as weapons of mass destruction are generally divided into four groups:

- **Viruses** — Viruses are living organisms that are much smaller than most bacteria **(Figure 4.12)**. Viruses are not capable of the basic metabolic functions necessary for independent growth. They require living cells in which to replicate. Viruses include:

  — Smallpox

  — Venezuelan equine encephalitis (VEE)

  — Viral hemorrhagic fever (VHF)

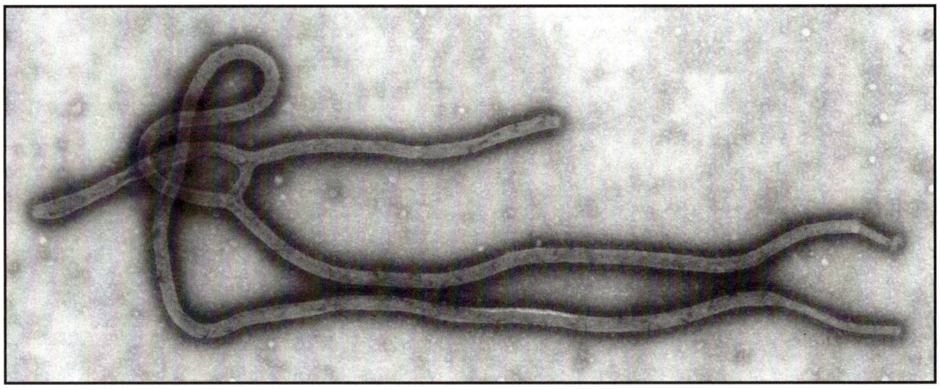

Figure 4.12 Viruses are living organisms that can be killed.

- **Bacteria** — Bacteria are single-celled living organisms capable of independent growth. They usually do not require a living host in which to replicate and many can be cultured in a lab. Some, like anthrax, have the ability to form spores that can survive for long periods of time in conditions that would otherwise kill the bacteria. Bacteria can also produce extremely potent toxins inside the body. Examples of bacteria include:

  — Anthrax

  — Brucellosis

  — Cholera

  — Plague

  — Tularemia

Biological Agent — Viruses, bacteria, or their toxins which are harmful to people, animals, or crops. When used deliberately to cause harm, may be referred to as a Biological Weapon.

Virus — Simplest type of microorganism that can only replicate itself in the living cells of its hosts. Viruses are unaffected by antibiotics.

Bacteria — Microscopic, single-celled organisms.

Rickettsia — Specialized bacteria that live and multiply in the gastrointestinal tract of arthropod carriers, such as ticks and fleas.

Toxin — Substance that has the property of being poisonous.

- **Rickettsia** – Rickettsias are specialized bacteria. The important difference between rickettsias and other types of bacteria is the vector: arthropods. Like viruses, rickettsias grow only within living cells. An example of a rickettsia is Q fever.
- **Toxins** — Toxins are nonliving chemical compounds — potent poisons produced by a variety of living organisms including bacteria, plants, and animals **(Figure 4.13)**. They produce effects similar to those caused by chemical agents. Biological toxins are far more toxic than most industrial chemicals. Examples of toxins are botulinum toxin (botulism), ricin, abrin, saxitoxin, staphylococcal enterotoxin B, and trichothecene mycotoxins.

Figure 4.13 Toxins are nonliving materials produced by living organisms that can damage living tissues.

Biological agents and toxins can enter the body through most normal routes of entry. Much depends on the particular agent, how it is disseminated, and whether exposure is to the agent itself (toxins) or to the resulting disease (bacteria/virus) in an infected person.

Toxins are not contagious. Some bacteria and viruses are contagious, but many are not. Three in particular that can be transmitted from one person to another to the degree that they require more than universal precautions are pneumonic plague, smallpox, and viral hemorrhagic fevers.

Federal Assistance for WMD/Terrorism Incidents
The scope of biological weapons is generally broader than most hazmat technicians will be able to evaluate during an incident. These categories are largely developed from WMD/terrorism incidents. When responding to these incidents, technicians should follow established rules and guidelines to determine what the hazards are and how to mitigate them effectively. Technicians are also encouraged to contact the FBI headquarters to discuss effective use of resources with the WMD coordinator. In the U.S., contacting the FBI WMD Coordinator is usually accomplished via local law enforcement.

A technician must understand the principles of **toxicology** because chemicals can have such a profound effect on the human body. Toxic substances can be categorized in nine categories:

- Asphyxiants
- Irritants
- Carcinogens
- Mutagens
- Infectious substances
- Corrosives
- Sensitizers
- Neurotoxins
- Teratogens

Chemicals in each of these categories have specific hazards and can have catastrophic effects on the human body. Other elements of toxicology that hazmat technicians must understand include the concepts of **dose** and concentration.

Toxicology — Study of the adverse effects of chemicals on living organisms.

Dose — Quantity of a chemical material ingested or absorbed through skin contact for purposes of measuring toxicity.

Toxicity — Degree to which a substance (toxin or poison) can harm humans or animals. Ability of a substance to do harm within the body.

Dose-Response Relationship — Comparison of changes within an organism per amount, intensity, or duration of exposure to a stressor over time. This information is used to determine action levels for materials such as drugs, pollutants, and toxins.

CAUTION
Understanding hazards is important for preventing and minimizing exposure.

Dose-Response Relationships

Coming in contact with a hazardous material will often result in some type of physical harm. The ability of a chemical to cause harm via interference with or destruction of individual cells is referred to as its **toxicity**. The quantitative relationship between a dose of a chemical and the biological effects produced by that chemical is known as the **dose-response relationship** (Figure 4.14). Within sample populations, the effects of the dose will be similar. However, there may be a wide range of responses within that population where some people are susceptible to the dose but others are resilient.

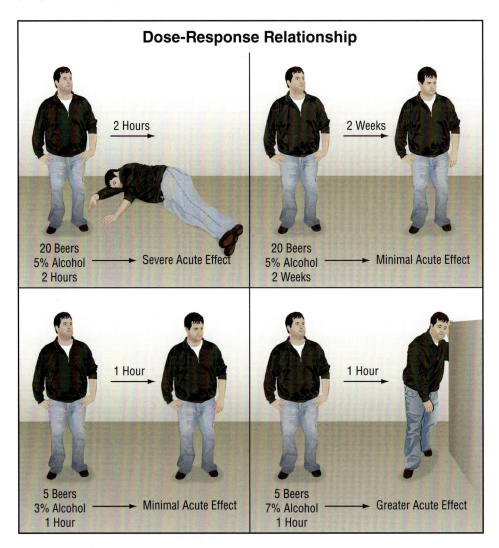

Figure 4.14 Contact with a set quantity of material over a set period of time may be different per material.

Measuring and Expressing Dose and Concentration

Concentrations of a substance can be expressed in the following terms:

- **Parts-per-million (ppm)** — Parts-per-million (ppm) is an expression of the concentration of a gas or vapor that can be found in air. It can be expressed as parts (by volume) of the gas or vapor in a million parts of air. It can also define the concentration of a particular substance in a liquid or solid.

- **Parts-per-billion (ppb)** — Parts-per-billion (ppb) is also an expression of the concentration of a gas or vapor that can be found in air. It can be expressed as parts (by volume) of the gas or vapor in a billion parts of air. It is usually used to express extremely low concentrations of unusually toxic gases or vapors. It will also define the concentration of a particular substance in a liquid or solid.

- **Milligrams per cubic meter (mg/m³)** — Milligrams per cubic meter are used to express the concentration of substances including dusts or mists in air.

Dosages can be expressed in the following ways:

- **Milligram per kilogram (mg/kg)** — Denotes quantity (in milligrams of weight) of a substance per kilogram of body weight

- **Milligram per square centimeter (mg/cm²)** — Denotes quantity (in milligrams of weight) of a substance per square centimeter of body surface area

- **Milligram per square meter (mg/m²)** — Denotes quantity (in milligrams of weight) of a substance per square meter body surface area

Toxicity

One way to study the effects of a toxic substance on a population is to study how that substance affects the population's demographics. By analyzing the birth and death rates post exposure, scientists can obtain a better understanding of how a certain product will affect a population. Some simple measures of toxicity use **bioassay** (biological assays) to measure the death rate in a sample population. The population measured is usually laboratory rats or mice. Common bioassays used to accomplish these tests and the results are referred to as LC_{50} and LD_{50}. While these measures of toxicity are important to understand, they are not always applicable for responders at an emergency incident.

Lethal dose 50 (LD_{50}) can be defined as the dose of a solid or liquid toxic substance, measured in milligrams of toxic substance per kilogram of body weight, that would kill 50 percent of the exposed sample population **(Figure 4.15)**. **Lethal concentration 50 (LC_{50})** can be defined as the concentration of a toxic gaseous substance, measured in parts per million (ppm) in air, that could kill 50 percent of the exposed sample population.

> **Bioassay** — Scientific experiment in which live plant or animal tissue or cells are used to determine the biological activity of a substance. *Also known as* Biological Assessment *or* Biological Assay.

> **Median Lethal Dose, 50 Percent Kill (LD_{50})** — Concentration of an ingested or injected substance that results in the death of 50 percent of the test population. LD_{50} is an oral or dermal exposure expressed in milligrams per kilogram (mg/kg); the lower the value, the more toxic the substance.

> **Median Lethal Concentration, 50 Percent Kill (LC_{50})** — Concentration of an inhaled substance that results in the death of 50 percent of the test population. LC_{50} is an inhalation exposure expressed in parts per million (ppm), milligrams per liter (mg/liter), or milligrams per cubic meter (mg/m³); the lower the value, the more toxic the substance.

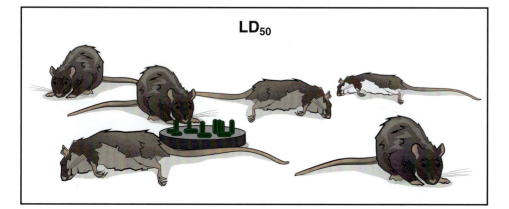

Figure 4.15 Lethal dose 50 (LD_{50}) refers to a concentration of a material that will kill 50% of the sample population.

Organophosphate Pesticides — Chemicals that kill insects by disrupting their central nervous systems; these chemicals deactivate acetylcholinesterase, an enzyme which is essential to nerve function in insects, humans, and many other animals. Because they have the same effect on humans, they are sometimes used in terrorist attacks.

Pesticides and Agricultural Chemicals

Pesticide is an umbrella term for a family of chemicals designed to kill specific target organisms. Many chemicals are toxic, but most were not intended to kill the way that **organophosphate pesticides** are. Even though pesticides are designed to kill specific target organisms (e.g., insects, rodents, or weeds), many pesticides can be toxic to both humans and the environment **(Table 4.5)**.

Other terms are used to categorize pesticides based on how they are designed to function **(Table 4.6)**. For example, pesticides are sometimes referred to as "crop protection chemicals" to reduce the negative connotation, but changing the name does not change the toxicity or other hazards associated with these materials.

Table 4.5 Types of Pesticides and Their Target Organisms

Type of Pesticide	Target Organisms
Algicides	Algae
Antimicrobials	Microorganisms (such as bacteria and viruses)
Biocides	Microorganisms
Fungicides	Fungi (including blights, mildews, molds, rusts)
Herbicides	Weeds
Insecticides	Insects and other arthropods
Miticides (Acaricides)	Mites (acarids)
Molluscicides	Snails and slugs
Nematocides	Nematodes (worm-like organisms that feed on plant roots)
Ovicides	Eggs of insects and mites
Rodenticides	Mice and other rodents

Table 4.6 Pesticide Categories and Functions

Type of Pesticide	Functional Design
Antifouling Agents	Kill or repel organisms that attach to underwater surfaces, such as boat bottoms.
Attractants	Attract pests (for example, to lure into a trap)
Defoliants	Cause leaves or other foliage to drop from a plant (e.g., to facilitate harvesting)
Desiccants	Promote drying of living tissues (e.g., unwanted plant tops)
Disinfectants and Sanitizers	Kill or inactivate disease-producing microorganisms on inanimate objects
Fumigants	Produce gas or vapor to destroy pests in buildings or soil
Insect Growth Regulators	Disrupt growth process of insects
Pheromones	Disrupt the mating behavior of insects
Plant Growth Regulators	Alter expected growth, flowering, or reproduction rate of plants (excludes fertilizers and other plant nutrients)
Repellants	Repel pests rather than destroy them (e.g., insect repellant)

Pesticides are transported in volumes ranging from large rail cars to small pickup trucks driven by pest control workers. Pesticides may be transported and dispersed via aircraft. Some pesticides are so toxic (via chemical composition, concentration, or both) that they are available only to trained and licensed applicators. The majority of pesticides can be found in occupancies throughout the community, including:

- Commercial and agricultural warehouses
- Farms and farm supply stores
- Nurseries and greenhouses
- Supermarkets
- Hardware and discount stores
- Residences

Radioactive and Nuclear

Incidents involving radioactive materials (RAM) are uncommon, but the hazmat technician may come in contact with radioactive materials and radiation almost anywhere. This section expands on information presented earlier on this chapter, with more of a focus on radiation and nuclear materials at incidents and delivered via weapons of mass destruction.

NOTE: Chapter 6 includes more information on detection and monitoring of radioactive materials.

Radioactive Material Exposure and Contamination

Radiation **exposure** occurs when a person is near a radiation source and is exposed to the energy from that source. Exposure and damage are not necessarily related. Depending on the length of exposure, energy, and type of radiation (alpha, beta, gamma, neutron), the victim may receive a dose during an incident involving radioactive material.

Exposure to a radioactive **contaminant** does not make a person or object radioactive. Radioactive **contamination** occurs when radioactive material is deposited on surfaces, skin, clothing, or any place where it is not desired **(Figure 4.16)**.

> **Exposure** — (1) Contact with a hazardous material, causing biological damage, typically by swallowing, breathing, or touching (skin or eyes). Exposure may be short-term (acute exposure), of intermediate duration, or long-term (chronic exposure). (2) People, property, systems, or natural features that are or may be exposed to the harmful effects of a hazardous materials emergency.

> **Contaminant** — Foreign substance that compromises the purity of a given substance.

> **Contamination** — Impurity resulting from mixture or contact with a foreign substance.

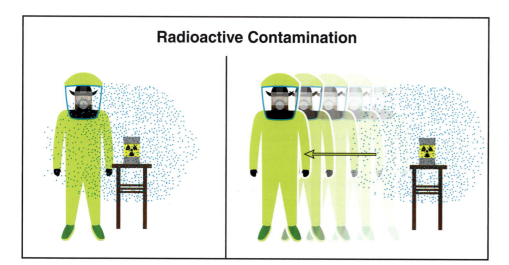

Figure 4.16 Exposure to radioactive material does not contaminate you.

> **Radiation Absorbed Dose (rad)** — English System unit used to measure the amount of radiation energy absorbed by a material; its International System equivalent is gray (Gy).

Damage is often discussed in terms of the **radiation absorbed dose (rad)**. A technician must know the dose-response relationship to evaluate whether the present types of radiation will cause damage, and what proximity and level of exposure will cause what kinds of harm.

CAUTION
Radiation does not spread; radioactive materials and contaminants spread.

Contamination only occurs when the radioactive material remains on a person or their clothing after coming into contact with the contaminant. A person can become contaminated externally, internally, or both. Radioactive material can enter the body via one or more routes of entry. An unprotected person contaminated with radioactive material receives radiation exposure until the source of radiation (radioactive material) is removed.

Technicians should keep the following considerations in mind:

- A person is externally contaminated (and receives external exposure) when radioactive material is on the skin or clothing.
- A person is internally contaminated (and receives internal exposure) when radioactive material is breathed, swallowed, or absorbed through wounds.
- The environment is contaminated when radioactive material spreads or is unconfined. Environmental contamination is another potential source of external exposure.
- Radioactive contamination is best determined by using a radiation detector that can detect alpha and beta contamination.

NOTE: Some contamination such as alpha contamination often requires the detector to be as close as 1 centimeter away from the source.

Ionizing Radiation for Technician Response

During hazmat responses, technicians will need to know specific information about the different types of ionizing particles:

- **Alpha (α) particles** — Energetic, positively charged alpha particles (helium nuclei) commonly emitted in the radioactive decay of the heaviest radioactive elements such as uranium and radium as well as by some manmade elements. Alpha-emitting radioisotopes are not considered a hazard outside the body. In addition, alpha particles:
  - Are usually blocked by the outer, dead layer of the human skin, but can be harmful if ingested or inhaled.
  - Do not travel far in open air.
  - Detection distances may not be significant; equipment may need to be near the source to detect particles.
  - Lose energy rapidly when passing through matter, and can be stopped completely with paper.

- **Beta (β) particles** — Fast-moving, positively or negatively charged electrons emitted from the nucleus during radioactive decay. Beta particles are emitted from manufactured and natural sources such as tritium, carbon-14, and strontium-90. Shielding beta emitters with dense metals can result in the release of X-rays (Bremsstrahlung radiation). Compared to alpha radiation, beta particles are more penetrating but less damaging over equally traveled distances. In addition, beta particles:
  — Are capable of penetrating the skin and causing radiation damage, but are generally more hazardous when inhaled or ingested.
  — Travel appreciable distances in air.
  — Detection distances for beta particles will vary based on the activity of the source.
  — Can be reduced or stopped by a layer of clothing, a thin sheet of metal, or thick Plexiglass™.
- **Neutron particles** — Particles that have a physical mass but have no electrical charge. Neutron radiation is generally encountered in large-scale operations including research laboratories, operating nuclear power plants, and soil moisture density gauges at construction sites. Fission reactions produce neutrons and gamma radiation. In addition, neutron particles:
  — May induce radioactivity in materials they encounter, including human skin. Neutron particles are particularly damaging to soft tissues.
  — Will travel through most materials.
  — Detection distances will vary based on the hardware and software available because noncharged particles are difficult to detect directly. Neutron radiation can be measured in the field using specialized equipment.
  — Shielding from neutron radiation requires materials with high amounts of hydrogen including hydrocarbons, water, and concrete.

Technicians will also need to know specific information about the different types of high-energy electromagnetic ionizing waves:

- **Gamma (γ) rays** — Gamma rays often accompany the emission of alpha or beta particles from a nucleus as high-energy **photons**. Gamma rays have neither a charge nor a mass but are penetrating. One source of gamma radiation in the environment is naturally occurring potassium-40. Common industrial gamma emitting sources include cobalt-60, iridium-192 and cesium-137. In addition, gamma rays:
  — Can easily pass completely through the human body or be absorbed by tissue.
  — Travel will vary depending on the isotope and activity levels.
  — Detection equipment includes a variety of **scintillators**.
  — Can be slowed or stopped via concrete, earth, and lead. Standard fire fighting protective clothing provides no protection against gamma radiation.
- **X-rays** — For the purposes of this manual, X-rays and gamma rays are identical and should be treated the same. The chances of encountering X-rays at a hazardous materials incident are remote because this type of radiation is only generated via a powered machine typically found in medical facilities and airports.

Photon — Weightless packet of electromagnetic energy, such as X-rays or visible light.

Scintillator — Material that glows (luminesces) when exposed to ionizing radiation.

Radiation Health Hazards

The effects of ionizing radiation occur at the cellular level. The human body is composed of many organs, and each organ of the body is composed of specialized cells. Ionizing radiation can affect the normal operation of these cells.

Radiation may cause damage to any material by ionizing the atoms in that material – changing the material's atomic structure. When atoms are ionized, the chemical properties of those atoms are altered. Radiation can damage a cell by ionizing the atoms and changing the resulting chemical behavior of the atoms and/or molecules in the cell. One way this occurs is by the release of free radicals, producing hydrogen peroxide at the cellular level. This results in DNA damage. If a person receives a sufficiently high dose of radiation and many cells are damaged, this may cause observable health effects including genetic mutations and cancer.

The biological effects of ionizing radiation depend on how much and how fast a radiation dose is received. There are two categories of radiation doses: acute and chronic.

Acute Doses. Exposure to radiation received in a short period of time is an **acute** dose. Acute exposures are usually associated with large doses. Some acute doses of radiation are permissible and have no long-term health effects. However, high levels of radiation received over a short time can produce serious health effects that include reduced blood count, hair loss, nausea, vomiting, diarrhea, and fatigue. Extremely high levels of acute radiation exposure, such as those received by victims of a nuclear bomb, can result in death within a few hours, days, or weeks.

> **Acute** — Characterized by sharpness or severity; having rapid onset and a relatively short duration.

> **Chronic** — Marked by long duration; recurring over a period of time.

Chronic Doses. Small amounts of radiation received over a long period of time are a **chronic** dose. The body is better equipped to handle a chronic dose of radiation than it is an acute dose because the body has enough time to replace dead or nonfunctioning cells with healthy ones. Chronic doses do not result in the same detectable health effects seen with acute doses. However, studies have confirmed that chronic exposure to ionizing radiation does cause cancer (World Health Organization). Examples of chronic radiation doses include the everyday doses received from natural background radiation and those received by workers in nuclear and medical facilities.

The exposures likely to be encountered by responders at most hazmat incidents are unlikely to cause any health effects, especially if proper precautions are taken. Even at terrorist incidents, responders are unlikely to encounter dangerous or lethal doses of radiation.

Protection from Radiation

Because radiation is invisible, it may be difficult to determine whether it is involved in an incident. Radiation monitoring should be conducted at any incident with an explosion or suspected terrorism, especially if responders note the presence of Class 7 radioactive materials packaging **(Figure 4.17)**. While most incidents involving radioactive materials present minimal risks to emergency responders, it is still necessary to take appropriate precautions to prevent unnecessary exposures.

Figure 4.17 The presence of packaging indicating Class 7 radioactive materials should be noted.

The ALARA (As Low As Reasonably Achievable) method or principle to limit exposure to radiation includes **(Figure 4.18)**:

- **Time** — Decrease the amount of time spent in areas where there is radiation. At a minimum, the time required is:
  - Time it takes to enter the incident
  - Time spent within the zone
  - Time required to exit the zone

- **Distance** — You have to know your dose rate to know the safe distances from the radioactive material. Increase the distance from a radiation source. Doubling the distance from a point source divides the dose by a factor of four. This calculation is sometimes referred to as the **inverse square law**. When the radius doubles, the radiation spreads over four times as much area, so the dose is only one-fourth as much **(Figure 4.19)**. If sheltered in a contaminated area, keep a distance from exterior walls and roofs. This calculation is only a rule of thumb, and the information must be supplemented with information from your radiation meter.

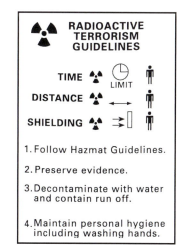

Figure 4.18 Time, distance, and shielding must be used strategically to minimize the dose received.

Inverse Square Law — Physical law that states that the amount of radiation present is inversely proportional to the square of the distance from the source of radiation.

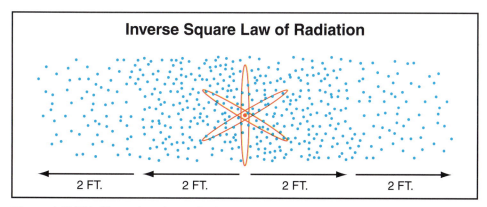

Figure 4.19 This illustration shows the inverse square law as an indicator of radiation received at increasing distance.

- **Shielding** — Create a barrier between responders and the radiation source with a building, earthen mound, or vehicle. Buildings – especially those made of brick or concrete – provide considerable shielding from radiation. For example, exposure from fallout is reduced by about 50 percent inside a one-story building and by about 90 percent at a level belowground.

Chapter 4 • Analyzing the Incident: Understanding Common Families and Special Hazards of Hazardous Materials/WMD

> **CAUTION**
> Use shielding to limit the dose.

Explosives and Incendiaries

An explosive is any substance or article that is designed to function by explosion or that, by chemical reaction within itself, is able to function in a similar manner. Explosives range from being very sensitive to shock, heat, friction, or contamination, to being relatively insensitive, needing an initiating device to function. Most explosives are used in mining operations, with a small percentage used for construction and related purposes. In recent decades, first responders have become increasingly aware of the potential threat of explosives. Improvised explosive devices (IEDs) include pipe bombs and mail bombs.

> **CAUTION**
> Every response involving an unknown powder is potentially a homemade explosive incident.

Explosive Ordnance Disposal (EOD) — Emergency responders specially trained and equipped to handle and dispose of explosive devices. *Also called* Hazardous Devices Units *or* Bomb Squad.

Common explosives include:

- Dynamite
- Ammonium nitrate and fuel oil (ANFO)
- Kenepac
- Homemade explosives (HME)
- Black powder

Several factors that affect an explosives and incendiaries response include:

- Regardless of the presence or absence of a written threat, the material being handled should be treated as an explosive. The **explosives ordnance disposal (EOD) team** will be responsible to determine how to address a lab that has an identified device, or components of a device, associated with it.
- The use of technology to identify the material should NOT be the goal. Field screening techniques should be used before other technology. After field screening is conducted, the material can be taken for further analysis if more information is needed.
- The color or materials used only comes into play with information confirmed, such as printed small jars with material in it.

Understanding the Danger

One reason explosives are so dangerous is that they contain both an oxidizer and a fuel component in their structures **(Figure 4.20)**. For example, trinitrotoluene [$C_6H_2CH_3(NO_2)_3$], commonly known as TNT, begins as toluene (C_7H_8), which provides the fuel component. Three hydrogen atoms are replaced with three nitro groups (NO$_2$), which are strong oxidizers. With two sides of the fire triangle already complete, all it takes is the introduction of energy in the form of heat, shock, or friction to cause a reaction.

Emergency responders must be particularly cautious with older explosives, which can be even more unstable than those recently manufactured, and those that have already been stressed by heat, shock, or friction.

WARNING!
Explosive materials that have been exposed to heat, shock, or friction may detonate a significant amount of time after exposure without apparent cause.

Figure 4.20 Explosive materials do not require an additional source of oxygen to ignite and burn.

Danger at Any Division

The six **division numbers** of explosives identified in U.S. **Code of Federal Regulations (CFR)**, Title 49, were established based on how explosives are expected to behave under normal conditions of transport. They do not necessarily reflect how these materials will behave when exposed to fire. The six divisions are:

- **Division 1.1** — Articles and substances having a mass explosion hazard
- **Division 1.2** — Articles and substances having a projection hazard, but not a mass explosion hazard
- **Division 1.3** — Articles and substances having a fire hazard, a minor blast hazard, and/or a minor projection hazard, but not a mass explosion hazard
- **Division 1.4** — Articles and substances presenting no significant hazard (explosion limited to package)
- **Division 1.5** — Very insensitive substances having a mass explosion hazard
- **Division 1.6** — Extremely insensitive articles which do not have a mass explosion hazard

Many emergency responders have been killed or injured because they underestimated the hazard potential of relatively insensitive explosives. However, when explosives are exposed to fire, there is little difference between the six divisions. While the degree of risk varies, all Class 1 explosive materials can be deadly if they explode.

NOTE: Explosive material divisions are explained in more detail in the IFSTA manual, **Hazardous Materials for First Responders**.

Types of Explosives

Explosives are classified based on their detonation velocity (**Figure 4.21, p. 118**). **Low explosives** are placarded as Division 1.4. These materials **deflagrate**. Examples include black powder, smokeless powder, and solid rocket fuel. Low explosives have more of a "propelling" effect than a "shattering" effect, so they are used primarily for propulsion. Low explosives can be initiated by a

> **Division Number** — Subset of a class within an explosives placard that assigns the product's level of explosion hazard.
>
> **Code of Federal Regulations (CFR)** — Rules and regulations published by executive agencies of the U.S. federal government. These administrative laws are just as enforceable as statutory laws (known collectively as federal law), which must be passed by Congress.
>
> **Low Explosive** — Explosive material that deflagrates, producing a reaction slower than the speed of sound.
>
> **Deflagrate** — To explode (burn quickly) at a rate of speed slower than the speed of sound.

Chapter 4 • Analyzing the Incident: Understanding Common Families and Special Hazards of Hazardous Materials/WMD   **117**

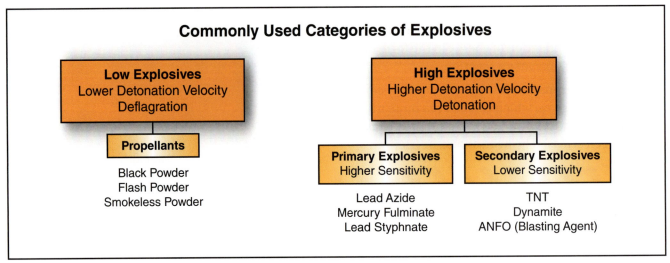

Figure 4.21 Categories of explosives indicate how quickly they burn, not how dangerous they are.

High Explosive — Explosive that decomposes extremely rapidly (almost instantaneously) and has a detonation velocity faster than the speed of sound.

Detonate — To explode or cause to explode. The level of explosive capability will directly affect the speed of the combustion reaction.

Primary Explosive — High explosive that is easily initiated and highly sensitive to heat; often used as a detonator. *Also known as* Initiation Device.

Secondary Explosive — High explosive that is designed to detonate only under specific circumstances, including activation from the detonation of a primary explosive. *Also known as* Main Charge Explosive.

Homemade Explosive (HME) — Explosive material constructed using common household chemicals. The finished product is usually highly unstable.

simple flame reaction. They can also be initiated by shock or friction, but do not require the shock of a blasting cap.

High explosives are placarded as Division 1.1. These materials **detonate** with a shattering effect. High explosives are further categorized based on their sensitivity to shock, friction, flame, heat, or any combination of these factors.

Primary explosives are extremely sensitive and hazardous to handle. They are sometimes called *detonators* or *initiation devices* because they are mostly used to initiate secondary explosives. Examples of primary high explosives include:

- Blasting caps
- Safety fuses filled with black powder
- Mercury fulminate
- Lead styphnate
- Lead azide

Secondary explosives are much less sensitive, but far more powerful than primary explosives. They are sometimes called *main charge explosives* because they do the bulk of the work after being initiated by a shock from a primary explosive. Examples of secondary explosives include:

- Dynamite
- Nitroglycerin
- TNT
- Cylconite (RDX)
- Pentaerythrital tetranitrate (PETN)

Homemade Explosives/Improvised Explosive Devices

Homemade explosive (HME) materials are typically made by combining an oxidizer with a fuel **(Figure 4.22)**. Many of these materials are fairly simple to make and require little technical expertise or specialized equipment. However, the explosive materials created are often highly unstable.

Components of Improvised Explosives

🔥 Potential Fuels + **🔥 Potential Oxidizers** = **💥 Explosive Blends (Oxidizer + Fuel)**

Hydrocarbons:
Alcohol
Carbon Black
Charcoal
Dextrin
Diesel
Ethylene Glycol
Gas
Kerosene
Naphtha
Rosin
Sawdust
Shellac
Sugar
Vaseline
Wax/Parfin

Energetic Hydrocarbons:
Nitrobenzene
Nitromethane
Nitrocellulose

Elemental "Hot" Fuels:
Powdered Metals
- Aluminum
- Magnesium
- Zirconium
- Copper
Phosphorus
Sulfur
Antimony Trisulfide

Oxidizers:
Perchlorate
Chlorate
Hypochlorite
Nitrate
Peroxide
Iodate
Chromate
Dichromate
Permaganate
Sodium Chlorate
Potassium Chlorate
Ammonium Nitrate
Potassium Nitrate
Hydrogen Peroxide
Barium Peroxide
Ammonium Perchlorate
Calcium Hypochlorite
Nitric Acid
Lead Iodate
Sodium Chlorate
Potassium Permanganate
Lithium Chromate
Potassium Dichromate

Nitrate Blends:
ANFO (Ammonium Nitrate + Diesel Fuel)
ANAI (Ammonium Nitrate + Aluminum Powder)
ANS (Ammonium Nitrate + Sulfur Powder
ANIS (Ammonium Nitrate + Icing Sugar)
Black Powder (Potassium Nitrate + Charcoal + Sulfur)

Chlorate/Perchlorate Blends:
Flash Powder (Potassium Chlorate/Perchlorate + Aluminum Powder + Magnesium Powder + Sulfur)
Poor Man's C-4 (Potassium Chlorate + Vaseline)
Armstrong's Mixture (Potassium Chlorate + Red Phosphorus)

Liquid Blend:
Hellhoffite (Nitric Acid + Nitrobenzene)

Common Precursors Used To Make Explosives

Precursors:
Hydrogen Peroxide
Sulfuric Acid (battery acid)
Nitric Acid
Hydrochloric Acid (muriatic acid)
Urea
Acetone
Methyl Ethyl Ketone
Alcohol (Ethyl or Methyl)
Ethylene Glycol (antifreeze)
Glycerin(e)
Hexamine (camp stove tablets)
Citric Acid (sour salt)

Nitrated Explosives:
Nitroglycerine (Glycerine + Mixed Acid [Nitric Acid + Sulfuric Acid])
Ethylene Glycol Dinitrate (EGDN) (Ethylene Glycol + Mixed Acid [Nitric Acid + Sulfuric Acid])
Methyl Nitrate (Methyl Alcohol [methanol] + Mixed Acid [Nitric Acid + Sulfuric Acid])
Urea Nitrate (Urea + Nitric Acid)
Nitrocotton (Gun Cotton) (Cotton + Mixed Acid [Nitric Acid + Sulfuric Acid])

Peroxide Explosives:
Triacetone Triperoxide (TATP) (Acetone + Hydrogen Peroxide + Strong Acid [Sulfuric, Nitric, or Hydrochloric])
Hexamethylene Triperoxide Diamine (HMDT) (Hexamine + Hydrogen Peroxide + Citric Acid)
Methyl Ethyl Ketone Peroxide (MEKP) (Methyl Ethyl Ketone + Hydrogen Peroxide + Strong Acid [Sulfuric, Nitric, or Hydrochloric])

Figure 4.22 Homemade explosive materials can include a variety of fuels and oxidizers.

Responders typically stage 1,000 feet (300 meters) away from a suspected explosive material incident. Technicians may do some field screening of potential HMEs. Detection of explosives will be addressed in greater detail in later chapters of this manual.

Many oxidizers and fuels can be combined to form improvised explosive materials. The sections that follow will discuss peroxide-based explosives, potassium chlorate, and urea nitrate. These do not represent a comprehensive list.

Acetone Peroxide (TATP) — Triacetonetriperoxide (TATP) is typically a white crystalline powder with a distinctive acrid (bleach) smell and can range in color from a yellowish to white color. Similar to Hexamethylene triperoxide diamine (HMTD).

Hexamethylene triperoxide diamine (HMTD) — Peroxide-based white powder high explosive organic compound that can be manufactured using nonspecialized equipment. Sensitive to shock and friction during manufacture and handling. *Similar to* acetone peroxide (TATP).

Peroxide-Based Explosives

Peroxide-based explosives include **acetone peroxide (TATP)** and **hexamethylene triperoxide diamine (HMTD)**. Peroxide-based explosives can be made by mixing concentrated hydrogen peroxide, acetone, and either hydrochloric or sulfuric acid. Both TATP and HMTD are dangerous to make and handle because they are unstable, both during the manufacturing process and as a finished product **(Figure 4.23)**. Specialized equipment is not needed to manufacture TATP and HMTD so they can be made almost anywhere. TATP is typically a white crystalline powder with a distinctive acrid smell. TATP can range in color from a yellowish to white color **(Figure 4.24)**.

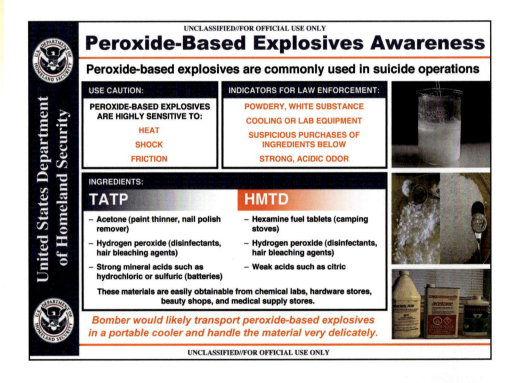

Figure 4.23 TATP and HMDT are dangerous at all stages of their manufacture, including as initial ingredients.

Figure 4.24 TATP as a finished material may look deceptively benign.

120 Chapter 4 • Analyzing the Incident: Understanding Common Families and Special Hazards of Hazardous Materials/WMD

Chlorate-Based Explosives

Chlorate-based oxidizers may be used in IEDs. They are commonly seen as a white crystal or powder that must be mixed with a fuel source. Chlorates are a common ingredient in some fireworks and can be purchased in bulk at fireworks/chemical supply houses. Chlorates are also used in printing, dying, steel, weed killer, matches, and the explosive industry.

Nitrate-Based Explosives

Nitrate-based oxidizers may be used in IEDs. Some nitrate sources may already have a fuel source included, such as black powder and smokeless powder. Others may require the addition of a separate fuel source. Nitrates are commonly found in ammonium nitrate and fertilizers.

Improvised Explosive Devices

Improvised explosive devices (IEDs) are standalone devices that may include a homemade explosive. These items are not commercially manufactured but may use items that were scavenged from commercially manufactured materials. They are usually constructed for a specific target and can be contained within almost any object **(Figure 4.25)**. They are relatively easy to make and can be constructed in virtually any location or setting.

> **Improvised Explosive Device (IED)**— Any explosive device constructed and deployed in a manner inconsistent with conventional military action.

Figure 4.25 IEDs can be contained in almost any object.

Inexperienced designers may create IEDs that fail to detonate, or in some cases detonate during the building process or when being moved or placed. Bomb makers who specialize in IED manufacture often make more sophisti-

Munitions — Military reserves of weapons, equipment, and ammunition.

cated varieties. Some groups are known to produce sophisticated devices that are constructed with components scavenged from conventional **munitions** and standard consumer electronics components, such as speaker wire, cellular phones, or garage door openers **(Figure 4.26)**. IEDs often contain nails, tacks, broken glass, bolts, and other items that will cause additional shrapnel damage and fragmentation injuries.

Figure 4.26 Simple electronic components can be used to create complex improvised explosives. *Courtesy of the U.S. Army, photo by Spc. Ben Brody.*

What This Means To You
Targeted Attacks
Targeted attacks in the modern world routinely use unique combinations of known materials. As in any other kind of threat, first responders should consider the whole scene while evaluating whether an item may be a disguised danger. If you didn't place an item, you should be suspicious of it.

Identification of IEDs
The bomber's imagination is the only limitation to the design and implementation of IEDs **(Figure 4.27)**. Responders should be cautious of any item(s) that attract attention because they seem out of place, anomalous, out of the ordinary, curious, suspicious, out of context, or unusual.

Figure 4.27 IEDs can be disguised to look like common goods.

IEDs may be placed anywhere. Usually, bombers try to avoid detection when placing IEDs. The level of security and awareness of the public, security forces, and employees affect where and how a terrorist will place an IED.

IED Types Categorized by Containers

IEDs are typically categorized by their container (such as pipe, backpack or vehicle bomb) and the way in which they are initiated. Bomb types are typically categorized based on the outer container.

- **Pipe bombs** — The most common type of IED found in the United States **(Figure 4.28, p. 124)**. Characteristics of the pipe bomb and those of the pipe bomb manufacturers include the following:
  - Length ranges from 4 to 14 inches (102 mm to 356 mm).
  - Steel or polyvinyl chloride (PVC) pipe sections are filled with explosives and the ends are capped or sealed.
  - Easily obtained materials such as black powder or match heads.
  - Filled or wrapped with nails or other materials that will become shrapnel when the bomb detonates.
  - Can throw shrapnel up to 300 feet (90 m) with lethal force.

- Detonate with a homemade fuse or with commercially available fuses.
- Explosive filler can get into the pipe threads making the device extremely sensitive to shock or friction.

Figure 4.28 Pipe bombs are the most common type of IED in the U.S.

Figure 4.29 Briefcases may disguise an explosive device.

- **Satchel, backpack, knapsack, duffle bag, briefcase, or box bombs** — Some terrorists may fill these bags with explosives or an explosive device **(Figure 4.29)**. Terrorists use these devices because it is common to see people carrying backpacks or other types of bags. These IEDs may include electronic timers or radio-controlled triggers so there may be no external wires or other items visible. These bombs come in any style, color, or size (even as small as a cigarette pack).
- **Plastic bottle bombs** — Manufacturers of these bombs fill plastic soda bottles (or any size of plastic drink bottles) with a material (such as dry ice) or combination of reactive materials that will expand rapidly, causing the container to explode. The Internet lists many variations of plastic bottle bombs. Be careful around plastic containers containing multilayered liquids and containers with white or gray liquids with cloudy appearances. Do not attempt to move or open plastic bottle bombs. Once initiated, they can detonate at any time. Manufacturers may use materials such as:

  — Pool chemicals
  — Alcohol
  — Aluminum
  — Drain

  — Dry ice
  — Acid
  — Toilet bowl
  — Driveway cleaners

- **Fireworks** — Some IED manufacturers may modify and/or combine legally obtained fireworks to form more dangerous explosive devices.
- **M-Devices** — Small devices constructed of cardboard tubes (often red) filled with flash powder and sealed at both ends and are ignited by fuses. The most common are M-80s, which measure 5/8 × 1½ inches (16 mm by 38 mm). At one time, M-80s were available in the U.S. as commercial fireworks. However, for safety reasons, they were made illegal in 1966.
- **Carbon Dioxide (CO_2) grenades** — These devices are made by drilling a hole in and filling used CO_2 containers (such as those used to power pellet pistols) with an explosive powder; they are usually initiated by a fuse. Shrapnel may be added to the outside of the container. These devices, also known as *crickets*, have a small range but will create a great deal of destruction within that range.
- **Tennis ball bombs** — A tennis ball may be filled with an explosive mixture and ignite using a simple fuse.
- **Other existing objects** — Items that seem to have an ordinary purpose can be substituted or used as the bomb container. Examples include fire extinguish-

WARNING!
Do not move, handle, or disturb an IED when found!

ers, propane bottles, trash cans, gasoline cans, and books.

Mail, Package, or Letter Bombs

A package or letter may be used to conceal the explosive device or material. Opening the package or letter usually triggers the bomb.

Package or letter bomb indicators include **(Figure 4.30)**:

- Package or letter that has no postage, noncancelled postage, or excessive postage.
- Parcels may be unprofessionally wrapped with several combinations of tape to secure them and endorsed: *Fragile, Handle with Care, Rush,* or *Do Not Delay.*
- Sender is unknown, no return address is available, or the return address is fictitious.
- Mail may bear restricted endorsements such as *Personal* or *Private*. These endorsements are particularly important when addressees do not usually receive personal mail at their work locations.
- Postmarks may show different locations than return addresses.

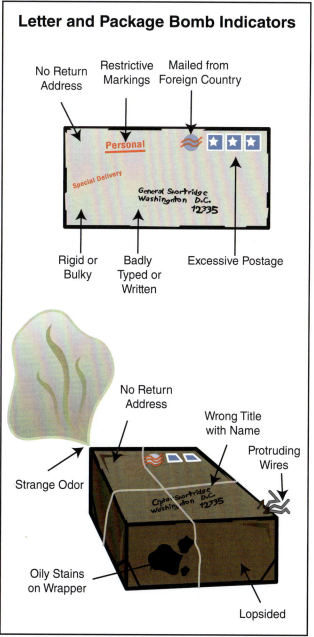

Figure 4.30 Explosives delivered via mail will often have telltale indicators.

- Common words are misspelled on mail.
- Mail may display distorted handwriting, or the name and address may be prepared with homemade labels or cut-and-paste lettering.
- Package emits a peculiar or suspicious odor.
- Mail shows oily stains or discoloration.
- Letter or package seems heavy or bulky for its size and may have an irregular shape, soft spots, or bulges.
- Letter envelopes may feel rigid or appear uneven or lopsided.
- Mail may have protruding wires or aluminum foil.
- Package makes ticking, buzzing, or whirring noises. Unidentified person calls to ask if a letter or package was received.

Person-Borne Improvised Explosive Devices (PBIED)

Person-Borne Improvised Explosive Devices (PBIEDs) typically consist of bombs worn or carried by a suicide bomber **(Figure 4.31)**. Suicide bombers wear the PBIED in the form of vests with many pockets sewn into them to hold explosive materials. Terrorists may also carry PBIEDs or they may be attached to coerced or unwilling victims.

Person-Borne Improvised Explosive Device (PBIED) — Improvised explosive device carried by a person. This type of IED is often employed by suicide bombers, but may be carried by individuals coerced into carrying the bomb.

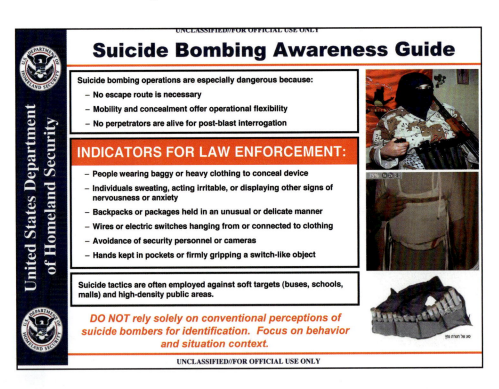

Figure 4.31 Potential suicide bombers may show some indicators ahead of the fact.

Individuals carrying briefcases or packages are inherently suspicious to security forces, particularly in locations where personnel have to pass through a security checkpoint. In addition to bags, packages, and cases, clothing can conceal a bomb. The contours of bulky suicide vests or belts may be visible prior to detonation **(Figure 4.32)**. Terrorists might also wear unseasonable or atypical attire, such as a coat (which may conceal a suicide belt) during warm weather. Wires or other materials exposed on or around the body could also be an indication of a bomb.

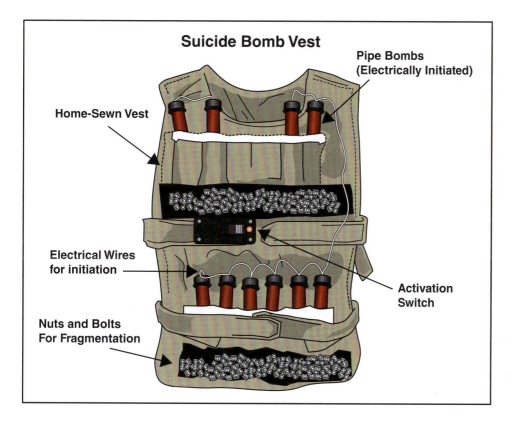

Figure 4.32 Common components of a suicide vest include explosives and materials to increase the potential damage.

Behavioral indicators of potential suicide bombers include the following:

- Fear, nervousness, or overenthusiasm
- Profuse sweating
- Keeping hands in pockets
- Repeated or nervous touching or patting of clothing
- Slow-paced walking while constantly shifting eyes to the left and right
- Significant attempts to avoid security personnel
- Obvious or awkward attempts to blend in with a crowd
- Obvious disguising of appearance
- Actions indicating a strong determination to get to a target
- Repeated visits to a high-risk location during the recon/target acquisition phase
- Placing items in locations that seem out of place or arouse curiosity

ALERT

The FBI uses the acronym ALERT to designate indicators of a possible suicide bomber:

- **A**lone and nervous
- **L**oose and/or bulky clothing
- **E**xposed wires (possibly through a sleeve)
- **R**igid mid-section (explosives device or a rifle)
- **T**ightened hands (may hold a detonation device)

Figure 4.33 Bomb robots are designed to handle some types of explosive materials.

Vehicle-Borne Improvised Explosive Device (VBIED) — An improvised explosive device placed in a car, truck, or other vehicle. This type of IED typically creates a large explosion.

Never approach a suspected or confirmed suicide bomber who is injured or deceased. If there are several strong indicators that there is a suicide bomber, the first priority is to clear and isolate the area and observe the bomber with binoculars or spotting scopes. The first approach must be conducted via trained personnel from an equipped explosive ordnance disposal (EOD) team, or using a bomb disposal robot **(Figure 4.33)**.

Vehicle-Borne Improvised Explosive Devices (VBIEDs)

Vehicle-Borne Improvised Explosive Devices (VBIEDs) may contain many thousands of pounds (kilograms) of explosives that can cause massive destruction. The explosives can be placed anywhere in a vehicle. When using small vehicles, such as passenger cars, the explosives are often concealed in the trunk.

Indicators of a possible VBIED include:

- Preincident intelligence or 9-1-1 calls leading to the suspected vehicle
- Vehicle parked suspiciously for a prolonged amount of time in a central location, choke point, or other strategic location
- Vehicle abandoned in a public assembly, tourist area, pedestrian area, retail area, or transit facility
- Vehicle parked between, against, or close to the columns of a multi-story building
- Vehicle which appears to be weighted down or sits unusually low on its suspension
- Vehicle with stolen, nonmatching plates, or no plates at all
- Wires, bundles, electronic components, packages, unusual containers, liquids or materials visible in the vehicle
- Unknown liquids or materials leaking under vehicle
- Unusually screwed, riveted, or welded sections located on the vehicle's bodywork
- Unusually large battery or extra battery found under the hood or elsewhere in the vehicle
- Blackened windows or covered windows
- The hollows of front or rear bumpers have been sealed, taped, or otherwise made inaccessible
- Tires that seem solid instead of air-inflated
- Bright chemical stains or unusual rust patches on a new vehicle
- Chemical odor present or unusual chemical leak beneath vehicle
- Wiring protruding from the vehicle, especially from trunk or engine compartment
- Wires or cables running from the engine compartment, through passenger compartment, to the rear of vehicle

- Wires or cables leading to a switch behind sun visor
- The appearance or character of the driver does not match the use or type of vehicle
- The driver seems agitated, lost, and unfamiliar with vehicle controls
- Anything that seems out of place, unusual, abnormal, or arouses curiosity

WARNING!
Never approach a suspicious vehicle once an indicator of possible VBIED has been noticed.

Response to Explosive/IED Events

All operations must be conducted within an incident command system and determined by the risk/benefit analysis. In addition, do the following:

- Follow designated SOP/Gs.
- ALWAYS proceed with caution, especially if an explosion has occurred or it is suspected that explosives may be involved in an incident.
- Understand that secondary devices may be involved.
- Request EOD (bomb squad) personnel, hazmat, and other specialized personnel as needed.
- Treat the incident scene as a crime scene until proven otherwise.
- NEVER touch or handle a suspected device, even if someone else already has. Only certified, trained bomb technicians should touch, move, defuse, or otherwise handle explosive devices **(Figure 4.34)**.
- Do not use two-way radios, cell phones, **mobile data terminals (MDT)** within a minimum of 300 feet (90 m) of any device or suspected device. The larger the suspicious device, the larger the standoff distance.
- Use intrinsically safe communications equipment within the isolation zone.
- Note unusual activities or persons at the scene and report observations to law enforcement.
- Limit personnel exposure until the risk of secondary devices is eliminated.

Figure 4.34 Only people who are trained in the proper treatment of explosives should attempt to handle potential IEDs and HMEs.

Mobile Data Terminal (MDT) — Mobile computer that communicates with other computers on a radio system.

Train with Specialized Bomb Disposal Personnel

If there are bomb squads in your area, ask for their assistance with your training and planning. Most bomb technicians will be glad to provide your agency with training on their procedures and equipment, since they will require your support during an incident. One key issue for fire and EMS departments is to become familiar with your local bomb squad operations and entry suits so you will know how to remove protective clothing and equipment from an injured bomb technician in case of emergency.

When technicians respond to a bombing incident, a primary search will have to be completed. In these situations, local protocol must be followed. Regardless of who completes the primary search and subsequent rescue operations when required, the Incident Commander should limit the number of personnel in the blast area to the minimum number of personnel required to carry out critical lifesaving operations. When determining risk, evaluate both the potential for additional explosions and structural stability.

WARNING!
Avoid staging near gardens, garbage bins, or other vehicles that could conceal explosive or incendiary devices. Limit exposure until the secondary device risk is eliminated.

Chapter Notes

World Health Organization (WHO). 2016. "Ionizing radiation, health effects and protective measures." Last modified April 2016. http://www.who.int/mediacentre/factsheets/fs371/en/

Chapter Review

1. What are some hazards posed by hydrocarbons and hydrocarbon derivatives besides flammability?
2. How do oxidizers differ from flammable material?
3. What are some specific hazards presented by reactive materials?
4. How do acids and bases differ when they come into contact with living tissue?
5. Describe different ways in which radiation can be measured.
6. What are some common symptoms for the different types of chemical warfare agents?
7. How is dosage expressed as compared to concentration?
8. What kinds of protective measures should be taken for the different types of radiation?
9. What type of conditions does the CFR assume about how the six explosive divisions are supposed to behave?
10. What kinds of items can improvised explosive devices be disguised as?

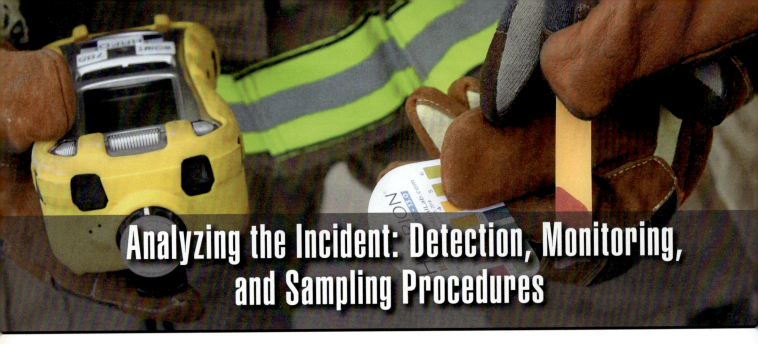

Analyzing the Incident: Detection, Monitoring, and Sampling Procedures

Chapter Contents

Detection and Monitoring Operations 135
- Fundamentals of Monitoring 135
- Calibration 139
- Action Levels 143
- Individual Monitoring and Area Monitoring ...144
- Confined Spaces 146
- Cargo Containers 146

Sampling Techniques 146
- Sample Collection for Hazard Identification ...147
- Field Screening Samples 147

Evidence Collection Techniques 150
- Evidence Collection 151
- Chain of Custody for Evidence 152
- Evidence Collection and Preservation 152
- Decontaminating Samples and Evidence ...152

Maintenance of Monitoring, Detection, and Sampling Equipment 152
Chapter Review 153
Skill Sheets 154

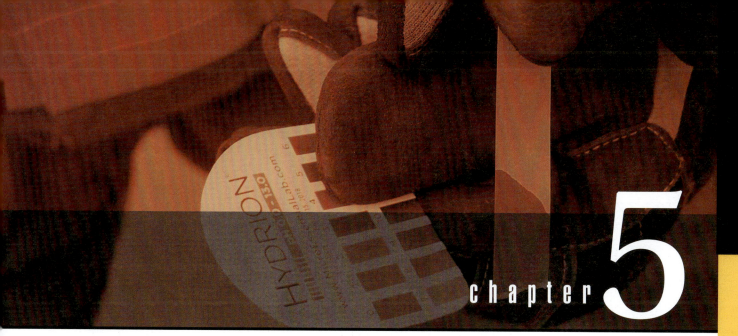

chapter 5

Key Terms

Calibrate 139	Confined Space 146
Calibration 139	Detection Limit.. 137
Calibration Test... 140	Direct-Reading Instrument 138
Canadian Transportation Emergency Centre (CANUTEC) 142	Evidence.. 147
Chain of Custody...................................... 150	Flame Ionization Detector (FID) 147
Chemical Transportation Emergency Center (CHEMTREC®)............................ 142	Instrument Response Time 138
	Purge 136
	Saturation.. 137

Analyzing the Incident: Detection, Monitoring, and Sampling Procedures

JPRs Addressed In This Chapter

7.2.1, 7.2.5, 7.3.4

Learning Objectives

After reading this chapter, students will be able to:

1. Describe detection and monitoring operations at a hazardous materials incident. [NFPA 1072 7.2.1, 7.2.5, 7.3.4]
2. Describe sampling techniques. [NFPA 1072, 7.2.1]
3. Describe evidence collection techniques. [NFPA 1072, 7.2.1, 7.3.4]
4. Recognize the importance of monitoring, detection, and sampling equipment maintenance. [NFPA 1072, 7.2.1]
5. Skill Sheet 5-1: Collect samples of hazardous material solid, liquid, or gas. [NFPA 1072, 7.2.1]
6. Skill Sheet 5-2: Collect samples of hazardous material solid, liquid, or gas. [NFPA 1072, 7.2.1]
7. Skill Sheet 5-3: Perform maintenance and testing on monitoring equipment, test strips, and reagents. [NFPA 1072, 7.2.1]

Chapter 5
Analyzing the Incident: Detection, Monitoring, and Sampling Procedures

This chapter addresses the following topics:
- Detection and monitoring operations
- Sampling techniques
- Evidence collection techniques
- Maintenance of monitoring, detection, and sampling equipment

Detection and Monitoring Operations

Detection and monitoring equipment helps to provide additional or confirming information so that emergency responders can employ the safest approach in a risk-based response to a product release. For example, detection and monitoring results can help determine appropriate PPE, safe areas and evacuation zones, and control tactics. Monitoring must be an ongoing process.

No single instrument will completely identify or quantitate compounds that are identified. Remember that at the levels that may be toxic, enough material may not be present to register on a device. Regardless, equipment can be used to:

- Help estimate the concentrations of a known product
- Help identify unknown products
- Identify hazards
- Assist in evaluating the health and safety effects of a chemical
- Evaluate exposures during medical surveillance

The following sections describe instrument usability and other factors. They will also address many of the monitors available to the hazmat technician. Much of the information presented is generic in nature, based on the instrument's technology. Specific information about an instrument should be obtained from the instrument's manufacturer.

Fundamentals of Monitoring

Monitoring and detection instruments are often designed with certain features that make them safe and user-friendly in the field. Some features that technicians should consider when evaluating detection and monitoring equipment include:

- **Portability and durability** — Instruments should be portable and rugged. They should be able to withstand exposures to the elements and have an internal power supply.

- **Cost** — In addition to the initial cost of the device, there may be other expenses to consider. These can include maintenance costs and calibration gases, sensor replacements, library updates, and training.

- **Ease of operation** — Operation while wearing PPE should be a consideration for monitoring and detection equipment **(Figure 5.1)**.

Figure 5.1 While choosing PPE, consider your ability to perform necessary tasks while wearing the equipment properly. *Courtesy of the U.S. Air Force, photo by Taylor Marr.*

Purge — To expel an inert gas through a device's hosing and/or intake system to remove any residual contaminants.

- **Usability in a hazardous environment** — Instruments should be explosion proof, intrinsically safe, or **purged**.

A device will not contribute to ignition if it follows these general guidelines:

- It is certified or listed by an agency (such as Underwriters Laboratories) as explosion-proof, intrinsically safe, or purged for a given hazard class
- It is used, maintained, and serviced according to the manufacturer's instructions

However, the device is not certified for use in atmospheres other than those recommended by the manufacturer. All certified devices must be marked to show class, division, and group.

NOTE: Hazard classes, placarding, and labeling are briefly discussed in Chapter 1 of this manual. More information may be found in the IFSTA manual, **Hazardous Materials for First Responders**.

Technology

A technician must be able to read, interpret, and communicate data from the detection and monitoring equipment used at a hazardous materials incident. Several factors are important in evaluating the reliability and usefulness of a monitor and the data it generates. These factors include:

- **Instrument response time** — The interval between the time an instrument senses a contaminant and when it generates data. Response time is important for producing reliable and useful field results. Response times for direct-reading instruments may range from a few seconds to several minutes. Response time relies on:
  — Test(s) to be performed
  — Wait time between sample periods (the time for analysis, data generation, and data display)
  — Sensitivity of the instrument

- **Interference** — Some types of equipment may be subject to some types of interference, such as fluorescence in Raman and FTIR spectroscopy, which may impair the credibility of the equipment's readings.

- **Sensitivity** — The ability of an instrument to accurately measure changes in concentration. Sensitivity is important when slight concentration changes can be dangerous. The lower detection limit is the lowest concentration to which an instrument will respond, known as the lower **detection limit**. The upper detection limit is defined by the **saturation** concentration.

- **Selectivity** — The ability of an instrument to detect and measure a specific chemical or group of similar chemicals. Additionally, selectivity depends upon interfering compounds that may produce a similar response. Selectivity and sensitivity must be reviewed and interpreted together. Interferences can affect the accuracy of the instrument reading.

- **Specificity** — A comparison of the relationship between an antigen and an antibody. Specificity measures the portion of negative samples that are correctly identified as negative. This metric is only applicable to handheld assays.

- **Operating range** — The range of concentration accurately measured by the instrument.

- **Amplification** — The instrument's ability to increase small electronic signals emanating from the detector to the readout. Changing the amplification of the detector does not change its sensitivity. However, it may be useful in calibration. Instruments with amplifier circuits can be affected by radio frequency from pulsed DC or AC power lines, transformers, generators, and radio wave transmitters such as portable radios **(Figure 5.2)**.

> **Detection Limit** — The smallest quantity of a material that is identifiable within a stated confidence level.
>
> **Saturation** — The concentration at which the addition of more solute does not increase the levels of dissolved solute.

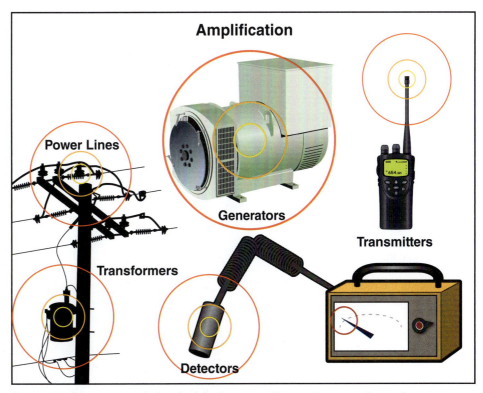

Figure 5.2 Some types of electrical devices can affect an instrument's readings.

- **Accuracy** — The relationship between a true value and the instrument reading. Technicians must understand the error factors of the devices that they use.

- **Reliability** — The ability of a device to provide consistently accurate readings.

Another consideration is that the instrument should be direct-reading, with little or no need to interpolate, integrate, or compile large amounts of data. **Direct-reading instruments** offer results at the time of sampling. While no instrument will give results instantaneously, direct-reading instruments can detect a contaminant concentration as low as 1 part per million (PPM) without the need for samples to be sent to a laboratory for analysis. This real-time sampling is beneficial in many aspects of planning and response at a hazardous materials incident. When selecting an instrument, compare the desired sensitivity, range, accuracy, selectivity, and ability in order to vary amplification of detector signals with the available instrument characteristics.

Instrument Response Time

The **instrument response time** is the interval between the time an instrument senses a chemical or contaminant and the moment it produces data **(Figure 5.3)**. Response times may depend on:

- Type of test to be performed
- Dead time required for the instrument to analyze and display the data
- Sensitivity of the instrument

Some monitoring instruments may be passive, meaning they will rely on normal air movements and currents to obtain the sample. Other instruments may utilize a pump and draw tube to obtain the sample. There may be significant variations to the response time based on the length of the draw tube.

Some instruments may respond as quickly as 10 seconds, while others may take more than one minute. Emergency responders should consult the instrument operation manual and allow for the appropriate time for the instrument to respond before recording any readings.

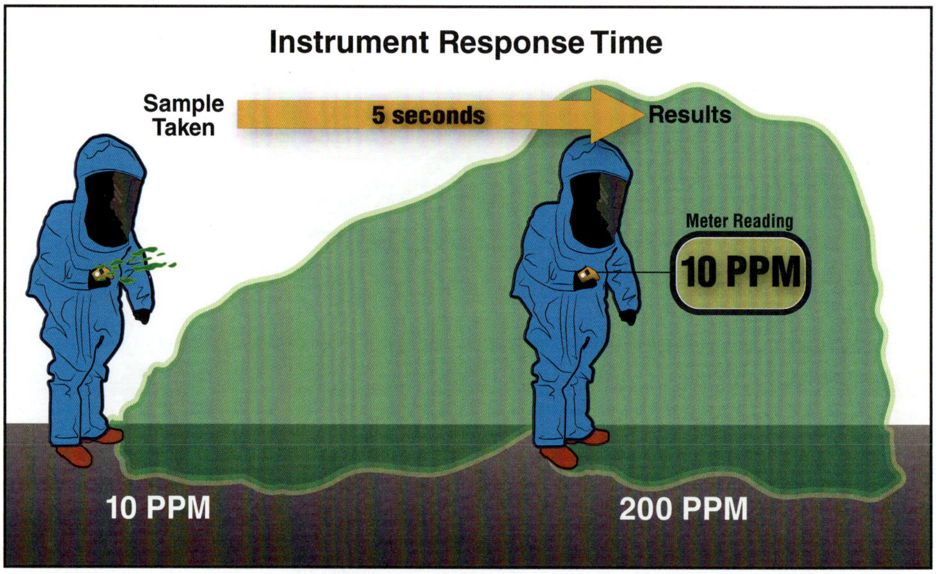

> **Direct-Reading Instrument** — A tool that indicates its reading on the tool itself, without requiring additional resources. Each instrument is designed for a specific monitoring purpose.

> **Instrument Response Time** — Elapsed time between the movement (drawing in) of an air sample into a monitoring/detection device and the reading (analysis) provided to the user. *Also known as* Instrument Reaction Time.

Figure 5.3 A technician must understand how the time interval required for an instrument to deliver a result will affect the significance of the reading.

Automatic Pumps for Manual Devices

Monitoring equipment furnished with automatic pumps can be a time saver for the hazmat technician. However, the monitor's response time may be delayed based on the length of the sample draw tube. Most instruments with integrated pumps will require 1 to 2 seconds per foot of sample draw tube. If a meter is equipped with a 10-foot (3 m) sample draw tube, there may be as much as a 20-second delay in obtaining a stable reading. While this may be a sizable delay in obtaining the readout, automated pumps will deliver a result more quickly and reliably than a hand-operated pump.

Calibration

Monitoring instruments are **calibrated** at the factory to respond accurately to a given vapor or gas within a specific concentration range. Technicians should check correct instrument response before and after each use, and at intervals established by the AHJ. Check the instrument response against the required calibration gas (or a check gas if the calibration gas is not available or is too dangerous) to verify that the instrument is operating and responding correctly **(Figure 5.4)**. If the readings are consistent then it may be assumed that the instrument is operating properly. If the calibration checks are outside the manufacturer-specified range, any information obtained may be inaccurate.

Calibrate — Operations to standardize or adjust a measuring instrument.

Calibration — Set of operations used to standardize or adjust the values of quantities indicated by a measuring instrument.

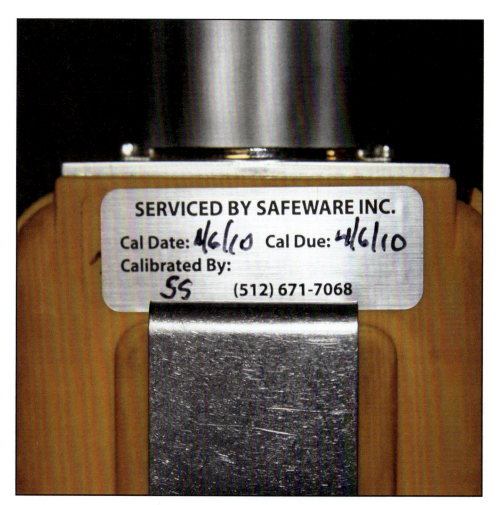

Figure 5.4 Calibration checks must be performed at intervals established by the AHJ.

Chapter 5 • Analyzing the Incident: Detection, Monitoring, and Sampling Procedures

The manufacturer's operating manual will provide specific instructions for instrument calibration. The appropriate check standards, regulators, and calibration gas must be available for instrument calibration. All calibration checks must be documented.

Bump Test

> **Calibration Test** — Set of operations used to make sure that an instrument's alerts all work at the recommended levels of hazard detected. *Also known as* Bump Test *and* Field Test.

A **calibration test** (bump test) verifies the instrument's calibration by exposing it to a known concentration of test gas. The instrument reading should be compared to the actual quantity of gas present. This quantity will be indicated on the calibration gas cylinder. This test will determine if the sensors are operating properly and that the instrument sounds alarms at the set action levels. A bump test is performed immediately prior to entering the hazard area to ensure the instrument is accurate and precise. This test is not intended to reset the calibration, but simply to determine the correct settings in the existing calibration.

If the instrument's response is within an acceptable range of the specified concentration, the calibration has been verified. Always follow the manufacturer's instructions regarding device calibration.

Relative Response/Correction Factors

Each instrument responds to a vapor or gas as if it is detecting its calibration gas. Although the instrument will give a reading, the data presented may be higher or lower than the actual concentration that is present. The instrument's *relative response* is the instrument's response to a reading relative to its calibration gas **(Table 5.1)**.

**Table 5.1
Comparison of Actual LEL and Gas Concentrations with Typical Instrument Readings**

Gas Type	Actual % LEL	Actual Gas Concentration	Typical Display Reading (% LEL)
Pentane	50%	0.07%	50%
Methane	50%	2.50%	100%
Propane	50%	1.05%	63%
Styrene	50%	0.55%	26%

Source: *Courtesy of MSA*

Correction factors or relative response curves can be used to convert the instrument reading to the true concentration of the vapor or gas present. This will only work if the vapor or gas can be identified. For example, a CGI calibrated to pentane will give a higher instrument reading when testing methane than the actual concentration. The relative response of an instrument to different chemicals can be calculated by dividing the instrument reading by the actual concentration. This response is expressed as a ratio or as a percent. Note that for the calibration standard the relative response should be 1.00 or 100 percent.

If the instrument is being used for a chemical that is different from the calibration standard, it may be possible to find a correction factor in the instrument's technical manual **(Table 5.2)**. The actual concentration can then be calculated. For example, if the instrument's relative response for xylene is 0.27 (27 percent) and the reading is 100 ppm, then the actual concentration is 370 ppm (100 ppm/0.27 = actual concentration). If the instrument has adjustable settings and a known concentration is available, the instrument may be adjusted to read directly for the chemical. Because recalibration takes time, this is usually done only if the instrument is going to be used for many measurements of the specific chemical.

NOTE: Not every jurisdiction will have SOP/Gs for response at incidents with unknown materials.

Table 5.2
Sample Correction Factors

Gas or Vapor	Factor
Hexane	0.68
Hydrogen	0.39
Isopropyl Alcohol	0.73
Methyl Ethyl Ketone	0.90
Methane	0.38
Methanol	0.58
Mineral Spirits	1.58
Nitro Propane	0.95
Octane	1.36
Pentene	0.86
Iso-Pentene	0.86
Isoprene	0.58
Propane	0.56
Styrene	1.27
Vinyl Acetate	0.70
Vinyl Chloride	1.06
O-Xylene	1.36

> **WARNING!**
> Any indication of the presence of a material is an actionable response.

Chemical Transportation Emergency Center (CHEMTREC®) — Center established by the American Chemistry Council that supplies 24-hour information for incidents involving hazardous materials.

Canadian Transportation Emergency Centre (CANUTEC) — Canadian center that provides fire and emergency responders with 24-hour information for incidents involving hazardous materials; operated by Transport Canada, a department of the Canadian government.

What This Means To You

Resources for More Information

As you work to identify hazards, you may find that an instrument correction factor is unavailable, even with a known chemical. You still must make a risk-based response. Resources you may be able to draw from include:

- AHJ SOP/Gs
- Known facts for the incident to which you are responding
- Science including molecular structure and equipment output
- Factors (circumstances) unique to this specific incident
- Similar incidents and known trends (experience), algorithms based on known information including calibration gases (e.g., methane)
- Outside resources to contact such as:
  - Specialists
  - Chemical producers
  - Emergency response centers including **Chemical Transportation Emergency Center (CHEMTREC®)** and **Canadian Transportation Emergency Centre (CANUTEC).**

Cross-Sensitivities and Interference

Some hazardous vapors and gases (chemicals) may interfere with proper operation of an instrument. Interferences can result in either decreased instrument sensitivity or false readings. Hazmat technicians frequently encounter these sensitivities when monitoring in high concentrations of product.

Cross-sensitivities and interference may occur on many different types of instrumentation. For example, acetylene and hydrogen can interfere with a CO sensor and give false readings. High levels of carbon dioxide may damage an oxygen sensor if the sensor is exposed to the gas for an extended period of time. Silicone sprays may damage combustible gas instrument sensors. Water vapor may interfere with the photoionization detector. Filters may be available to prevent cross-sensitivities and/or interference.

Instrumentation manufacturers typically supply information about interferences with their products. Technicians should consult this information before evaluating instrumentation readings. Unexpected interference, such as cross-sensitivity readings, provides extremely valuable information that can affect the response.

NOTE: Equipment, and the manufacturer's guidelines, will not provide all of the information needed for an effective response. Technicians should consult other resources as necessary to compare cross-sensitivities.

Interference may also come from environmental factors. Certain instruments include a correction chart.

An example of environmental interference is humidity. If instruments are stored in a climate-controlled atmosphere, the instrument should be allowed to adapt to the test atmosphere before operation. It is possible that a sensor may "fog up" similar to a camera lens when exposed to a humid environment. Sunlight and radio waves may also cause interference.

Simply checking the calibration of an instrument may not reveal some interference. The hazmat technician should determine if interfering components are present. This can be done by comparing the readings of one instrument with that of another instrument that will monitor for the same chemicals **(Figure 5.5)**.

Figure 5.5 Compare the readings of two types of detection instruments to verify proper functioning.

Action Levels

The AHJ should establish action levels. Action levels can be defined as a response to known or unknown chemicals or products that will trigger some action. When an action level (or action point) is reached, it may trigger:

- Removal of unprotected or unnecessary personnel
- Additional monitoring
- Alteration or adjustment of PPE
- Total area evacuation

Other factors that may influence action levels include manufacturer's recommendations. When considering action levels for LEL, it must be determined if the product is known or unknown. If the product is known and a relative response conversion is available, it may be possible to have high LEL action levels. If the product is unknown, technicians must take a more conservative approach.

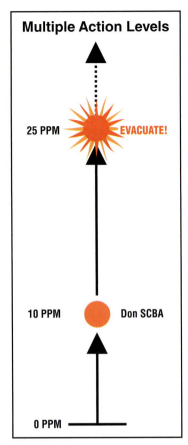

Figure 5.6 Hazards often intensify at higher concentrations.

There may be multiple action levels established for a response. As readings change, technicians must reassess the situation and take actions based on the new readings **(Figure 5.6)**.

In a risk-based response, action levels can vary depending on the mission. For example, for a rescue, high LEL levels may be acceptable. However, for tasks such as leak control, a much lower LEL will be used.

CAUTION
Be prepared to take action before you reach an action level.

Individual Monitoring and Area Monitoring

There are two main methods of sampling to determine exposures to chemical hazards: individual monitoring and area monitoring. When conducting individual monitoring, the emergency responder carries an instrument to evaluate the work area. Area monitoring is performed as a team by sampling the area in an organized manner.

Individual Monitoring

Individual monitoring is usually conducted with handheld equipment. Hazmat technicians must consider many factors when conducting this type of monitoring. These factors may influence not only the results of the sample, but also the safety of the technician. Some factors to consider include:

- The location of the area to be monitored.
- The proximity of the product within the area. Vapor density will dictate where the material may be located within a given area **(Figure 5.7)**.

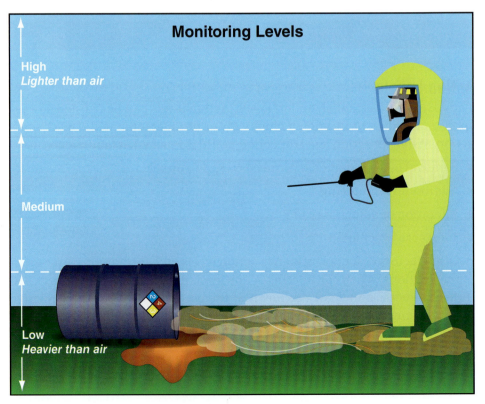

Figure 5.7 The properties of the material will affect concentrations in specific areas.

144 Chapter 5 • Analyzing the Incident: Detection, Monitoring, and Sampling Procedures

- The oxygen concentrations within the sample area.
- The specific instrument response to a given substance.

Sometimes the hazmat technician will know exactly what chemicals to monitor. In other cases the substance is unknown. In all cases, it is important that the hazmat technician utilize a risk-based response when conducting individual monitoring operations. The use of monitoring equipment will either help to confirm the concentrations of a known product or alert the responders to the presence of a potential hazard.

The responder should carefully plan monitoring to avoid wasting time and resources while gathering information. Monitoring locations should be chosen based on environmental conditions and any information that may be known about the hazard. If an interior location or confined space is to be monitored, it is important to consider vapor density. Whether the product hazards are known or unknown, the majority of chemicals that may be found during a response have a vapor density greater than one. Monitor all areas of a given location to determine the presence or absence of a hazardous product. If the monitoring operation is to be conducted outside, the technician must consider wind direction and speed when selecting sampling locations.

Area Monitoring

Area monitoring is performed as a team at planned locations within a given area where chemical dusts, vapor, or fumes are generated. The objective of area monitoring is to determine the presence and concentration of product in an area. In addition, the IC may ask the hazmat team to develop an average exposure potential for a defined area in a given zone. The team may use portable equipment in area monitoring; however, for long-term incidents, fixed equipment should be used to monitor for specific chemicals at defined grid positions **(Figure 5.8)**.

Figure 5.8 Fixed equipment will minimize set-up time and give more predictable results over the course of the incident.

Confined Space — Space or enclosed area not intended for continuous occupation, having limited (restricted access) openings for entry or exit, providing unfavorable natural ventilation and the potential to have a toxic, explosive, or oxygen-deficient atmosphere.

Photoionization detectors, oxygen sensors, explosivity meters, and chemical sensors can be combined in different ways to measure air concentrations at a hazmat scene. These instruments are useful for continuous monitoring during operations or for perimeter monitoring to warn responders and the community of increases in airborne contamination. Some units can be radio-linked to a central command post to provide real-time monitoring of conditions at multiple locations.

Hazmat teams should preplan and prepare area monitoring operations on an area map or drawing to define the exact locations that sampling will be needed. The IC should review and approve area monitoring plans prior to implementation.

Confined Spaces

Monitoring of **confined spaces** is critical for a safe operation. For example, the atmosphere must be tested for hazardous materials before responders enter the confined space. When working in a confined space, hazmat technicians should follow agency SOP/Gs prior to and during operations in the area **(Figure 5.9)**. For example, confined spaces may contain physical safety hazards such as loose wires, standing water, and unmarked openings. Responders should be trained to the level deemed appropriate by the AHJ. However, it is recommended that responders engaged in confined space activities be formally trained in this response area. For hazmat response, the hazards remain the same, and technicians should perform monitoring activities as with any other response.

Cargo Containers

Hazardous materials are sometimes transported without knowledge of port personnel **(Figure 5.10)**. When working in and around containers, hazmat responders must take extreme care because of the unknown conditions. Given the right tools, product identification may be possible. Technicians should attempt air monitoring through any open vents. Once it is deemed safe to enter, hazmat teams should use a risk-based response when working with the container's contents.

Sampling Techniques

To protect the public, hazmat technicians may be required to collect samples of contaminants or suspected contaminants to support medical treatment, determine how to best mitigate the situation, and determine the type of decontamination required. However, when technicians respond to incidents of a criminal nature, sampling concerns extend beyond public health and the samples become **evidence**. This evidence is crucial

Figure 5.9 Confined spaces must be monitored with a calibrated direct-reading instrument.

Figure 5.10 Not all cargo containers will provide a visual indication of their contents.

to law enforcement personnel for criminal investigation and prosecution. In fact, at criminal hazmat/WMD incidents, evidence preservation and sampling is one of the most important considerations. Steps for collecting samples of hazardous material solids, liquids, and gases are presented in **Skill Sheet 5.1**.

Evidence — Information collected and analyzed by an investigator.

Sample Collection for Hazard Identification

Samples can come in many forms. The particular type of material present and the amount of the material will determine the sampling method and equipment required. Public safety samples should be collected per AHJ requirements and may later be used as evidence. Steps for performing public safety sampling and evidence preservation are presented in **Skill Sheet 5.2**.

Field Screening Samples

Technicians must conduct field screens to eliminate specific hazards before the samples can be sent to a laboratory. These tests are necessary to ensure safety for individuals involved with packing, transporting, and performing lab tests on the samples. Once field screening has ruled out these hazards, the materials may be moved and transported to the appropriate laboratory.

Before field screening, responders should consider whether potential explosives are present, and should ensure that bomb squad personnel have cleared any potential explosives. Technicians should also check for crystallized materials around caps and containers as these are indicators of potentially shock-sensitive explosives and reactive chemicals.

CAUTION
Always screen before sampling.

To field screen, technicians will need to establish a suitable work area that is well-ventilated, preferably outdoors. When using destructive field screening techniques, be mindful of the need to leave enough material to be tested by the laboratory for evidentiary purposes.

Field screen samples to test for at a minimum include:

- **Corrosivity** — Check for corrosives on liquids using pH paper.
- **Explosivity** — Have bomb squad personnel check for explosive materials/devices.
- **Flammability** — Use a combustible gas meter to check flammability.
- **Oxidizer** — Use potassium iodide (KI) paper.
- **Radioactivity** — Check for alpha, beta, and gamma radiation.
- **Volatility** — Use a photoionization detector (PID) and **flame ionization detector (FID)** to detect volatile organic compounds (VOCs).

Flame Ionization Detector (FID) — Gas detector that oxidizes all oxidizable materials in a gas stream, and then measures the concentration of the ionized material.

Sampling Plans

Hazmat technicians should collect all samples in accordance with the sampling plan. Although sample plans will vary (following local or federal protocols), most will include all of the following sampling steps:

1. Preparing containers before entering the exclusion zone, using the system agreed upon by the AHJ and the receiving laboratory
2. Recording the sample location, conditions, and other pertinent information in a field notebook
3. Confirming that the sample container number agrees with the overpack container number
4. Wrapping the sample container in absorbent material
5. Placing the sample in an overpack container and sealing it with tamper-proof tape
6. Completing the chain-of-custody form **(Figure 5.11)**
7. Placing the sample and chain-of-custody form in an approved transport container

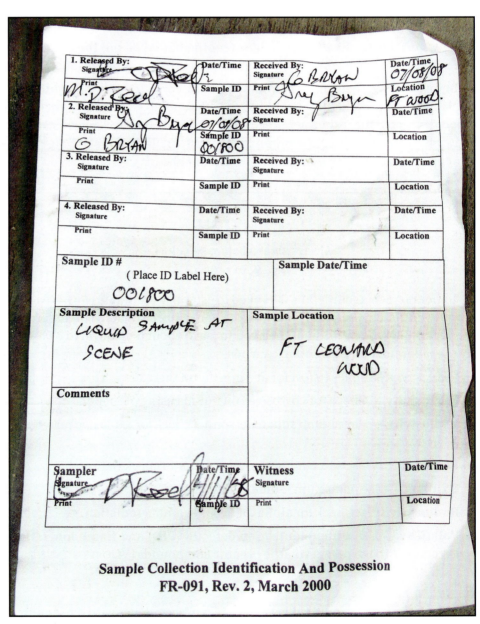

Figure 5.11 Be careful to track where a material sample came from and who handled it.

The sample plan should also include sampling protocols for:

- Protecting samples and evidence
- Field screening samples
- Labeling and packaging
- Decontaminating samples and evidence

A minimum of two individuals are recommended for a sampling team:

- A person (sampler) who takes the samples and handles all pertinent equipment (sampling tools, container, and sample) **(Figure 5.12)**
- An assistant who handles only clean equipment and provides it to the sampler when needed **(Figure 5.13)**

When possible, a third individual can be added to the sampling team. This individual also handles clean equipment and can provide assistance by documenting, photographing, and acting as an Assistant Safety Officer in the hot zone **(Figure 5.14)**. When conducting operations in a crime scene, each individual who enters the scene should have a specific task.

Figure 5.12 One member of a sampling team handles the material to be tested. *Courtesy of the United States Air Force.*

Figure 5.13 A second member of a sampling team handles only clean materials and supplies them as needed. *Courtesy of the United States Air Force.*

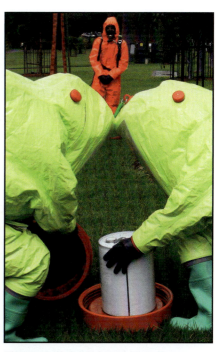

Figure 5.14 A third member of a sampling team may act as the Assistant Safety Officer in the hot zone. *Courtesy of Steven Baker, New South Wales Fire Brigades.*

Safe external packaging and transportation of evidence is the responsibility of the law enforcement AHJ in cooperation with the receiving laboratory and the operator of the transport vehicle. Technicians must be trained in sampling and evidence collection methods and the equipment used during sampling and collection activities. Many states have documents that provide guidance on evidence collection and sampling, and this information is available on the Internet.

Chain of Custody — Continuous changes of possession of physical evidence that must be established in court to admit such material into evidence. In order for physical evidence to be admissible in court, there must be an evidence log of accountability that documents each change of possession from the evidence's discovery until it is presented in court.

Chain of Custody for Sampling

The practice of tracking an item from the time it is found until it is ultimately disposed of or returned to the owner is known as **chain of custody** (**Figure 5.15**). Chain of custody is primarily used during evidence collection activities.

NOTE: In Canada, chain of custody is known as *continuity*.

For the purpose of sampling activities, chain of custody may be used for the following purposes:

- Public safety sampling
- Environmental
- Documentation
- Public health
- Identification

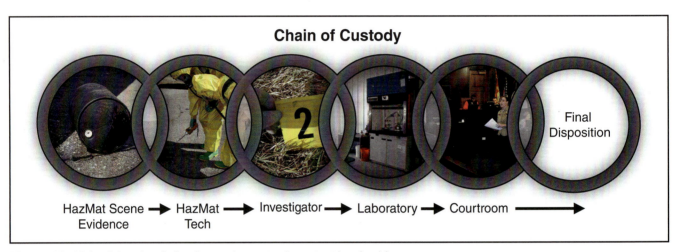

Figure 5.15 Maintaining the chain of custody creates important legal evidence.

Figure 5.16 The correct handling of evidence can assist in preventing future incidents.

Evidence Collection Techniques

Following proper procedures enables technicians to collect evidence that can help apprehend suspects and prevent incident recurrences (**Figure 5.16**). Anything at the scene may be considered evidence and may not be recognized as such until trained evidence technicians arrive and evaluate the situation.

Hazmat crime scenes generally fall under two categories:

1. A traditional crime where hazardous materials are involved; for example, a home invasion where a hazardous material was used to destroy or prevent the discovery of evidence.

2. A hazardous materials WMD or CBRNE event where the intent is to cause harm to persons, property, or the environment.

When technicians observe indicators of criminal activity, they must work with various law enforcement agencies to preserve and protect any potential evidence. Technicians may also be called to assist the law enforcement agency in the collection of evidence. Hazmat technicians assigned to perform evidence preservation and sampling activities must be trained to work under direction of law enforcement in a unified command structure and in accordance with the requirements of their jurisdiction.

Evidence Collection

Processing physical evidence at the crime scene is one of the most important parts of the investigation. Evidence must be documented, collected, preserved, and packaged with careful attention to scene integrity and protection from contamination or harmful change. During the processing of the scene and following documentation, technicians should appropriately package, label, and maintain evidence in a secure manner in accordance with law enforcement directives **(Figure 5.17)**.

During the response, investigators will perform the following duties:

- Maintain scene security throughout processing and until the scene is released.
- Lay out the investigative evidence collection grid system.
- Document the collection of evidence by recording its location at the scene, date of collection, and who collected it.
- Collect each item identified as evidence.
- Establish chain of custody.
- Obtain control samples.
- Provide blank samples.
- Immediately secure electronically recorded evidence (e.g., answering machine tapes, surveillance camera videotapes, computers) and remove them to a secure area.
- Identify and secure evidence in containers at the crime scene (e.g., label, date, initial container).
  — Different types of evidence require different containers (for example, porous, nonporous, and crushproof).
  — Follow the requirements of the lab that will be analyzing the samples with regard to the types of containers and the amount of sample to collect.
- Package items to avoid contamination and cross-contamination.
- Document the condition of firearms/weapons prior to rendering them safe for transportation and submission.
- Maintain evidence at the scene in a manner designed to diminish degradation or loss.
- Transport and submit evidence items for secure storage.

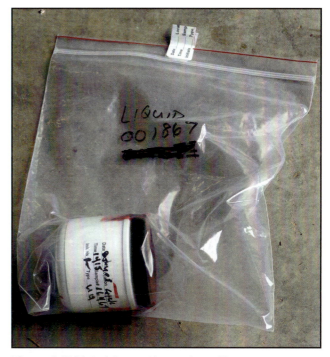

Figure 5.17 Materials must be packaged in a manner compatible with the AHJ including law enforcement.

Chapter 5 • **Analyzing the Incident: Detection, Monitoring, and Sampling Procedures**   **151**

Chain of Custody for Evidence

As a legal document, the chain of custody is a written history that must include each person who maintains physical control over the item throughout the process. Each person in the chain of custody is a candidate for court subpoena. All evidence must be handled and moved to an evidence custodian for documentation into the evidence chain in accordance with the AHJ's chain-of-custody procedures. Chain-of-custody requirements may be mandated at a municipal or federal level.

Evidence Collection and Preservation

Evidence collection will determine the sampling method and equipment required. In the case of suspicious letters and packages, the entire letter, envelope, or package (as well as the hazardous materials contained within) should be treated as physical evidence. Evidence collection should be collected per AHJ requirements.

Initial monitoring results should indicate the general characteristics of the potential threat and type of contaminant that is present (radiological, biological, chemical, or combination). With this information, investigators can better determine the correct sample method.

Wipe samples are used when contaminants are visible or suspected on surfaces. Typical equipment found in a wipe sample kit will vary based on AHJ requirements and the preferences of the laboratory that will process the samples.

The procedures for obtaining solid samples on surfaces are similar to those for wipe samples. Smaller amounts of powder may be collected using a swab, while a scoop or spatula is better for collecting granular solids. Scalpels are used for scraping wood or paint that may have absorbed a contaminant, and scissors are useful for cutting material or fabric that may have absorbed a vapor or liquid contaminant.

Decontaminating Samples and Evidence

Decontaminating evidence removes contamination from the exterior evidence packaging only. Technicians should not open exterior evidence packaging to decontaminate interior evidence packaging. Preserving evidence includes careful handling during decontamination, such as preserving fingerprints. Many evidence containers will be double-bagged or placed inside multiple containers for protection of the samples and the safety and health of the people handling them. Follow laboratory instructions and procedures for decontamination of evidence packages.

Maintenance of Monitoring, Detection, and Sampling Equipment

Hazmat technicians must know how to properly operate monitoring equipment and apply the information to response objectives and tactics. Responders must also know how to maintain and test their equipment. Steps for performing maintenance and testing on monitoring equipment, test strips, and reagents are presented in **Skill Sheet 5.3**.

In the field, hazmat technicians may be forced to work in proximity to a variety of hazards. They must understand the hazards and what bearing these dangers have on the detection and monitoring equipment that may be in use **(Figure 5.18)**. Hazardous products can potentially:

- Damage the detection equipment
- Prevent accurate readings
- Cause total failure of the equipment

NOTE: Maintaining responder safety takes priority over potential damage to equipment.

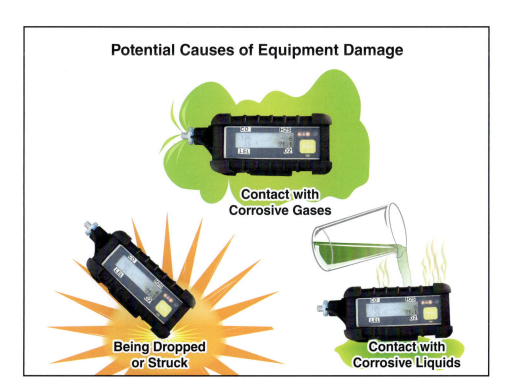

Figure 5.18 Equipment must be protected against preventable damage as much as possible to preserve their proper function.

Chapter Review

1. What are some factors that are important in evaluating the reliability and usefulness of a monitor and the data it generates?
2. At a minimum, what should field screen samples be tested for?
3. What are some evidence collection duties that investigators will perform during a response?
4. How can hazardous products affect monitoring and detection equipment?

SKILL SHEETS

5-1

Perform sampling techniques to identify hazards associated with solid, liquid, and gaseous substances. [NFPA 1072, 7.2.1]

WARNING: If this skill involves the use of actual hazardous material samples, hazardous materials can cause serious injury or fatality. Appropriate personal protective equipment (PPE) must be worn and safety precautions must be followed. The following skill sheet demonstrates general steps; specific hazmat incidents may differ in procedure. Always follow the AHJ procedures for specific incidents.

Liquid

- **Step 1:** Ensure proper detection, monitoring, or sampling method and equipment is chosen.
- **Step 2:** Ensure that all responders are wearing appropriate PPE.
- **Step 3:** Collect a sample of the material.

- **Step 4:** Test the substance using the instrument(s) provided by the AHJ.
- **Step 5:** Read, interpret, and report the results of the test.
- **Step 6:** Dispose of sample in accordance with appropriate regulations.
- **Step 7:** Decontaminate equipment and return to operational state per manufacturer's instructions.
- **Step 8:** Complete required reports and supporting documentation.

Solid

- **Step 1:** Ensure proper detection, monitoring, or sampling method and equipment is chosen.
- **Step 2:** Ensure that all responders are wearing appropriate PPE.
- **Step 3:** Collect a sample of the material.

- **Step 4:** Test the substance using the instrument(s) provided by the AHJ.
- **Step 5:** Read, interpret, and report the results of the test.
- **Step 6:** Dispose of sample in accordance with appropriate regulations.
- **Step 7:** Decontaminate equipment and return to operational state per manufacturer's instructions.
- **Step 8:** Complete required reports and supporting documentation.

Gas

- **Step 1:** Ensure proper detection, monitoring, or sampling method and equipment is chosen.
- **Step 2:** Ensure that all responders are wearing appropriate PPE.
- **Step 3:** Collect a sample of the material.

- **Step 4:** Test the substance using the instrument(s) provided by the AHJ.
- **Step 5:** Read, interpret, and report the results of the test.
- **Step 6:** Dispose of sample in accordance with appropriate regulations.
- **Step 7:** Decontaminate equipment and return to operational state per manufacturer's instructions.
- **Step 8:** Complete required reports and supporting documentation.

5-2
Collect samples of a hazardous material solid, liquid, or gas. [NFPA 1072, 7.2.1]

WARNING: If this skill involves the use of actual hazardous material samples, hazardous materials can cause serious injury or fatality. Appropriate personal protective equipment (PPE) must be worn and safety precautions must be followed. The following skill sheet demonstrates general steps; specific hazmat incidents may differ in procedure. Always follow the AHJ procedures for specific incidents.

Public Safety Sampling

Step 1: Develop appropriate sampling collection plan for hazard identification.

Step 2: Ensure any materials and equipment used for collection are certified "clean" or "sterile" and kept sealed until used.

Step 3: Collect the solid, liquid, and/or gas samples in the appropriate containers as per AHJ SOPs.

Evidence Preservation

Step 1: Develop appropriate evidence collection plan according to AHJ SOPs.

Step 2: Ensure any materials and equipment used for collection are certified "clean" or "sterile" and kept sealed until used.

Step 3: Collect the solid, liquid, and/or gas samples in the appropriate containers as per AHJ SOPs.

Step 4: Follow all chain-of-custody procedures established by the AHJ.

Step 4: Follow all chain-of-custody procedures established by the AHJ.

SKILL SHEETS

Perform maintenance and testing on monitoring equipment, test strips, and reagents.
[NFPA 1072, 7.2.1]

5-3

WARNING: If this skill involves the use of actual hazardous material samples, hazardous materials can cause serious injury or fatality. Appropriate personal protective equipment (PPE) must be worn and safety precautions must be followed. The following skill sheet demonstrates general steps; specific hazmat incidents may differ in procedure. Always follow AHJ procedures for specific incidents.

Step 1: Identify the device or item that requires testing or maintenance.

Step 2: Perform testing and/or maintenance as identified in the manufacturer's instructions and AHJ SOPs.

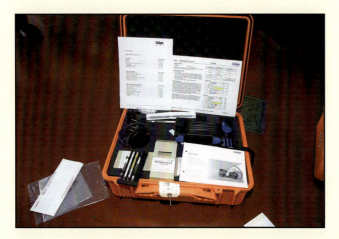

Step 3: Properly document any maintenance or testing performed.

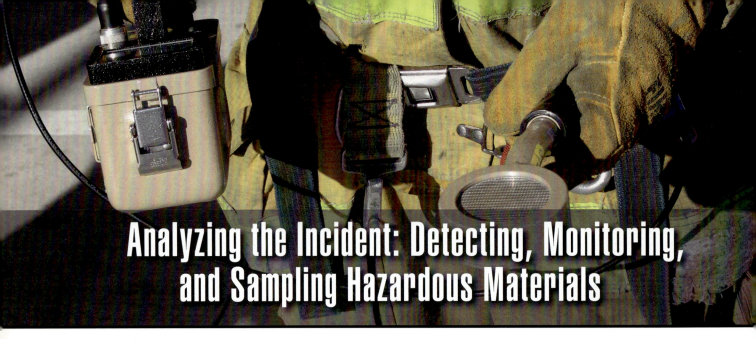

Analyzing the Incident: Detecting, Monitoring, and Sampling Hazardous Materials

Chapter Contents

Exposure **161**	Thermal Imagers 188
Routes of Entry..161	Infrared Thermometers 188
Contamination versus Exposure 162	**Other Detection Devices** **189**
Acute versus Chronic Exposure 163	Halogenated Hydrocarbon Meters 189
Radiological and Biological Exposures ... **164**	Flame Ionization Detectors..................................190
Exposure Limits...164	Gas Chromatography..190
Radiological Exposures 168	Mass Spectroscopy 191
Biological Exposures 169	Ion Mobility Spectrometry 192
Sensor-Based Instruments and	Surface Acoustic Wave..192
Other Devices **170**	Gamma-Ray Spectrometer 193
Oxygen Indicators 170	Fourier Transform IR 194
Combustible Gas Indicators (CGIs) 172	Infrared Spectrophotometers...............................194
Correction Factors 173	Raman Spectroscopy IR......................................195
Electrochemical Cells 175	Mercury Detection 197
Metal Oxide Sensors 175	Wet Chemistry 198
pH Meters 176	DNA Fluoroscopy 199
Multi Sensor Instruments...................................176	Polymerase Chain Reaction Devices.....................199
Colorimetric Methods..176	**Chapter Notes** **200**
Radiation Detection Instruments 183	**Chapter Review** **200**
Photoionization Detectors (PIDs)........................185	**Skill Sheets**.................................. **201**
Biological Immunoassay Indicators.....................187	

158 Chapter 6 • Analyzing the Incident: Detecting, Monitoring, and Sampling Hazardous Materials

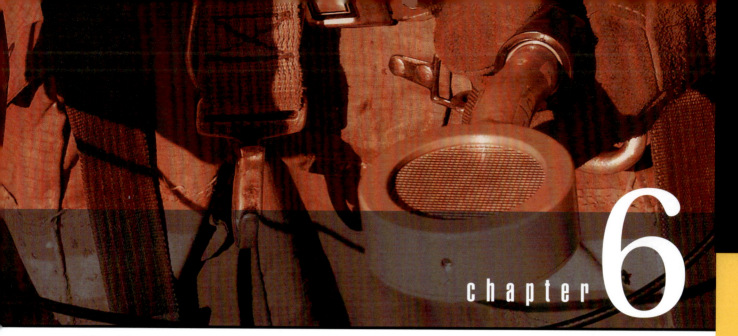

Chapter 6

Key Terms

- Adsorb 198
- Antibody 187
- Antigen 187
- Carcinogen 189
- Colorimetric Indicator Tube 177
- Combustible Gas Indicator (CGI) 172
- Correction Factor 169
- Counts per Minute (CPM) 169
- Dosimeter 168
- Electrochemical Gas Sensor 175
- Emissivity 189
- Exposure Limit 164
- Flame Ionization Detector (FID) 190
- Fluorimeter 199
- Fourier Transform Infrared (FT-IR) Spectroscopy 194
- Gamma-Ray Spectrometer 193
- Gas Chromatograph (GC) 190
- Geiger-Mueller (GM) Detector 183
- Geiger-Mueller (GM) Tube 184
- Immediately Dangerous to Life and Health (IDLH) 164
- Immunoassay (IA) 187
- Infrared 188
- Infrared Thermometer 188
- Ion Mobility Spectrometry (IMS) 192
- Ionization Potential (IP) 186
- Mass Spectrometer 191
- Multi-Use Detectors 176
- Nondispersive Infrared (NDIR) Sensor .. 176
- pH Indicator 176
- Polymerase Chain Reaction (PCR) 199
- Raman Spectrometer 195
- Reagent 181
- Roentgen (R) 169
- Roentgen Equivalent in Man (rem) 169
- Route of Entry 161
- Self-Reading Dosimeter (SRD) 184
- Sievert (Sv) 169
- Spectrometer 191
- Spectrophotometer 194
- Spectroscopy 191
- Surface Acoustic Wave (SAW) Sensor .. 192
- Thermal Imager 187
- Wet Chemistry 187

Chapter 6 • Analyzing the Incident: Detecting, Monitoring, and Sampling Hazardous Materials   159

Analyzing the Incident: Detecting, Monitoring, and Sampling Hazardous Materials

JPRs Addressed In This Chapter

7.2.1, 7.2.5

Learning Objectives

After reading this chapter, students will be able to:

1. Explain critical factors in preventing exposure to hazardous materials. [NFPA 1072, 7.2.1]
2. Explain methods for quantifying radiological and biological exposures. [NFPA 1072, 7.2.1, 7.2.5]
3. Describe types of sensor-based instruments and other devices used for detecting hazards. [NFPA 1072, 7.2.1]
4. Describe other detection devices that can be used for detecting hazards.
5. Skill Sheet 6-1: Demonstrate proper use of pH meters to identify hazards. [NFPA 1072, 7.2.1]
6. Skill Sheet 6-2: Demonstrate proper use of a multi-gas meter (carbon monoxide, oxygen, combustible gases, multi-gas, and others) to identify hazards. [NFPA 1072, 7.2.1]
7. Skill Sheet 6-3: Demonstrate proper use of colorimetric tubes to identify hazards. [NFPA 1072, 7.2.1]
8. Skill Sheet 6-4: Demonstrate proper use of pH paper to identify hazards. [NFPA 1072, 7.2.1]
9. Skill Sheet 6-5: Demonstrate proper use of reagent test paper to identify hazards. [NFPA 1072, 7.2.1]
10. Skill Sheet 6-6: Demonstrate proper use of radiation detection instruments to identify and monitor hazards. [NFPA 1072, 7.2.1]
11. Skill Sheet 6-7: Demonstrate proper use of dosimeters to identify personal dose received. [NFPA 1072, 7.2.1]
12. Skill Sheet 6-8: Demonstrate proper use of photoionization detectors to identify hazards. [NFPA 1072, 7.2.1]
13. Skill Sheet 6-9: Demonstrate the use of a noncontact thermal detection device to identify hazards. [NFPA 1072, 7.2.1]

Chapter 6
Analyzing the Incident: Detecting, Monitoring, and Sampling Hazardous Materials

This chapter will discuss of several factors of a successful response operation including:

- Exposure
- Radiological and biological exposures
- Sensor-based instruments and other devices
- Other detection devices

Exposure

Even the most cautious emergency responder has the potential of being exposed to a chemical at the scene of an emergency. No defined order exists for detection and monitoring activities, nor for applying the findings/results to the next operational tactics. Technicians must maintain risk-based response to get a clear picture of the entire incident profile. Regardless of the hazards that are immediately apparent, responders should also evaluate other risks. The best way to protect against toxic chemicals is to prevent exposure to them. The following sections explain three critical factors in preventing exposure.

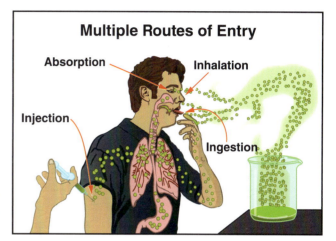

Figure 6.1 Hazardous chemicals can enter the body in four common ways.

Routes of Entry

Technicians should maintain the mindset of risk-based response when taking information presented by detection tools (including meters) into account. For example, always consider whether the PPE you are wearing will keep you safe against the types of risks that you are responding to, or that may be in the area. The following are the main **routes of entry** through which hazardous materials can enter the body and cause harm **(Figure 6.1)**:

- **Inhalation** — Process of taking in materials by breathing through the nose or mouth. Hazardous vapors, smoke, gases, liquid aerosols, fumes, and suspended dusts may be inhaled into the body. When a hazardous material presents an inhalation threat, respiratory protection is required. Inhalation is the most common route of exposure for responders and civilians.

- **Ingestion** — Process of taking in materials through the mouth by means other than simple inhalation. Taking a pill is a simple example of how a chemical might be deliberately ingested. However, poor hygiene after handling a hazardous material can lead to accidental ingestion. Examples include:

> **Route of Entry** — Pathway via which hazardous materials get into (or affect) the human body.

— Chemical residue on the hands that can be transferred to food and then ingested while eating; hand washing is very important to prevent accidental ingestion of hazardous materials

— Particles of insoluble materials that can become trapped in the mucous membranes and ingested after being cleared from the respiratory tract

- **Absorption** — Process of taking in materials through the skin or eyes. Some materials pass easily through the mucous membranes or areas of the body where the skin is the thinnest, providing the least resistance to penetration. The eyes, nose, mouth, wrists, neck, ears, hands, groin, and underarms are areas of particular concern. Many poisons are easily absorbed into the body. Others can enter the body through the eye such as when a contaminated finger touches the eye.

- **Injection** — Process of taking in materials through a puncture of the skin. Protection from injection must be a consideration when dealing with any sort of contaminated (or potentially contaminated) objects easily capable of cutting or puncturing the skin. Such items include: broken glass, nails, sharp metal edges, and tools like utility knives.

Contamination versus Exposure

Previous chapters introduced the concept that a chemical can react with air, water, light, and other chemicals. However, hazardous materials technicians need to understand what can happen when they come in contact with a chemical and what can happen when the chemical reacts with the emergency responder.

An individual can be exposed to a product in a number of settings without being aware of the exposure. For example, a building with a malfunctioning furnace can expose the occupants to the toxic properties of carbon monoxide. If individuals remain in the building over a period of time, they may start to feel the effects of this toxic chemical. By leaving the structure, they are no longer exposed to the hazard **(Figure 6.2)**. Exposure is said to occur when a toxic or infectious substance directly contacts the human body or is introduced into the body.

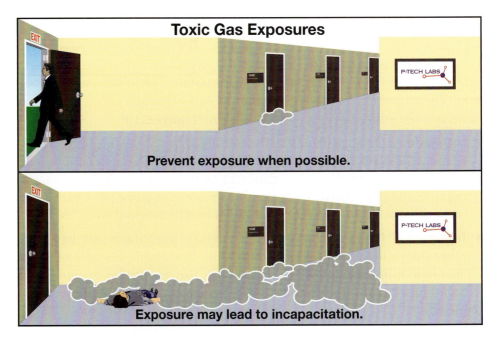

Figure 6.2 Exposure occurs when a toxic substance contacts or enters the body.

Contamination is the result of direct contact with a product, and it occurs when a product adheres to or wets any kind of exposure. A person is considered to be contaminated if in the course of working with a liquid chemical or a solid product, the product somehow spills and contacts clothing or skin.

Understanding the nuances of these terms and concepts is important in maintaining both a safe response and a safe work environment. For example, in an incident including only radioactive materials, a person in close proximity to a radioactive source can be exposed without being contaminated. Once the responder leaves the area, he or she is no longer exposed and has not been contaminated.

A responder working within a hot zone can also be contaminated without being exposed. This can occur if the responder is in approved and appropriate chemical protective clothing and the CPC is saturated with a chemical.

Acute versus Chronic Exposure

If an individual comes in contact with a chemical, several variables may influence the effects of the exposure. For example, responders may not have to come into contact with a product; merely being in proximity to the product may have a detrimental effect. When the toxicity of a chemical is discussed in relation to its effects on the body, how long a person was exposed must be evaluated. An acute exposure may be a single exposure over a relatively short period of time to a high level of a dangerous product. A chronic exposure may be an exposure over a longer duration. To put this in perspective, if a person were exposed to a high level of a radioactive source for a period of two minutes, it could be considered an acute exposure. If a person worked in a radiology department of a hospital and had prolonged exposures to X-rays over the course of his or her career, this would be considered a chronic exposure **(Figure 6.3)**.

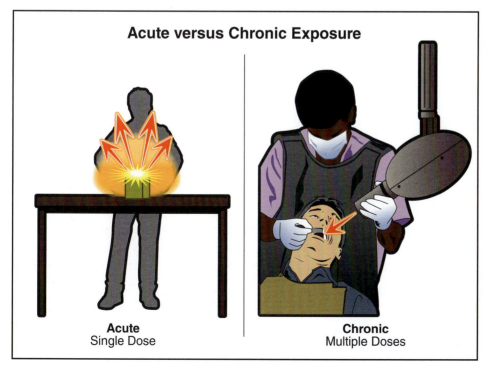

Figure 6.3 Acute and chronic doses are differentiated by the frequency of the incident.

Radiological and Biological Exposures

There may be a possibility for exposure even if the emergency responder operates safely on the scene of a hazmat emergency. For example, firefighters understand that in an ordinary structure fire, there is still the potential for exposure to toxic products of combustion long after the fire has been extinguished. Hazmat responders must also understand that even though they have taken precautions, exposure may still be possible. Because of this fact, they must thoroughly understand toxicology and exposures.

The ability to research a toxic material and fully understand the information provided in the research data is one of the most helpful aspects of a safe response. Research will help the responder determine the toxic properties of the products involved.

Exposure Limits

Exposure limits are used to estimate potential harm and outcomes at hazardous materials incidents. The National Institute for Occupational Safety and Health (NIOSH) Pocket Guide and other resources provide exposure limits for many materials. The following sections will provide definitions to several of the terms used to express these limits.

> **Exposure Limit** — Maximum length of time an individual can be exposed to an airborne substance before injury, illness, or death occurs.

> **Immediately Dangerous to Life and Health (IDLH)** — Description of any atmosphere that poses an immediate hazard to life or produces immediate irreversible, debilitating effects on health; represents concentrations above which respiratory protection should be required. Expressed in parts per million (ppm) or milligrams per cubic meter (mg/m^3); companion measurement to the permissible exposure limit (PEL).

Immediately Dangerous to Life and Health (IDLH)

The National Institute for Occupational Safety and Health (NIOSH) quotes the Occupational Safety and Health Administration (OSHA) standard number 29 CFR 1910.120 in one definition of an atmosphere considered to be **immediately dangerous to life and health (IDLH)**: "An atmospheric concentration of any toxic, corrosive, or asphyxiating substance that poses an immediate threat to life. It can cause irreversible or delayed adverse health effects and interfere with the individual's ability to escape from a dangerous atmosphere."

The Occupational Safety and Health Administration (OSHA) also further identifies an IDLH atmosphere in standard number 29 CFR 1910.134(b): "An atmosphere that poses an immediate threat to life, would cause irreversible adverse health effects, or would impair an individual's ability to escape from a dangerous atmosphere."

The exposure period is immediate. The NIOSH definition also states that the IDLH is the maximum concentration from which an unprotected person can expect to escape in a 30-minute period of time without suffering irreversible health effects.

Other Exposure Limits

Hazmat technicians commonly encounter other types of exposure limits when conducting research on hazardous materials. These limits typically come from industry, and responders must be familiar with the terminology **(Table 6.1)**.

NOTE: For more information on exposure limits, refer to Chapter 7, Air Monitoring.

Standards development entities such as American Conference of Governmental Industrial Hygienists® (ACGIH) may provide some data. Exposure limits separate from IDLH include:

Table 6.1
Exposure Limits Terminology

Term	Definition	Exposure Period	Organization
IDLH Immediately Dangerous to Life or Health	An atmospheric concentration of any toxic, corrosive, or asphyxiating substance that poses an immediate threat to life. It can cause irreversible or delayed adverse health effects and interfere with the individual's ability to escape from a dangerous atmosphere.*	Immediate (This limit represents the maximum concentration from which an unprotected person can expect to escape in a 30-minute period of time without suffering irreversible health effects.)	**NIOSH** National Institute for Occupational Safety and Health
IDLH Immediately Dangerous to Life or Health	An atmosphere that poses an immediate threat to life, would cause irreversible adverse health effects, or would impair an individual's ability to escape from a dangerous atmosphere.	Immediate	**OSHA** Occupational Safety and Health Administration
LOC Levels of Concern	10% of the IDLH		
PEL Permissible Exposure Limit**	A regulatory limit on the amount or concentration of a substance in the air. PELs may also contain a skin designation. The PEL is the maximum concentration to which the majority of healthy adults can be exposed over a 40-hour workweek without suffering adverse effects.	8-hours Time-Weighted Average (TWA)*** (unless otherwise noted)	**OSHA** Occupational Safety and Health Administration
PEL (C) PEL Ceiling Limit	The maximum concentration to which an employee may be exposed at any time, even instantaneously.	Instantaneous	**OSHA** Occupational Safety and Health Administration
STEL Short-Term Exposure Limit	The maximum concentration allowed for a 15-minute exposure period.	15 minutes (TWA)	**OSHA** Occupational Safety and Health Administration
TLV® Threshold Limit Value†	An occupational exposure value recommended by ACGIH® to which it is believed nearly all workers can be exposed day after day for a working lifetime without ill effect.	Lifetime	**ACGIH®** American Conference of Governmental Industrial Hygienists
TLV®-TWA Threshold Limit Value-Time-Weighted Average	The allowable time-weighted average concentration.	8-hour day or 40-hour work week (TWA)	**ACGIH®** American Conference of Governmental Industrial Hygienists
TLV®-STEL Threshold Limit Value-Short-Term Exposure Limit	The maximum concentration for a continuous 15-minute exposure period (maximum of four such periods per day, with at least 60 minutes between exposure periods, provided the daily TLV®-TWA is not exceeded).	15 minutes (TWA)	**ACGIH®** American Conference of Governmental Industrial Hygienists

Continued

Table 6.1 (continued)

Term	Definition	Exposure Period	Organization
TLV®-C Threshold Limit Value-Ceiling	The concentration that should not be exceeded even instantaneously.	Instantaneous	**ACGIH®** American Conference of Governmental Industrial Hygienists
BEIs® Biological Exposure Indices	A guidance value recommended for assessing biological monitoring results.		**ACGIH®** American Conference of Governmental Industrial Hygienists
REL Recommended Exposure Limit	A recommended exposure limit made by NIOSH.	10-hours (TWA) ††	**NIOSH** National Institute for Occupational Safety and Health
AEGL-1 Acute Exposure Guideline Level-1	The airborne concentration of a substance at or above which it is predicted that the general population, including "susceptible" but excluding "hypersusceptible" individuals, could experience notable discomfort. †††	Multiple exposure periods: 10 minutes 30 minutes 1 hour 4 hours 8 hours	**EPA** Environmental Protection Agency
AEGL-2 Acute Exposure Guideline Level-2	The airborne concentration of a substance at or above which it is predicted that the general population, including "susceptible" but excluding "hypersusceptible" individuals, could experience irreversible or other serious, long-lasting effects or impaired ability to escape. Airborne concentrations below AEGL-2 but at or above AEGL-1 represent exposure levels that may cause notable discomfort.	Multiple exposure periods: 10 minutes 30 minutes 1 hour 4 hours 8 hours	**EPA** Environmental Protection Agency
AEGL-3 Acute Exposure Guideline Level-3	The airborne concentration of a substance at or above which it is predicted that the general population, including "susceptible" but excluding "hypersusceptible" individuals, could experience life-threatening effects or death. Airborne concentrations below AEGL-3 but at or above AEGL-2 represent exposure levels that may cause irreversible or other serious, long-lasting effects or impaired ability to escape.	Multiple exposure periods: 10 minutes 30 minutes 1 hour 4 hours 8 hours	**EPA** Environmental Protection Agency
ERPG-1 Emergency Response Planning Guideline Level 1	The maximum airborne concentration below which it is believed nearly all individuals could be exposed for up to one hour without experiencing other than mild transient adverse health effects or perceiving a clearly defined objectionable odor.	Up to 1 hour	**AIHA** American Industrial Hygiene Association

Continued

Table 6.1 (concluded)

Term	Definition	Exposure Period	Organization
ERPG-2 Emergency Response Planning Guideline Level 2	The maximum airborne concentration below which it is believed nearly all individuals could be exposed for up to one hour without experiencing or developing irreversible or other serious health effects or symptoms that could impair an individual's ability to take protective action.	Up to 1 hour	**AIHA** American Industrial Hygiene Association
ERPG-3 Emergency Response Planning Guideline Level 3	The maximum airborne concentration below which it is believed nearly all individuals could be exposed without experiencing or developing life-threatening health effects.	Up to 1 hour	**AIHA** American Industrial Hygiene Association
TEEL-0 Temporary Emergency Exposure Limits Level 0	The threshold concentration below which most people will experience no appreciable risk of health effects.		**DOE** Department of Energy
TEEL-1 Temporary Emergency Exposure Limits Level 1	The maximum concentration in air below which it is believed nearly all individuals could be exposed without experiencing other than mild transient adverse health effects or perceiving a clearly defined objectionable odor.		**DOE** Department of Energy
TEEL-2 Temporary Emergency Exposure Limits Level 2	The maximum concentration in air below which it is believed nearly all individuals could be exposed without experiencing or developing irreversible or other serious health effects or symptoms that could impair their abilities to take protective action.		**DOE** Department of Energy
TEEL-3 Temporary Emergency Exposure Limits Level 3	The maximum concentration in air below which it is believed nearly all individuals could be exposed without experiencing or developing life-threatening health effects.		**DOE** Department of Energy

* The NIOSH definition only addresses airborne concentrations. It does not include direct contact with liquids or other materials.

** PELs are issued in Title 29 *CFR* 1910.1000, particularly Tables Z-1, Z-2, and Z-3, and are enforceable as law.

*** Time-weighted average means that changing concentration levels can be averaged over a given period of time to reach an average level of exposure.

† TLVs® and BEIs® are guidelines for use by industrial hygienists in making decisions regarding safe levels of exposure. They are not considered to be consensus standards by the ACGIH®, and they do not carry the force of law unless they are officially adopted as such by a particular jurisdiction.

†† NIOSH may also list STELs (15-minute TWA) and ceiling limits.

††† Airborne concentrations below AEGL-1 represent exposure levels that could produce mild odor, taste, or other sensory irritation.

- **Permissible Exposure Limit (PEL)** — OSHA's definition of a regulatory limit on the amount or concentration of a substance in the air. PELs may also contain a skin designation. The PEL is the maximum concentration to which the majority of healthy adults can be exposed over a 40-hour work week without suffering adverse effects. The PEL is reported as an 8-hour, time-weighted average (TWA) unless otherwise noted.

- **Permissible Exposure Limit – Ceiling (PEL-C)** — OSHA's definition as the maximum concentration to which an employee may be exposed at any time.

- **Recommended Exposure Limits (REL)** — Recommendations provided by NIOSH and based on a 10-hour, time-weighted average.

- **Threshold Limit Value® (TLV)** — Value provided by the American Conference of Governmental Industrial Hygienists® (ACGIH). TLV® is an occupational exposure value to which it is believed nearly all workers can be exposed day after day for a working lifetime without any ill effects.

- **Threshold Limit Value® – Ceiling (TLV-C)** — The ACGIH® has defined as a concentration that should not be exceeded for any time period, even instantaneously.

- **Threshold Limit Value® – Short Term Exposure Limit (TLV-STEL)** — The ACGIH® has defined as the maximum concentration for a continuous 15-minute exposure period. The TLV®-STEL is calculated on a daily basis as four 15-minute continuous work periods with a minimum of a 60-minute period between each exposure provided the TWA is not exceeded.

- **Threshold Limit Value®–Time Weighted Average (TLV-TWA)** — The ACGIH®defines the TLV®-TWA as the allowable time-weighted average concentration when an employee is exposed for 8 hours per day, 40 hours per week.

Radiological Exposures

Emergency responders should be familiar with the terms used to express radiation dose and exposure. These units may be used on radiation dose instruments, called **dosimeters** and *radiation survey meters* **(Figure 6.4)**. As with curies and becquerels, two systems of units are used to measure and express radiation exposure and radiation dose (energy absorbed from the radiation) **(Figure 6.5)**.

> **Dosimeter** — Detection device used to measure an individual's exposure to an environmental hazard such as radiation or sound.

Figure 6.4 Badge dosimeters typically display the absorbed radiation dose.

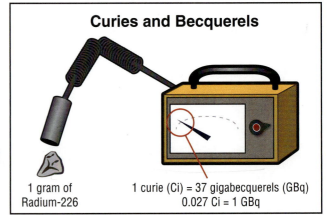

Figure 6.5 Curies (Ci) are used to indicate large amounts of activity, and becquerels (Bq) are used to indicate small amounts of energy.

The U.S. still commonly uses the English System, and these units are as follows:

- **Roentgen (R)** — Used for measuring exposure and is applied only to gamma and X-ray radiation. This is the unit used on most U.S. dosimeters. Roentgen is expressed in R per hour (R/hr) and is used on radiation survey meters.

- **Radiation absorbed dose (rad)** — Used to measure the amount of radiation energy absorbed by a material. This unit applies to any material and all types of radiation, but does not take into account the potential effect that different types of radiation have on the human body. For example, 1 rad of alpha radiation is more damaging to the human body than 1 rad of gamma radiation.

- **Roentgen equivalent in man (rem)** — Used for the absorbed dose equivalence as pertaining to a human body and is applied to all types of radiation. This unit takes into account the energy absorbed (as measured in rad) and the biological effect on the body due to different types of radiation. Rem is used to set dose limits for emergency responders. For gamma and X-ray radiation, a common conversion between exposure, absorbed dose, and dose equivalent is as follows: 1 R = 1 rad = 1 rem.

- **Counts per minute (cpm)** is a measure of radioactivity and is the number of atoms in a quantity of radioactive material that is detected to have decayed in one minute. CPM is an aggregate measurement of radioactive decay at a certain point. When used with a **correction factor** and the type of material is known, it may be converted into Roentgen.

- **Sievert** is an International System of Units (SI) term used for a derived unit or dose of radiation. This modern unit of radiation dose incorporates quality factors for the biological effectiveness of the different types of ionizing radiation.

> **Roentgen (R)** — English System unit used to measure radiation exposure, applied only to gamma and X-ray radiation; the unit used on most U.S. dosimeters.
>
> **Roentgen Equivalent in Man (rem)** — English System unit used to express the radiation absorbed dose (rad) equivalence as pertaining to a human body; used to set radiation dose limits for emergency responders. Applied to all types of radiation.
>
> **Counts per Minute (CPM)** — Measure of ionizing radiation in which a detection device registers the rate of returns over time. Primarily used to detect particles, not rays.
>
> **Correction Factor** — Manufacturer-provided number that can be used to convert a specific device's read-out to be applicable to another function. *Also known as* Conversion Factor, Multiplier *and* Response Curve.
>
> **Sievert (Sv)** — SI unit of measurement for low levels of ionizing radiation and their health effect in humans.

Biological Exposures

Biological exposures can be very difficult to mitigate based on the fact that the emergency responder may not know they are present. Biological and chemical warfare agents affect humans in different ways. Effects of exposure to chemical agents are almost always immediate. However, effects of exposure to biological agents may not become apparent for several days and can affect wide geographical areas. The emergency responder must be familiar with the following concepts:

- **Incubation period** — Defined as the time from the exposure to a biological agent to the appearance of symptoms in an infected person. Also known as Latency. This is not a measurable time period for field or clinical applications.

- **Infectious dose (ID)** — Amount of pathogen (measured in number of microorganisms) required to cause an infection in the host. Identification can vary according to the pathogenic agent and the affected person's age and overall health. This is not a measurable factor for field or clinical applications.

Sensor-Based Instruments and Other Devices

The following sections will detail some of the types of sensor-based instruments available to emergency responders. While this is not an all-inclusive list, it is a fair representation of common sensor-based instruments. The operation of each type of sensor will not be different between a standalone sensor versus a bundled detector, but other functionalities may be affected. For example, the sensor range may be significantly shorter in a bundled sensor device.

NOTE: Bundled detectors are included in more detail later in this chapter.

Limitations of New Technology

Manufacturers of detection and monitoring equipment may be able to offer new options for response applications. Technicians must exercise caution when replacing or purchasing new equipment, because they must know the uses (readings, interpretations) and limitations (scope of use) of all equipment that they will rely on during a response. When researching new equipment, technicians should examine many sources of information to determine whether the equipment performs its intended task under the applicable conditions.

Oxygen Indicators

OSHA defines action levels for specific oxygen concentrations. Risk-based response should focus on the specific incident and not limit evaluation to those numbers. Oxygen indicators are used to evaluate an atmosphere for the following:

- Oxygen concentration for respiratory purposes — Atmospheric oxygen levels should be monitored continuously during incidents. Any sustained change from normal or measured oxygen levels can represent a significant displacement of oxygen with some other product **(Figure 6.6)**. Anyone without supplied air respirators/SCBA in a reduced oxygen environment may suffer harmful exposure to that environment and the hazardous chemicals within it.

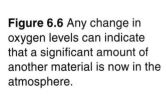

Figure 6.6 Any change in oxygen levels can indicate that a significant amount of another material is now in the atmosphere.

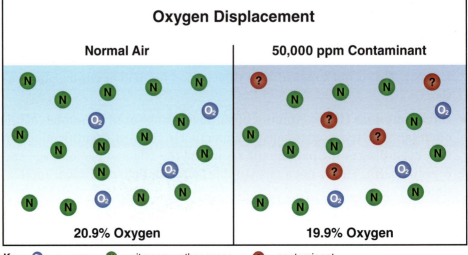

- Increased risk of flammability — High concentrations of oxygen, or any sustained elevations, increase the risk of material flammability.

Similar to an air freshener, O_2 sensors work continuously while activated. The sensor will last a shorter duration as more of it is used. To confirm the effectiveness of a detector, check it with a bump test. Then try calibrating. If neither test returns satisfactory results, replace the sensor. The owner's manual for the specific device should include guidance for frequency of bump testing and calibration.

CAUTION
A 1 percent drop in oxygen level indicates that 50,000 ppm of some other product has taken its place in the atmosphere.

WARNING!
Increases in the oxygen level may indicate that some type of reaction or displacement has taken place or is ongoing.

Oxygen sensors are often used in tandem with other sensors because instruments require sufficient oxygen for accurate results. Combustible gas indicators may not give reliable results if the oxygen concentration is below 10 percent **(Figure 6.7)**.

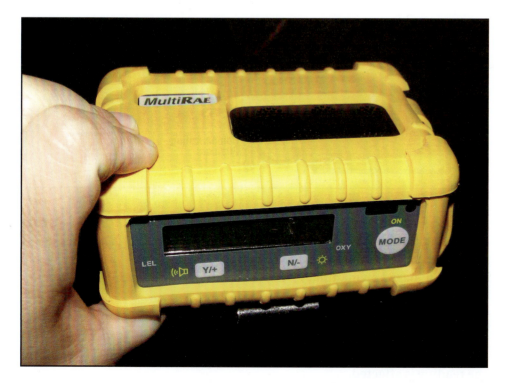

Figure 6.7 Combustible gas indicators may work reliably in oxygen-deficient atmospheres.

> **CAUTION**
> Breathing into a sensor may provide erroneous readings because of the CO_2 levels in exhaled air.

Between freezing and boiling temperature ranges, most sensors should function accurately. At temperature ranges beyond those extremes, the sensor may be damaged or have slower response times. It is recommended that the instrument be calibrated at the temperature at which it will be utilized.

NOTE: Training on each type of sensor/detector/monitor should include reference to the effects of temperature and humidity on the specific device and its operation.

Combustible Gas Indicators (CGIs)

Combustible gas indicators (CGIs) are available in many styles and configurations. Some CGIs have a pump to draw the air sample into the detector, while other units use ambient air that diffuse over the sensor. Many units are combination meters that contain an oxygen sensor along with other gas or vapor detecting sensors.

Combustible Gas Indicator (CGI) — Electronic device that indicates the presence and explosive levels of combustible gases, as relayed from a combustible gas detector.

Combustible gas indicators use a combustion chamber containing a filament that burns the flammable gas or an electronic sensor to measure the concentration of a flammable vapor or gas in air **(Figure 6.8)**. The resulting reading is displayed via the CGI as a percentage of the lower explosive limit (%LEL) of the calibration gas. Utility companies sometimes use CGIs that measure the actual flammable range, not only a percentage of the LEL.

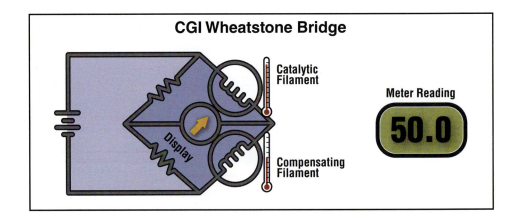

Figure 6.8 One type of combustible gas indicator uses a hot filament coated with a catalyst to burn the flammable gas and then detect the results.

A concentration greater than the LEL and lower than the UEL indicates that the ambient atmosphere is readily combustible, and hazmat technicians must understand how their meter will react in this situation. They should operate equipment based on manufacturers' instructions and the AHJ's policies and procedures.

Similar to the oxygen sensor, the response of some flammable gas sensors is temperature dependent. If the temperature at which the instrument was calibrated differs from the temperature in the sample area, the accuracy of the reading is affected.

Hotter temperatures can provide a higher-than-actual reading and cooler temperatures can reduce the reading. It is best to zero (bump) the instrument in the temperature it will be used.

Flammable gas sensors are intended for use only in normal oxygen atmospheres. Oxygen-deficient atmospheres will affect readings, potentially to the point of invalidating the information.

CAUTION
Oxygen-deficient atmospheres may affect the validity of meter readings.

The safeguards that prevent the combustion source from igniting a flammable atmosphere are not designed to operate in oxygen-enriched atmospheres. Organic lead vapors, sulfur compounds, and silicone compounds will also interfere with the operation of the filament. Acid gases may corrode the filament while silicone vapors even in small quantities may poison the sensor.

CGIs have specific calibration considerations. Common calibrant gases for CGIs include:

- Methane
- Pentane
- Propane
- Hexane

When responders use a CGI calibrated to one gas (such as methane) to measure other known flammable gases/vapors (such as propane), the actual LEL of the gas being measured may differ from the reading of the CGI. Examples of correction factors for various gases were included in Chapter 5 of this manual.

Responders using LEL meters must make allowances for these potential discrepancies in order to correctly interpret LEL readings. Manufacturers provide this additional data to work specifically with individual meters.

Correction Factors

Combustible gas indicators (CGIs) and photoionization detectors (PIDs) are most useful when monitoring two categories of known gases:

- The calibrant gas
- Gases specified in the manufacturer's instructions

In some cases, the gas expected at an incident will be more hazardous than may be available in the area selected for calibration. In this case, a correction factor may be chosen to calibrate the meter to a "surrogate" gas. Correction factors are scaling factors that offer a correction to the readout of the gas to which the sensor was calibrated. They do not make the sensor specific to gas other than the calibration gas. A correction factor can be defined as a measure of the sensitivity of the PID sensor to a specific gas.

The PID can be calibrated to any gas, but the correction factors for all manufacturers refer to isobutylene **(Table 6.2)**. The NIOSH Pocket Guide provides the ionization potential numbers for many materials.

	Table 6.2		
	Sample PID Correction Factors*		
	Correction Factor*		
Compound Name	**Lamp Strength 10.6**	**Lamp Strength 11.7**	**IP (eV)**
Acetone	1.1	1.4	9.71
Ethyl Ether	1.1	1.7	9.51
Methyl Bromide	1.7	1.3	10.54
Nitrogen Dioxide	16	6	9.75
Phosgene	No Response	6.8	11.2
Toluene	.50	.51	8.82

* For a RAE Systems, Inc. PID calibrated to isobutylene.
** Technicians must understand how to use the correction factor based on the manufacturer's instructions, for example, multiplying the instrument's reading by the correction factor.

Manufacturers supply correction factors relevant to the readings of the detectors. Correction factors are intended to be a resource that a responder in the field can use to translate a reading into a potential other answer. Correction factors should be maintained so they are immediately accessible in the field during detection/monitoring operations. Monitors with a built-in electronic dictionary of these factors should be updated per the manufacturer's recommendations.

Correction factors are only useful if the technician knows the material that the monitor is sampling, and that material is included in the library. Correction factors are available for many gases, but not for all. For example, most monitors do not have a correction factor for gasoline. Some manufacturers offer a wider variety of conversion or correction factors than others. Ultimately, technicians must remember that any reading indicates a hazard. At most incidents, the AHJ will define a safe action level as 10 percent of the LEL.

WARNING!
Any reading on a meter indicates the presence of a hazard.

What This Means To You
Limitations of Conversion/Correction Factors

Technicians should know how correction factors can be used to show hazards that their detectors may not specifically detect. Technicians can

> use a chemical formula to predict whether the meter reading has included known risk factors. The use of correction factors may or may not be approved for technicians per their respective AHJ.
>
> Reasons technicians may not choose to use conversion/correction factors include:
>
> - Correction factors do not include a margin of error as a safety buffer, so the final range may not reflect safety considerations.
> - The system of converting between actual readings and expected readings can be confusing.
> - Technicians not familiar with the correction factors available in their meter may inadvertently choose the wrong one for the incident.
> - Technicians may not have access (correction factor numbers may not be calculated) for the chemical they want to calculate.

Electrochemical Cells

Electrochemical cell sensors are a type of fuel cell designed to produce a current that is precisely related to the amount of the target gas in the atmosphere. Measurement of the current gives a measure of the concentration of the target gas in the atmosphere. Essentially, the electrochemical cell consists of:

- Container
- Two electrodes
- Connection wires
- Electrolyte (typically sulfuric acid)

In the case of carbon monoxide (CO), the CO is oxidized at one electrode to carbon dioxide while oxygen is consumed at the other electrode. For CO detection, the electrochemical cell has advantages over other technologies. For example, CO detectors that use an electrochemical cell require a relatively small amount of power, operate at room temperature, and display CO readings as well as cross-sensitivities to some chemicals. The operational life spans of these sensors can vary based on:

- Manufacturer
- Frequency of use
- Concentrations of detected materials during use

> **Electrochemical Gas Sensor** — Device used to measure the concentration of a target gas by oxidizing or reducing the target gas and then measuring the current.

CAUTION
Be careful to choose a detector that can detect the target material.

Metal Oxide Sensors

Metal oxide sensors are used to detect gases such as natural gas and hydrogen sulfide. Inside the sensor, gases interact with a thin film of metal oxide. Depending on their properties, the gases are either reduced or oxidized. This reduction or oxidation causes an increase or decrease in the conductivity of the metal oxide film. The degree of change in sensor resistance is translated into the concentration of the gas detected.

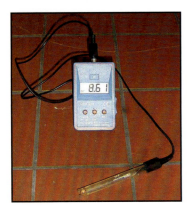

Figure 6.9 pH meters must be calibrated for accuracy.

pH Indicator — Chemical detector for hydronium ions ($H3O^+$) or hydrogen ions (H^+). Indicator equipment includes impregnated papers and meters.

Multi-Use Detectors — Device with several types of equipment in one handheld device. Used to detect specific types of materials in an atmosphere. *Also known as* Multi-Gas Meter.

Nondispersive Infrared (NDIR) Sensor — Simple spectroscope that can be used as a gas detector.

pH Meters

Testing the acidity of liquids has been performed since the early 1900s. The use of pH paper is a common means of testing acidity and will be discussed later in this chapter **(Figure 6.9)**. The basic principle of using **pH indicators** is to measure the concentration of hydrogen ions. Acids dissolve in water forming positively charged hydrogen ions. The greater this concentration of hydrogen ions, the stronger the acid. Similarly, alkalis or bases dissolve in water forming negatively charged hydrogen ions. The stronger the alkali or base in the solution, the higher the concentration of negatively charged hydrogen ions.

All pH meters must be calibrated. Technicians can achieve calibration by dipping the meter's probe into a buffer solution of a known pH and following the manufacturer's recommended steps for the specific pH meter in use.

Disadvantages of these meters include:

- They must be calibrated prior to each use.
- They are easily damaged.
- They require manual handling.

Because pH paper and strips are reliable, they are the primary means of determining pH levels in the field. Occasionally, pH meters are encountered and used in hazmat response. Steps for demonstrating proper use of pH meters to identify hazards are presented in **Skill Sheet 6-1**.

Multi Sensor Instruments

Many sensor-based instruments are equipped to measure multiple specific gases, including oxygen. **Multi-use detectors** may include five or six different types of sensors. This setup may include a variety of configurations. When purchasing multi-sensor instruments, hazmat teams have the ability to specify the configuration of their sensors based upon their needs. For example, a **nondispersive infrared (NDIR) sensor** may be packaged onto other multi-use sensors to detect carbon dioxide. Steps for demonstrating proper use of a multi-gas meter (carbon monoxide, oxygen, combustible gases, multi-gas, and others) to identify hazards are presented in **Skill Sheet 6-2**.

Types of sensors that are commonly bundled include a PID plus four or five other sensors including an:

- Oxygen sensor
- LEL sensor
- CO sensor
- Hydrogen sulfide (H_2S) sensor

Like other sensor-based instruments, these detection monitors are subject to interference from a variety of sources such as other gases, vapors, temperature, and barometric pressures. Sensors must be replaced periodically and calibrated with gas specific to the sensor.

Colorimetric Methods

Colorimetric detection methods can help indicate and/or identify the presence of a chemical through a chemical reaction that results in the color change of the test medium. Technicians may view the color change visually or via more sophisticated equipment. Hazmat response professionals use colorimetric methods widely. These methods have been found to be reliable, though some have a wide variance of accuracy called *standard deviation*. The following sections introduce the different colorimetric detection methods that are available to the emergency responder. Technicians can obtain more specific information on these methods directly from the manufacturer of these instruments. There may be multiple tubes with multiple ranges available for the same chemical.

Technicians would be wise to research the gaps that may exist between the detection capabilities of their equipment. For example, if a flame ionization detector (FID) is not available, a tube that can improve the detection range for a photoionization detector (PID) may be useful. It may be less expensive for a department to collect a set/box of tubes than a standalone instrument if the gap in detection devices is wide enough for chemicals that the technician frequently needs to detect.

> **Colorimetric Indicator Tube** — Small tube filled with a chemical reagent that changes color in a predictable manner when a controlled volume of contaminated air is drawn through it. *Also known as* Detector Tube.

Colorimetric Tubes/Chips

Colorimetric indicator tubes consist of a glass tube impregnated with an indicating chemical. The tube is connected to a piston or bellows-type pump **(Figure 6.10)**. The pump pulls a known volume of contaminated air at a predetermined rate through the tube. The contaminant reacts with the indicator chemical in the tube, producing a change in color where the length of the color change in the tube is proportional to the contaminant concentration. Steps for demonstrating proper use of colorimetric tubes to identify hazards are presented in **Skill Sheet 6-3**.

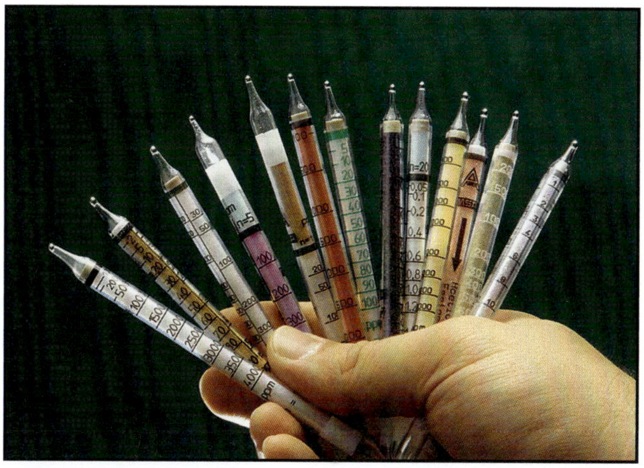

Detector tubes are normally chemical-specific. Some manufacturers do produce tubes for groups of gases, such as aromatic hydrocarbons or alcohols. Concentration ranges on the tubes may be in ranges defined by parts per million or a percent range. A preconditioning filter may precede the indicating chemical in order to:

Figure 6.10 Colorimetric tubes will show different colors based on the detection of specific chemicals. *Courtesy of MSA.*

- Remove contaminants (other than the one in question) that may interfere with the measurement
- Remove humidity
- React with a contaminant to change it into a compound that reacts with the indicating chemical

Colorimetric indicator tube kits are available from several different manufacturers. These kits identify or classify the contaminants as a member of a chemical group such as acid gas, halogenated hydrocarbon, and others. This identification is done by sampling with certain combinations of tubes at the same time using a special multiple tube holder or by using tubes in a specific sampling sequence. There are also screening kits designed specifically to detect weapons of mass destruction.

Due to gaps in technology, detector tubes are sometimes the only means of monitoring under some conditions. Detector tubes are not always as accurate or precise as other monitoring equipment, but they can be more accurate than LEL sensors.

The chemical reactions involved in the use of the tubes are affected by temperature. Cold temperatures slow the reactions and thus the response time. To reduce this problem, it is recommended that the tubes be kept warm (for example, inside a coat pocket) until used. Warm temperatures increase the reaction and can cause a problem by discoloring the indicator when a contaminant is not present. This discoloration can happen even in unopened tubes. Therefore, tubes should be stored at a moderate temperature.

Some tubes do not have a pre-filter to remove humidity and may be affected by high humidity. The manufacturer's instructions usually indicate whether humidity is a problem and list any correction factors to use if humidity affects the tube.

The chemicals used in the tubes deteriorate over time. For this reason, the tubes are assigned a shelf life. This varies from one to three years depending on the tube's manufacturer and chemistry. A tube is only certified to be accurate until the expiration date.

NOTE: Tubes that are out of date can be used for training exercises, but they may give false readings.

CAUTION
Storing colorimetric indicator tubes in a hot compartment or area can cause them to fail prematurely.

An advantage that detector tubes have over some other instruments is the possibility of selecting a tube specific to a chemical. However, some tubes will respond to interfering compounds. Fortunately, manufacturers provide information about interfering gases and vapors with the tubes.

Interpretation of results can be difficult. Because the tube's length of color change indicates the contaminant concentration, the user must be able to see the end of the stain **(Figure 6.11)**. Color changes may not be precise or clear at times. Some stains are diffused and not clear-cut; others may have an uneven endpoint. The length of the stain may change over time and some tubes need longer reaction times before the color change is obvious. When in doubt, use the highest value that would be obtained from reading the different aspects of the tube.

The total volume to be drawn through the tube varies with the tube type and the manufacturer. The volume needed is given as the number of pump strokes required, i.e., the number of times the technician manipulates the piston or bellows. Also, the air does not instantaneously pass through the tube. It may take 1 to 2 minutes for each volume (stroke) to be completely drawn. Therefore, sampling times can vary from 1 to 100 minutes or

Figure 6.11 The length of the stained section will indicate the concentration of the detected contaminant.

more per tube. This can make the use of detector tubes time-consuming. In addition, the responder must be stationary while conducting the test. Battery-operated pumps can make these tests easier to conduct.

Due to these many considerations, it is important to read the instructions provided with, and are specific to, a set of tubes **(Figure 6.12)**. The information should include:

- The number of pump strokes needed
- Time for each pump stroke
- Interfering gases and vapors
- Effects of humidity and temperature
- Shelf life
- Proper color change
- Reusability of the tube

While there are many limitations and considerations for using detector tubes, they are versatile in being able to measure a wide range of chemicals with a single pump. Also, there are some chemicals for which detector tubes are the only direct-reading indicators.

Closely associated with colorimetric tubes technology are colorimetric chips. They may be referred to as a *chip measurement system (CMS)*. CMSs use chemical-specific measuring chips with an electronic analyzer. These chips have small tubes, sometimes referred to as capillaries, that are filled with a reagent system for the designated chemical.

Most CMSs are considered direct-reading instruments. These electronic analyzing instruments offer a highly reliable measurement for specific gases and vapors in a digital readout format. CMSs tend to offer a quick response, sometimes accurate to within 7 percent of measured values for some products. In addition to these features, CMSs are simple to use.

pH Paper/Strips

pH is a logarithmic scale; the difference between each pH unit is a factor of 10-fold difference in concentration. For each whole pH value below 7, the representative material is 10 times more acidic than the next higher value. For example, a pH of 4 is 10 times more acidic than a pH of 5 and 100 times (10 × 10) more acidic than a pH of 6. The same holds true for pH values above 7, each of which is 10 times more alkaline than the value below it. Pure water is neutral, with a pH of 7.0 **(Figure 6.13)**.

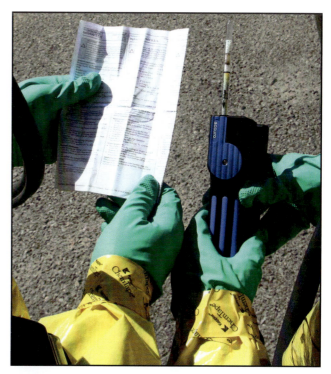

Figure 6.12 Operating instructions may vary by manufacturer and device.

pH Concentration

Concentration of Hydrogen Ions Compared to Distilled Water	pH
10,000,000	0
1,000,000	1
100,000	2
10,000	3
1,000	4
100	5
10	6
1	7
1/10	8
1/100	9
1/1,000	10
1/10,000	11
1/100,000	12
1/1,000,000	13
1/10,000,000	14

Figure 6.13 The pH of water is 7.0, at the center of the scale.

pH paper is easy to use and interpret. It is used to visually determine whether a liquid or solution is acidic or basic. Acids turn it red, and bases turn it blue **(Figure 6.14)**. pH paper can be used for both liquids and vapors. Steps for demonstrating proper use of pH paper to identify hazards are presented in **Skill Sheet 6-4**.

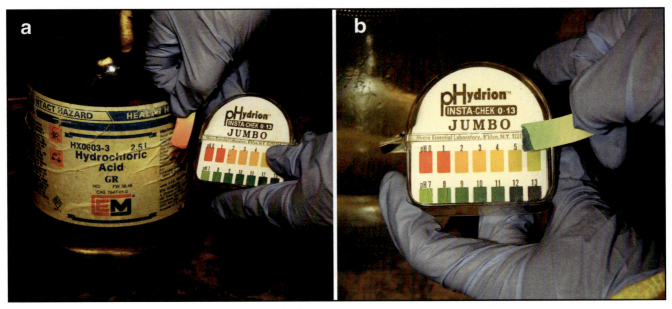

Figure 6.14 Acids turn pH paper red, and bases turn it blue.

There are several varieties of pH paper available with different pH ranges **(Figure 6.15)**. pH paper is highly sensitive to differences in acid or base concentration. Since pH is a logarithmic scale, pH paper can be used to measure the acidity or alkalinity of liquid and solid samples by mixing the solid with water and measuring the pH of the resulting solution.

NOTE: The purity of the water, whether distilled water or tap water, may not make a significant difference for this application. Follow the AHJ's and manufacturers recommendations.

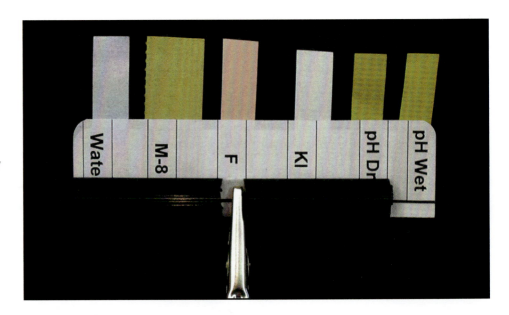

Figure 6.15 A "bear claw" can be constructed with some type of horizontal holder and a variety of detection papers.

180   Chapter 6 • **Analyzing the Incident: Detecting, Monitoring, and Sampling Hazardous Materials**

When attempting to determine the corrosivity of an unknown atmosphere, the entry team may elect to tape a strip of pH paper to their outermost face shield. This will allow technicians to view the strip of pH paper while keeping their hands free for other tasks. Some manufacturers make pH paper with an adhesive backing to allow it to be taped to a suit to determine whether acids or bases are present in the air. Some vapors may not produce as quick of a color change when a preapplied adhesive is used. Attaching a strip with adhesive tape may be a better solution.

NOTE: In corrosive atmospheres, flammable gases may be released. In these instances, hazmat technicians can use pH paper in conjunction with other monitoring equipment to identify corrosive/flammable atmospheres.

CAUTION
All basic (alkaline) corrosive gases will burn in a flame; not all flame-burning gases are corrosive.

CAUTION
pH paper will not detect materials it is not designed to detect. Use multiple sensors to monitor an environment during hazardous materials response.

Reagent Papers

A variety of reagent and water finding papers are available for specific ions typically found in water. Chemical test strips are a type of **reagent** paper. Test strips used by technicians can include:

- pH
- Fluoride
- Chloride
- Nitrates
- Peroxides
- Oxidizers

These are colorimetric tests and are typically used for measuring water for these components at the low parts-per-million range. The presence of water is not necessarily a hazard in itself, but the ability to test for water is helpful during field screening at an unknown hazard response because it helps minimize errors in some types of equipment. Use of reagent papers coincides with pH paper and other risk-based response detection devices. These strips are used with pipettes to identify unknown liquids. Steps for demonstrating proper use of reagent test paper to identify hazards are presented in **Skill Sheet 6-5**.

Several types of multi-functional reagent papers are available. These typically will contain tests for pH, oxidizers, and others. These test strips must be

> **Reagent** — Chemical that is known to react to another chemical or compound in a specific way, often used to detect or synthesize another chemical.

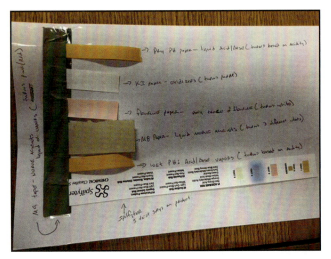

Figure 6.16 This array of test strips includes a wastewater detector strip. *Courtesy of Scott Kerwood.*

stored in a climate-controlled environment to prevent deterioration and most have a limited shelf life. Types of Spilfyter® test strips include wastewater classifiers and chemical classifiers **(Figure 6.16)**.

M8, M9, and M256 Chemical Agent Detection Papers

M8 and M9 chemical agent detection papers change color in the presence of liquid nerve agents or blistering agents. M8 paper is dipped into the suspect liquid and the color change is compared to a color chart to identify specific chemical warfare agents. M9 detects nerve and blister agents as droplets in the air. Both types of paper are easy to use but will only show that there is a probable toxic agent (gas) present and will not display the level of material present. They are both extremely useful in that they respond to a chemical warfare agent immediately. Both papers will also react to some industrial chemicals.

NOTE: M8 paper can also identify whether a liquid is organic or water. If the liquid is organic, the paper will absorb it. Water will bead on the paper.

The M256A1 Detector Kit is a handheld set of colorimetric tests that uses several different chemistries including immunoassay techniques **(Figure 6.17)**. The M256A1 Detector Kit may indicate the presence of an agent when it is actually not present because of some types of interference.

This kit can detect the presence of airborne WMD agents including:

- **Nerve agents** – G and V series organophosphate agents
- **Blister agents** – Sulfur Mustard (H), Nitrogen Mustard (HN), Phosgene Oxime (CX), and Lewisite (L)
- **Blood agents** – Hydrogen Cyanide (AC) and Cyanogen Chloride (CK)

Figure 6.17 The M256A1 Detector Kit can be used to detect airborne WMD agents.

The tests can take up to 30 minutes to perform and is effectively a two-person task since the testing requires specific steps be done at specific times. This is most frequently done by having one person read the instructions and monitor the time while a second person performs the test.

Radiation Detection Instruments

Detection and measurement of radioactive contamination can help responders avoid hazardous exposure at the scene of a nuclear incident. Proper detection equipment is also needed to verify that victims are free from radiological contamination.

Monitoring instruments are used to detect the presence of radiation by collecting charged particles (ions). The radiation measured is usually expressed as exposure per unit time using various units of measure. These units include the curie (Ci), the becquerel (Bq), and counts per minute (CPM). A variety of instruments are available for detecting and measuring radiation.

Radiation detection and measurement instruments are used routinely to monitor personnel working around or with radiation sources. They are also used to check for any leakage of radiation from containers used to store or transport radioactive materials. Hazmat technicians should always utilize radiation detection equipment. Some types are small enough to clip onto an SCBA strap. Steps for demonstrating proper use of radiation detection instruments to identify and monitor hazards are presented in **Skill Sheet 6-6**.

Geiger-Mueller

With a gas-filled detector, radiation ionizes the gas inside the detection chamber and the instrument's electronics measure the quantity of ions created. Ion chambers and **Geiger-Mueller (GM) detectors** are common examples of gas-filled detectors (**Figure 6.18**).

Geiger-Mueller (GM) Detector — Detection device that uses GM tubes to measure ionizing radiation. *Also known as a* Geiger Counter.

Figure 6.18 Geiger counters are used to detect radiation.

An ion chamber is a simple type of gas-filled detector that often uses ambient air as the detection gas, which can cause the chamber to be affected by temperature and humidity. Ion chambers are often calibrated so that the response is directly related to the intensity of the radiation, making them reliable instruments when encountering radiations with varying energies.

GM detectors are sealed from outside air and are not typically affected by temperature or humidity. **GM tubes** with a very thin window may be capable of detecting alpha, beta, and gamma radiation, making them useful for detecting radiological contamination. GM tubes that use a sealed metal body are better suited for measuring penetrating gamma radiation, a possible external exposure hazard. The metal case makes this type of probe less suitable for use in detecting radiological contamination.

> **Geiger-Mueller (GM) Tube** — Sensor tube used to detect ionizing radiation. This tube is one element of a Geiger-Mueller detector.

Scintillation Detectors

With scintillation detectors, radiation interacts with a crystal such as sodium iodide, cesium iodide, or zinc sulfide to produce a small flash of light. The electronics of the instrument amplify this light pulse thousands of times in order to produce a signal that can be processed. A device called a photomultiplier tube that is attached to the crystal provides the light amplification. Some scintillation detectors have a thin Mylar® covering over the crystal, making them useful for detecting radiological contamination. Generally speaking, scintillation detectors are most useful when detection of small amounts of radiation is required.

Scintillation crystals that are sealed in a metal body are better suited for measuring penetrating gamma radiation. Scintillation detector probes are usually larger than gas-filled detector probes because of the photomultiplier tube. They are also susceptible to breakage if not handled properly. Dropping the instrument can shatter the crystal, photomultiplier tube, or both.

Dosimeters and Badges

Dosimetry devices are useful for keeping track of the wearer's total accumulated radiation dose. The dosimeter measures the total radiation dose an individual has received. Several different types of dosimeters are available. Dosimeters may be included as a feature in some commercial detectors. Steps for demonstrating proper use of dosimeters to identify personal dose received are presented in **Skill Sheet 6-7**.

Self-reading dosimeters (SRDs) are commonly used, partly because they do not require processing at a lab to retrieve dose information. Generally speaking, SRDs only measure gamma and X-ray radiation. An SRD measures the radiation dose in:

- Roentgens (R)
- Milliroentgens (mR)
- Microsieverts
- Milisieverts
- Gray (Gy)

> **Self-Reading Dosimeter (SRD)** — Detection device that displays the cumulative reading without requiring additional processing. *Also known as* Direct-Reading Dosimeters (DRDs) *and* Pencil Dosimeters.

Responders should record the SRD reading before entering a radiation field (hot zone). While working in the hot zone, responders should read the SRD at 15- to 30-minute intervals and again upon exit from the hot zone. If a reading is higher than expected or if the SRD reading is off-scale, the AHJ should determine the appropriate action levels based on the hazards and the mission. As always, the IC, local SOPs, and the circumstances of the incident determine the appropriate response. Using the ALARA (As Low as Reasonably Achievable) principle, it should be possible to conduct many types of operations, including rescues, below the 5 rem (0.05 Sv) limit.

Photoionization Detectors (PIDs)

A photoionization detector (PID) detects concentrations of flammable/combustible gases and vapors in air by using an ultraviolet light source to ionize the airborne contaminant **(Figure 6.19)**. PIDs can detect levels as low as 0.1 part per million and in some instances even parts per billion. If a PID presents readings, a technician will know that some substance in the atmosphere will burn. Steps for demonstrating the proper use of photoionation detectors to identify hazards are presented in **Skill Sheet 6-8**.

Figure 6.19 Photoionization detectors (PIDs) use a lamp to ionize contaminants.

PIDs use a fan or pump to draw air into the instrument's detector where the contaminants are exposed to UV light. The resulting charged particles (ions) are collected and measured **(Figure 6.20)**. Once the gas or vapor is ionized in

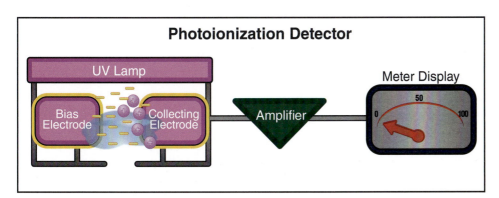

Figure 6.20 The components of a photoionization detector (PID) can be simplified as a lamp, electrodes, amplifier, and display. Other components can be added to increase the efficiency and range of the device.

the instrument, it can be detected and measured. Photoionization detectors are also used in gas chromatographs.

Since the ability to detect a chemical depends on the ability to ionize it, technicians must compare the **ionization potential (IP)** of a chemical to be detected to the energy generated by the UV lamp of the instrument **(Table 6.3)**. There is a limit imposed by the components of air. That is to say, if the lamp is too energetic, oxygen and nitrogen will ionize and interfere with the readings for contaminants.

> **Ionization Potential (IP)** — Energy required to free an electron from its atom or molecule.

Table 6.3
Ionization Potential and Lamp Strength

Chemical Name	Ionization Potential (IP)	Necessary Lamp Strength*
Isopropyl Ether	9.20 eV	9.5eV - 11.8eV
Methyl Ethyl Ketone	9.54 eV	10.0 eV - 11.8eV
Cyclohexane	9.88 eV	10.0 eV - 11.8eV
Ethyl Alcohol	10.47 eV	10.6 eV - 11.8eV
Acetylene	11.40 eV	11.7eV or 11.8eV
Hydrogen Cyanide	13.60 eV	Undetectable
Hydrogen Fluoride	15.98 eV	Undetectable

*Typical Lamp Strengths: 9.5eV, 10.0 eV, 10.2 eV, 10.6 eV, 11.7eV, 11.8eV; lamp strength must be higher than IP for the material to be detected

WARNING!
Getting no response or no change on a meter does NOT mean that a hazard is not present.

Some limitations of photoionization detectors include:

- Dust may collect on the lamp and block the transmission of the UV light.
- Condensation may form on the lamp and reduce the available light.
- Moisture may reduce the ionization of the chemical and reduce the overall readings.
- As the lamp ages, output intensity will decrease.

NOTE: One way to field check the sensor is to cover or block the inlet with a gloved hand. If the sensor continues to show a reading, the bulb needs to be cleaned.

PIDs are calibrated to a single chemical. The instrument's response to chemicals other than the calibration gas/vapor can vary. In some cases, the instrument response can decrease at high concentrations.

Although PIDs are calibrated to a single chemical, it is still possible to gain an accurate reading of a target product even though the instrument was not calibrated to that specific product. The use of correction factors can help the responder determine that information based on the operator's manual.

Photoionization detectors, oxygen sensors, explosive detection meters, and chemical sensors can be grouped in different combinations in field instruments to measure air concentrations at a hazmat scene. These instruments are useful for continuous monitoring during operations or for perimeter monitoring to warn responders and the community of airborne contamination increases. Some units can be radio-linked to a central command post to provide real-time monitoring of conditions at multiple locations. An advantage of the PID is that it does not destroy the test sample and also allows for the sample to be retained for later analysis. A bigger advantage is that it reads materials that flame-burn (burn with a visible flame) at any detectable concentration.

Immunoassay (IA) — Test to measure the concentration of an analyte (material of interest) within a solution.

Antibody — Specialized protein produced by a body's immune system when it detects antigens (harmful substances). Antibodies can only neutralize or remove the effects of their analogous antigens.

Antigen — Toxin or other foreign substance that triggers an immune response in a body.

Biological Immunoassay Indicators

Biological **immunoassay (IA)** indicators (also known as assays) are identification systems that detect the effects of a biological agent by relying on the evidence of a reaction, in the form of **antibodies** (**Figure 6.21**). IAs can detect certain **antigens** (biological agents) by targeting proteins unique to that agent.

During risk-based response, biological assays can supplement other detection methods to help the hazmat technician provide the Incident Commander with information needed for critical decision-making. If malicious intent is suspected, technicians may use the IA (if available) to screen for the presence of a biological agent.

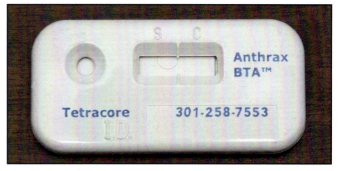

Figure 6.21 Immunoassay indicators are used to detect a reaction to a biological agent.

Technicians can also use **wet chemistry** in conjunction with IA indicators. If a material is a solid or a solid in a solution, and has a neutral pH value, the next step is to apply a protein test to the substance. Most materials will never create a positive protein test. If the protein test indicates a positive result, the application of an IA test would be appropriate. Wet chemistry is discussed in greater length later in this chapter.

Immunoassay methods are based on the response of specific antibodies to toxic materials. Living organisms, when exposed to toxic materials, will generate antibodies (proteins and enzymes) to respond to the attack as part of the organism's immune system. During subsequent exposures, these proteins and enzymes react to the presence of the toxic material rapidly and at low concentrations.

IA tests may offer quick results. One limitation of these tests is that they may indicate biological agents where none exist.

Wet Chemistry — Branch of analysis with a focus on chemicals in their liquid phase.

Thermal Imager — Electronic device that forms images using infrared radiation. *Also known as* Thermal Imaging Camera.

Thermal Imagers

A **thermal imager (TI)**, also known as a thermal imaging camera (TIC), is a type of thermographic imaging device used widely by firefighters. By rendering

Figure 6.22 In addition to helping firefighters find victims, thermal imagers can be used to detect hot spots and temperature differences in hazmat containers.

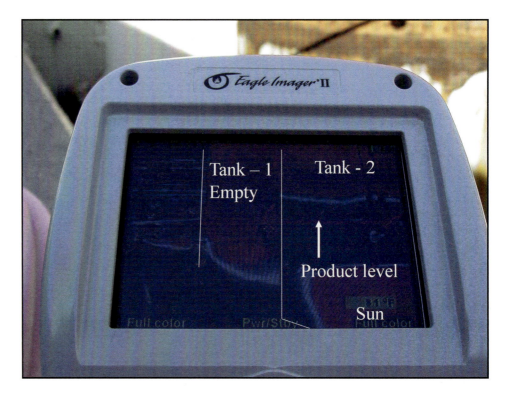

Infrared — Invisible electromagnetic radiant energy at a wavelength in the visible light spectrum greater than the red end but lower than microwaves.

infrared radiation as visible light, such imagers allow firefighters to see areas of heat through smoke, darkness, or heat-permeable barriers **(Figure 6.22)**. TIs are typically handheld, but may be helmet-mounted. They are constructed using heat- and water-resistant housings, and are designed to withstand the hazards of fireground operations.

A TI consists of five components:

- Optic system
- Amplifier
- Display
- Detector
- Signal processing

These parts work together to render infrared radiation, such as that given off by warm objects or flames, into a visible light representation in real time. The imager display shows infrared output differentials, so two objects with the same temperature will appear to be the same "color." Many thermal imagers use grayscale to represent normal temperature objects, but highlight dangerously hot surfaces in different colors.

Hazmat responders use TIs to look for hot spots or differences in temperature. Using a TI, the responder may identify a chemical reaction, such as a polymerization reaction, occurring in a container. They can also be used to determine the liquid level in uninsulated containers or flow in a transfer hose. Some TIs are sensitive enough to differentiate cold temperatures and are useful in identifying leaks involving products such as propane and ammonia. They also are useful for identifying differences in temperature between containers.

Infrared Thermometers

Infrared Thermometer — Noncontact measuring device that detects the infrared energy emitted by materials and converts the energy factor into a temperature reading. *Also known as* Temperature Gun.

Infrared thermometers detect temperature using a portion of the thermal radiation (sometimes called blackbody radiation) emitted by the object of measurement **(Figure 6.23)**. When IR thermometers are pointed at the object to be measured, the device displays a temperature. By knowing the amount of infrared energy emitted by the object and its **emissivity**, technicians can often determine the object's temperature. Infrared thermometers are a subset of devices more precisely called *thermal radiation thermometers* or *radiation thermometers* for short. Infrared thermometers are typically very accurate. Steps for demonstrating the use of a noncontact thermal detection device to identify hazards are presented in **Skill Sheet 6-9**.

The most basic infrared thermometer design consists of a lens that focuses the infrared thermal radiation onto a detector. The detector converts the radiant power to an electrical signal that can be displayed in units of temperature after being compensated for ambient temperature. This configuration facilitates temperature measurement from a distance without technicians having to come in contact with the target object. The infrared thermometer is useful for measuring temperature under circumstances where thermocouples or other probe type sensors cannot be used or do not produce accurate data.

Figure 6.23 Infrared thermometers can sense temperature without contacting the material.

Emissivity — Measure of an object's ability to radiate thermal energy.

Technicians can use infrared thermometers to serve a wide variety of temperature monitoring functions such as:

- Determining the surface temperature of a container
- Determining the temperature of a material
- Determining if a chemical reaction is taking place in and outside the container
- Monitoring materials in the process of a chemical reaction

Other Detection Devices

To this point, this chapter has concentrated on the typical meters and instrumentation available to most hazardous materials technician teams. However, there are many nontraditional instruments that can be useful when attempting to mitigate a hazmat incident. The following sections address some instrumentation that technicians should consider when developing a cache of detection devices. Some of these devices may already be available to the emergency responder, while some may have to be researched and purchased. In any case, having the proper equipment available will not only assist in mitigation but also increase the safety of the response.

Halogenated Hydrocarbon Meters

Halogenated hydrocarbons are **carcinogens** and can pose a serious health risk. Technicians can detect halogenated hydrocarbons by using colorimetric tubes and chips and other more advanced methods. Many "tic tracer" leak detectors are capable of monitoring for halogenated hydrocarbons. These portable devices are capable of detecting vapor concentrations of 25 parts per million by volume (ppmv) and indicating a concentration of 25 ppmv or greater. The devices do this by emitting an audible or visual signal that varies as the concentration changes. A disadvantage when compared to other detection devices is that these meters do not provide a numerical value. Another disadvantage is that they are not typically intrinsically safe.

Carcinogen — Cancer-producing substance.

Flame Ionization Detector (FID) — Gas detector that oxidizes all oxidizable materials in a gas stream, and then measures the concentration of the ionized material.

Advantages of halogenated hydrocarbon meters include:

- Simple to use.
- Chemical family specific.
- Used to verify the readings from other detection devices.

Flame Ionization Detectors

Figure 6.24 Flame ionization detectors can return results on nearly all organic compounds.

Flame ionization detectors (FIDs) use a hydrogen flame (combustion) to ionize organic airborne contaminants in order to detect carbon-based flammable gas **(Figure 6.24)**. Once they are ionized, many contaminants can be detected and measured. The FID responds to virtually all organic compounds.

CAUTION
The recommended warm-up times are essential for proper functioning of a FID.

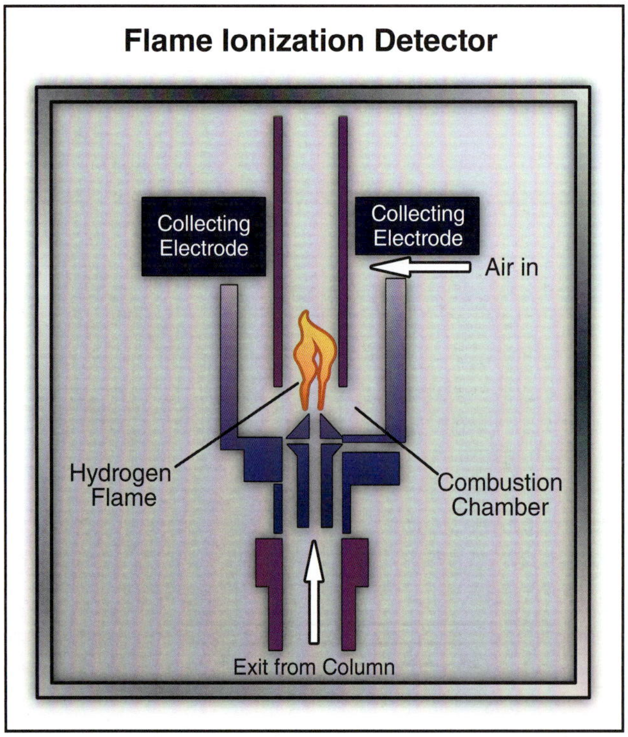

FIDs primarily detect carbon (C) and hydrogen (H). This generalized sensitivity is due to the breaking of chemical bonds, which requires a set amount of energy and is a known reproducible event **(Figure 6.25)**. FIDs are the most sensitive detector for saturated hydrocarbons, alkanes, and unsaturated hydrocarbon alkenes. Substances that contain substituted functional groups, such as hydroxide (OH) and chloride (Cl), tend to reduce the detector's sensitivity. Overall, however, the detectability remains sound.

FIDs respond only to organic compounds. Thus, they do not detect inorganic compounds like chlorine or ammonia. FIDs may detect some halogen compounds, but only if those compounds are present in large quantities.

FIDs are complementary for use with PIDs. Using these devices together is helpful because they expand the range of products that can be detected. Combination FID/PID meters are also commercially available. FIDs require some advanced knowledge for operation.

Figure 6.25 Despite its use of a flame, an FID may be intrinsically safe.

Gas Chromatography

A **gas chromatograph (GC)** is a tool used extensively to measure the presence and identity of chemicals in air, water, and soil. The role of GC is mainly to separate and measure the various components of the mixture of chemicals. This technique can be exceptionally sensitive and specific depending on the product to be measured.

GC is commonly used in many research and industrial labs. However, it does have some relevance in the hazmat response field. Technicians can analyze

Gas Chromatograph (GC) — Apparatus used to detect and separate small quantities of volatile liquids or gases via instrument analysis. *Also known as* Gas-Liquid Partition Chromatography (GLPC).

a broad variety of samples as long as the compounds are sufficiently stable when exposed to high temperatures.

GC takes into account three different factors between the sample and a stationary phase:

- Affinity
- Volatility
- Molecular weight

During operation, a sample is injected into the column. An inert gas such as helium helps push the sample through. After the sample passes through the column, it reaches the detector, commonly a thermal conductivity detector, which will measure the thermal conductivity of the carrier gas. This will generate an electrical signal that transforms into a chromatogram.

Gas chromatography equipment requires additional training. Technicians should have an idea of what materials they expect to find, via use of a PID or FID, before using a **spectrometer**.

A significant issue with gas chromatographic data is the possible misidentification of the impurity peak due to interferences in the chromatogram. It is possible to report the presence and concentration of a product in air, water, or soil when the product may not be present or may be present at a much lower level. One of the most significant contributors of interference in a device's readings may be as simple as the software or hardware updates required to keep it current and functioning optimally.

Mass Spectroscopy

Mass spectrometers are generally the most sensitive detectors for GC and may measure components in the low parts per million or even parts per billion ranges. Use of this technology can aid the responder in identification of unknown materials.

Mass spectrometers also have a two-fold advantage:

- Being able to break down individual compounds into fragments
- Comparing those fragments with the fragmentation pattern of a known sample of the component being measured for identification purposes

Mass spectrometers ionize samples in order to determine their composition. Like other **spectroscopy** devices, they compare test results to a library of known measurements in order to make a positive identification.

The operator of this technology must be a highly trained technician who can deliver viable samples to the equipment. New generations of GC/MSs are field portable, handheld, and easier to operate, but still require a high level of skill to interpret data **(Figure 6.26)**. The GC/MS technology is best used to identify volatile and semi-volatile organic compounds including chemical warfare agents and toxic industrial chemicals.

> **Spectrometer** — Apparatus used to measure the intensity of a given sample based on a predefined spectrum such as wavelength or mass.
>
> **Mass Spectrometer** — Apparatus used to ionize a chemical and then measure the masses within the sample.
>
> **Spectroscopy** — Study of the results when a material is dispersed into its component spectrum. *Also known as* Spectrography.

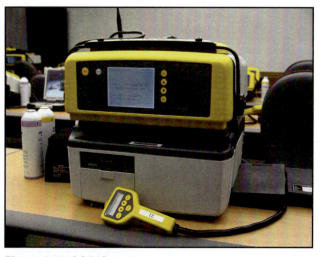

Figure 6.26 GC/MS equipment has become more portable over time, but the results must still be analyzed and interpreted.

Ion Mobility Spectrometry

Ion mobility spectrometry (IMS) is the basis of many chemical agent monitors used in emergency response **(Figure 6.27)**. This technology is deployed in many airports and is used for detection and screening of explosives, chemical warfare agents, and illicit drugs. In an industrial setting, ion mobility is used to detect toxic or volatile chemicals that are otherwise difficult to detect based on their behaviors. A primary limitation of this technology is that analysis is dependent on vapor pressure, which may limit the detection of some explosives with low vapor pressure.

Chemical agent monitors (CAMs), improved chemical agent detectors (ICADs), automatic chemical agent detector alarms (ACADAs), and joint chemical agent detectors (JCADs) are gross-level detectors of nerve, blood, blister, and choking agents. They provide a visible and audible warning of agent doses, but only when the concentration is above the initial effects dose. These are handheld, battery-operated, post-attack devices for monitoring chemical agent contamination on personnel or equipment. Ion mobility detection may not yield accurate readings if certain organic solvents are present.

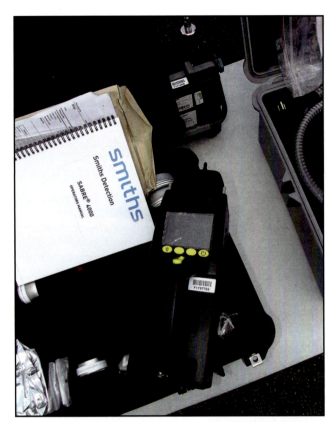

Figure 6.27 Ion mobility spectrometry can help identify some materials that are difficult to detect via other methods.

A Nickel-63 source ionizes gases from the sample which form a cluster of ions. A gate circuit opens and ions of a specific polarity enter a chamber called a drift tube or drift region. The clusters of ions are focused and accelerated by an electromagnetic field moving through the drift tube in approximately 10-20 milliseconds. The ions arrive at an electrode and produce an electrical current proportional to the relative number of ions present. The combination of the drift time and charge-current then causes an alarm to sound, indicating the presence of a nerve or blister agent.

Ion mobility tends to be fast and reliable instrumentation that has been tested under many circumstances and conditions. In most instances, the user can easily maintain this technology, which only requires periodic checks and yearly maintenance. Ion mobility will work efficiently for particulate analysis as well using wipe sampling technology. Ion mobility will attempt to offer a name for the detected material, as opposed to a library of potential options.

Surface Acoustic Wave

Surface acoustic wave (SAW) sensors are based on technology that was developed in the early 20th century. SAW technology, used in meters such as MSA SAW MiniCAD®, may potentially be used in the detection of nerve and blister agents **(Figure 6.28)**.

The MiniCAD® uses a pair of SAW microsensors that are extremely sensitive to small changes in the mass of the surface coatings that act as sponges for chemical warfare agents. A small pump collects the vapor samples, concentrates them, and passes them over the microsensors. An onboard microcomputer analyzes the responses to determine if a hazard exists. Analysis time is typically one minute.

> **Ion Mobility Spectrometry (IMS)** — Technique used to separate and identify ionized molecules. The ionize molecules are impeded in travel via a buffer gas chosen for the type of detection intended. Larger ions are slowed more than smaller ions; this difference provides an indication of the ions' size and identity.
>
> **Surface Acoustic Wave (SAW) Sensor** — Device that senses a physical phenomenon. Electrical signals are transduced to mechanical waves, and then back to electrical signals for analysis.

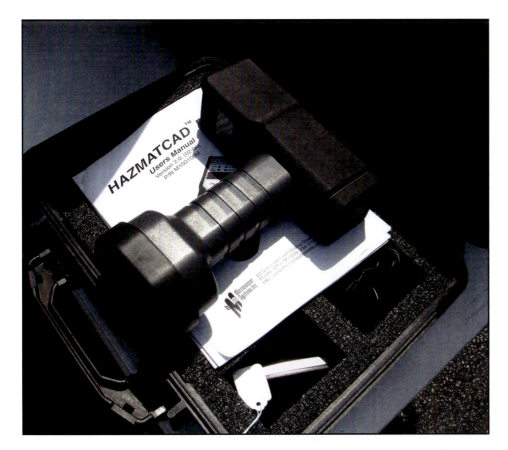

Figure 6.28 Surface acoustic wave sensors may be helpful in detecting nerve and blister agents.

> **Gamma-Ray Spectrometer** — Apparatus used to measure the intensity of gamma radiation as compared to the energy of each photon.

Gamma-Ray Spectrometer

A **gamma-ray spectrometer** determines the energies of the gamma-ray photons emitted by the source **(Figure 6.29)**. Radioactive nuclei (radionuclides) commonly emit gamma rays corresponding to the typical energy levels in nuclei with reasonably long lifetimes. Such sources typically produce gamma-ray "line spectra" (i.e., many photons emitted at discrete energies).

Most radioactive sources produce gamma rays of various energies and intensities. When technicians collect and analyze these emissions with a gamma-ray spectroscopy system, the result can be a gamma-ray energy spectrum. A detailed analysis of this spectrum is typically used to determine the identity and quantity of gamma emitters present in the source. The gamma spectrum is characteristic of the gamma-emitting nuclides contained in the source. Just as in optical spectroscopy, the optical spectrum is characteristic of the atoms and molecules contained in the sample.

Gamma spectroscopy can serve a beneficial purpose in the hazmat response field if this technology is available to the responder. Gamma spectroscopy offers reliable quantitative results for gamma-emitting isotopes. It is also a fast, reliable, and sensitive means for the complete assessment of a gamma radiation field at a fixed location.

After using risk-based response to determine a name, gamma spectroscopy can differentiate between commercial/medical or illicit materials. Keep in mind that if the product is not in the purchased isotope library for the device, it will be unable to detect it.

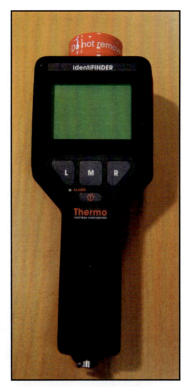

Figure 6.29 A gamma-ray spectrometer identifies the activity level of a source.

Fourier Transform Infrared (FT-IR) Spectroscopy — Device that uses a mathematical process to convert detection data onto the infrared spectrum.

Spectrophotometer — Apparatus used to measure the intensity of light as an aspect of its color.

Fourier Transform IR

Fourier Transform Infrared (FT-IR) spectroscopy is similar to infrared spectroscopy and the principle of operation is also similar. The main difference is in the way the signal is received and processed. A simple optical device called an *interferometer* presents a unique type of signal which has all infrared frequencies encoded into it. The use of the interferometer allows for a quick response.

FT-IR spectroscopy was developed to overcome the limitations that may be found in dispersive instruments. In other IR technology, only one signal is received and processed to indicate the results. In FT-IR, many signals (potentially more than one thousand) are received and then averaged. This result offers much more sensitivity and accuracy.

Fourier Transform IR tends to be extremely accurate for specific materials that are included in the library. Types of materials that are detected reliably via FT-IR include:

- Solids, liquids, gels, and pastes
- Explosives
- Organic compounds
- Mineral acids
- Substances in aqueous solutions
- Crystalline semi-metals
- Fluorescent materials

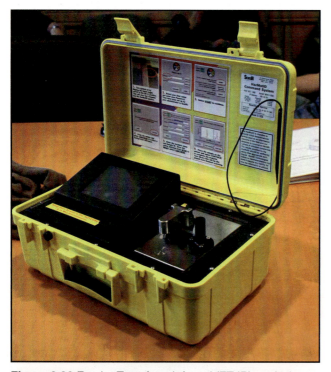

Figure 6.30 Fourier Transform Infrared (FT-IR) analysis requires controlled conditions for accurate results and interpretation.

FT-IR analysis should not take place in the field. This technology uses a limited number of moving parts, is internally calibrated, and must be returned to the factory for calibration **(Figure 6.30)**. It may be necessary to consult the equipment's manufacturer in order to analyze the spectra. This approach, called "reach back" must be established via AHJ policies because it requires specialized interpretation of the equipment's readings. The product must be physically handled in order to obtain a sample. The presence of water may affect the results.

FT-IR technology cannot detect (see) through glass containers. The following chemicals or products can be difficult or near impossible to identify:

- Ionic substances
- Metals
- Elemental substances
- Complex mixtures
- Aqueous solutions
- Some strong acids

Infrared Spectrophotometers

A **spectrophotometer** is a type of photometer that can measure intensity of a beam of light as a function of the light source wavelength (color). Technicians cannot carry this equipment into the field for use. They may stage it in a mobile trailer unit, and a small number of features may be programmed after it is moved. Changes may easily unbalance its calibration. Other field-use instruments available can monitor and accurately identify thousands of compounds in their libraries.

The spectrophotometer operates by passing a beam of light through a given sample and measuring the strength of the light that reaches the detector. Organic compounds absorb infrared radiation at frequencies that are characteristic and different for various molecules, allowing for infrared analysis. Infrared absorbance patterns are distinct enough that a well-trained chemist can often interpret the patterns and identify the characteristics of the molecule, sometimes even identifying the specific molecule and identifying one isomer out of several possible structures. The amount of infrared energy that a sample absorbs can be related directly to the amount of a specific component in that sample.

Infrared instruments specifically designed to analyze samples in the field are available from a number of companies. A significant problem with the analysis of both liquid and solid samples is the impact of water contamination on the ability to obtain useable, interpretable spectra. Water has such a strong, intense infrared spectrum that a significant presence of water in the sample may obscure critical bands and make it impossible for the software to interpret the bands and make identification.

Raman Spectrometer — Apparatus used to observe the absorption, scattering, and shifts in light when sent through a material. The results are unique to the molecule.

Some of these instruments allow a change in the compound being monitored, presuming that the compound is already in the instrument's database. In most cases, setting up the instrument to measure a compound that is not already in the database is difficult, if not impossible, in the field and will often require laboratory preparation of standard gas mixtures to allow the choice of monitoring frequencies and calibration factors.

Gas sampling is typically done by pumping ambient air through a transparent chamber between the infrared light source and a detector. An infrared scan is a plot of the amount of infrared radiation absorbed by the sample as a function of the wavelength of the IR radiation. The IR pattern is sometimes called a "fingerprint" because it is characteristic of a molecule. Similar molecules will display similar, but not identical, IR results.

Raman Spectroscopy IR

In a **Raman spectrometer**, a sample is exposed to a high intensity light **(Figure 6.31)**. The light reflected by, or scattered by, the sample is analyzed for absorbance characteristics of vibrational, rotational, stretching, or bending of the various bonds in the molecule. These energy characteristics are associated with the types of bonds and so they have the same values as the infrared absorbance. Raman spectroscopy is similar to infrared spectroscopy technology.

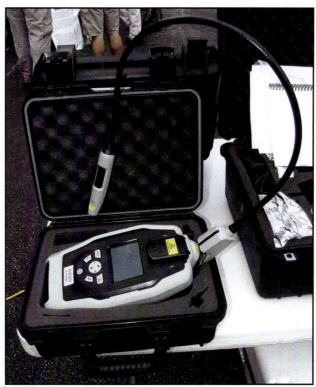

Figure 6.31 Raman spectroscopy uses a high intensity light, such as a laser, to identify materials.

In a Raman spectrometer, a high intensity light irradiates the sample (liquid or solid), and the light reflected or scattered by the sample is analyzed. Some of the energy from the light source is taken up by the sample and is "redistributed" between the various rotational, vibrational, stretching, and bending modes of the molecule. The net result is that the light reflected or scattered by the sample has "absorbance" bands characteristic of that specific molecule. Since these are the same energy absorbance as the infrared spectra, the interpretation of the Raman spectrum is quite similar to the IR. However, some of the vibrational, rotational, bending, and stretching frequencies present in the infrared spectrum will be missing in the Raman spectrum. The theoretical explanation as to why all of the same frequencies are not present in both infrared and Raman spectra is far beyond the level of this text. Raman spectra also tend to be "less intense" than the corresponding infrared spectra.

Raman spectroscopy has one significant advantage over infrared in that water does not have a Raman absorption spectrum. This is significantly different from infrared spectroscopy where water is a strong absorber — often absorbing so much of the infrared energy that no interpretable spectrum can be observed by infrared. Raman spectroscopy can often accurately analyze solid samples containing significant amounts of water and aqueous solutions. Using Raman technology in concert with FT-IR technology will aid the identification process.

NOTE: Field screening should detect water before a material is tested via Raman spectroscopy.

On the negative side, Raman spectroscopy is limited to the detection of visible quantities of liquids or solids. Depending on the model, most products using this technology will only detect substances in clear containers. Opaque containers such as brown bottles will not work. The use of a laser in this technology will also offer some limitations. The unshielded laser light may be an eye hazard and black or dark materials will not be able to be identified because they will absorb the laser light. Also, the laser may ignite flammables. Reducing the sample size and lowering the laser output and scan delay can help to reduce this risk.

NOTE: The Raman spectrometer will provide an estimated time to analyze the sample. The longer this sample estimate, the less potential for an accurate spectra.

CAUTION
Risks associated with Raman spectrometry include eye hazards from unshielded lasers and ignition of flammable materials.

Hazards of Raman Spectroscopy
Explosive materials may ignite or explode when exposed to a laser. If the equipment allows, technicians should apply the lowest laser setting along with a scan delay. This procedure will increase the necessary number of scans to the process, but will protect the hazmat technician from delivering too much energy that could cause ignition or detonation.

Mercury Detection

Mercury is a liquid metal that behaves like a solid in terms of vapor pressure. During risk-based response, technicians may detect mercury as a solid.

Mercury releases into the atmosphere primarily via burning fossil fuels. Once in the atmosphere, it is deposited on land and sea, where it eventually enters the food chain in fish or shellfish. Mercury is then transformed to methyl mercury and preconcentrated to potentially dangerous levels. Mercury can be found in industry and has a presence in laboratories. Mercury is used in thermometers and barometers, and in some floats and switches.

Mercury Incident, Williamson County, Texas, December 2014

On the afternoon of December 15, a resident called the regional office of the Texas Commission on Environmental Quality. The resident had spilled mercury pellets in his garage a few months earlier, and wanted to be sure that his attempt at cleaning the spill was effective. The spilled amount was equivalent to 5-6 thermometers in volume, leftover from his father's gold mining operation.

He was asked to restrict access to the garage, and to expect an EPA team the next morning. On further interactions, the resident indicated that a few weeks after the spill, his three children had shown signs of fifth disease, a common and mild viral infection. One child had more persistent symptoms with no apparent cause, and at the beginning of the month, started having seizures. In the past few days, all three children were ultimately hospitalized with symptoms of chronic mercury poisoning, and responded well to treatment. On the day of the call, the children had been tested for mercury, with higher than normal results. Neither of the parents showed symptoms.

On December 16, the house was tested for mercury vapor, and readings were well above the allowable reading of one microgram per cubic meter ($\mu g/m^3$) [.1 ppb] on the Jerome 505 Mercury Analyzer. In one ice chest, the readings were over 200 $\mu g/m^3$ (20 ppb) and most of the house was over 7 $\mu g/m^3$ (.7 ppb). The Texas Commission on Environmental Quality (TCEQ) and Environmental Protection Agency (EPA) barred the family from the house while clean-up work was underway, and worked out a plan to conduct the work without billing the family. The rest of the day was spent removing items from the garage to see if the readings would come down.

On December 17, the EPA brought more sensitive equipment. The Mercury Vapor Analyzer (Lumex RA-915+) could not show readings higher than 50 $\mu g/m^3$ (5 ppb) but was able to help identify locations of the remaining mercury contamination. The first focus was the wall separating the garage and the house. With all items removed from the garage, the readings were still over 50 $\mu g/m^3$ (5 ppb).

On December 18, the base plate of the garage wall was identified as the place with the highest readings. A firefighter crew was called in to evaluate the structural stability of the wall. They braced the wall so the base plate could be removed. Under the baseplate, the clean-up crew found a silver dollar sized puddle of mercury. When it was removed, the readings were still over 10 $\mu g/m^3$.

Over the 19th through 23rd, work continued with the garage warming to the temperatures that would be needed for a valid mercury reading: around

> 85° Fahrenheit (29 °C). Items were sorted for return to the family versus disposal based on their readings. During this process, a drain was identified as having the highest remaining readings: over 16 µg/m³ (1.6 ppb) where the rest of the house was closer to 1.6 µg/m³ (.16 ppb). A decontamination powder, HgX®, was left in the drains over the holiday break.
>
> On December 27, the drain was treated again with HgX®, and this time the whole house tested under 1 µg/m³ (.1 ppb) at temperatures between 80° and 90° Fahrenheit (27 and 32 °C). Results were finally good enough that the EPA was called to conduct their eight-hour test the next day. Results from that test were within acceptable parameters, and the house was returned to the residents on December 30.
>
> **Lessons Learned:** Mercury is a highly potent toxin, and chronic exposure may not be immediately obvious. Mercury is also difficult and time-consuming to clean up, even with the correct materials and resources.
>
> **Source:** Williamson County Hazardous Materials Response Team

An advantage to mercury detection is that is has a wide detection range. This technology is prone to few interferences. If technicians suspect the presence of mercury, field screening can be helpful. Two methods of field screening for mercury include the use of a silica optical cell, or a gold film detector.

Technicians can detect mercury concentration using an optical cell made of high purity fused silica. In this process, a sample is drawn continuously through the detector, and ultraviolet light is absorbed and measured. This so-called "cold vapor" measuring method is extremely sensitive for mercury determination and has been used successfully for many years. Sensitivity of cold vapor detections is about 0.1 µg/m³ (0.01 ppb).

> **Adsorb** — The collection of a liquid or gas on the surface of a solid in a thin layer.

Another method of detecting mercury is by the use of gold film detectors. The gold film sensor is inherently stable and selective to mercury. When the sample cycle activates, the internal pump draws a precise volume of air over the sensor. Mercury in the sample **adsorbs** on the surface of the gold film, and the sensor registers it as proportional change in electrical resistance of the gold film. Sensitivity of the detector is about 0.003 mg/m³ (3 µg/m³ [.3 ppb]).

The gold film becomes saturated over time, and may be contaminated by touching surfaces. Gold film sensors may be cleaned (regenerated) by heating the mercury off of the sensor. This process is called regeneration. The process may be time-consuming, and it requires the use of AC power that may not be available at the emergency scene.

Wet Chemistry

Wet chemistry encompasses many types of kits and strips, some of which have already been discussed in this chapter. Some wet chemistry kits contain portable chemistry sets designed to enable logical and progressive testing of a sample in order to identify a chemical or product.

Wet chemistry predates FT-IR and Raman spectroscopy in terms of resources available to field technicians. A number of field test kits are available, including test strips specific for different potential contaminants. Depending on the system being used, it may be possible to identify hazardous substances and common products. Responders must be aware that most tools and testing equipment may not readily identify unknown materials.

The AHJ must determine, based on risk-based response, what functions of wet chemistry should be used within their system. Some kits require extensive training. Depending on the AHJ, wet chemistry approaches that may be used include:

- Reagent tests
- Chemical test strips
- pH paper
- Multifunctional test strips
- HAZCAT
- HEINZ 5 STEP™ Kit
- EPA Field Screening Kit

DNA Fluoroscopy

DNA fluoroscopy devices are used in the detection of biological agents. They are designed to detect specific DNA sequences that are helpful in the detection of protein-based biological substances.

Fluorimeters can be used to determine whether a sample contains DNA and RNA because the pyrimidine and purine ring structures in these molecules are fluorescent.

A sample that contains baby powder (talc), corn starch, or other material that does not contain DNA will not fluoresce. A sample of biologically active material will contain DNA or RNA. This equipment is not portable in a conventional sense. It cannot be manually carried into the field.

NOTE: Biotoxins such as ricin or botulinus toxin will not contain DNA or RNA and will not show up in a fluorimeter analysis.

Polymerase Chain Reaction Devices

Technicians can use a **polymerase chain reaction (PCR)** to quickly amplify or copy segments of DNA, as well as identify DNA. Devices using PCR can detect and identify biological agents and toxins.

PCR is an automated process that amplifies a segment of DNA. The sample segment is heated until it separates (denatures) into two pieces. This process results in a duplication of the original DNA sample with each of the samples containing a new and old strand of DNA. Each of these strands can be used to create new samples, and the process can go on exponentially. This automated process is relatively quick and is conducted by a machine called a thermocycler. The thermocycler is programmed to alter the temperature of the reaction every few minutes to allow the DNA denaturing and synthesis.

Polymerase chain reaction devices have evolved as a widely used methodology for biological investigation because they can detect and quantify trace samples of DNA. While accurate, its field use is limited. In addition, it takes an average of 30 minutes to process the test samples.

> **Fluorimeter** — Device used to detect the fluorescence of a material, especially as pertains to the fluorescent qualities of DNA and RNA.
>
> **Polymerase Chain Reaction (PCR)** — Technique in which DNA is copied to amplify a segment of DNA to diagnose and monitor a disease or to forensically identify an individual.

Chapter Notes

National Institute for Occupational Safety and Health (NIOSH). December 1994. "Immediately Dangerous to Life or Health Concentrations (IDLH)." Immediately Dangerous to Life or Health (IDLH). http://www.cdc.gov/niosh/idlh/idlhintr.html

Occupational Safety and Health Administration (OSHA). (n.d.). Regulations (Standards- 29 CFR). https://www.osha.gov/pls/oshaweb/owadisp.show_document?p_table=STANDARDS&p_id=12716

Chapter Review

1. What is the difference between exposure and contamination?
2. What are exposure limits?
3. What two hazards are oxygen indicators designed to detect?
4. How does temperature affect the calibration of flammable gas sensors?
5. Correction factors for PIDs are useful only when what two variables are known?
6. Name four types of colorimetric detection methods and describe their basic functions.
7. Name three types of radiation detection instruments and describe their methods of detection.
8. What are some limitations of photoionization detectors?
9. What do biological assays test for?
10. What are four temperature monitoring functions that infrared thermometers can perform?
11. What are some other detection devices available and what hazards are they designed to detect?

6-1

Demonstrate proper use of pH meters to identify hazards. [NFPA 1072, 7.2.1]

WARNING: If this skill involves the use of actual hazardous material samples, hazardous materials can cause serious injury or fatality. Appropriate personal protective equipment (PPE) must be worn and safety precautions must be followed. The following skill sheet demonstrates general steps; specific hazmat incidents may differ in procedure. Always follow the AHJ's procedures for specific incidents.

Step 1: Ensure proper detection, monitoring, or sampling method and equipment is chosen.
Step 2: Ensure that all responders are wearing appropriate PPE.
Step 3: Perform initial inspection to ensure device is serviceable.
Step 4: Turn on the pH meter.
Step 5: Remove the protective cap from the electrode.

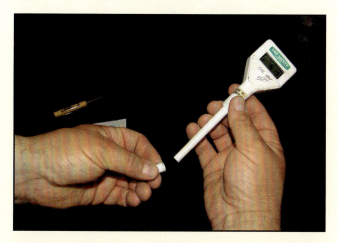

Step 6: Calibrate the pH meter in a test solution with a known pH per manufacturer's instructions.

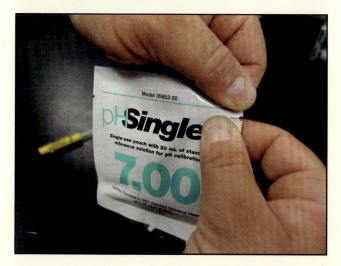

Step 7: Once calibrated, rinse and return the electrode to its operational state per manufacturer's instructions.

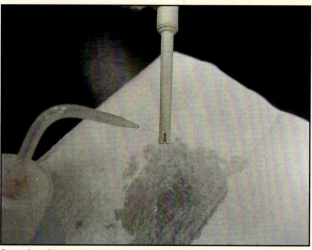

Step 8: Place the electrode in the liquid to be tested and make note of the reading.

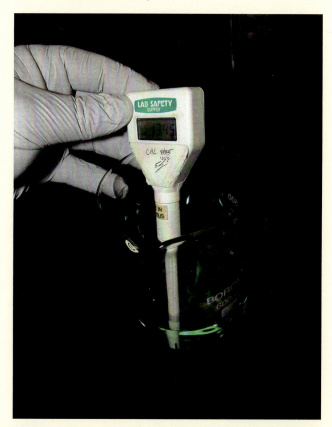

Step 9: Report readings according to AHJ's procedures.
Step 10: Remove electrode from the liquid, rinse, and return to operational state per manufacturer's instructions.
Step 11: Replace the protective cap on the electrode.
Step 12: Turn off the pH meter.
Step 13: Decontaminate equipment and return to operational state per manufacturer's instructions.
Step 14: Complete required reports and supporting documentation.

SKILL SHEETS 6-2

Demonstrate proper use of a multi-gas meter (carbon monoxide, oxygen, combustible gases, multi-gas, and others) to identify hazards. [NFPA 1072, 7.2.1]

WARNING: If this skill involves the use of actual hazardous material samples, hazardous materials can cause serious injury or fatality. Appropriate personal protective equipment (PPE) must be worn and safety precautions must be followed. The following skill sheet demonstrates general steps; specific hazmat incidents may differ in procedure. Always follow the AHJ's procedures for specific incidents.

NOTE: Specific procedures will vary depending on the equipment used. Refer to the manufacturer's instructions for complete directions.

Step 1: Ensure proper detection, monitoring, or sampling method and equipment is chosen.

Step 2: Ensure that all responders are wearing appropriate PPE.

Step 3: Perform initial inspection to ensure device is serviceable.

Step 4: Select the monitor and identify the gases it will detect.

Step 5: Perform a bump test to ensure the meter is functioning properly.

Step 6: Perform a "fresh air" calibration of the monitor prior to entry.

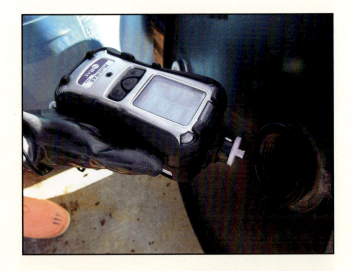

202 Chapter 6 • Analyzing the Incident: Detecting, Monitoring, and Sampling Hazardous Materials

6-2

Demonstrate proper use of a multi-gas meter (carbon monoxide, oxygen, combustible gases, multi-gas, and others) to identify hazards. [NFPA 1072, 7.2.1]

Step 7: Properly monitor the area per AHJ's requirements.

Step 8: Report results according to AHJ's requirements.
Step 9: When monitoring is complete, turn off the instrument.
Step 10: Decontaminate equipment and return to operational state per manufacturer's instructions.
Step 11: Complete required reports and supporting documentation.

SKILL SHEETS

6-3

Demonstrate proper use of colorimetric tubes to identify hazards. [NFPA 1072, 7.2.1]

WARNING: If this skill involves the use of actual hazardous material samples, hazardous materials can cause serious injury or fatality. Appropriate personal protective equipment (PPE) must be worn and safety precautions must be followed. The following skill sheet demonstrates general steps; specific hazmat incidents may differ in procedure. Always follow the AHJ's procedures for specific incidents.

NOTE: Specific procedures will vary depending on the equipment used. Refer to the manufacturer's instructions for complete directions.

Tube System

Step 1: Ensure proper detection, monitoring, or sampling method and equipment is chosen.

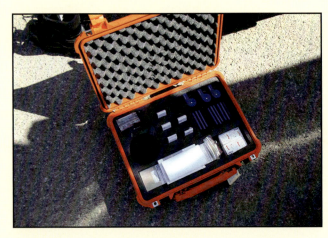

Step 2: Ensure that all responders are wearing appropriate PPE.

Step 3: Use the manufacturer's instruction manual to select the proper colorimetric tube for sampling and check expiration dates for the tube.

Step 4: Perform a functional test to check device per manufacturer's instructions to ensure correct operation.

Step 5: Reset the counter.

Step 6: Properly break both ends off of the tube(s) using the provided tube cutter.

Step 7: Insert the tube into the hand pump in the proper direction.

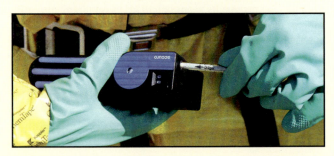

Step 8: Hold the tip of the tube an appropriate distance away from the product or container opening, taking care not to come into contact with any solid or liquid product.

6-3
Demonstrate proper use of colorimetric tubes to identify hazards. [NFPA 1072, 7.2.1]

Step 9: Sample the product based on manufacturer's instructions.

Step 10: Remove the tube from the pump and read, interpret, and record the results per manufacturer's instructions.

Step 11: Dispose of sampling tube in accordance with appropriate regulations.

CAUTION: Used tubes may be a hazardous waste and/or sharps hazard.

Step 12: Decontaminate equipment and return to operational state per manufacturer's instructions.

Step 13: Complete required reports and supporting documentation.

SKILL SHEETS 6-4

Demonstrate proper use of pH paper to identify hazards. [NFPA 1072, 7.2.1]

WARNING: If this skill involves the use of actual hazardous material samples, hazardous materials can cause serious injury or fatality. Appropriate personal protective equipment (PPE) must be worn and safety precautions must be followed. The following skill sheet demonstrates general steps; specific hazmat incidents may differ in procedure. Always follow the AHJ's procedures for specific incidents.

Step 1: Ensure proper detection, monitoring, or sampling method and equipment is chosen.

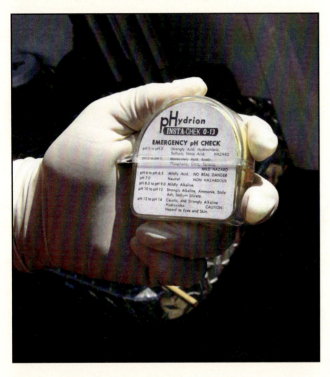

Step 2: Ensure that all responders are wearing appropriate PPE.
Step 3: Inspect the paper to ensure it has not been exposed or expired.
Step 4: Remove a piece of appropriate size pH paper from the roll or remove strip from the container and secure the paper.

Step 5: Sample the product.
NOTE: If monitoring vapors or gases, pH paper should be wetted.

Step 6: Compare results to pH paper color scale to determine if the product is an acid, a base, or neutral. Record results.

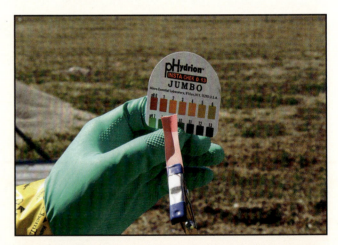

NOTE: Confirmation of a corrosive atmosphere will eliminate the use of electronic meters for further testing.

Step 7: Report results according to AHJ's procedures.
Step 8: Dispose of pH papers in accordance with appropriate regulations.
Step 9: Decontaminate equipment and return to operational state per manufacturer's instructions.
Step 10: Complete required reports and supporting documentation.

6-5
Demonstrate proper use of reagent test paper to identify hazards. [NFPA 1072, 7.2.1]

WARNING: If this skill involves the use of actual hazardous material samples, hazardous materials can cause serious injury or fatality. Appropriate personal protective equipment (PPE) must be worn and safety precautions must be followed. The following skill sheet demonstrates general steps; specific hazmat incidents may differ in procedure. Always follow the AHJ's procedures for specific incidents.

Step 1: Ensure proper detection, monitoring, or sampling method and equipment is chosen.

Step 2: Ensure that all responders are wearing appropriate PPE.
Step 3: Inspect the test paper to ensure it has not been exposed or expired.
Step 4: Remove a piece of appropriate size reagent test paper.

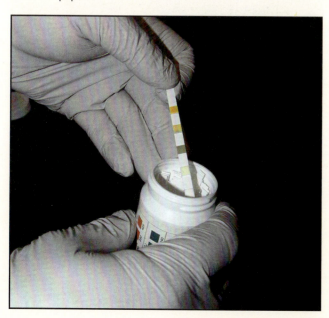

Step 5: Sample the product.
NOTE: Reagent test paper should be wetted per manufacturer's recommendations.

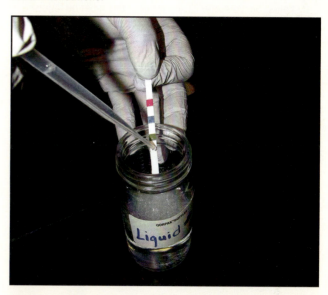

Step 6: Identify any color changes to the reagent test paper and compare them with the provided reference.

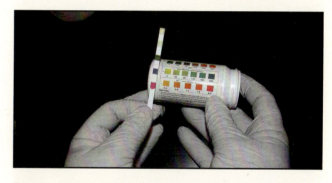

Step 7: Report results according to AHJ's procedures.
Step 8: Dispose of reagent test paper in accordance with appropriate regulations.
Step 9: Decontaminate equipment and return to operational status per manufacturer's instructions.
Step 10: Complete required reports and supporting documentation.

SKILL SHEETS

6-6

Demonstrate proper use of radiation detection instruments to identify and monitor hazards. [NFPA 1072, 7.2.1]

WARNING: If this skill involves the use of actual hazardous material samples, hazardous materials can cause serious injury or fatality. Appropriate personal protective equipment (PPE) must be worn and safety precautions must be followed. The following skill sheet demonstrates general steps; specific hazmat incidents may differ in procedure. Always follow the AHJ's procedures for specific incidents.

NOTE: Specific procedures will vary depending on the equipment used. Refer to the manufacturer's instructions for complete directions.

Step 1: Ensure proper detection, monitoring, or sampling method and equipment is chosen.

Step 2: Ensure that all responders are wearing appropriate PPE.
Step 3: Select the appropriate monitor for the potential hazard(s).
Step 4: Perform initial inspection to ensure device is serviceable.

Step 5: Ensure that the monitor has been maintained and appropriately calibrated according to AHJ's SOPs and manufacturer's instructions.

Step 6: Turn on the meter and test detector against check source.
Step 7: Acquire background radiation levels.

208 Chapter 6 • Analyzing the Incident: Detecting, Monitoring, and Sampling Hazardous Materials

6-6

Demonstrate proper use of radiation detection instruments to identify and monitor hazards. [NFPA 1072, 7.2.1]

Step 8: Properly monitor the area per AHJ's requirements.

Step 9: Determine the presence of ionizing radiation.

Step 10: Compare radiation values to AHJ's SOPs. Record results.

Step 11: Report results according to AHJ's requirements.

Step 12: When monitoring is complete, turn off the instrument.

Step 13: Decontaminate equipment and return to operational state per manufacturer's instructions.

Step 14: Complete required reports and supporting documentation.

SKILL SHEETS 6-7

Demonstrate proper use of dosimeters to identify personal dose received. [NFPA 1072, 7.2.1]

WARNING: If this skill involves the use of actual hazardous material samples, hazardous materials can cause serious injury or fatality. Appropriate personal protective equipment (PPE) must be worn and safety precautions must be followed. The following skill sheet demonstrates general steps; specific hazmat incidents may differ in procedure. Always follow the AHJ's procedures for specific incidents.

Step 1: Ensure proper detection, monitoring, or sampling method and equipment is chosen.

Step 2: Ensure that all responders are wearing appropriate PPE.

Step 3: Perform initial inspection to ensure device is serviceable.
Step 4: Ensure the dosimeter is properly calibrated.
NOTE: The dosimeter should be logged to you.

Step 5: Ensure the dosimeter reads zero.
Step 6: Don the dosimeter per manufacturer's instructions.

Step 7: Perform mission activity.

Step 8: Doff the dosimeter.
Step 9: Follow manufacturer's instructions and AHJ's procedures regarding dosimeter analysis.
Step 10: Report results per AHJ's procedures.
Step 11: Decontaminate equipment and return to operational state per manufacturer's instructions.
Step 12: Complete required reports and supporting documentation.

6-8

Demonstrate proper use of photoionization detectors to identify hazards. [NFPA 1072, 7.2.1]

WARNING: If this skill involves the use of actual hazardous material samples, hazardous materials can cause serious injury or fatality. Appropriate personal protective equipment (PPE) must be worn and safety precautions must be followed. The following skill sheet demonstrates general steps; specific hazmat incidents may differ in procedure. Always follow the AHJ's procedures for specific incidents.

NOTE: Specific procedures will vary depending on the equipment used. Refer to the manufacturer's instructions for complete directions.

Step 1: Ensure proper detection, monitoring, or sampling method and equipment is chosen.

Step 2: Ensure that all responders are wearing appropriate PPE.

Step 3: Perform initial inspection to ensure device is serviceable.

Step 4: Perform a "fresh air" calibration.

Step 5: Operate the device per manufacturer's instructions and AHJ's procedures.

Step 6: Properly monitor the area per AHJ's requirements.

Step 7: Identify correction factors and apply them as necessary.

Step 8: Report results per AHJ's requirements.

Step 9: When monitoring is complete, turn off the device.

Step 10: Decontaminate equipment and return to operational state per manufacturer's instructions.

Step 11: Complete required reports and supporting documentation.

6-9
Demonstrate the use of a noncontact thermal detection device to identify hazards.
[NFPA 1072, 7.2.1]

WARNING: If this skill involves the use of actual hazardous material samples, hazardous materials can cause serious injury or fatality. Appropriate personal protective equipment (PPE) must be worn and safety precautions must be followed. The following skill sheet demonstrates general steps; specific hazmat incidents may differ in procedure. Always follow the AHJ's procedures for specific incidents.

Step 1: Ensure proper detection, monitoring, or sampling method and equipment is chosen.

Step 2: Ensure that all responders are wearing appropriate PPE.

Step 3: Perform initial inspection to ensure device is serviceable.

Step 4: Insert appropriate batteries and turn on the thermal detector. Select Celsius or Fahrenheit if applicable.

Step 5: Standing as close as safely possible, use the aiming laser to assess the target from the lowest point on the target and scan up slowly and methodically.

Step 6: Record results.

Step 7: When monitoring is complete, turn off the device.

Step 8: Decontaminate equipment and return to operational state per manufacturer's instructions.

Step 9: Return to proper storage per manufacturer's instructions.

Step 10: Complete required reports and supporting documentation.

Analyzing the Incident: Collecting and Interpreting Hazard and Response Information

Chapter Contents

Introduction to the Technical Research Job Function 217
    Product Information and Behavior Prediction 217
    Data Interpretation for PPE Selection 218
    Technical Information and Research Documentation 220

Research 221
    Hazardous Materials Profiling 221
    Chemical Assessment 224

Written Technical References 225
    Reference Manuals 225
    Safety Data Sheet (SDS) 231
    Shipping Papers 232
    Facility Documents 233

Electronic Technical Resources 234
    Computer-Aided Management of Emergency Operations (CAMEO) 234
    Wireless Information System for Emergency Responders (WISER) 236
    Palmtop Emergency Action for Chemicals (PEAC®-WMD) 237
    E-Plan for First Responders 237
    Chemical Hazards Emergency Medical Management (CHEMM) 237
    Mobile Applications (Apps) 237
    National Pipeline Mapping System (NPMS) 238
    Weather Data 239
    Plume Modeling 239
    Other Internet Resources 240

Technical Information Centers and Specialists 240
    Emergency Response Centers 240
    Poison Control Centers 242
    Chlorine Emergency Plan (CHLOREP) 242
    Interagency Radiological Assistance Plan (IRAP) ... 243
    U.S. WMD-Civil Support Teams 243
    EPA Emergency Response Team (ERT) 243
    Explosive Ordnance Disposal (EOD) 243
    FBI WMD Coordinator 243
    Other Federal Resources 244

Chapter Notes 244
Chapter Review 244
Skill Sheets 245

Chapter 7

Key Terms

- CAS® Number 224
- Chemical Assessment 224
- Chemical Inventory List (CIL) 233
- CHLOREP 242
- Computer-Aided Management of Emergency Operations (CAMEO) 234
- Datasheet 234
- Emergency Response Guidebook (ERG) 226
- Geographic Information Systems (GIS) 240
- Globally Harmonized System of Classification and Labeling of Chemicals (GHS) 224
- Hazardous Materials Profile 221
- Local Emergency Planning Committee (LEPC) 233
- Metadata 238
- Pipeline and Hazardous Materials Safety Administration (PHMSA) 232
- Risk-Based Response 217
- Standard Transportation Commodity Code (STCC) 225
- UN/NA Number 224
- Wireless Information System for Emergency Responders (WISER) 236

Analyzing the Incident: Collecting and Interpreting Hazard and Response Information

JPRs Addressed In This Chapter

7.2.2

Learning Objectives

After reading this chapter, students will be able to:

1. List responsibilities of the technical research specialist during a hazmat incident. [NFPA 1072, 7.2.2]
2. Describe types of research undertaken by hazardous material technicians. [NFPA 1072, 7.2.2]
3. Describe written reference materials available to aid in hazardous material research. [NFPA 1072, 7.2.2]
4. Describe electronic technical resources available to aid in hazardous material research. [NFPA 1072, 7.2.2]
5. Describe technical information centers and specialists available to aid in hazardous material research. [NFPA 1072, 7.2.2]
6. Skill Sheet 7-1: Use approved reference resources to interpret hazard and response information. [NFPA 1072, 7.2.2]

Chapter 7
Analyzing the Incident: Collecting and Interpreting Hazard and Response Information

This chapter introduces reference resources that are typically available to technicians at hazmat incidents:

- Introduction to technical research
- Research
- Written technical references
- Electronic technical resources
- Technical information centers and specialists

Introduction to Technical Research

Decision-making at hazardous materials incidents requires precise information regarding the products involved, the container involved, and the environment in which the incident is occurring. The research specialist and the hazmat technician in charge of resource management must obtain, organize, and disseminate information to the Incident Commander (IC) and the other decision-makers within the incident's Command structure. Timely and accurate delivery of important information must follow a **risk-based response**.

The role of the research specialist (also called Science Officer, Research Officer, or Technical Specialist) is important during a hazmat response. In this role, you must properly identify the product at the emergency scene and document the product's chemical and physical properties so that responders can initiate a proper response. Steps for using approved reference resources to interpret hazard and response information are included in **Skill Sheet 7-1**.

The research specialist gathers information needed to understand the hazards presented by the incident. The Incident Commander can then use this information to formulate a risk-based response **(Figure 7.1)**. Risk-based response is discussed further in Chapter 9 of this manual.

Risk-Based Response — Method using hazard and risk assessment to determine an appropriate mitigation effort based on the circumstances of the incident.

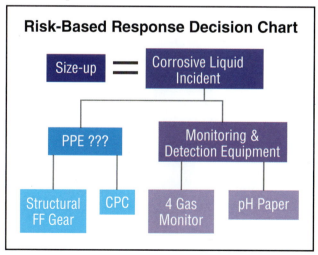

Figure 7.1 Risk-based response decisions must include an evaluation of the conditions that are present at an incident.

Product Information and Behavior Prediction

Responders must understand the chemicals and products involved in a hazmat incident before they can successfully mitigate the incident. Obtaining precise and accurate information on the products involved allows the hazmat technician to make well-informed decisions based on the product's known chemical and physical properties.

Understanding the physical properties of the chemical is extremely important for behavior prediction. It is equally important to understand where the hazmat tech can find that information. Many different chemical references are available, as explained later in this chapter. While most have similar information, some sources specialize in certain chemicals.

U.S. Hazard Communication

U.S. OSHA issues regulations relating to worker safety under Title 29 *CFR*. OSHA regulations of interest to first responders include:

- The HAZWOPER regulation (29 *CFR* 1910.120[q]) which was introduced in Chapter 1 of this manual
- The Hazard Communication regulation (29 *CFR* 1910.1200)
- The Process Safety Management of Highly Hazardous Chemicals regulation (29 *CFR* 1910.119)

The Hazard Communication Standard (HCS) is designed to ensure that information regarding chemical hazards and associated protective measures is disseminated to workers and employers. This dissemination is accomplished by requiring chemical manufacturers and importers to evaluate the hazards of the chemicals they produce or import and provide information about the chemical through labels on shipped containers and safety data sheets (SDSs, formerly called material safety data sheets or MSDSs).

NOTE: SDSs are developed for industry and may not always be useful for emergency response.

Data Interpretation for PPE Selection

There is no single chemical protective clothing (CPC) type or fabric that will protect emergency responders from every chemical **(Figure 7.2)**. CPC is subject to factors such as permeation, penetration, and degradation, and this information is provided in each product's documentation. However, CPC documentation does not provide information on other hazards that can damage the material, such as thermal and mechanical stressors. When taking a risk-based approach for PPE selection, all hazards present must be considered.

WARNING!
Chemical protective clothing (CPC) cannot protect against all hazards.

Figure 7.2 Chemical protective clothing is designed for specific hazards, and will not protect against some other types of hazards.

At an incident, when using risk-based response, technicians need to assess many factors to determine the appropriate PPE. For example, consider factors including **(Figure 7.3)**:

Figure 7.3 PPE selection should be based on a number of factors anticipated at the incident scene, including ambient conditions and considerations affecting the material itself.

- Mission type and duration
- Environment
- Product's state of matter (solid, liquid, or gas)
- Product's temperature
- Product attributes, such as:
  — Toxicity by skin absorption
  — Corrosivity
  — Vapor pressure
  — Flammability
  — Reactive properties
  — Oxidation
  — Radioactive contamination
  — CPC compatibility

These considerations will help determine both the level and type of PPE needed. Performing product monitoring, researching the material's chemical and physical properties, and contacting a technical specialist can help provide additional information.

CAUTION
It is not always possible to determine the product's exact identity.

Technical Information and Research Documentation

Hazmat technicians can find a wide range of useful reference materials. Though all may not be written with the technician in mind, valuable information may be found in professional publications such as:

- Chemical engineers
- Industrial hygienists
- United States Coast Guard
- Department of Defense
- The Association of American Railroads

During the incident's chemical research phase, documentation must be accurate and technicians must take an organized approach to researching chemicals and compounds. Charts and worksheets are excellent tools for organizing the data and ensuring thorough and systematic analysis.

Chemical research documentation will be used for:

- Conducting the hazard assessment concerning the overall health of the general public and the emergency responders
- Determining PPE and respiratory protection
- Aiding decisions about chemical monitoring

Research

Quality research will improve safety at an incident because hazmat technicians will gain a better understanding of the product's physical and hazardous properties and will make well-informed decisions about protective actions. Research can help determine what effects the product will have on humans and the environment, and it allows responders to ensure that the general public is protected. Research also can show the effect that the product will have on chemical protective clothing. Chemical research will ensure that technicians select the proper suit materials and respiratory devices for use at the incident. Two important research elements include a hazardous materials profile and a chemical assessment.

Hazardous Materials Profiling

Conducting a **hazardous materials profile** (recon assessment) is an efficient tool for size-up and hazard analysis. A hazard profile provides the hazmat technician with an organized approach to understanding the physical and hazardous characteristics of a product and allows for a response that is well-informed and risk-based. This information is the core of risk-based response. When profiling a chemical during a hazmat response, the process will be mostly the same regardless whether the chemical is known or unknown.

When the chemical is known, technicians may extract baseline information from reference manuals. Baseline information should be similar to the factors considered while selecting PPE, such as:

- State of matter
- Flash point
- Toxicity
- Vapor density
- Expansion ratio
- Specific gravity
- Polymerization potential
- Explosive limits
- Reactivity
- Vapor pressure
- Ionization potential

Incorporating monitoring and detection equipment into the decision-making process will help validate findings from the reference manuals as well as serve as a guide in a response when a chemical cannot be definitively identified. As long as responders use proper PPE, toxicity concerns are minimized. However, the technician must consider other individuals who may have been exposed to the product.

When relying on technology for hazmat response, the AHJ must consider the broad spectrum of response data. Include the following monitoring and detection devices when conducting a reconnaissance sweep or validating a chemical during profiling (**Figure 7.4**):

- **Radiation detector** — Any radiation reading above normal background levels is a concern. Retreat immediately if elevated levels of radiation are detected.
- **pH paper** — pH paper can be used to detect corrosive gases (methyl amines, a base) and vapors (hydrochloric acid).
- **Reagent paper** — Used to detect numerous substances.

> **Hazardous Materials Profile** — A chemical size-up based upon the suspected identity, or not, of a chemical hazard. This is validated with monitoring and detection equipment upon performing a reconnaissance entry. Profiling allows the hazmat technician to predict hazards and validate the actual entry conditions even if the product is not positively identified. *Also known as* Hazard Profile.

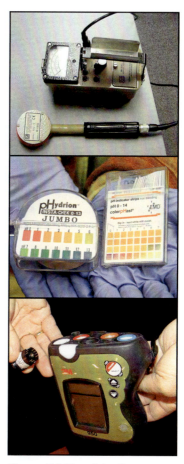

Figure 7.4 Use detection and monitoring equipment to investigate an environment or product.

- **Combustible gas detector (LEL meter)** — LEL meters will detect a flammable atmosphere. LEL sensors are typically contained within a traditional four- or five-gas meter with other oxygen and toxicity sensors.
- **Thermal detector** — Used to determine if a chemical reaction or polymerization is taking place.
- **Photoionization detector** — Used to detect some gases and vapors.

Hazmat profiling is an ongoing process. As new information is gathered, technicians should develop response objectives based upon the profiled information.

While the hazmat technician must collect quality information for decision-making purposes, care must be taken to prevent swamping the decision-making process with unnecessary information **(Figure 7.5)**. Instead, the hazmat technician should be equipped with monitoring and detection technology that can validate and update the known variables as the response continues.

Figure 7.5 A hazardous materials technician must be able to determine the response objectives based on quality information.

Risk-based response must be followed when responding to both known and unknown materials. If a product is known, technology should be used to establish and confirm baseline information during the recon process. If the product is unknown, response personnel must ascertain all possible information, and verify it as possible with technology.

> **CAUTION**
> Safety must always take priority in a hazmat response. Build decisions from validated information only.

It is possible that containers can be mismarked. Shipping or facility documents can have errors or misspellings. In these cases, the information derived from the reference manuals will give the responders incorrect information, a potentially deadly error. It is essential that emergency response crews profile the correct information **(Figure 7.6)**.

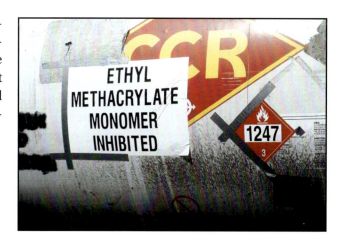

Figure 7.6 This container likely contains ethyl methacrylate monomer (EMA), despite the UN Number placard reference to methyl methacrylate monomer (MMA).

Methyl Methacrylate (MMA) versus Ethyl Methacrylate (EMA)

Methyl methacrylate (MMA) and ethyl methacrylate (EMA) vary somewhat from each other **(Figure 7.7)**. The differences include the size of the molecule, with EMA being larger and less able to penetrate body tissue. The monomers are used for similar purposes. In cosmetics, EMA is used for acrylic fingernail extensions (FDA 2016). In medical application, MMA is used as bone cement, and EMA is used in dental work. In industry, MMA polymer is known as Plexiglas and Lucite, among other trademarked applications.

Figure 7.7 MMA and EMA are different molecules.

Methyl methacrylate is commonly used as an example of a monomer that can create hazardous conditions if it is permitted to polymerize under uncontrolled conditions. Hazards include a rapid increase in temperature that can lead to the catastrophic failure of the monomer's container if the heat is not dissipated adequately.

NOTE: Some chemical reactions can double for every interval of increase in temperature or pressure.

Chemical Assessment — An organized approach at quantifying the risks associated with the potential exposure to the chemical.

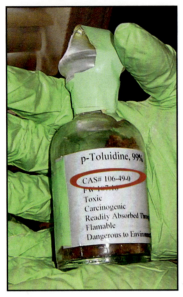

Figure 7.8 The CAS® registry number is a unique identifier for a particular substance.

UN/NA Number — Four-digit number assigned by the United Nations to identify a specific hazardous chemical. North America (DOT) numbers are identical to UN numbers, unless the UN number is unassigned.

CAS® Number — Number assigned by the American Chemical Society's Chemical Abstract Service that uniquely identifies a specific compound.

Globally Harmonized System of Classification and Labeling of Chemicals (GHS) — International classification and labeling system for chemicals and other hazard communication information, such as safety data sheets.

Chemical Assessment

Decision-making at a hazmat incident requires information regarding the product (or products), the container, and the environment in which the incident is occurring. The research specialist must manage all the information resources available when producing a comprehensive **chemical assessment**. Given the urgency to determine the hazards present, the research specialist must have a system to gather information.

The following items should be analyzed to accomplish a rapid and organized chemical assessment:

- **Product identification** — The first thing you should do is identify the product involved in the release or impending release. The following information assists in locating product hazard information:
  — Product name
  — Synonyms
  — Transportation placards and labels
  — **UN/NA number**
  — **CAS® number** (Figure 7.8)
  — Safety data sheets
  — Shipping papers
  — **Globally Harmonized System (GHS)** information
  — **Standard Transportation Commodity Code (STCC)**

- **State of matter** — Due to dispersion characteristics, this has a significant effect on personal protection and tactical operations.

- **Toxic inhalation hazards** — Due to their toxicity and dispersion characteristics, these substances present a substantial life-safety threat and must be protected against accordingly.

- **Flammability hazards** — A substance's flammability will have a significant effect on personal protection and tactical options. During research, obtain the following critical information:
  — Flash point
  — Ignition temperature
  — Flammable or explosive limits

- **Toxicity hazards** — Evaluating the toxicity hazard presented by a substance requires the consideration of the routes of entry and the degree of toxicity, as well as the physical characteristics affecting dispersion. Toxicity will also affect decisions about PPE and tactical options. During research, obtain the following critical information:
  — Carcinogen
  — Teratogen
  — Mutagen
  — Route of entry
  — Exposure limits
  — Corrosivity/contact hazard

- **Dispersion characteristics** — The dispersion method and pattern will affect exposure protection, selection of PPE, and control tactics. Key dispersion characteristics that should be researched include:
  — Vapor pressure
  — Expansion ratio
  — Boiling point
  — Vapor density
  — Specific gravity
  — Water solubility
- **Reactivity characteristics** — Substances may react with other materials present at the incident or with those introduced by response personnel as part of a tactical operation. For example, inserting a wood plug in a vessel containing an oxidizer can cause a reaction. Key reactivity characteristics that should be researched include:
  — Air reactivity
  — Water reactivity
  — Incompatibility
  — Polymerization
  — Temperature sensitivity
- **Radiation** — Is radiation present and at what levels? What is the product's state of matter? Is it removable?

> **Standard Transportation Commodity Code** — Numerical code on the waybill used by the rail industry to identify the commodity. *Also known as* STCC Number.

Written Technical References

Written technical references provide technical data and may also suggest response options. During the response, the responsible party may provide additional reference materials particular to the product. Reference materials provide information such as:

- Chemical technical data
- Physical and chemical properties of the material
- Treatment methods
- Decontamination options

  Types of written technical references include:
- Reference manuals
- Safety data sheets (SDSs)
- Shipping papers
- Facility documents

Reference Manuals

The following sections will introduce reference manuals that technicians commonly use at hazmat incidents. The references listed are commonly encountered in the field. These manuals are either available commercially or through government sources. In addition to print versions, these resources are increasingly available as Internet accessible PDFs via their publishers. This list is not intended to be all-inclusive; the reference manuals used by hazmat technicians will vary depending on the preferences of the AHJ.

Similarities among reference materials often include:

- Basic description of how the manual is organized
- References
- Acronyms
- Glossary of terms

References are often organized for a specialized use. Knowing the intended use will aid a technician in identifying the proper usage of that source at a hazardous materials incident. Be careful to use any reference's most current edition, especially when researching hygienic standards or toxicology data. Materials' physical and chemical properties do not change over time and should remain consistent edition to edition. However, different testing procedures can result in different values. For example, the flash point of a material may be measured via multiple techniques and devices, and each may yield a different result. Use multiple sources to inform your understanding of a topic. The AHJ will dictate the minimum number and type of sources used. Responders should use due diligence to ensure that they thoroughly research a product.

Emergency Response Guidebook (ERG)

The **Emergency Response Guidebook (ERG)** is primarily designed for use at transportation incidents involving hazardous materials. The *ERG* is intended to be a starting point for research, and it only provides general information on included chemicals. Once the chemical is identified, responders should use other references that provide more specific chemical and toxicological information.

The *ERG* provides:

- A quick guide to assist emergency responders in identifying the chemical.
- General information on how responders can protect themselves and others while acquiring more specific information.
- Special information such as polymerization hazards.
- Cross-references based on the UN/NA numbers or the actual spelling of the chemical.
- A primary identification tool that responders should use only at the onset of an incident.

The *ERG* may be the quickest way to determine if a product will polymerize, and can provide an indication whether a material in transport is a radioactive material or a corrosive gas. The following are examples of the chemical groups designated in the *ERG* and their associated guide numbers:

- **Explosives** — Guide numbers: 112, 113, and 114
- **Corrosive Gases** — Guide numbers: 118, 123, 124, and 125
- **Radioactive** — Guide numbers: 161 though 166

The *ERG* also establishes initial isolation distances based on the involved product's state of matter:

Emergency Response Guidebook (ERG) — Manual that aids emergency response and inspection personnel in identifying hazardous materials placards and labels; also gives guidelines for initial actions to be taken at hazardous materials incidents. Developed jointly by Transport Canada (TC), U.S. Department of Transportation (DOT), the Secretariat of Transport and Communications of Mexico (SCT), and with the collaboration of CIQUIME (Centro de Información Química para Emergencias).

- **Solids** – 75 feet (25 m)
- **Liquids** – 150 feet (50 m)
- **Gases** – 330 feet (100 m)

Emergency Handling of Hazardous Materials in Surface Transportation

The *Emergency Handling of Hazardous Materials in Surface Transportation* manual was developed by the Association of American Railroads (AAR) and the Bureau of Explosives (BoE). The manual is divided into two major sections. The first section addresses general information on approaching a hazardous materials incident, product identification, and response information for each U.S. DOT hazard class. The second section consists of commodity-specific emergency response information for each hazardous material regulated by the U.S. DOT, Transport Canada, and the International Maritime Organization. The manual does not provide toxicity data.

The manual contains multiple cross-references. The first contains a list of the DOT UN/NA numbers and page references for many of the commonly shipped materials. The second list is the seven-digit product code that may be referred to as the Standard Transportation Commodity Code (STCC). Finally, the manual includes a cross-reference for all seven-digit product codes, a description of the product assigned to that code, and the page number where it may appear.

NOTE: This reference manual is no longer available new; however, many hazmat teams still use it.

Chemical Hazards Response Information System (CHRIS)

The *Chemical Hazards Response Information System (CHRIS)* manual is designed for U.S. Coast Guard personnel responsible for overseeing incidents involving waterways. This reference system contains multiple manuals, Volume 1 and Volume 2.

Volume 1, The Condensed Guide to Chemical Hazards, is a good resource for first responders, containing information specific to each hazardous material that may be encountered. However, responders must properly identify the material before this information is useful.

Volume 2, The Hazardous Substances Data Manual, may be one of the most useful references when responding to water-based spills and waste sites. Port security personnel use this reference regularly. It contains information on hazardous materials that are commonly shipped in large volume over water. The information is easily understood and may be used as a quick reference to determine the needed actions to safeguard life and property.

NOTE: The U.S. Coast Guard is no longer updating the CHRIS manual. Existing versions may be found on the Internet as PDFs.

NIOSH Pocket Guide to Chemical Hazards (NPG)

The National Institute for Occupational Safety and Health (NIOSH) produces the *NIOSH Pocket Guide to Chemical Hazards (NPG)* **(Figure 7.9)**. This document is based on NIOSH guidelines for chemical hazards and recognized references in the field of industrial hygiene, occupational medicine, and toxicology.

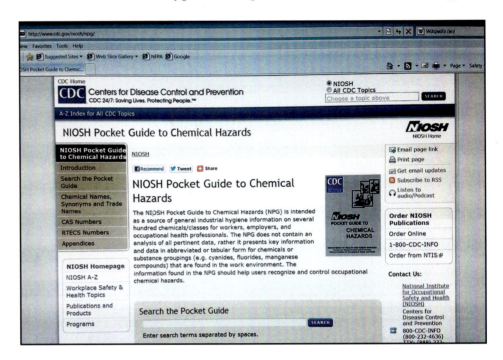

Figure 7.9 The *NIOSH Pocket Guide* is organized to provide quick information about known materials.

This guide is presented in a table format and is intended to provide quick information on general industrial hygiene practices. The *NIOSH Pocket Guide* includes the following information:

- Chemical names, synonyms, trade names, CAS®, RTECS, and DOT ID and Guide numbers
- Chemical formula, conversion factors
- NIOSH Recommended Exposure Limits (RELs)
- OSHA Permissible Exposure Limits (PELs)
- NIOSH Immediately Dangerous to Life and Health (IDLH) values
- Physical description and chemical and physical properties of agents
- Measurement methods
- Personal protection and sanitation recommendations
- Respirator selection recommendations
- Agent incompatibility and reactivity
- Exposure routes, symptoms, target organs, and first aid information

NFPA Fire Protection Guide to Hazardous Materials

The *Fire Protection Guide to Hazardous Materials* produced by the National Fire Protection Association (NFPA) provides good information on the properties of hazardous chemicals. This reference manual is important for those who use chemicals in industry and those who respond to chemical emergencies including fires, accidental spills, and transportation accidents.

Six NFPA documents compose the current edition and can be used to identify not only common chemicals that may be found in transport and industry but other chemicals that may only be found in a laboratory setting. The documents are as follows:

- **NFPA 49, Hazardous Chemicals Data** — Provides information on over 300 chemicals. In addition to the NFPA 704 hazard index markings, it also provides specific information on health, fire, and reactivity data. All chemicals in this section are arranged alphabetically by DOT shipping names.

- **NFPA 325, Guide to Fire Hazard Properties of Flammable Liquids, Gases, and Volatile Solids** — Provides fire hazard properties for over 1,300 flammable substances.

- **NFPA 491, Guide to Hazardous Chemical Reactions** — Includes information on over 3,500 mixtures of two or more chemicals that may be potentially dangerous at moderate or elevated temperatures.

- **NFPA 704, Standard System for the Identification of the Hazards of Materials for Emergency Response** — Provides the degree of health, flammability and reactivity for many common industry chemicals. The newest edition also includes information on aerosol products.

- **NFPA 497, Recommended Practice for the Classification of Flammable Liquids, Gases, or Vapors and of Hazardous (Classified) Locations for Electrical Installations in Chemical Process Areas** — Includes information on certain flammable gases and vapors, flammable liquids, and combustible liquids for purposes of selecting appropriate electrical equipment in hazardous locations.

- **NFPA 499, Recommended Practice for the Classification of Combustible Dusts and of Hazardous (Classified) Locations for Electrical Installations in Chemical Process Areas** — Provides information about certain combustible dusts for purposes of selecting appropriate electrical equipment in hazardous locations.

Hawley's Condensed Chemical Dictionary (CCD)

Hawley's Condensed Chemical Dictionary (CCD) is a compilation of technical information and descriptions that cover thousands of chemicals. The *CCD* is organized to offer three types of information:

- Technical descriptions of chemicals and processes
- Expanded definitions of chemicals and terminology
- Descriptions and/or identification of trade names

Jane's CBRN Response Handbook

Jane's CBRN Response Handbook, an expanded version of *Jane's Chem-Bio Handbook* (which is no longer in publication), is a field reference guide that allows responders to quickly reference chemicals, biological, or radiological weapons and their effects. Its contents include information about the following:

- Preincident planning
- Postincident management
- Biological agents and treatment
- On-scene procedures
- Chemical agents and treatment
- Radiological hazards and treatment

Merck Index

The *Merck Index* is a publication of the Royal Society of Chemistry that is intended to serve chemists, pharmacists, physicians, and other allied health professionals. This resource offers the emergency responder vital information on chemicals, drugs, and biological information. The *Merck Index Online* can be searched for over 500 organic named reactions **(Figure 7.10)**.

Figure 7.10 The *Merck Index Online* can be searched quickly using various types of information.

The *Merck Index* supplies concise information on thousands of chemicals and compounds. Records in the *Merck Index* contain the following information:

- Chemical, common, generic, and systematic names
- Trademarks and associated companies
- CAS® registry numbers
- Molecular formulas, weights, and percentage compositions
- Capsule statements identifying compound classes and scientific significance
- Chemical, biomedical, and patent literature references
- Physical and toxicity data
- Therapeutic and commercial uses
- Caution and hazard information

Sax's Dangerous Properties of Industrial Materials

Sax's Dangerous Properties of Industrial Materials is a multiple volume reference set that provides hazard information for up to 28,000 chemicals. It provides relevant information on toxicology, reactivity, explosive potential, and regulatory information.

The first volume contains a CAS® number and synonym cross index. The remaining volumes contain the following descriptive information:

- Dangerous Properties of Industrial Materials (DPIM) codes
- Hazard ratings
- Entry name
- CAS® numbers
- DOT (UN/NA) numbers
- Molecular formula
- Line structure formula (a graphic representation of the molecular structure)

- Material descriptions
- Toxicology data

Symbol Seeker: Hazard Identification Manual

Symbol Seeker: Hazard Identification Manual helps responders identify many international hazard markings **(Figure 7.11)**. This global reference provides a means to identify hazardous materials in any transportation mode or on any fixed facility. This reference does not offer any specific information on chemicals, substances, or products beyond identification systems. Systems within this reference include:

- Hazmat marking systems
- Nuclear and radioactive marking systems
- Biohazard marking systems
- Ordnance and weapon marking systems
- Marking systems specific to individual countries

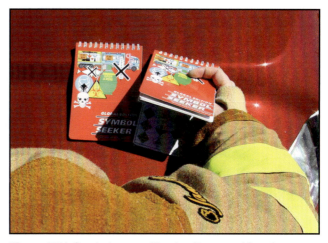

Figure 7.11 Symbol recognition is often considered an Awareness Level skill, but this resource may help a technician identify markings not commonly encountered.

Threshold Limit Values & Biological Exposure Indices (TLVs® & BEIs®)

The *Threshold Limit Values and Biological Exposure Indices (TLVs® & BEIs®)* are published by the American Conference of Governmental Industrial Hygienists (ACGIH®). Designed for use by industrial hygienists and emergency responders, the indices provide occupational exposure guidelines for hundreds of chemical substances and physical agents.

Cryogenic Heat Transfer

Heat transfer associated with cryogenic liquid spills must be well understood while responding to incidents involving cryogenic containers. Current computer-aided design models can yield data specific to one material at a time (Barron).

Safety Data Sheet (SDS)

A safety data sheet (SDS) is a detailed information bulletin prepared by a chemical's manufacturer or importer that provides specific information about the product. Similar to a material safety data sheet (MSDS), SDSs are formatted according to Globally Harmonized System (GHS) specifications. Sample templates for safety data sheets can be found online. While both the U.S. and Canada mandate SDSs, responders may still see MSDSs developed to American National Standards Institute (ANSI) standards, OSHA MSDS standards, Canadian MSDS standards, or the new GHS standard. SDSs are introduced in Chapter 2 of this manual.

SDSs are often good sources of detailed information about a particular material to which emergency responders have access. The sheets can be acquired from the material's manufacturer, the supplier, the shipper, an emergency response center such as CHEMTREC®, or the facility hazard communication plan. SDSs are sometimes attached to shipping papers and containers as well.

The GHS for Hazard Classification and Communication specifies minimum information to be provided on SDSs. These sheets are used worldwide. SDSs must include the following sections:

1. Identification
2. Hazard(s) identification
3. Composition/information on ingredients
4. First aid measures
5. Fire fighting measures
6. Accidental release measures
7. Handling and storage
8. Exposure controls/personal protection
9. Physical and chemical properties
10. Stability and reactivity
11. Toxicological information
12. Ecological information
13. Disposal considerations
14. Transport information
15. Regulatory information
16. Other information

Shipping Papers

Shipping papers or bills of lading are required to accompany a product that is in transit, regardless of the means of transportation **(Figure 7.12)**. Shipping papers can be a reliable means of product identification for chemicals in transit. The information can be provided on a bill of lading, waybill, or similar document. The exact location of the documents varies. Hazardous waste shipments must be accompanied by the Environmental Protection Agency's (EPA) Uniform Hazardous Waste Manifest document which is typically attached to the shipping papers.

Figure 7.12 Shipping papers may include a number of documents that together meet the requirements of the AHJ.

Per the **Pipeline and Hazardous Materials Safety Administration (PHMSA)**, the Basic Description provided in shipping papers will follow a sequence indicated by the acronym, ISHP:

- I = Identification Number
- S = Proper Shipping Name
- H = Hazard Class or Division
- P = Packing Group

For products shipped by rail, the Standard Transportation Commodity Code (STCC) must be included. If a material is hazardous, the number 49 will precede the STCC number. If a material is considered hazardous waste, the number 48 will precede the STCC number. This numbering system will not appear on the tank, but will be included on the cargo manifest or waybill.

Pipeline and Hazardous Materials Safety Administration (PHMSA) — Branch of the U.S. Department of Transportation (DOT) that focuses on pipeline safety and related environmental concerns.

Bulk and facility containers may contain a lower volume or pressure of a product than the containers can hold, for a number of reasons including prior use from that container or to meet weight requirements. To determine an accurate pressure measurement, containers may be equipped with self-reading sensors. For some containers, the level of lading may be measured using a thermal imager.

Paperless Hazard Communication Program

PHMSA is currently developing a program for paperless communications, referred to as HM-ACCESS (hazardous materials automated cargo communications for efficient and safe shipments). The intent of this program is to improve the speed and accuracy of product shipment updates and alerts. Ultimately, the program will affect transportation of materials via roadway, rail, maritime, and air. The program is still in pilot testing as of this manual's printing.

Source: http://www.phmsa.dot.gov/portal/site/PHMSA/

Facility Documents

OSHA's Hazard Communication Standard (HCS) requires U.S. employers to maintain **Chemical Inventory Lists (CILs)** of all their hazardous substances **(Figure 7.13)**. Because CILs usually contain information about the locations of materials within a facility, they can be useful tools in identifying containers that may have damaged or missing labels or markings, such as a label or marking made illegible because of fire damage.

Chemical Inventory List (CIL) — Formal tracking document showing details of stored chemicals including location, manufacturer, volume, container type, and health hazards.

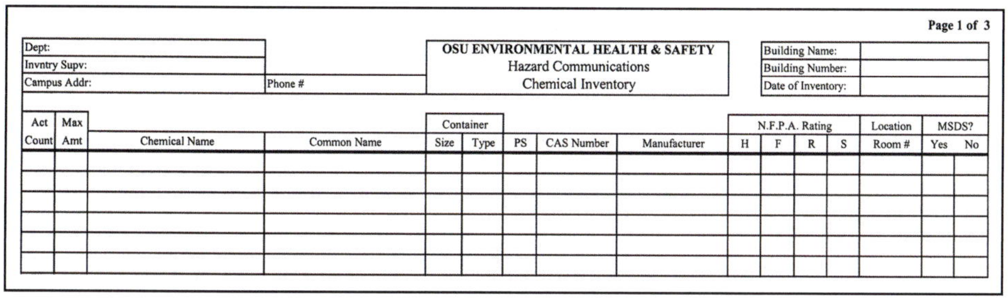

Page 1 of 3

Dept:
Invntry Supv:
Campus Addr: Phone #

OSU ENVIRONMENTAL HEALTH & SAFETY
Hazard Communications
Chemical Inventory

Building Name:
Building Number:
Date of Inventory:

Act Count	Max Amt	Chemical Name	Common Name	Container Size	Container Type	PS	CAS Number	Manufacturer	N.F.P.A. Rating H	N.F.P.A. Rating F	N.F.P.A. Rating R	N.F.P.A. Rating S	Location Room #	MSDS? Yes	MSDS? No

Figure 7.13 The chemical inventory list in a fixed facility will help technicians identify a product.

Several other documents and records may provide information about hazardous materials at a facility such as the following:

- Shipping and receiving documents
- Inventory records
- Risk management and hazardous communication plans
- Chemical inventory reports (known as Tier II reports)

The **Local Emergency Planning Committee (LEPC)** is another potential source of information. Emergency response plans developed by LEPCs are a good source of information for emergency responders.

Local Emergency Planning Committee (LEPC) — Community organization responsible for local emergency response planning. Required by SARA Title III, LEPCs are composed of local officials, citizens, and industry representatives with the task of designing, reviewing, and updating a comprehensive emergency plan for an emergency planning district; plans may address hazardous materials inventories, hazardous material response training, and assessment of local response capabilities.

LEPCs were designed to provide a forum for emergency management agencies, responders, industry, and the public to accomplish the following tasks:

- Work together to evaluate, understand, and communicate chemical hazards in the community
- Develop appropriate plans in case these chemicals are accidentally released

Electronic Technical Resources

Technical resources and references have advanced with technology. Many common written resources and references are now available in an electronic format. Electronic resources have the added benefit of search features that allow technicians to access information in a more efficient manner than with print resources. When looking for electronic resources, keep in mind that the availability of resources will depend upon the brand and type of device used. Some resources may be available only on certain operating systems such as Windows, OS X (Macintosh), or Linux, or on certain mobile devices such as iOS (iPhone or iPad) or Android.

Local electronic resources may include GIS information and maps. Such information may include pipeline and utility locations, street maps, building blueprints, and topography maps.

Technicians can also access many references on smartphones and mobile devices **(Figure 7.14)**. While electronic resource use is increasing, it may still be necessary to have print resources available should there be issues in accessing the electronic data.

Figure 7.14 Smart phones and mobile devices are increasingly used to access data.

Computer-Aided Management of Emergency Operations (CAMEO)

Computer-Aided Management of Emergency Operations (CAMEO) is a resource designed by the National Oceanic and Atmospheric Administration (NOAA). CAMEO is a system of software applications that helps emergency responders develop safe response plans. It can be used to access, store, and evaluate information critical in emergency response.

CAMEO has several tools to assist a variety of users **(Figure 7.15)**. The database and information management tools include modules ranging from facility and contact information along with SARA Title III Tier 2 submission forms. CAMEO Chemicals has an extensive chemical database with critical response information for thousands of chemicals. The chemical database includes two primary types of **datasheets**: the chemical datasheet and the UN/NA datasheet.

The chemical datasheet provides the following information:

- Physical properties
- Health hazards
- Air and water reactive information
- Fire fighting information

The UN/NA datasheet provides information from the *ERG* and shipping information from the Hazardous Materials Table (49 *CFR* 172.101). CAMEO also has the capability to predict what hazards might occur if the chemicals in a specific collection are mixed together.

> **Computer-Aided Management of Emergency Operations (CAMEO)** — A system of software applications that assists emergency responders in the development of safe response plans. It can be used to access, store, and evaluate information critical in emergency response.
>
> **Datasheet** — Document that includes important information regarding a specific utility or resource in a standardized format.

Figure 7.15 CAMEO is designed to help emergency responders develop safe response plans.

CAMEO includes a mapping application and a dispersion modeling program. MARPLOT (Mapping Applications for Response, Planning and Local Operational Tasks) is a mapping application that allows users to visualize their data and display the information on computer maps. These maps may be printed either individually or on area maps. This is an effective tool when working to determine what areas will be affected by either a real-time or potential chemical release.

ALOHA is CAMEO's atmospheric dispersion modeling program used for evaluating the release of hazardous chemical vapors. ALOHA allows the user to estimate the potential harm to an area based on:

- Toxicological and physical characteristics of a chemical
- Atmospheric conditions
- Any other specific circumstances that may be affecting the release

By acquiring this information, technicians can establish and plot threat zones using the MARPLOT functions. The threat zones are established by using the type of chemical release such as leaks, fires, or explosions.

Wireless Information System for Emergency Responders (WISER)

The **Wireless Information System for Emergency Responders (WISER)** is an electronic resource that may be downloaded free of charge **(Figure 7.16)**. This system brings a wide range of information to the hazmat responder including:

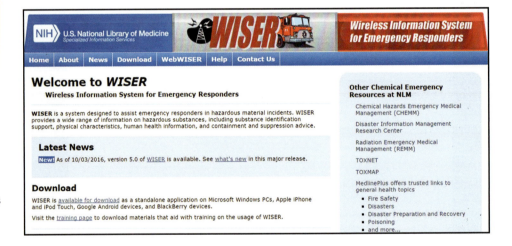

> **Wireless Information System for Emergency Responders (WISER)** — This electronic resource brings a wide range of information to the hazmat responder such as chemical identification support, characteristics of chemicals and compounds, health hazard information, and containment advice.

Figure 7.16 WISER consolidates data from a range of reference manuals.

- Chemical identification support
- Characteristics of chemicals and compounds
- Health hazard information
- Containment advice

Similar to CAMEO, WISER develops its database from a series of relevant sources. The Hazardous Substance Data Bank provides end users with toxicology data from the National Library of Medicine. Chemical data referenced in WISER are consolidated from a core set of reference manuals including:

- *Emergency Response Guidebook*
- Chemical Hazards Response Information System
- NFPA *Fire Protection Guide to Hazardous Materials*
- *NIOSH Pocket Guide to Chemical Hazards*
- EPA Integrated Risk Information System (IRIS)
- POISINDEX® Information System
- TOMES® Information System

- ACGIH® Guidelines for Selection of Chemical Protective Clothing
- ACGIH® TLVs® & BEIs®
- Emergency Handling of Hazardous Materials in Surface Transportation

Palmtop Emergency Action for Chemicals (PEAC®)-WMD

PEAC®-WMD (Palmtop Emergency Action for Chemicals), by AristaTek, includes information on over 100,000 chemicals, synonyms, and trade names. The software provides mapping and dispersion modeling. The system can quickly calculate standoff distances and exclusion zones at the incident, using the actual weather and topography conditions present.

E-Plan for First Responders

E-Plan for First Responders provides users with Tier II hazardous chemical information for fixed facilities around the U.S. E-Plan provides the only electronic access to this information; the information is only available in paper form otherwise and is typically not accessible in the initial stages of an incident. Access to this information can be very beneficial to first responders.

Chemical Hazards Emergency Medical Management (CHEMM)

Chemical Hazards Emergency Medical Management (CHEMM) is a website operated by the U.S. Department of Health and Human Services **(Figure 7.17)**. CHEMM allows responders easy access to information that can be used to plan, prepare, and respond to mass-casualty incidents involving chemicals. CHEMM offers guidance on chemical identification, acute patient treatment, patient management, and other useful tools. Some information on the CHEMM website can be downloaded in advance and saved onto a computer, allowing access to the information without the need for an Internet connection.

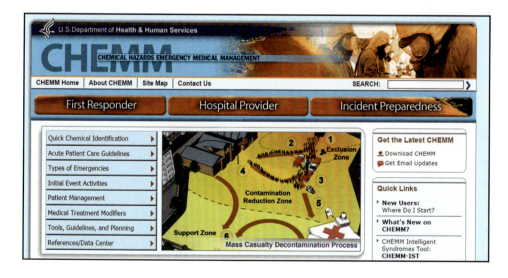

Figure 7.17 CHEMM provides information geared toward helping responders address health concerns at an incident.

Mobile Applications (Apps)

While it is beyond the scope of this manual to provide a comprehensive list and description of all the mobile applications (apps) currently used by hazmat responders, the following may be of use:

- WISER
- HazRef Lite
- ChemAlert
- Periodic Table
- Electronic *ERG*
- Pipeline Emergencies
- Chem Safety
- HazMat Segregation
- Google Earth
- QRG
- Nooly
- AskRail
- Hazmat IQ
- Hazmat Load
- iTriage
- iRPG
- HazMatch (Kappler)
- ChemHazards
- HazMat Training
- SAFER Mobile
- Mobile REMM
- RadResponder
- Weather Apps
- Chemical Companion
- Radiation Emergency Medical Management (REMM)

National Pipeline Mapping System (NPMS)

The National Pipeline Mapping System (NPMS) is a geographic information system (GIS) created by the U.S. Office of Pipeline Safety (OPS) in cooperation with other federal and state governmental agencies and the pipeline industry. Types of data collected via the NMPS includes:

- Geospatial data
- Attribute data
- Public contact information

Metadata within the NPMS system includes:

- Interstate and intrastate hazardous liquid trunklines
- Hazardous liquid low-stress lines
  — Gas transmission pipelines
  — Liquefied natural gas (LNG) plants
  — Hazardous liquid breakout tanks.

Attributes in the NPMS pipeline data layer include:

- Operator identification number
- Operator name
- System name
- Subsystem name
- Diameter (voluntary data element)
- General commodities transported
- Interstate/intrastate designation
- Operating status (in service, abandoned, retired)
- Geospatial accuracy estimate

The NPMS does NOT contain information on pipeline features such as:

- Interconnects
- Pump and compressor stations

> **Metadata** — Information that provides background and detail about other types of information.

- Valves
- Capacity
- Operating pressure
- Direction of flow
- Throughput
- Distribution and gathering pipelines

Weather Data

Having access to accurate weather data is just as critical as having access to chemical reference manuals. Safe mitigation of an emergency incident will require the correct interpretation of weather data. Humidity, atmospheric pressure, temperature, precipitation, wind speed, and wind direction can have a profound effect on a hazardous materials incident. Having access to these data is critical for a safe response.

Weather data is readily available online. Local news stations and national outlets have weather information available on the Internet. Numerous radar images are also available and allow for a reasonable cross-sectional view of an affected area for the purpose of precipitation prediction. Responders should look ahead to the weather forecast, especially during extended incidents. Weather conditions can deteriorate quickly and cause unsafe conditions for responders.

Many hazmat teams use portable weather stations for more exact data of the microclimate of a response area **(Figure 7.18)**. Portable weather stations allow for more exact weather information within the Command Post. Many of the common portable weather stations can provide the following information:

- Barometric pressure
- Temperature
- Wind speed and direction
- Relative humidity

Figure 7.18 Weather stations collect data in a specific area.

Plume Modeling

A variety of plume modeling programs are available to emergency responders. As discussed earlier, CAMEO has a comprehensive and popular plume modeling program. The U.S. EPA also has plume modeling software available named Visual Plumes. This software application is a comprehensive tool for exposure assessment **(Figure 7.19)**. Many plume modeling software applications provide the user with

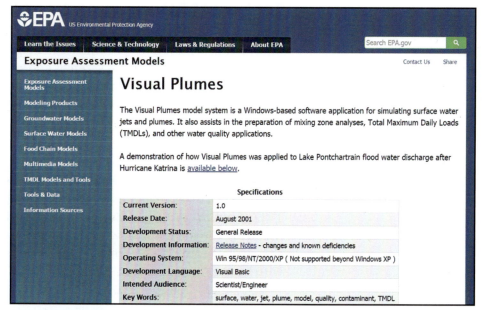

Figure 7.19 Plume modeling software can show the effects of an infusion of salt water into a freshwater lake.

Chapter 7 • Analyzing the Incident: Collecting and Interpreting Hazard and Response Information 239

> **Geographic Information Systems (GIS)** — Computer software application that relates physical features on the earth to a database to be used for mapping and analysis. The system captures, stores, analyzes, manages, and presents data that refers to or is linked to a location.

the ability to plot the plume footprint on a map so that the affected areas and levels of concentration can be readily identified. Technicians can use many of these applications in conjunction with preexisting **GIS** mapping systems.

While plume modeling applications can be beneficial to the emergency responder, they are not without limitations. When using plume models, the model must be matched with the material. For example, using a non dense gas model with a dense gas will cause large errors. Because of this, forecasts from even sophisticated dispersion modeling applications may have single event errors.

Other Internet Resources

As technology improves, the number of online references and resources will continue to expand and diversify. Many Internet resources deliver similar information regarding chemicals and chemical compounds.

NOTE: Internet addresses change frequently and therefore are not provided in this text. A simple Internet search of the listed databases will provide the most current information.

Some online resources are free and some require a subscription. The following online databases are available:

- NIST Chemistry WebBook
- ChemSpider
- CHEMnetBASE
- Toxicology Data Network (TOXNET)
- Agency for Toxic Substances and Disease Registry (ATSDR)
- Radiation Emergency Medical Management (REMM)
- Pesticide Action Network (PAN) Pesticide Database
- Ansell Chemical Resistance Guide

Technical Information Centers and Specialists

In addition to the resources and references discussed so far, other technical information centers and specialists can help the emergency responder. Some resources offer a wide array of services and specialties; others have a narrower focus. Hazardous materials technicians and response organizations should keep these valuable resources and contacts available so they may be utilized on-scene.

Emergency Response Centers

The chemical industry in the United States established CHEMTREC® as a public service hotline for emergency responders **(Figure 7.20)**. The CHEMTREC® hotline provides information and assistance for emergency incidents involving chemicals and hazardous materials. These centers are staffed with experts who provide 24-hour assistance to emergency responders dealing with hazmat emergencies.

The Canadian Transport Emergency Centre (CANUTEC) is operated by Transport Canada **(Figure 7.21)**. This national, bilingual (English and French) advisory center is part of the Transportation of Dangerous Goods Directorate.

Figure 7.20 CHEMTREC is often accessed via telephone but also has an Internet resource available.

Figure 7.21 CANUTEC offers French and English language resources for hazardous materials relevant to Canada.

CANUTEC has a scientific database on chemicals manufactured, stored, and transported in Canada and is staffed by professional scientists who specialize in emergency response and are experienced in interpreting technical information and providing advice.

Mexico has two emergency response centers: National Coordination of Civil Protection (CECOM) and Emergency Transportation System for the Chemical Industry (SETIQ), which are operated by the National Association of Chemical Industries **(Figure 7.22, p.242)**.

The *ERG* provides a list of emergency response centers and their telephone numbers. The responder should provide the center as much of the following information as possible:

- Caller's name, callback telephone number, and FAX number
- Location and nature of problem (spill, fire)
- Name and identification number of material(s) involved
- Shipper/consignee/point of origin

Figure 7.22 SETIQ is one Spanish language resource available in Mexico.

- Carrier name, railcar reporting marks (letters and numbers), or truck number
- Container type and size
- Quantity of material transported/released
- Local conditions (weather, terrain, proximity to schools, hospitals, waterways)
- Injuries, exposures, current conditions involving spills, leaks, fires, explosions, and vapor clouds
- Local emergency services that have been notified

The emergency response center will do the following:

- Confirm that a chemical emergency exists.
- Record details electronically and in written form.
- Provide immediate technical assistance to the caller.
- Contact the shipper of the material or other experts.
- Provide the shipper/manufacturer with the caller's name and callback number so that the shipper/manufacturer can deal directly with the party involved.

Poison Control Centers

Poison control centers are located around the United States. These technical information centers provide free 24-hour advice to anyone located within the fifty states, Puerto Rico, Micronesia, American Samoa, and Guam. Poison control centers provide poison expertise and treatment advice by telephone. All poison control centers can be reached at one phone number: 1-800-222-1222.

The American Association of Poison Control Centers has developed a series of guidelines to assist poison control center staff in the appropriate management of patients with suspected exposures to certain poisons. These centers can be helpful to a responder considering both victim and responder health and safety.

CHLOREP — Program administered and coordinated by The Chlorine Institute to provide an organized and effective system for responding to chlorine emergencies in the United States and Canada, operating 24 hours a day/ 7 days a week with established phone contacts.

Chlorine Emergency Plan (CHLOREP)

The Chlorine Institute developed the chlorine emergency plan known as **CHLOREP** to assist in the handling of chlorine emergencies in the United

States and Canada. The system is operated through CHEMTREC®. Upon receipt of an emergency notification, CHEMTREC® will notify the nearest manufacturer in accordance with a mutual aid plan. The manufacturer will contact someone within the Command structure of the incident to determine the need for a response team. Each manufacturer has trained personnel and equipment available to respond.

Interagency Radiological Assistance Plan (IRAP)

The Interagency Radiological Assistance Plan (IRAP) is designed to assist when responding to radiological emergencies. IRAP operates under the U.S. Federal Department of Energy and works closely with other federal, state, military and regional agencies. If deemed necessary, IRAP will contact and work with the Nuclear Regulatory Commission. The main responsibilities of IRAP response teams are to assess the hazards, keep the public informed, and recommend emergency action to the Command team to minimize hazards.

U.S. WMD-Civil Support Teams

Weapon of Mass Destruction-Civil Support Teams (WMD-CSTs) operate at the direction of state governors within the United States. The mission of U.S. WMD-CSTs is to support civil authorities at domestic CBRN incident sites. WMD-CSTs identify CBRN agents/substances, assess current and projected consequences, advise on response measures, and assist with requests for additional support.

EPA Emergency Response Team (ERT)

The EPA Emergency Response Team (ERT) was established under the National Contingency Plan. The ERT objective is to act as or advise the On-Scene Coordinator and Regional Response Teams on environmental issues related to spill containment, cleanup, and damage assessment. The team provides expertise in biology, chemistry, hydrology, and engineering of environmental emergencies. The ERT provides support to the full range of emergency response actions, including unusual or complex emergency incidents such as underwater releases. In such cases, the ERT can bring in special equipment and experienced responders and can provide the On-Scene Coordinator or lead responder with advice.

Explosive Ordnance Disposal (EOD)

Explosive Ordnance Disposal (EOD), Public Safety Bomb Disposal (PSBD) and Bomb Squad teams provide expertise in identifying, disabling, and disposing of items involving explosives. These teams may be military or civilian based, and technicians should be aware that their response could take time to arrive on scene.

FBI WMD Coordinator

The U.S. Federal Bureau of Investigation's Weapons of Mass Destruction Coordinator (WMDC) is a special agent who is charged with developing relationships with federal, tribal, state, and local communities (public safety, academia, industry). The WMD Coordinator acts as a conduit to FBI headquarters and the Federal Government for technical information, advice (pre-event planning and prevention), and assistance during potential WMD or chemical, biological, radiological, nuclear, high-yield explosive (CBRNE) incidents. WMD Coordinators organize investigative activities between the FBI and public safety during acts of terrorism, especially those involving weapons of mass destruction or CBRNE materials. Each FBI Field Office has at least one WMDC.

Other Federal Resources
Several federal resources are available in the United States to assist in hazardous materials incidents. These include:
- Department of Homeland Security (DHS)
- Department of Energy (DOE)
- Department of Justice (DOJ)
- Military assets

Chapter Notes
Barron, Randall, Gregory F. Nellis. *Cryogenic Heat Transfer*, Second Edition. CRC Press; Boca Raton, 2016.

U.S. Food & Drug Administration (FDA). 'Methacrylate Monomers in Artificial Nails ("Acrylics").' Nail Care Products. U.S. Department of Health and Human Services. Last modified October 2016. http://www.fda.gov/Cosmetics/ProductsIngredients/Products/ucm127068.htm

Chapter Review
1. What are the main responsibilities of the technical research job function?
2. What type of research does a hazardous materials technician have to do at an incident?
3. List some hazardous materials reference manuals and their specific advantages.
4. What types of applications can electronic resources provide?
5. What are some technical information centers and specialists available to hazmat technicians?

7-1

Use approved reference resources to interpret hazard and response information.
[NFPA 1072 7.2.2]

WARNING: If this skill involves the use of actual hazardous material samples, hazardous materials can cause serious injury or fatality. Appropriate personal protective equipment (PPE) must be worn and safety precautions must be followed. The following skill sheet demonstrates general steps; specific hazmat incidents may differ in procedure. Always follow the AHJ's procedures for specific incidents.

Step 1: Collect hazard and response information.

Step 2: Interpret hazard and response information.

Step 3: Identify signs and symptoms of exposures (including target organ effects).

Step 4: Determine radiation exposure rates from containers.

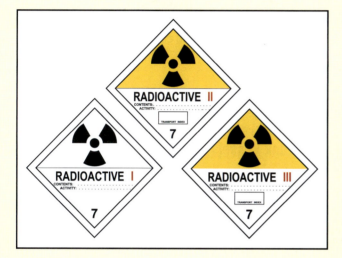

Step 5: Communicate hazards and response information.

Chapter 7 • Analyzing the Incident: Collecting and Interpreting Hazard and Response Information

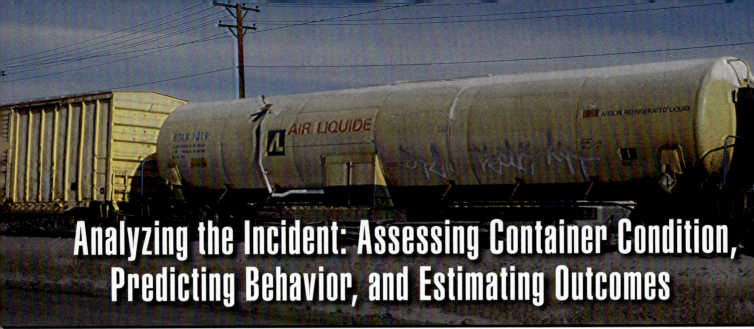

Analyzing the Incident: Assessing Container Condition, Predicting Behavior, and Estimating Outcomes

Photo courtesy of Rich Mahaney.

Chapter Contents

Non-Bulk Containers 249
    Bags 250
    Carboys and Bottles 251
    Drums and Pails 252
    Cylinders 253
    Other Common Small Containers 256
Intermediate Bulk Containers (IBCs) 256
Ton Containers (Pressure Drums) 258
Highway Cargo Containers 260
    Tank Markings 261
    Specification Plates 261
    Nonpressure Cargo Tanks 263
    Low-Pressure Cargo Tanks 267
    Corrosive Liquid Tanks 270
    High-Pressure Cargo Tanks 273
    Cryogenic Tanks 276
    Tube Trailers 279
    Dry Bulk Carriers 281
Railway Tank Cars 282
    Tank Car Markings 282
    Tank Car Structure 285
    Safety Features of Railway Tank Cars 287
    Tank Car Fittings 289
    General Service (Nonpressure) Railway Tank Cars .. 291
    Pressure Railway Tank Cars 293
    Cryogenic Railway Tank Cars 295
    Specialized Cars 296
Intermodal Containers 297
    Intermodal Tanks 299
    Intermodal Tanks for Liquids and Solid Hazardous Materials 300

Intermodal Tanks for Non-Refrigerated Liquefied Compressed Gases 300
Intermodal Tanks for Refrigerated Liquefied Gases . 301
Marine Tank Vessels 302
    Tankers 302
    Cargo Vessels 304
    Barges 305
Air Freight Cargo 305
Pipelines 306
    Principles of Pipeline Operation 306
    Basic Identification 306
    Construction Features 307
Fixed Facility Containers 309
    Nonpressure Tanks 309
    Pressure Tanks 312
    Cryogenic Tanks 313
Other Storage Facility Considerations 313
    Laboratories 313
    Batch Plants 314
    Non-Regulated and Illicit Container Use 314
Radioactive Materials Packaging 314
    Excepted 314
    Industrial 314
    Type A Packaging 315
    Type B Packaging 315
    Type C Packaging 316
    Descriptions and Types of Radioactive Labels 317
Chapter Notes 317
Chapter Review 318
Skill Sheets 319

Chapter 8

Key Terms

Air Bill 306	Low-Pressure Storage Tank 312
American Society of Mechanical Engineers (ASME) 263	Manway 289
Baffle 266	Maximum Allowable Working Pressure (MAWP) 260
Beam 298	Other Regulated Material (ORM) 250
Bill of Lading 266	Pressure Relief Valve 291
Boiling Liquid Expanding Vapor Explosion (BLEVE) 293	Pressure Storage Tank 312
Capacity Stencil 284	Pressure Vessel 312
Coffer Dam 303	Radiation 314
Consist 283	Railcar Initials and Numbers 282
Continuous Underframe Tank Car 286	Safety Relief Device 291
Cryogen 276	Specification Marking 283
Cryogenic Liquid Storage Tank 313	Supervisory Control and Data Acquisition (SCADA) 308
Cylinder 253	T-Code 298
Dewar 255	Thermal Insulation 288
Excepted Packaging 314	Thermal Protection 288
Frameless Tank Car 285	Toxic Inhalation Hazard (TIH) 252
Head Shield 287	Transport Index (TI) 317
Heat Induced Tear 293	Type A Packaging 315
Hydrostatic Test 253	Type B Packaging 315
Industrial Packaging 314	Type C Packaging 316
Intermediate Bulk Container (IBC) 257	Vacuum Relief Valve 291
Intermodal Container 297	Valve 289
Liquefied Natural Gas (LNG) 276	Waybill 292
Liquefied Petroleum Gas (LPG) 293	

Chapter 8 • Analyzing the Incident: Assessing Container Condition, Predicting Behavior, and Estimating Outcomes 247

Analyzing the Incident: Assessing Container Condition, Predicting Behavior, and Estimating Outcomes

JPRs Addressed In This Chapter

7.2.1, 7.2.3

Learning Objectives

After reading this chapter, students will be able to:

1. Identify characteristics of non-bulk containers. [NFPA 1072, 7.2.3]
2. Identify characteristics of intermediate bulk containers. [NFPA 1072, 7.2.3]
3. Identify characteristics of ton containers. [NFPA 1072, 7.2.3]
4. Identify characteristics of highway cargo containers. [NFPA 1072, 7.2.3]
5. Identify characteristics of railway tank cars. [NFPA 1072, 7.2.3]
6. Identify characteristics of intermodal containers. [NFPA 1072, 7.2.3]
7. Identify characteristics of marine tank vessels. [NFPA 1072, 7.2.3]
8. Identify characteristics of air freight cargo. [NFPA 1072, 7.2.3]
9. Identify characteristics of pipelines. [NFPA 1072, 7.2.3]
10. Identify characteristics of fixed facility containers. [NFPA 1072, 7.2.3]
11. Identify characteristics of other storage facilities. [NFPA 1072, 7.2.3]
12. Identify characteristics of radioactive materials packaging. [NFPA 1072, 7.2.1, 7.2.3]
13. Skill Sheet 8-1: Evaluate the condition of a hazardous materials container. [NFPA 1072, 7.2.3]

Chapter 8
Analyzing the Incident: Assessing Container Condition, Predicting Behavior, and Estimating Outcomes

This chapter introduces common containers and their features. Chapter 11 will address containers with an emphasis on how to perform product control once the product has been released. Steps involved in evaluating the condition of a hazardous materials container are presented in **Skill Sheet 8-1**. Common containers include:

- Non-bulk containers
- Ton containers (pressure drums)
- Railway tank cars
- Marine tank vessels
- Pipelines
- Radioactive materials packaging
- Intermediate bulk containers (IBCs)
- Highway cargo containers
- Intermodal containers
- Air (space) freight cargo
- Fixed facility containers
- Other storage facility considerations

Non-Bulk Containers

Hazardous material shipments may be found in all types of containers, both large and small. The U.S. Department of Transportation (DOT) divides these containers into bulk and non-bulk packaging (see information box). Steps for evaluating the condition of a hazardous materials container are presented in **Skill Sheet 8-1**.

NOTE: Many types and details of containers and packaging are discussed in the IFSTA manual, **Hazardous Materials for First Responders**.

Non-bulk packaging is widely used across the United States and may be found in any transportation port or location. Because of the way the material is packaged, it may be difficult to determine if the shipment is hazardous. The emergency responder should not completely rely on markings to determine the hazard. Some containers may not be marked or may be marked inappropriately. The shape and size of a container can aid in determining potential hazards.

CAUTION
Contained materials may displace oxygen.

> **CAUTION**
> Monitor the atmosphere before and while opening any container.

Bulk and Non-Bulk Packaging

The U. S. Department of Transportation (DOT) defines the capacities of bulk and non-bulk packaging. For more information on these defined capacities, refer to the Code of Federal Regulations, Title 49.

Bulk packaging is defined as a container that has no intermediate form of containment and has:

1. A maximum capacity greater than 119 gallons (450 L) as a receptacle for a liquid;
2. A maximum net mass greater than 800 pounds (400 kg) and a maximum capacity greater than 119 gallons (450 L) as a receptacle for a solid; or
3. A water capacity greater than 1,000 pounds (454 kg) as a receptacle for a gas.

Non-bulk packaging is defined as a packaging which has:

1. A maximum capacity of 119 gallons (450 L) or less as a receptacle for a liquid;
2. A maximum net mass of 800 pounds (400 kg) or less and a maximum capacity of 119 gallons (450 L) or less as a receptacle for a solid; or
3. A water capacity of 1,000 pounds (454 kg) or less as a receptacle for a gas

Other Regulated Material (ORM) — Material, such as a consumer commodity, that does not meet the definition of a hazardous material and is not included in any other hazard class but possesses enough hazardous characteristics that it requires some regulation; presents limited hazard during transportation because of its form, quantity, and packaging.

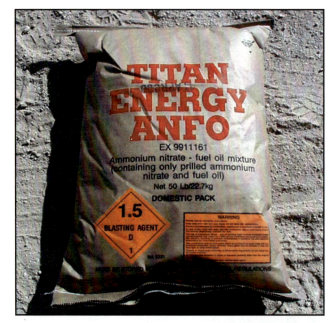

Figure 8.1 Bags are commonly used for dry goods that are not highly shock-reactive. *Courtesy of the U.S. Bureau of Alcohol, Tobacco, Firearms, and Explosives, and the Oklahoma Highway Patrol.*

Bags

Bags come in a variety of material and may hold a wide array of contents. A bag may be constructed out of plastic, multilayered paper, or may be a paper bag that is layered with plastic.

NOTE: *49 CFR* defines bags but does not specify a weight limitation.

Typical contents of the bag may include:

- Dry corrosive
- Blasting agents **(Figure 8.1)**
- Explosives
- Flammable solids
- Oxidizers or organic peroxide
- Poisons
- Pesticides
- **Other regulated materials (ORM)**

Non-bulk bags can be stacked on pallets with enough material to constitute a bulk shipment. When working with bags, hazmat technicians should consider hazards including:

- The type of materials used in bag construction causes them to be fragile and prone to damage and the release of contents.
- The structure of bags makes them susceptible to environmental conditions **(Figure 8.2)**.
- Contamination and spread of material is easy with bags.
- Forklifts may easily breach a bag without the operator noticing.
- Containment and confinement techniques may require atypical combinations of response techniques.
- Flammable dust may complicate the incident based on the material, location, and quantity.

Figure 8.2 This bag is designed to be durable but may still lose structural integrity from photodegredation. *Courtesy of Barry Lindley.*

CAUTION
Bags are easily damaged and should be handled carefully.

Carboys and Bottles

Similar to jerry cans, carboys are a limited-use, non-bulk container used almost exclusively for the shipment of hazardous materials. Carboys and bottles are types of containers that may be constructed of plastic, glass, or steel. Both types of containers have a narrow neck and a larger internal capacity. Bottles may hold as little as a few milliliters or as much as 1 gallon (4 L). Typical carboy sizes range between 6 gallons (25 liters) and 16 gallons (60 L) (Soraka 2008).

Typical contents of carboys include:

- Class 3 (Flammables)
- Class 4 (Solids dissolved or suspended in solvents)
- Class 5 (Oxidizers)
- Class 6 (Poisons)
- Class 8 (Corrosives)

Due to their construction, carboys are a relatively safe mode of transportation for hazardous materials and typically are not prone to the same type of damage and corrosion that may be found on other types of containment devices. However, if the carboy is constructed of glass or plastic or if the outer packaging is damaged or not sized correctly, the internal containment device (which is protected by an outer cushion container made of wood, cardboard, or Styrofoam™) may be damaged.

Hazmat technicians should be aware of the following considerations related to carboys and bottles:

- Colored glass often indicates that the material is photosensitive **(Figure 8.3)**.
- Documentation, labels, and placards may not represent the true contents.
- Placards and labels may not be required for carboys, so unknown flammable or combustible products may be present.

Figure 8.3 Materials that are sensitive to light may be protected using tinted glass.

- Outer packaging may react with contents if a breach occurs.
- Crystals on the rim of the carboy indicate that something has changed inside the container, potentially making the product sensitive to vibration, friction, or temperature changes. When crystals are present, the integrity of the container and stability of the product may be in doubt. If exposed to vibration, friction, or temperature changes, the product might explode.

Drums and Pails

Drums are a commonly used container type for a variety of materials, with the exception of compressed gases and etiological (infectious) agents. Drums may be constructed out of the following materials:

- **Metal** — Steel and stainless steel drums are commonly used to carry materials that are flammable and solvents. Stainless steel drums typically carry materials that are highly corrosive and are **toxic inhalation hazards (TIHs)**.
- **Fiberboard** — Used for solid materials but may contain liquids or slurries in bags. Fiberboard drums may or may not be lined with an inner containment material. Pesticides, food-grade products, and corrosive solids are common materials found in these drums.
- **Plastic** — Typically carry corrosive products but may also carry flammables and solvents. These drums may also incorporate an inner containment material.

> **Toxic Inhalation Hazard (TIH)** — Volatile liquid or gas known to be a severe hazard to human health during transportation.

Figure 8.4 A drum intended to contain heavy materials may have parallel support rings so it can be rolled. *Courtesy of Rich Mahaney.*

Drums may be configured with either an open head or closed head (**Figure 8.4**). An open-head drum has a top lid that can be fully removed, whereas a closed-head drum will only have smaller openings such as a bung hole. Drums may have two-threaded openings. Drums can hold up to 119 gallons (450 L) liquid capacity (*49 CFR 173.3*). The most common capacity of drums is 55 gallons (220 L) (Soraka, 2008).

Drums are sometimes prone to damage depending on the construction and may leak from the seams or bung openings. Metal drums may be prone to corrosion or rot. Mechanical damage is a concern for all types of drum material. Punctures, tears, and overpressure are also causes of drum damage.

Pails may be considered a type of drum, but with a lower content capacity. Pails have a wide variety of uses and are found in all types of locations. Pails may be constructed of metal, fiberboard, or plastic and may hold anywhere from 1 to 13 gallons (3 L to 50 L) or more of material (Soraka, 2008). Pails may be prone to the same types of leakage and damage as drums because of similar construction materials.

Materials commonly found in pails may include:

- Flammable solids
- Flammable or combustible liquids
- Oxidizers or organic peroxide
- Poisons

When dealing with drums or pails, the hazmat technicians should:

- Consider the integrity of the drum, especially if it has been reused or abandoned.
- Look for indicators of potential hazards such as discoloration, staining, and scoring.
- Assume that empty drums have residual product or vapor until proven otherwise.
- Note that drums with multiple roll rings may be carrying a heavier than normal material or may be reconditioned and of questionable integrity.
- Look for bulges that indicate that there is a pressure buildup inside the container. Vacuums are also possible in these containers.
- Remember that drums are also used for salvage and cleanup and may contain materials that are not appropriate for the container.
- Be aware that 55 gallons (220 L) drums are commonly used to contain hazardous waste. The waste may be from multiple locations, have multiple layers, and contain multiple hazard classes.

Cylinders

A **cylinder** is a pressurized vessel that may contain compressed or liquified gases, flammable or combustible liquids, poisons, corrosives, or radioactive materials. All these materials can be contained at potentially high pressures, and capacities up to 450 L (119 gallons). Cylinders can be found in a wide variety of locations including:

- Manufacturing facilities
- Medical facilities
- University labs
- Construction sites
- Emergency responder staging

Basic Identification

While not an industry standard, cylinders often display characteristics relevant to their contents. A cylinder that is short and broad will have a lower pressure than cylinders that are long and thin **(Figure 8.5)**. Cylinders with a weld seam on the long axis will not be used for high-pressure containment.

The U.S. DOT establishes regulations for the care, maintenance, and manufacture of cylinders in the United States. DOT specification markings must be located near the top of the cylinder and indicate:

- **Hydrostatic test** dates
- Manufacturer
- DOT specification
- Serial numbers
- Inspector marks

> **Cylinder** — Enclosed container with a circular cross-section used to hold a range of materials. Uses include compressed breathing air, poisons, or radioactive materials. *Also known as* Tank *or* Bottle.

> **Hydrostatic Test** — Testing method that uses water under pressure to check the integrity of pressure vessels.

Figure 8.5 Cylinder shapes are often customized to their contents' characteristics. *Courtesy of Rich Mahaney.*

Figure 8.6 Use labels to identify a material regardless of any color coding that may be present. *Courtesy of Rich Mahaney.*

The Compressed Gas Association has recommended a color-coding system for cylinders, which many medical gas manufacturers follow. However, this color-coding system is not required by law; therefore, emergency responders cannot rely on this as a means of identifying the contents of a compressed gas cylinder. Technicians should use labels to properly identify the contents of all cylinders **(Figure 8.6)**.

> **WARNING!**
> Marking and color-coding of cylinders is not an industry standard and cannot be relied on for identification purposes. Use labels to identify cylinder contents.

Construction Features

The construction of cylinders uses materials with a high tensile strength. Steel is the most common of those materials, but others may include aluminum, carbon fiber, and polycarbonate.

A cylinder will include valve devices that are specific to the product contained in the cylinder. Stop angle valves are a common feature of most cylinders, except for cylinders designed for unique uses such as medical grade gases that have valves that cannot be adapted to any other type of gas.

Pressure relief devices are safety devices that work in tandem with the valve. If the pressure of the cylinder exceeds the rated pressure of the relief device, the pressure relief device will activate and relieve the excess pressure. In most cases, once a pressure relief device activates, it cannot be reset and must be replaced.

Pressure relief devices may be a simple burst disc type. This device is installed in the back of the valve and is nothing more than a small metal gasket that will rupture at a predetermined pressure. A low melting point metal may comprise the pressure relief device so that in case of fire impingement or temperature increase the relief device will activate and prevent catastrophic failure of the cylinder. Not all cylinders incorporate such safety devices.

Cylinders are an inherently strong type of containment vessel. Although leaks are uncommon in a well-maintained cylinder, mechanical damage may reduce the overall strength of the cylinder. Leaks may occur at the threaded connections for the valve assembly or within the valve assembly itself. Based on the orientation of the cylinder, the leak may either be a gaseous leak or a liquid leak. Repair kits and recovery vessels may be used with leaking chlorine containers. Hazmat technicians should carefully inspect the cylinder and review the hydrostatic test dates.

Cryogenic Cylinders

Cryogenic cylinders are vessels that are designed and manufactured to store super-cooled materials. Cryogenic cylinders must be able to accommodate the material at both its gaseous state (at normal atmospheric pressure) and its liquid state (when autorefrigerated). Cryogenic cylinders vary in capacity. Pressure in cryogenic cylinders includes low and high ranges, depending on the intended use of the vessel. The valve assemblies on a cryogenic cylinder will be constructed to dispense both a gas and a liquid.

A **dewar** flask is a nonpressurized, insulated container that has a vacuum space between the outer shell and the inner vessel. The dewar flask is designed for the storage and dispensing of cryogenic materials such as liquid nitrogen, liquid oxygen, and helium. Dewars have a bulky appearance due to the insulation that is used to keep the cryogenic material at the desired temperature.

Dewar — Container designed for the movement of small quantities of cryogenic liquids within a facility; not designed or intended to meet Department of Transportation (DOT) requirements for the transportation of cryogenic materials.

Cryogenic Materials Release

A fire department received a call that a pregnant adult had tripped and fallen at a fast food restaurant. The fire department responded and found the victim still lying at the top of a staircase leading to the basement. She was incoherent, anxious, and breathing heavily.

Thinking that she may have tripped or fallen on something on the staircase, two firefighters went to investigate. As they moved down the stairs, one noticed an odd smell. Another felt a small burning in his nose and throat. They noticed paint cans and some chemical cylinders in the basement and decided to leave because they both suspected something was wrong. When the firefighters reached the top of the stairs again, both felt lightheaded. One firefighter collapsed, but did not lose consciousness.

Realizing that something more than a simple trip and fall incident was occurring, the fire department removed the pregnant victim from the area and evacuated the restaurant. A hazmat team was dispatched to the scene.

The hazmat team's initial recon showed a 17 percent oxygen deficiency in the basement coupled with an alarm on their natural gas meter. They did not suspect a natural gas leak, so this information complicated the response, and they withdrew from the basement. Utilities were secured, and eventually it was determined that the leak was from a cryogenic dewar cylinder containing carbon dioxide (CO_2). The hazmat team located small thumb valves on the top of the tank and closed them, hoping to stop the leak. The team then ventilated the space with a confined space ventilation fan, and eventually the area was declared safe.

Further research revealed that the chemical and physical properties of CO_2 mimicked those of natural gas in their particular brand of natural gas meter. This resulted in the false positive their meter registered in the basement during the initial recon.

Lessons Learned:

Historically, restaurants have used compressed gas CO_2 cylinders to carbonate their beverages. However, more and more are switching to a new system that utilizes liquefied CO_2 in cryogenic dewar cylinders **(Figure 8.7)**. The expansion ratio of liquid CO_2 is around 670:1, and as it is released it turns into a colorless gas that quickly displaces the O_2 in the atmosphere, potentially creating a very dangerous situation. The transition to this container type for this product has created a new hazard at restaurants and other businesses and industries that use carbon dioxide systems inside their buildings.

Figure 8.7 Cryogenic containers for carbon dioxide (CO_2) are increasingly used in restaurants as part of beverage carbonation systems. *Courtesy of Rich Mahaney.*

Figure 8.8 Y cylinders contain compressed gases. *Courtesy of Barry Lindley.*

Figure 8.9 These cylinders operate in a cascade system.

Figure 8.10 Corrugated fiberboard boxes may contain almost any classification of hazardous material. *Courtesy of Rich Mahaney.*

Y Cylinders

Y cylinders are a type of compressed gas cylinder that can be bulk or non-bulk. A typical "Y" ton container will have a specification such as DOT 3AA-2400 or DOT3AA-480 (pressure is dependent on product) **(Figure 8.8)**. These containers are typically 7 ft (2115 mm) long, 2 ft (600 mm) in diameter, have a wall thickness of about 0.6 inches (15 mm), and, when empty, weigh about 1,200 lbs (600 kg). These containers have a water capacity of approximately 120 gallons (480 L). Often used for refrigerants, they typically operate in a cascade system **(Figure 8.9)**.

Two specifications of Y cylinders are defined based on size (*49 CFR 178.37*):

- The DOT–3AA cylinder is a seamless steel cylinder with a water capacity (nominal) of not over 1,000 lbs (500 kg) and a service pressure of at least 150 psig (1 136 kPa).

- The DOT–3AAX cylinder is a seamless steel cylinder with a water capacity of not less than 1,000 lbs (500 kg) and a service pressure of at least 500 psig (227 kPa).

Other Common Small Containers

Wood and fiberboard boxes may be used as primary packaging devices or as cases for smaller inner containers such as carboys. Boxes may carry an array of hazardous materials, and proper labeling must be used for identification purposes. Wooden boxes may be used to carry every classification of hazardous material including compressed gas cylinders. Fiberboard boxes may be used to carry every classification of hazardous material except compressed gases and poisons **(Figure 8.10)**. While boxes cannot carry compressed gases themselves, they can carry products such as aerosol cans which pose a pressurized hazard.

Multicell packaging is a packaging device that is form-fitted to other containers. The multicell packaging can serve as a protective device for the container. The U.S. DOT limits their capacity to no more than six cells and a maximum of four liters.

Intermediate Bulk Containers (IBCs)

Intermediate bulk containers (IBCs), typically called totes, are designated by the U.S. DOT as either rigid or flexible portable packaging (other than a cylinder or portable tank) designed for mechanical handling. Design standards for IBCs in the U.S., Canada, and Mexico are based on United Nations Recommendations on the

Transportation of Dangerous Goods (UN Recommendations).

IBCs are authorized to transport a wide variety of materials and hazard classes:

- Aviation fuel (turbine engine)
- Gasoline
- Alcohols
- Toluene
- Corrosive liquids
- Solid materials in powder, flake, or granular forms

> **Intermediate Bulk Container (IBC)** — Rigid (RIBC) or flexible (FIBC) portable packaging, other than a cylinder or portable tank, that is designed for mechanical handling with a maximum capacity of not more than 3 cubic meters (3,000 L, 793 gal, or 106 ft^3) (*49 CFR 178.700*).

Flexible intermediate bulk containers (FIBCs) are sometimes called bulk bags, bulk sacks, supersacks, big bags, tote bags, or totes. These containers are flexible, collapsible bags or sacks that are used to carry solid materials, and their designs vary greatly **(Figure 8.11)**. FIBCs may be constructed of multiwall paper or other textiles. A common-sized supersack FIBC can carry 2000 pounds (1 000 kg) and can be stacked one on top of another depending on design and the material inside. FIBCs are sometimes transported inside a rigid exterior container made of corrugated board or wood.

Figure 8.11 Bulk bags may carry solid materials, depending on design and construction. *Courtesy of Rich Mahaney.*

Rigid intermediate bulk containers (RIBCs) are typically made of steel, aluminum, wood, fiberboard, or plastic and are often designed to be stacked. RIBCs can contain both solid materials and liquids **(Figure 8.12)**. Some liquid containers may look like smaller versions of intermodal nonpressure tanks with metal or plastic tanks inside rectangular box frames. Other RIBCs may be large, square or rectangular boxes or bins. Rigid portable tanks may be used to carry various liquids, fertilizers, solvents, and other chemicals.

Hazmat technicians should be aware of the following considerations related to IBCs:

- They share many of the same issues as barrels and drums.
- Some have valves that have the potential to leak and sometimes are difficult to access.
- Some have containment vessels incorporated into the container but may not be able to contain the entire volume of the container.
- Specific products have specifically designed containers.
- Depending on the protective housing, patching and plugging operations can be difficult.
- Supersacks can be extremely difficult to handle because there is no easy

Figure 8.12 Rigid IBCs are often designed to be stacked. *Courtesy of Rich Mahaney.*

way to pick up or move if it comes off the pallet. Supersacks also have some of the same environmental considerations as bags **(Figure 8.13)**.

Figure 8.13 These RIBCs are holding materials that are dangerous when wet. *Courtesy of Rich Mahaney.*

Ton Containers (Pressure Drums)

The U.S. DOT refers to ton containers as multi-unit tank car tanks (DOT 110 and DOT 106). These containers are typically stored on their sides, and the ends (heads) are convex or concave.

Ton containers have two valves in the center of one end, one above the other. One valve connects to a tube going into the liquid space, and the other valve connects to a tube going into the vapor space above **(Figure 8.14)**. Some of these containers also have a pressure-relief device in case of fire or exposure to elevated temperatures. They may also have fusible plugs that can melt and relieve pressure in the container.

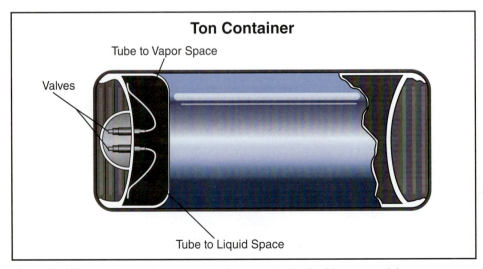

Figure 8.14 Ton containers have valves that connect to the liquid space and the vapor space.

Ton containers commonly contain chlorine and are often found at locations such as water treatment plants and commercial swimming pools. They may also contain materials such as sulfur dioxide, anhydrous ammonia, or Freon® refrigerant.

Ton containers are an extremely rigid type of containment device. Leaks in this type of container often occur at the valves. Based on the orientation of the cylinder, the leak may either be a gaseous leak or a liquid leak. Specialized repair kits for chlorine and sulfur dioxide are available and should be utilized should a leak in either the valves or fusible links occur. More information on these kits and on leaks involving ton containers is provided in Chapter 13.

Chlorine Emergency Kits

The Chlorine Institute developed specialized repair kits for emergency response to hazmat incidents involving chlorine (Indian Springs Manufacturing). Three kits were developed for use on different containers:

- Emergency Kit "A" ("A Kit") for compressed gas cylinders **(Figure 8.15)**

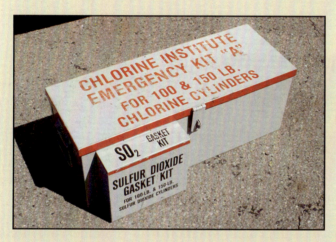

Figure 8.15 The A kit is intended for use with compressed gas cylinders. *Courtesy of Rich Mahaney.*

- Emergency Kit "B" ("B Kit") for use on chlorine ton containers **(Figure 8.16)**

Figure 8.16 The B kit is intended for use with chlorine ton containers. *Courtesy of Rich Mahaney.*

- Emergency Kit "C" ("C Kit") for use with chlorine tank cars and intermodal portable tanks **(Figure 8.17)**

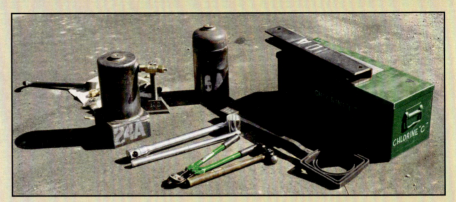

Figure 8.17 The C kit is intended for use with chlorine tank cars and intermodal portable tanks. *Courtesy of Rich Mahaney.*

In addition, a hazmat team may deploy chlorine recovery vessels to contain a release, per the AHJ. Ammonia recovery vessels may also be available in some jurisdictions.

Highway Cargo Containers

Highway vehicles that transport hazardous materials include cargo tanks (also called tank motor vehicles and tank trucks), dry bulk containers, compressed gas tube trailers, and mixed load containers (also called box trucks or dry van trucks). These vehicles transport all types of hazardous materials in a wide range of quantities.

Cargo tank trucks are recognizable because they have construction features, fittings, attachments, or shapes that are characteristic of their uses. If emergency responders recognize one of the cargo tank trucks described in this section, they must base their determination of the container contents on placards, shipping papers, or other formal sources of information.

Cargo tank trucks are commonly used to transport bulk amounts of hazardous materials by road. Most cargo tank trucks that haul hazardous materials are designed to meet government tank-safety specifications. The government specifications establish minimum tank construction material thicknesses, required safety features, and **maximum allowable working pressure (MAWP)**. The two specifications currently in use are the motor carrier (MC) standards and DOT/TC standards. Cargo tank trucks built to a given specification are designated using the MC or DOT/TC initials followed by a three-digit number identifying the specification (such as MC 306 and DOT/TC 406). Some cargo tanks have multiple compartments. Each compartment is considered a separate tank and may contain different products.

Tanks not constructed to meet one of the common MC or DOT/TC specifications are commonly referred to as *nonspec tanks*. Nonspec tanks may haul hazardous materials, such as molten asphalt and sulfur, if the tank was designed for a specific purpose and exempted from the MC or DOT/TC requirements **(Figure 8.18)**.

> **Maximum Allowable Working Pressure (MAWP)** — A percentage of a container's test pressure. Can be calculated as the pressure that the weakest component of a vessel or container can safely maintain.

Figure 8.18 Trucks that do not meet DOT/TC specifications may still be used to carry some types of hazardous materials. *Courtesy of Rich Mahaney.*

Nonhazardous materials may be hauled in either non-spec cargo tank trucks or cargo tank trucks that meet a designated specification.

Tank Markings

When making response-based decisions at an incident involving a cargo tank, technicians must know the tank's contents. Many highway cargo vehicle tanks will display a number of markings. Some of these markings may directly correlate to the contents, while others will not help identify the product.

By DOT regulations, all compressed gases (both flammable and nonflammable) and cryogenic liquids must have their shipping name displayed on the tank. The markings must be located on both sides of the tank and at both ends. Highway cargo tanks are frequently marked with the product's brand name **(Figure 8.19)**.

Specification Plates

It is the manufacturer's responsibility to certify that each tank has been designed, constructed, and tested in accordance with cargo tank truck requirements. All tanks that meet established specifications are required to carry a variety of information that may be helpful to the emergency responder.

Each cargo tank must carry two types of plates **(Figure 8.20)**:

Figure 8.19 Some markings, such as the brand name of the transportation company, will aid a technician in determining the contents of the tank. *Courtesy of Bill Hand, Houston Fire Department (ret).*

Figure 8.20 The information included on a nameplate and specification plate meet separate requirements.

- Nameplate **(Table 8.1, p. 262)**
- Specification plate **(Table 8.2, p. 263)**

Specification plates must be corrosion-resistant and permanently attached to the cargo tank truck or its integral supporting structure. Plates must be permanently and plainly marked in English by stamping or embossing, and must be affixed to the left side of the vehicle near the front of the cargo tank truck in a place that is readily accessible for inspection.

Table 8.1
Nameplate Information

DOT Requirement	Plate Abbreviation	Units
DOT-specification number	DOT XXX **NOTE:** *"XXX"* is replaced with the applicable specification number. For cargo tanks having a variable specification plate, the DOT specification number is replaced with the words "See variable specification plate."	n/a
Original test date	Orig. Test Date	month and year
Tank maximum allowable working pressure	Tank MAWP	psig
Cargo tank test pressure	Test P	psig
Cargo tank design temperature range	Design temp. range	_ °F to _ °F
Nominal capacity	Water cap.	gallons
Maximum design density of lading	Max. lading density	pounds per gallon
Material specification number—shell	Shell matl, yyy*** **NOTE:** *"yyy"* is replaced by the alloy designation and *"***"* by the alloy type	n/a
Material specification number—heads	Head matl, yyy*** **NOTE 1:** *"yyy"* is replaced by the alloy designation and *"***"* by the alloy type **NOTE 2:** When the shell and heads materials are the same thickness, they may be combined (Shell& head matl, *yyy****)	n/a
Weld material	Weld matl.	n/a
Minimum thickness—shell	Min. shell-thick **NOTE:** When minimum shell thicknesses are not the same for different areas, show top _, side _, bottom _	inches
Minimum thickness—heads	Min. heads thick.	inches
Manufactured thickness—shell	Mfd. shell thick. (top _, side _, bottom _) **NOTE:** Required when additional thickness is provided for corrosion allowance.	inches
Manufactured thickness—heads	Mfd. heads thick. **NOTE:** Required when additional thickness is provided for corrosion allowance.	inches
Exposed surface area	n/a	square feet

Table 8.2
Specification Plate Information

DOT Requirement	Plate Abbreviation
Cargo tank motor vehicle manufacturer	CTMV mfr.
Cargo tank motor vehicle certification date (if different from the cargo tank certification date.)	CTMV cert. date
Cargo tank manufacturer	CT mfr.
Cargo tank date of manufacture (month and year)	CT date of mfr.
Maximum weight of lading (in pounds)	Max. Payload
Maximum loading rate in gallons per minute	Max. Load rate, GPM
Maximum unloading rate in gallons per minute	Max. Unload rate
Lining material (if applicable)	Lining
Heating system design pressure in psig (if applicable)	Heating sys. press.
Heating system design temperature in °F (if applicable)	Heating sys. temp.

Insulated tanks and some certification tanks may have multiple plates. One plate may be located on the left front of the tank permanently mounted under the insulation. These certification plates may only be viewed if the insulation is removed. The other certification plate is typically located on the outside of the insulating jacket on the front left side of the tank or within the operator's cabinet.

In addition to the DOT specification/certification plate, some tanks that are certified to the **American Society of Mechanical Engineers (ASME)** code for pressure carriers must also carry a separate certification plate. ASME certification plates can be identified by the embossed "U" on the upper left-hand corner of the plate. A vacuum-loaded cargo tank must have an ASME code-stamped specification plate marked with a minimum internal design pressure of 25 psig (274 kpa) and be designed for a minimum external design pressure of 15 psig (205 kpa).

American Society of Mechanical Engineers (ASME) — Voluntary standards-setting organization concerned with the development of technical standards, such as those for respiratory protection cylinders.

Nonpressure Cargo Tanks

Nonpressure cargo tanks may carry any product from food-grade liquids to petroleum products such as gasoline and fuel oil. These cargo tanks carry the MC 306 designation or the DOT 406 designation and make up more than half of the total fleet of tankers on the road today **(Figure 8.21, p. 264)**. These tanks are designed to accommodate pressures not exceeding 3 psig (122 kPa). Nonpressure cargo tanks often comprise more than one compartment. Common products shipped in these tanks may include:

- Gasoline
- Alcohols
- Flammable and combustible liquids
- Fuel oil
- Food-grade liquids

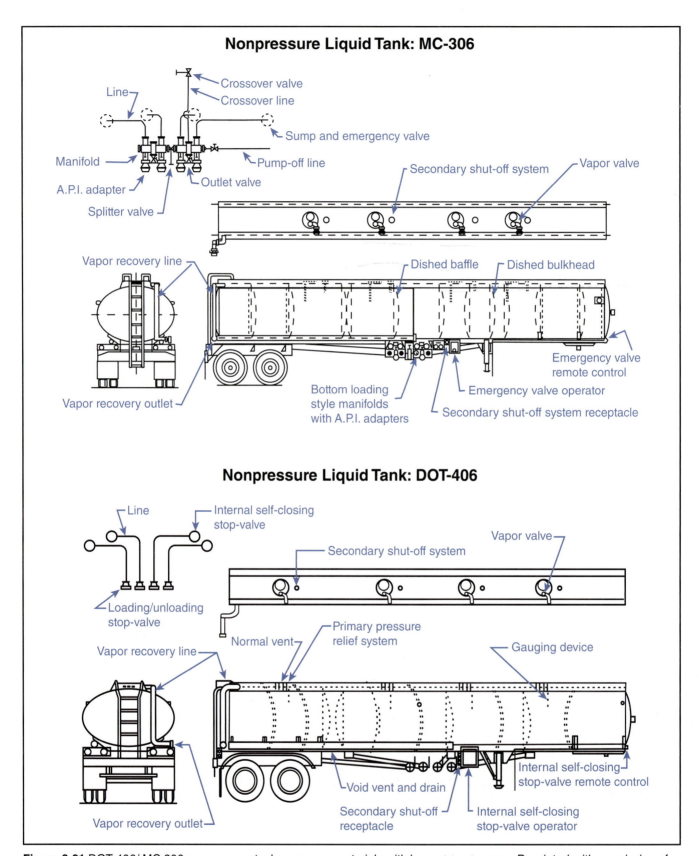

Figure 8.21 DOT 406/ MC 306 nonpressure tank cars carry materials with low vapor pressure. *Reprinted with permission of the Hazardous Materials Response Handbook, Sixth Edition, copyright © 2013. NFPA, Quincy, MA.*

Table 8.3
Nonpressure Liquid Tank

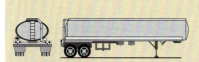

**Nonpressure Liquid Tank
DOT406, TC406, SCT-306
(MC306, TC306)**

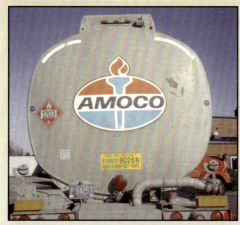

- Pressure less than 4 psi (28 kPa)
- Typical maximum capacity: 9,000 gallons (34 069 L)
- New tanks made of aluminum
- Old tanks made of steel
- Oval shape
- Multiple compartments
- Recessed manways
- Rollover protection
- Bottom valves
- Longitudinal rollover protection
- Valve assembly and unloading control box under tank
- Vapor-recovery system on curb side and rear, if present
- Manway assemblies, and vapor-recovery valves on top for each compartment
- Possible permanent markings for ownership that are locally identifiable

Carries: Gasoline, fuel oil, alcohol, other flammable/combustible liquids, other liquids, and liquid fuel products

Basic Identification

The nonpressure cargo tank can be identified by its elliptical cross-section and nearly flat heads **(Table 8.3)**. The owner's name is usually permanently marked on these oval tanks. These tanks are commonly top loaded and unloaded through discharge valves located at the bottom of the tank. The number of discharge valves is often a good indicator of how many compartments are located within the tank. Nonpressure tanks typically have the following:

- Rollover protection running the length of the tank
- Multiple compartments
- A separate manhole for each compartment
- An emergency shutoff on driver's side front

MC 306 and DOT 406 tankers may carry a wide variety of product quantities. This size tanker typically carries 9,000 gallons (36 000 L), but may also carry significantly more or less volume. Each compartment may have a different volume of product. The volumes must be marked on each compartment. The emergency responder should refer to the **bill of lading** for the exact quantity of product being hauled.

Construction Features

MC 306 cargo tanks are usually constructed of aluminum. However, tanks constructed prior to August 31, 1995, may be constructed of carbon steel. These tanks are often steel-walled and insulated and may be compartmentalized. Each compartment has its own manhole assembly located at the top of the tank. Large compartments may have more than one manhole assembly.

If a tank is not compartmentalized, it will have **baffles** to help control liquid movement. Baffles are similar to the separators that divide tanks, but they are equipped with holes to allow the liquid to move through the baffle at a suppressed rate making the transport of the liquid safer **(Figure 8.22)**. A marking indicating that the tank has baffle holes may be visible on the rear of the tank **(Figure 8.23)**. The marking of the baffle holes will be helpful if the tank needs to be drilled open for off-loading operations. Vapor recovery lines are an integral part of this type of tank but are not an indicator of how many compartments the tank contains.

> **Bill of Lading** — Shipping paper used by the trucking industry (and others) indicating origin, destination, route, and product; placed in the cab of every truck tractor. This document establishes the terms of a contract between a shipper and a carrier. It serves as a document of title, contract of carriage, and receipt for goods. *Similar to* Waybill.

> **Baffle** — Partition placed in vehicular or aircraft water tanks to reduce shifting of the water load when starting, stopping, or turning.

Figure 8.22 Baffles are holes strategically placed in tank barriers to slow the rate of movement of a liquid within the tank. *Courtesy of Barry Lindley.*

Figure 8.23 The locations of baffle holes may be marked on the rear exterior of the tank.

MC 306 tanks are equipped with rollover protection that may run the entire length of the tank. Emergency shutoffs are usually manual and may be located at the front of the tank. This tank may also include a fusible link.

The DOT 406 will have a thicker shell than MC 306 tanks and allows for a maximum pressure of 4 psig (129 kPa). The manways of the DOT 406 tank truck must be able to withstand higher pressures and are rated to be leak free at 36 psig (350 kPa).

The most common leak point of nonpressure cargo tanks is through the manholes and dome covers **(Figure 8.24)**. However, additional points may appear if the cargo tank has been subjected to mechanical damage. The lower discharge valves are traditionally equipped with "shear" type leak protection if the tank is subject to a motor vehicle accident. Even though the shear protection is in place and has activated, the discharge piping may still contain a significant volume of product.

Figure 8.24 Access entryways are common leak points in nonpressure cargo tanks.

Low-Pressure Cargo Tanks

Low-pressure cargo tanks, also known as low-pressure chemical tanks, make up approximately 20 percent of the tanker fleet on the road. These tanks carry the MC 307 or DOT 407 designation **(Figure 8.25, p. 268)**. Low-pressure cargo tanks transport liquids that may have a higher vapor pressure than those products carried in their nonpressure counterparts. Typical contents carried in the low-pressure tanker may include:

- Flammable liquids
- Combustible liquids
- Mild corrosives
- Poisons

Basic Identification

Low-pressure liquid chemical cargo tanks will have a circular cross-section with flat heads. Viewed from behind, insulated tanks may have a horseshoe

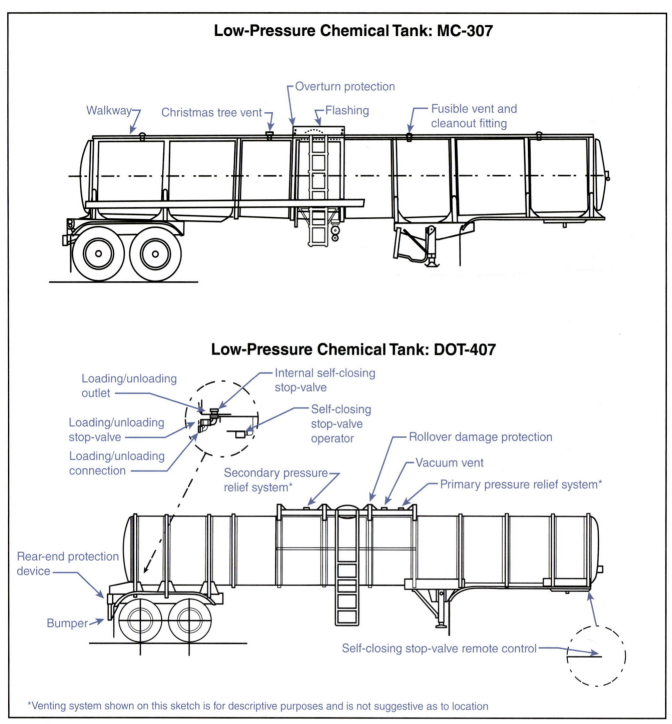

Figure 8.25 DOT 407/MC 307 low pressure chemical tanks are designed to carry materials with higher vapor pressure than atmospheric pressure tanks. *Reprinted with permission of the Hazardous Materials Response Handbook, Sixth Edition, copyright © 2013. NFPA, Quincy, MA.*

shape **(Table 8.4)**. Low-pressure cargo tanks will have a manway at the top but may be loaded from either the manway or through the internal valves located at the bottom of the tank. These tanks usually have a single compartment. The off-loading valve is typically located in the rear of the tank. The manway rollover protection and ladder are typically in the center of the tank when viewed from the side.

Table 8.4
Low-Pressure Chemical Tank

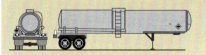

Low-Pressure Chemical Tank
DOT407, TC407, SCT-307
(MC307, TC307)

- Pressure under 40 psi (172 kPa to 276 kPa)
- Typical maximum capacity: 7,000 gallons (26 498 L)
- Rubber lined or steel
- Typically double shell
- Stiffening rings may be visible or covered
- Circumferential rollover protection
- Single or multiple compartments
- Single- or double-top manway assembly protected by a flash box that also provides rollover protection
- Single-outlet discharge piping at midship or rear
- Fusible plugs, frangible disks, or vents outside the flash box on top of the tank
- Drain hose from the flash box down the side of the tank
- Rounded or horse shoe-shaped ends

Carries: Flammable liquids, combustible liquids, acids, caustics, and poisons

The MC 307 tanker typically carries 6,000 gallons (28 000 L) in a capacity tank. In some locations the DOT 407 may be rated with a higher capacity due to its thicker design. The emergency responder should refer to the bill of lading for the exact quantity of product being shipped.

Construction Features

The low-pressure cargo tank may be constructed of aluminum, mild steel, or stainless steel. This tank will have rollover protection around the manway area. Another feature of this cargo tank is the use of stiffening rings to increase the tank's structural integrity. The stiffening rings may or may not be visible based on the design of the tank. These tanks may have an incorporated heating system. The DOT 407 cargo tank will have a thicker shell and material. Some 407 tanks allow for vacuum service. Emergency shutoffs are located on the driver's side front of the tank. These tanks typically have 3-inch (75-mm) cleanout fittings.

A small percentage of low-pressure cargo tanks have multiple compartments. Each compartment of a compartmentalized tank is considered a separate tank. Compartmentalized tanks can carry more than one class of product at a time. Safety features of low-pressure cargo tanks include a fusible cap.

As with most cargo tanks, the manways and valves are a common point for leakage **(Figure 8.26)**. Tank leaks may be difficult to locate due to the presence of insulation. Inspection is paramount for this type of tank.

Figure 8.26 As with other types of tanks, the access entryways of low-pressure tanks are a common leak point.

Corrosive Liquid Tanks

The corrosive liquid tank, also called a corrosive cargo tank, is a uniquely shaped tanker that transports heavy, high density liquids and toxic inhalation hazards (TIH). These tanks account for around 12 percent of containers on the road. Corrosive tanks will have either a MC 312 or DOT 412 designation **(Figure 8.27)**. These tanks typically carry materials that are corrosive in nature like sodium hydroxide, hydrochloric acid, and sulfuric acid. MC 312 cargo tanks are used as vacuum trucks and may carry products besides corrosives.

Basic Identification

The corrosive liquid tank typically features the manway and valves located in the rear and discharge lines located in the top rear of the tank **(Table 8.5, p. 272)**. Because corrosives are usually heavy, the overall volume that can be carried in this type of tanker is typically lower than that of other types. Because of its relatively small capacity, the tank will appear to be small in diameter and have convex heads. External stiffening rings are a trademark of corrosive tanks. These tanks can also be insulated and/or heated.

Corrosive cargo tankers are traditionally single tanks with no compartmentalization. For the exact quantity being transported, the emergency responder should refer to the bill of lading.

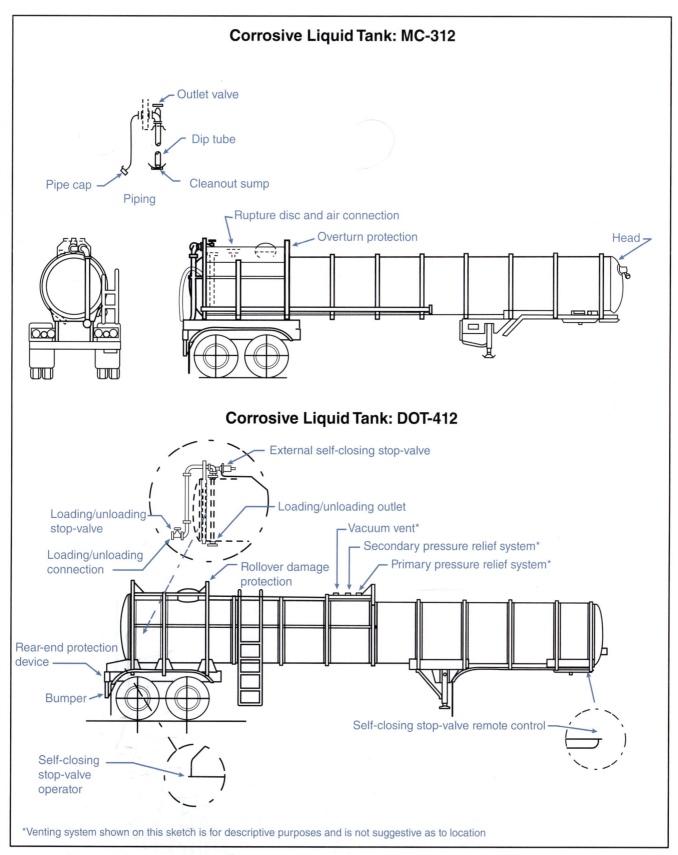

Figure 8.27 DOT 412/MC 312 corrosive liquid tanks carry materials with high density and high toxic inhalation hazard.
Reprinted with permission of the Hazardous Materials Response Handbook, Sixth Edition, copyright © 2013, NFPA, Quincy, MA.

Table 8.5
Corrosive Liquid Tank

 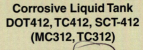

**Corrosive Liquid Tank
DOT412, TC412, SCT-412
(MC312, TC312)**

- Pressure less than 75 psi (517 kPa)
- Typical maximum capacity: 7,000 gallons (26 498 L) [per NFPA]
- Rubber lined or steel
- Typically single compartment
- Small-diameter round shape
- Exterior stiffening rings may be visible on uninsulated tanks
- Typical rear top-loading/unloading station with exterior piping extending to the bottom of the tank
- Splashguard serving as rollover protection around valve assembly
- Flange-type rupture disk vent either inside or outside the splashguard
- May have discoloration around loading/unloading area or area painted or coated with corrosive-resistant material
- Permanent ownership markings that are locally identifiable

Carries: Corrosive liquids (usually acids)

Construction Features

The medium-pressure or corrosive cargo tank is typically made of stainless steel or carbon steel and may be lined with several different materials. Chlorobutyl or butyl rubber is a typical lining for corrosives. Corrosive tanks can also be made of aluminum or fiberglass reinforced plastic (FRP). Tanks can be insulated or noninsulated. Tanks contain stiffening rings that are located closer together to help increase the tank's integrity. If it is noninsulated, these rings are visible. These tanks will also feature rollover and splash protection around the manway and fittings. Typically, these tanks are top unloading and the piping comes from the top of the tank to the ground level.

Typically the fittings for the piping and valves are flanged and include gaskets between the flanges. Most of these tanks do not have emergency shutoffs and the valves must be manually opened and closed. These tanks were originally designed to carry corrosives. They are often discolored around the loading and unloading areas and may be painted with corrosive-resistant paint. Today many of these tanks carry noncorrosive toxic inhalation hazard (TIH) materials.

As with most cargo tanks, the manways and valves are common leakage points. Because of the nature of the contents, the tank may be prone to leakage and failure if the product leaks through its liner. Inspection is paramount for this type of tank.

High-Pressure Cargo Tanks

High-pressure cargo tanks transport liquefied gases and high vapor pressure materials. The contents must remain under pressure in order to maintain a liquid state. These tanks will carry the designation of MC 331 **(Figure 8.28)**.

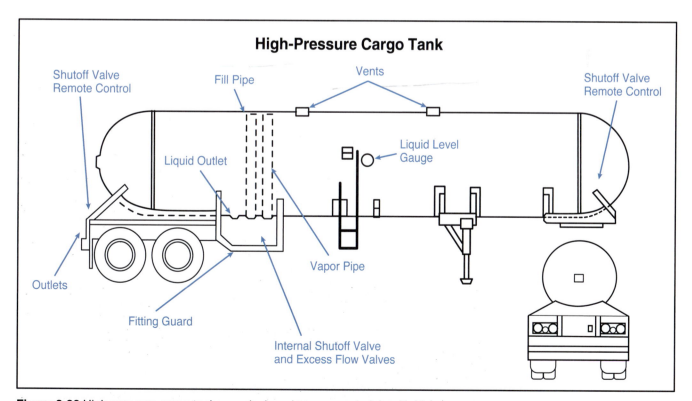

Figure 8.28 High-pressure cargo tanks are designed to carry materials with high vapor pressures.

High-pressure cargo tanks account for approximately 10 percent of the tanks on our roadways. Common products shipped in high-pressure cargo tanks may include:

- Anhydrous ammonia
- Propane
- Chlorine
- Other gases that have been liquefied under pressure

Basic Identification

High-pressure cargo tanks are round with protruding, rounded heads **(Table 8.6)**. The rounded head enables the tank to withstand these high internal pressures. While the MC 331 is considered a highway bulk tank, the propane "bobtail" truck is its intercity counterpart **(Figure 8.29)**. Bobtail-type vehicles carry substantially less than the typical capacity of high-pressure cargo tanks.

Figure 8.29 Bobtail-style trucks are designed to carry a smaller volume of high-pressure material in an urban environment.

The DOT requires that the upper two-thirds of noninsulated tanks be painted white or another highly reflective color to help decrease the tank's internal temperatures. Chlorine trucks are MC 331 tanks but look different from other high-pressure cargo tanks because they have a domed protective housing on the rear. The emergency responder should refer to the bill of lading for the exact quantity of product being transported. The liquid gauge can also indicate the amount of liquid in the tank.

Construction Features

High-pressure cargo tanks are constructed of steel or insulated aluminum and are not compartmentalized. All valves on the MC 331 tank must be labeled to indicate whether it will control liquid or vapor **(Figure 8.30)**. If the cargo tank has a water capacity below 3,500 gallons (14 000 L) it must have at least one emergency shutoff valve. Any MC 331 tank with a water capacity greater than 3,500 gallons (14 000 L) must have both mechanical and thermal dis-

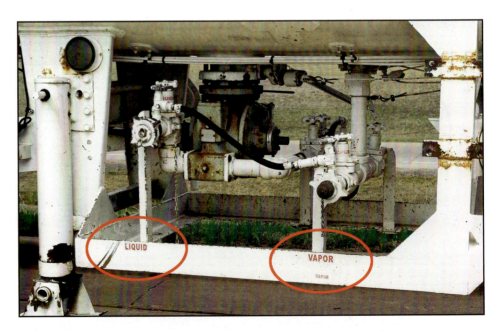

Figure 8.30 The valves on an MC 331 tank must be labeled to indicate whether it will control liquid or vapor.

Table 8.6
High-Pressure Cargo Tank

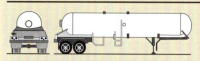

High-Pressure Tank
MC-331, TC331, SCT-331

- Pressure above 100 psi (689 kPa)
- Typical maximum capacity: 11,500 gallons (43 532 L)
- Single steel compartment
- Noninsulated
- Bolted manway at front or rear
- Internal and rear outlet valves
- Typically painted white or other reflective color
- Large hemispherical heads on both ends
- Guard cage around the bottom loading/unloading piping
- Uninsulated tanks, single-shell vessels
- Permanent markings such as the product name

Carries: Pressurized gases and liquids, anhydrous ammonia, propane, butane, and other gases that have been liquefied under pressure

High-Pressure Bobtail Tank

Used for local delivery of liquefied petroleum gas and anhydrous ammonia

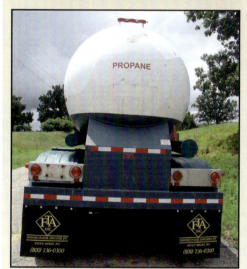

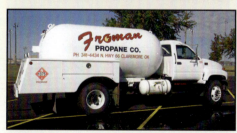

charge control valves located at each end of the tank. Safety valve thresholds must be set at 110 percent of the tank's overall design pressure. These tanks must include temperature and pressure gauges and may have liquid gauging devices **(Figure 8.31)**.

The MC 331 is a very rugged tank designed to protect its contents. The most common leak point for this type of tanker is typically within the valve mechanisms. Flanges of the valves and the plugs and caps are also common leak points.

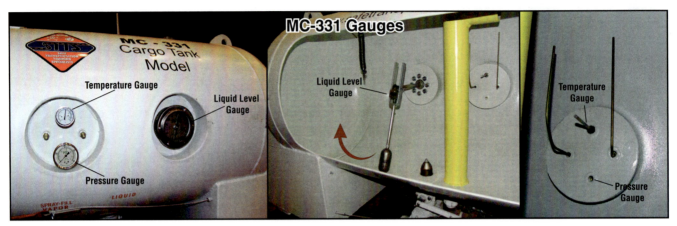

Figure 8.31 This high-pressure tank cutaway model shows gauges for temperature, pressure, and liquid levels.

Cryogenic Tanks

Cryogen — Gas that is converted into liquid by being cooled below -130°F (-90°C). *Also known as* Refrigerated Liquid *and* Cryogenic Liquid.

Liquefied Natural Gas (LNG) — Natural gas stored under pressure as a liquid.

As previously mentioned, cryogenic tankers are specially designed to carry gases that have been liquefied by reducing their overall temperature. The contents of these tanks will be extremely cold, which may pose more of a hazard than those associated with the product itself. Cryogenic tanks are classified as MC 338. Cryogenic cargo tanks make up a small percentage of the tanks that transport product over roadways. Cryogenic materials offer a unique safety hazard, and the tanks used to contain them have many features intended to safety control the product, but those features are limited. Most manufacturers of **cryogens** will have their own response teams.

Common products carried in cryogenic tanks include gases that have been liquefied by lowering the temperature, such as:

- Liquid oxygen
- Liquid nitrogen **(Figure 8.32)**
- Liquid carbon dioxide
- Liquid hydrogen
- **Liquefied natural gas (LNG)**

WARNING!
The rapid expansion of vapors from cryogens can quickly displace oxygen.

Figure 8.32 Cryogenic tanks are often labeled as a refrigerated liquid, as is this container placarded for liquid nitrogen. *Courtesy of Barry Lindley.*

Basic Identification

Because cryogenic liquids are transported at extremely cold temperatures, these tankers must be adequately insulated to protect their contents. This insulation will give the tank a bulky appearance **(Table 8.7, p. 278)**. This style of tank is deceptive because it holds far less product than it appears. The tank is round with flat ends. A loading/unloading station will be located either in the rear of the tank or just forward of the rear wheels **(Figure 8.33)**. To determine the actual quantity of product being transported, the emergency responder should refer to the bill of lading.

Figure 8.33 This cryogenic tank has a loading/unloading station at the rear of the tank.

Construction Features

The cryogenic tank is constructed of aluminum or stainless steel and has flat heads. The material is based on its service temperature rating. It is comprised of a welded inner tank that holds the product surrounded by a vacuum space that contains insulating material. The MC 338 has a final outer shell made of steel.

Table 8.7
Cryogenic Liquid Tank

**Cryogenic Liquid Tank
MC338, TC338, SCT-338
(TC341, CGA341)**

- Well-insulated steel or aluminum tank
- Possibly discharging vapor from relief valves
- Round tank with flat ends
- Large and bulky double shelling and heavy insulation
- Loading/unloading station attached either at the rear or in front of the rear dual wheels, typically called the doghouse in the field
- Permanent markings such as *REFRIGERATED LIQUID* or an identifiable manufacturer name

Carries: Liquid oxygen, liquid nitrogen, liquid carbon dioxide, liquid hydrogen, and other gases that have been liquefied by lowering their temperatures

A pressure gauge must be located so that the driver may view it from the cab. MC 338 tanks that carry products, such as oxygen, must have discharge precautions set at 110 percent of the design pressure of the tank. Additionally, a thermal closure must activate at a preset temperature.

The MC 338 is an extremely rugged tank designed and built as a tank within a tank. The valves are the most vulnerable to leakage. However, mechanical damage and stress can compromise the integrity of the tank.

A unique feature of the MC 338 is its ability to vent based on temperature and pressure. What may appear as a leak or the activation of a pressure relief device may actually be the result of properly working safety equipment **(Figure 8.34)**.

Figure 8.34 The MC 338 is designed to spontaneously vent, depending on temperature and pressure conditions.

Tube Trailers

The DOT does not classify compressed-gas tube trailers as cargo tanks. This unique carrier is actually a modified semi-trailer and is comprised of individual steel tubes that may be stacked and banded together **(Table 8.8, p. 280)**. The tubes may carry individual quantities of gas or may be linked together in a cascade-style system. The tubes have a high internal working pressure. Only one product can be carried in each tube at a time.

These tubes carry gas under pressure and occasionally liquefied gases such as anhydrous hydrochloric acid. Products commonly transported in tube trailers include:

- Argon
- Helium **(Figure 8.35)**

Figure 8.35 Compressed helium may be transported via tube trailers.

- Carbon dioxide
- Nitrogen
- Refrigerant gases
- Silicon tetraflouride

Table 8.8
Compressed-Gas/Tube Trailer

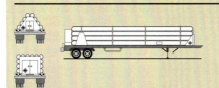

Compressed-Gas/Tube Trailer

- Pressure at 3,000 to 5,000 psi (20 684 kPa to 34 474 kPa) (gas only)
- Individual steel cylinders stacked and banded together
- Typically has over-pressure device for each cylinder
- Bolted manway at front or rear
- Valves at rear (protected)
- Manifold enclosed at the rear
- Permanent markings for the material or ownership that is locally identifiable

Carries: Helium, hydrogen, methane, oxygen, and other gases

Dry Bulk Carriers

Dry bulk carriers are nonspecification cargo tanks that are not regulated and do not conform to DOT specifications. Nonspecification cargo carriers are off-loaded through bottom ports. An auxiliary engine compressor may assist pneumatic off-loading. These unique tanks are distinguished by their large sloping V-shaped compartments, known as hoppers (**Table 8.9**).

Table 8.9
Dry Bulk Cargo Trailer

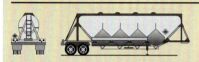

Dry Bulk Cargo Trailer

- Pressure usually between 15 psi (100 kPa) to 25 psi (170 kPa)
- Typically not under pressure
- Bottom valves
- Shapes vary, but has V-shaped bottom-unloading compartments
- Rear-mounted, auxiliary-engine-powered compressor or tractor-mounted power-take-off air compressor
- Air-assisted, exterior loading and bottom unloading pipes
- Top manway assemblies

Carries: Oxidizers, corrosive solids, cement, plastic pellets, and fertilizers

Common products transported may include:

- Fertilizers
- Corrosive solids
- Oxidizers
- Cement

Railway Tank Cars

NFPA 1072 discusses a wide variety of railway tank cars. Tank cars are classified according to their construction features, fittings, and function. Responders should evaluate the types of railway tank cars in their jurisdiction. **Table 8.10** provides the designated class numbers and appropriate approving authority for the different types of cars.

Table 8.10 Tank Car Class Numbers and Approving Authority	
Type of Car	**Approving Authority and Class Numbers**
Nonpressure Tank Cars	DOT-103
	DOT-111
	AAR-201
	DOT-104
	DOT-115
	AAR-203
	AAR-206
	AAR-211
Pressure Tank Cars	DOT-105
	DOT-114
	DOT-109
	DOT-120
	DOT-112
Cryogenic Liquid Tank Cars	DOT-113
	AAR-204W

NOTE: DOT = Department of Transportation, AAR = Association of American Railroads

Shipments of hazardous materials may also be transported in specialty railcars. Some specialty tank cars will be introduced later in this chapter. In addition to covering the different types of tank cars, this section will address tank car markings, structure, safety features, and fittings.

Tank Car Markings

A tank car must display several different markings. These markings may be helpful to hazmat technicians, enabling them to identify contents, shippers/car owners, and tank car capacities.

Three of these markings include:

- Reporting marks (initials and numbers)
- Specification marking
- Capacity stencil

Railcar initials and numbers (also known as reporting marks) are a way to identify the tank and its owner. Reporting marks are stenciled on the left

> **Railcar Initials and Numbers** — Combination of letters and numbers stenciled on rail tank cars that may be used to get information about the car's contents from the railroad's computer or the shipper. *Also known as* Reporting Marks.

side of the tank and on each end. Some shippers also stencil these numbers on the top of the car in case an accident turns the car on its side. Reporting marks include up to four letters indicating the tank's owner **(Figure 8.36)**. Additionally, the reporting marks may include up to six digits **(Figure 8.37)**. The combination of letters and numbers along with shipping papers such as the train **consist** (train list) and wheel report can identify the tank owner.

> **Consist** — Rail shipping paper that contains a list of cars in the train by order; indicates the cars that contain hazardous materials. Some railroads include information on emergency operations for the hazardous materials on board with the consist. *Also known as* Train Consist.

Figure 8.36 The four letters in this reporting mark identifies the tank's owner.

Figure 8.37 The six numbers in this reporting mark identifies the tank itself.

Specification markings are stenciled on the right side of the tank on the longitudinal side **(Figure 8.38, p. 284)**. These marks represent the DOT, TC, or American Association of Railroads (AAR) standards to which the tank car

> **Specification Marking** — Stencil on the exterior of a tank car indicating the standards to which the tank car was built; may also be found on intermodal containers and cargo tank trucks.

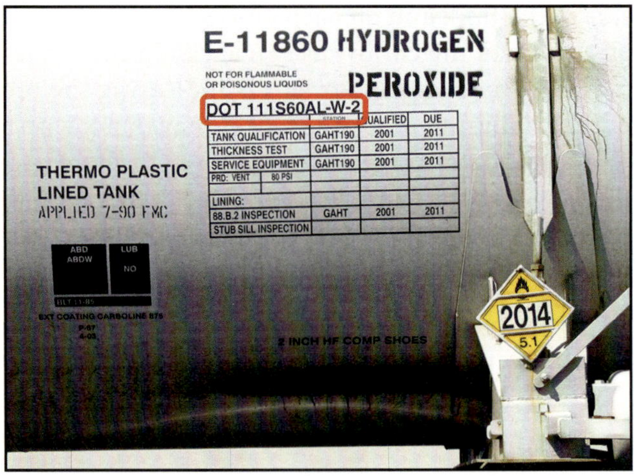

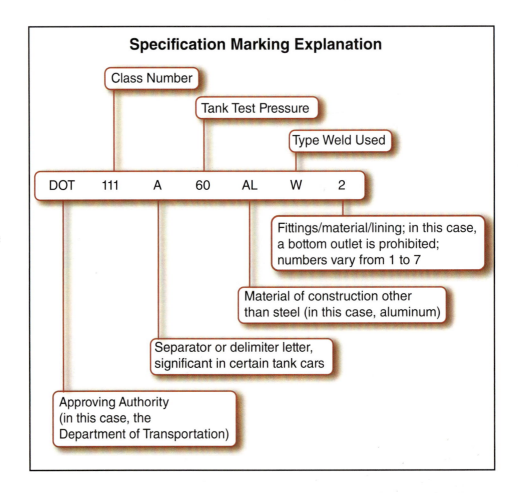

Figure 8.38 Specification markings are located on a predictable place on the tank itself.

Figure 8.39 Specification markings include some specific types of information unique to the container itself.

was constructed. The specification markings do not identify the tank's cargo **(Figure 8.39)**.

The **capacity stencil** shows the volume of the rail tank car. The volume in gallons (and sometimes liters) is stenciled on both ends of the car under the car's reporting marks. The volume in pounds (and sometime kilograms) is stenciled on the sides of the cars under the reporting marks **(Figure 8.40)**. For certain tank cars (DOT-111A100W4 cars), the water capacity (water weight) of the tank in pounds (and typically kilograms) is stenciled on the sides of the tank near the center of the car.

Capacity Stencil — Number stenciled on the exterior of a tank car to indicate the volume of the tank. *Also known as* Load Limit Marking.

Figure 8.40 The volume of a rail tank car in both weight and volume is stenciled on the side and both ends. *Courtesy of Rich Mahaney.*

New-style tank cars are equipped with two identical identification plates; all other cars have this identification stamped into the heads of the cars. These plates must be permanently mounted on the inboard surface of the tank's structure. Information included on the tank's identification plates will include the material from which the tank is constructed and specified equipment such as bottom and top shelf couplers, head shields, and any thermal protection.

Additionally, some materials shipped by rail must feature the name of that product stenciled on the side of the tank. Tank cars with stenciled markings are known as dedicated tank cars. These cars are allowed to carry only the product which is stenciled on the tank **(Table 8.11, p. 286)**. If another product is to be shipped in this container, the car must qualify for the new product and have new stenciling applied.

Tank Car Structure

A rail tank car includes the tank and the truck assembly (body) of the car. The truck assembly is similar to a chassis and includes the wheels, axles, truck bolsters, and bowl pins. It is possible to construct the car in several different ways.

The bottom of the tank may be **frameless**, called a *stub sill*, where all of the stresses of the railcar will be borne by the tank itself **(Figure 8.41)**. The stub sill is a short structural member welded to the end of the tank. The stub sill attaches the tank to the truck assembly and absorbs the forces of train movement.

> **Frameless Tank Car** — Direct attachment of a rail tank car to the truck assembly. This type of construction transfers all of the stresses of transport from the railcar to the stub sill assembly and the tank itself. *Known as* Stub Sill.

Figure 8.41 A stub sill is welded to the end of the tank and transfers the stress of the train's movement to the tank itself. *Courtesy of Rich Mahaney.*

Table 8.11
Stenciled Commodity Names

ACROLEIN, STABILIZED	HYDROGEN FLUORIDE, ANHYDROUS
AMMONIA, ANHYDROUS, LIQUEFIED	HYDROGEN PEROXIDE, AQUEOUS SOLUTIONS (greater than 20% hydrogen peroxide)
AMMONIA SOLUTIONS (more than 50% ammonia)	HYDROGEN PEROXIDE, STABILIZED
BROMINE *or* BROMINE SOLUTIONS	HYDROGEN PEROXIDE *and* PEROXYACETIC ACID MIXTURES
BROMINE CHLORIDE	NITRIC ACID (other than red fuming)
CHLOROPRENE, STABILIZED	PHOSPHORUS, AMORPHOUS
DISPERSANT GAS *or* REFRIGERANT GAS (as defined in §173.115 of this subchapter)	PHOSPHORUS, WHITE DRY *or* PHOSPHORUS, WHITE, UNDER WATER *or* PHOSPHORUS WHITE, IN SOLUTION, *or* PHOSPHORUS, YELLOW DRY *or* PHOSPHORUS, YELLOW, UNDER WATER *or* PHOSPHORUS, YELLOW, IN SOLUTION
DIVISION 2.1 MATERIALS (47 listed)	
DIVISION 2.2 MATERIALS (in Class DOT 107 tank cars only) (47 listed)	
DIVISION 2.3 MATERIALS (49 listed)	PHOSPHORUS WHITE, MOLTEN
FORMIC ACID	POTASSIUM NITRATE *and* SODIUM NITRATE MIXTURES
HYDROCYANIC ACID, AQUEOUS SOLUTIONS	POTASSIUM PERMANGANATE
HYDROFLUORIC ACID, SOLUTION	SULFUR TRIOXIDE, STABILIZED
HYDROGEN CYANIDE, STABILIZED (less than 3% water)	SULFUR TRIOXIDE, UNINHIBITED

Source: §172.330 Tank cars and multi-unit tank car tanks.

Continuous Underframe Tank Car — Construction of a rail tank car that includes full support of the tank car. The underframe rests on the truck assembly during transport. *Also known as* Full Sill.

Figure 8.42 A full sill runs the entire length of the tank and absorbs the stress of the train's movement. *Courtesy of Rich Mahaney.*

In contrast, a rail tank car may have a **continuous underframe**, called *full sill* (**Figure 8.42**). Tank cars manufactured with a continuous underframe feature a one-piece assembly that runs the length of the railcar. The continuous underframe absorbs all of the forces created by the train's movement. The topside of the underframe holds the tank in place. The bottom of the underframe rests on the truck assembly. A body bolster is a structural cross member mounted at a right angle to the underframe. In essence, it is a cradle for the tank. Tank cars equipped with body bolsters will have one bolster at each end of the railcar.

Safety Features of Railway Tank Cars

A railway tank car may be equipped with various features that can help increase the safety of product transportation. The following sections will discuss safety features that may be found on different railcars.

Head Shields

Head shields help protect the heads of a tank car when transporting hazardous materials. All pressure cars must have head shields. The head shields may or may not be visible. Head shields offer an extra layer of puncture protection on the ends of the tank. If required, newly constructed tanks will have full head shields **(Figure 8.43)**. Older tanks may have a "half head" or a trapezoidal plate of steel welded to the lower half of the tank ends **(Figure 8.44)**. Jacketed tank cars may incorporate a full plate that protects the entire head of the tank.

> **Head Shield** — Layer of puncture protection added to the head of tanks. Head shields may or may not be visible, depending on the construction of the tank and the type of protection provided.

Figure 8.43 Full head shields protect the entire surface of one end of a pressure car. *Courtesy of Steve George.*

Figure 8.44 Half head shields protect the bottom half of one end of a pressure car. *Courtesy of Rich Mahaney.*

Thermal Insulation — Materials added to decrease heat transfer between objects in proximity to each other.

Thermal Protection — Materials added to the shell of a railway tank car to increase the durability of the tank car against direct flame impingement or a pool of fire.

Insulation

Thermal insulation helps to protect a tank's cargo from outside temperatures. Insulation may be found on both pressure and nonpressure tank cars. Cryogenic tank cars always have insulation to protect their cold contents.

Fiberglass and polyurethane foams are the most common types of insulating materials used on railway tank cars. Perlite is used to insulate cryogenic products. The tank's outer jacket conceals the insulation.

Thermal Protection

In contrast to insulation, **thermal protection** is designed to protect a tank car from direct flame impingement or a pool of fire. Thermal ratings are as follows: pool fires for 100 minutes and torch fires for 30 minutes. This type of protection holds for tank cars shipping either a liquefied flammable gas such as propane or poisonous gases. Some cars incorporate both thermal protection and insulation to protect cargo such as ammonia.

The primary type of thermal protection used on tank cars is jacketed thermal protection. Jacketed thermal protection consists of mineral wool or other manufactured ceramic fibers in a blanket form which is held in place by the outer metal jacket. Spray-on thermal protection consists of a rough textured coating containing asbestos that is sprayed onto the tank. This product will expand when exposed to fire.

Figure 8.45 This tank has electric coils to maintain the temperature of its liquid contents. *Courtesy of Rich Mahaney.*

CAUTION
Older thermal protection materials may contain asbestos.

Heating Coils

Some tanks may be equipped with heating coils located either inside or outside the tank **(Figure 8.45)**. Steam, hot water, or heated oil from an external source can be used to heat thick or solidified materials such as asphalts or waxes. Heating coils may be in place to reheat materials such as metallic sodium and molten sulfur.

Outlets and inlets for interior coils must have caps in place during transport so that if there is an accident, no material will release into the environment. Caps are not required for exterior coils.

Figure 8.46 Tank cars that carry hazardous materials must be equipped with train car couplers that meet FRA/AAR Standards. *Courtesy of Rich Mahaney.*

Top and Bottom Shelf Couplers

Top and bottom shelf couplers are train car couplers with vertical restraint mechanisms that reduce the potential for coupler disengagement **(Figure 8.46)**. Tank cars that are transporting hazardous materials must have this safety equipment in place.

Skid Protection

Skid protection is a safety feature that prevents the loss of a tank car's contents in the event of a derailment. The skid plate attaches to the tank in the area of the bottom fittings. There is also top skid protection, which will help reduce the amount of mechanical stress on the tank and any fitting located on the bottom of the car.

Tank Car Fittings

Tank cars may be equipped with a variety of fittings that allow for loading and unloading of product, gauges to determine product levels and temperatures, and safety features such as pressure relief devices **(Figure 8.47)**. In order to effectively communicate with railway personnel on scene, responders should be able to recognize and understand the function of common features.

Figure 8.47 Fittings on rail cars include pressure regulation equipment on cryogenic or carbon dioxide carriers. *Courtesy of Rich Mahaney.*

Manways

Manways are the most obvious fittings found on most tank cars **(Figure 8.48)**. They are large openings located at the top of the cars and will usually have permanently attached cover plates. Manways allow access into the interior of the tank so that it may be cleaned, inspected, and repaired. They play an integral part in tank car identification.

Valves and Venting Devices

Valves are the fittings that allow product to flow in one direction or another and are the primary means of loading and unloading **(Figure 8.49, p. 290)**. Common valves found on railway tanks include plug type and ball type. An eduction pipe connects the valve to the bottom of the tank. If equipped, sumps

> **Manway** — Opening that is large enough to admit a person into a tank. This opening is usually equipped with a removable, lockable cover. *Also known as* Manhole.

> **Valve** — Mechanical device with a passageway that controls the flow of a liquid or gas.

Figure 8.48 Access points for cleaning, maintenance, and inspection are often large and easily noticeable. *Courtesy of Rich Mahaney.*

Figure 8.49 Valves are intended to allow the product to flow into or out of the tank. *Courtesy of Rich Mahaney.*

Figure 8.50 Pressure relief valves are set to release the contained gases until the internal pressure is reduced back to an acceptable level. *Courtesy of Rich Mahaney.*

Figure 8.51 Safety relief devices are set to rupture and vent the contained gases to prevent the tank itself from bursting. *Courtesy of Rich Mahaney.*

are located at the lowest point of the tank — even below the eduction pipe on a flat bottom tank. Product will flow into a sump and allow for total drainage of the tank. Tank cars often include safety features to protect valves and piping on top of the car to protect those features in the incident of a roll over. These safety features include valve protection and housings.

Safety Relief Devices

Safety relief devices allow the tank's internal pressure to be relieved. Most relief devices are spring-operated to allow the device to close when the tank's internal pressure is reduced to normal limits. Pressure relief devices and safety vents are two types of safety relief devices found on tank cars.

Pressure relief devices (PRDs), including **pressure relief valves**, are typically set to activate at 75 percent of the tank's test pressure **(Figure 8.50)**. The PRD may activate at 82.5 percent for some gases. PRDs are only effective when the tank is upright. Pressures can be identified by the stencil on the car.

Safety relief devices are different from PRDs in that a frangible disk will rupture at a predetermined temperature or pressure **(Figure 8.51)**. The predetermined pressure is typically 33 1/3 percent of the burst pressure of the tank. Unlike relief devices, once a safety vent opens it cannot be closed. Once this device has been activated, someone who has been properly trained must replace it.

Some vents are combination units that have a rupture disk or breaking pin that is spring operated. These are typically set at 75 percent of the test pressure of the tank. There is a telltale (indicator) valve between the rupture disk and the relief valve. The use of this telltale valve tells whether the rupture disk has activated.

Vacuum relief valves prevent internal vacuums from occurring in nonpressure tanks during normal temperature changes. The purpose of these valves is to introduce outside air into the tank. The vacuum relief valves close when there is no longer a vacuum. These valves are usually set around 0.75 psig (106 kPa). If operating properly, these devices should prevent a tank from collapsing during off-loading operations.

Other Fittings

Some types of fittings are provided to simplify evaluation of a railway tank's conditions. These fittings include:

- **Sample lines** — Allow a sample of the tank's product to be taken without unloading the car.

- **Thermometer wells** — Closed tubes filled with a liquid like ethylene glycol that extend into a tank to allow a thermometer to be introduced to sample the temperature of the product.
- **Gauging devices** — Tools used to measure the amount of product or vapor space in a tank.
- **Bottom outlet valves** — Fittings located at the bottom of the tank that can be used for either off-loading or cleaning.

General Service (Nonpressure) Railway Tank Cars

General service railway tank cars are the most common type of tank car in North America. These types of cars are commonly categorized as DOT 111, with some variation in the allowed parameters. These tanks carry a variety of liquids, solids, and molten solids, including materials such as:

- Crude oil
- Sulfuric acid
- Vegetable oils
- Wastes **(Figure 8.52)**
- Caustic soda
- Tallow
- Fruit juice
- Ethanol
- Molten sulfur
- Alcohols
- Wine
- Solvents
- Tomato paste

> **Safety Relief Device** — Device on rail car cargo tanks with an operating part held in place by a spring; the valve opens at preset pressures to relieve excess pressure and prevent failure of the vessel.
>
> **Pressure Relief Valve** — Pressure control device designed to eliminate hazardous conditions resulting from excessive pressures by allowing this pressure to release in manageable quantities.
>
> **Vacuum Relief Valve** — Pressure control device designed to introduce outside air into a container during periods of heating and cooling.

Figure 8.52 This nonpressure tank car is placarded for liquid hazardous wastes.

CAUTION
The appearance of general service tank cars may be slightly changed with the addition of housings intended to protect valves in the instance of a rollover.

Because general service tanks are so widely used by the railway industry, the only way to determine the tank's contents is by shipping papers, placards, or tank markings. Common hazardous materials transported by these tanks include:

- Flammable and combustible liquids
- Flammable solids
- Oxidizers and organic peroxides
- Liquid poisons
- Corrosives

Basic Identification

Nonpressure tank cars are cylindrical in shape with rounded heads. They will have at least one manway to access the interior of the tank. Fittings for loading and unloading and other hardware will typically be visible on the exterior of the tank. Some new DOT 111 tank cars now enclose fittings in a single protective housing, similar to a pressure tank car **(Figure 8.53)**. If a single protective housing is present on a tank car, check the specification marks to confirm if it is a pressure or nonpressure car.

Figure 8.53 Newer tank cars may use a housing to protect external fittings.

Nonpressure tank cars are designed for materials with vapor pressures of 25 psig (274 kPa) or less at 70° Fahreinheit (21° Celcius). Responders should refer to the **waybill** to determine the total contents of all compartments in the tank.

> **Waybill** — Shipping paper used by a railroad to indicate origin, destination, route, and product; a waybill for each car is carried by the conductor. *Similar to* Bill of Lading.

Construction Features

Most nonpressure tank cars are constructed from carbon steel. Aluminum and stainless steel may also be used. The tank car may be manufactured with a full or stub sill. If the tank car is not constructed with an underframe, the tank must be designed to bear the stress of all train movement.

Nonpressure tank cars may be compartmentalized. If the tank has compartments, each one must be used as a separate tank and have its own fittings and expansion dome if equipped **(Figure 8.54)**. All railway tankers must be built to mechanical standards designed for rail freight cars. Additionally, they must meet *49 CFR Part 179* and the *AAR Specifications for Tank Cars*.

Figure 8.54 The two compartments in this tank car have independent fittings. *Courtesy of Rich Mahaney.*

This style of tank car is most prone to leaks at the valves and fittings. Manways are the most common source of a leak. As with any tank, mechanical damage may occur in the event of a railway accident and may compromise the tank's integrity. Carefully evaluate the tank and its contents if involved in an accident.

A damaged tank may explosively fail via **heat induced tear** when temperatures are high enough. This type of container failure is common for low pressure tank cars transporting flammable/combustible liquids such as crude oil and ethanol. Fire causes the tank shell to tear and fail, often along a seam. Heat induced tears primarily occur in low-pressure containers in contrast to **boiling liquid expanding vapor explosions (BLEVEs)** which primarily occur in pressure containers.

Pressure Railway Tank Cars

Similar in design to nonpressure tank cars, the pressure tank car is able to carry liquids of a high vapor pressure or highly hazardous materials. Pressure tank car specifications include DOT 105, a common car for chlorine; DOT 112, a common tank car for **liquefied petroleum gas (LPG)** and ammonia; and DOT 114, a common tank for refrigerants. Common types of products transported in pressure tank cars include:

- Liquefied gases
- Flammables
- Water reactives
- Toxics
- Corrosives

Examples of specific products include:

- Acrolien
- Anhydrous hydrogen fluoride
- Bromine

Heat Induced Tear — Rupture of a container caused by overpressure, often along the seam. This type of failure primarily occurs in low-pressure containers transporting flammable/combustible liquids.

Boiling Liquid Expanding Vapor Explosion (BLEVE) — Rapid vaporization of a liquid stored under pressure upon release to the atmosphere following major failure of its containing vessel. Failure is the result of over-pressurization caused by an external heat source, which causes the vessel to explode into two or more pieces when the temperature of the liquid is well above its boiling point at normal atmospheric pressure.

Liquefied Petroleum Gas (LPG) — Any of several petroleum products, such as propane or butane, stored under pressure as a liquid.

- Ethylene oxide
- Pyrophoric liquids
- Sodium metal
- Chlorine **(Figure 8.55)**

Figure 8.55 The DOT 105 tank car is commonly used to transport chlorine. *Courtesy of Rich Mahaney.*

- Liquefied petroleum gas (LPG)
- Carbon dioxide
- Anhydrous hydrochloric acid
- Vinyl chloride
- Methyl amines
- Anhydrous ammonia

Basic Identification

Cylindrical in cross-section, the pressure tank car has an enclosed protective housing mounted on the pressure plate located around the center tank. These tanks may be insulated. Responders should refer to the waybill of the relevant tank because they may vary by the manufacture date, type of contents, and capacity.

Construction Features

Pressure tank cars are constructed of steel, stainless steel, or aluminum and have rounded heads. These tankers load in a standard way and typically have their fittings (including loading and unloading pressure relief engaging) inside a protective housing mounted on the pressure plate.

All railway tankers must be built to meet mechanical standards designed for rail freight cars as well as *49 CFR Part 179* and the *AAR Specifications for Tank Cars*. Pressure tankers must also meet or exceed ASME standards for pressure vessels.

This style of tank car is most prone to leaks in the valves, fittings, and attachments. As with any tank, mechanical damage may occur in the event of a railway accident and may compromise the tank's integrity. Carefully evaluate the tank and its contents if involved in an accident.

Cryogenic Railway Tank Cars

Cryogenic liquid rail tankers carry low-pressure refrigerated liquids. Cryogenic tank car specifications include DOT 113 and AAR 2014. These refrigerated liquids are transported at temperatures of -130°F (-90°C) and below. These products are gases in their natural state but have been cooled through refrigeration to become a liquid. Common products transported in cryogenic tank cars include:

- Argon
- Liquefied natural gas (LNG)
- Hydrogen
- Ethylene **(Figure 8.56)**
- Nitrogen
- Oxygen

Figure 8.56 This cryogenic tank car is labeled for ethylene. *Courtesy of Barry Lindley.*

Basic Identification

Cryogenic tank cars have a cylindrical cross-section with round heads. The size of the tank may not be representative of the amount of product carried. Cryogenic tank cars are traditionally manufactured as a tank within a tank to allow for the insulation needed to keep the product cold.

Products shipped in this type of car are normally gases in their natural state but have been cooled to become a liquid. These products have a large expansion ratio if released into the atmosphere. The emergency responder should refer to the waybill to determine the actual amount of product carried in the tank car.

Construction Features

This type of tank has a high alloy steel inner tank supported by a strong carbon steel outer tank. The space between the two tanks is a vacuum space that is heavily insulated to keep the shipment cold. The combination of the insulation and the vacuum reduces the loss of temperature and will help maintain temperature for up to 30 days.

Fittings for this type of railway cargo tank, including loading and unloading valves, will be kept in ground-level cabinets on both sides of the tank or in the center of one end of the car **(Figure 8.57)**. Cryogenic products may also be shipped in a tank located in a standard boxcar, referred to as an *XT boxcar*. This boxcar will look similar to ordinary boxcars but will carry a cryogenic tank within. All valves for the cryogenic tank inside an XT boxcar will be located inside the boxcar.

Figure 8.57 Fittings for cryogenic tank cars are protected in cabinets accessible from the ground. *Courtesy of Rich Mahaney.*

The valves and fittings are the most prone to leaks in the cryogenic tank car. Pressure plates are commonly the most probable cause of a leak. While most railway tank cars go through rigorous inspection, insulated tanks can be difficult to inspect due to the double tanks and insulation.

As with any tank, mechanical damage may occur in the event of a railway accident and may compromise the tank's integrity. Carefully evaluate the tank and its contents if involved in an accident.

Specialized Cars

The previous sections introduced the most common types of railway tank cars, but other types of railway cars can be encountered. While not specifically tank cars, these cars may present unique hazards. Because railcars do not usually possess any specific identifying features, it may be difficult to recognize a railway car that transports hazardous materials **(Figure 8.58)**. Boxcars may carry hazardous materials in drums, crates, bags, boxes, or liquid bladders.

Figure 8.58 This hopper car carries plastic pellets, which can compromise a natural habitat if they are released. *Courtesy of Barry Lindley.*

Pneumatically Unloaded Hopper Cars

Pneumatically unloaded hopper cars force their product out of the hopper using air pressure. These cars are not considered nonpressure tank cars, but they may be designed to withstand up to 80 psig (650 kPa). Some materials that may be carried in this type of car include caustic soda, calcium carbide, and other dry bulk products **(Figure 8.59)**.

Refrigerated Cars

Refrigerated cars can be classified in three different categories. These types of cars have some integrated hazards besides the contents. Insulated bunkerless cars may have heaters located at the top of the doorways. Mechanical refrigerated cars may have an electrical generator that will supply 220 volts to the refrigeration unit. The generator may carry between 500 and 550 gallons (2 000 L and 2 200 L) of fuel. Mechanical refrigerated cars may also contain refrigerant gases such as Freon®. The atmosphere inside refrigerator cars may not contain oxygen if they are filled with nonflammable gas to help preserve and chill products. These cars may also be fumigated and have toxic contents like phosphine.

Figure 8.59 Pneumatically unloaded hopper cars pressurize dry materials. *Courtesy of Rich Mahaney.*

Intermodal Containers

Intermodal containers appeal to shipping companies because they can be transferred between modes of transportation without being off-loaded **(Figure 8.60)**. As with other modes of transport, box containers may transport a variety of hazardous materials, including smaller hazmat containers within. Intermodal tank containers (also called portable tanks) are a tank or cylinder within a framelike structure. Because intermodal tank containers are handled more frequently than bulk cargo tanks, there may be a greater risk of damage or leakage. Intermodal containers may be shipped from (and manufactured) virtually anywhere in the world. Therefore, hazmat technicians must be prepared to deal with unusual containers and contents.

> **Intermodal Container** — Freight containers designed and constructed to be used interchangeably in two or more modes of transport. *Also known as* Intermodal Tank, Intermodal Tank Container, *and* Intermodal Freight Container.

Figure 8.60 Intermodal containers are durable and standardized to work with a range of transportation systems. *Courtesy of Rich Mahaney.*

Beam — Structural member subjected to loads, usually vertical loads, perpendicular to its length.

There are two major types of frame construction used for intermodal tank containers: box type and **beam** type. Box-type intermodal containers encase the tank within the framework of a box **(Figure 8.61)**. Beam-type intermodal containers will only have framework at the ends of the tank **(Figure 8.62)**. Intermodal containers may be refrigerated, heated, or lined.

Figure 8.61 Intermodal tanks may be contained in a box-type framework. *Courtesy of Rich Mahaney.*

Figure 8.62 Intermodal tanks may be protected by a framework at the ends of the tank only.

T-Code — Portable tank instruction code used to identify intermodal containers used to transport hazardous materials. This set of codes replaces the IMO type listings.

Tank instruction codes, or **T-Codes**, are included in each tank's identification records. This manual only references the new set of T-Codes. These codes correspond to certain design specifications and instructions **(Table 8.12)**. T-Codes are not required to appear on tank specification plates, but they are often included on the tank somewhere. IMO codes may still be referenced because T-Codes do not address all current regulations.

298 Chapter 8 • Analyzing the Incident: Assessing Container Condition, Predicting Behavior, and Estimating Outcomes

Table 8.12
Portable Tank Instruction Codes

T-Codes	Analogous Tanks	Likely to Contain
T1-T5	IMO Type 2 or IM 102	Nondangerous liquids
T6-T14	IMO Type 1 excluding codes T3 and T4	Dangerous liquids
T15-T22	IMO Type 5 or DOT Spec 51	Dangerous gases
T23	IMDG 4.2.5	Organic peroxides and self-reactive substances
T50	Spec 51, DOT 51, IMO Type 5	Liquified compressed gases
T75	IMO Type 7	Refrigerated liquefied gases

Intermodal Tanks

Intermodal tanks can be both pressurized and nonpressurized. Materials transported in these containers include:

- Liquid and solid hazardous materials (*49 CFR 178*)
- Non-refrigerated liquefied compressed gases (*49 CFR 178*)
- Refrigerated liquefied gases (*49 CFR 178*)

The DOT Hazardous Materials Table (*49 CFR 172.101*) assigns reference codes for each hazardous material that may be shipped in an intermodal container. Effective January 1, 2003, specifications for intermodal tanks include design, construction, inspection, and testing requirements. Tanks constructed prior to 2003 are still in use, and must meet current standards **(Table 8.13)**.

Table 8.13
Intermodal Tank Container Descriptions

Specification	Materials Transported	Capacity	Design Pressure
IM 101 Portable Tank	Hazardous and nonhazardous materials, including toxics, corrosives, and flammables with flash points below 32°F (0°C)	Normally range from 5,000 to 6,300 gallons (18 927 to 23 848 L)	25.4 to 100 psi (175 to 689 kPa) {1.75 to 6.89 bar}
IM 102 Portable Tank	Whiskey, alcohols, some corrosives, pesticides, insecticides, resins, industrial solvents, and flammables with flash points ranging from 32 to 140°F (0 to 60° C)	Normally range from 5,000 to 6,300 gallons (18 927 to 23 848 L)	14.5 to 25.4 psi (100 to 175 kPa) {1 to 1.75 bar}
Spec. 51 Portable Tank	Liquefied gases such as LPG, anhydrous ammonia, high vapor pressure flammable liquids, pyrophoric liquids (such as aluminum alkyls), and other highly regulated materials	Normally range from 4,500 to 5,500 gallons (17 034 to 0 820 L)	100 to 500 psi (689 to 3 447kPa) {6.89 to 34.5 bar}

Intermodal Tanks for Liquids and Solid Hazardous Materials

Tanks for liquid and solid hazardous materials are the most common intermodal tanks used in transportation **(Figure 8.63)**. These tanks may also carry nonhazardous liquids or solids. Tanks formerly known as IM 101 and IM 102 containers fall into this category.

Figure 8.63 Nonpressure intermodal tanks may contain solids or liquids. *Courtesy of Rich Mahaney.*

IM 102 tanks for liquid and solid hazardous materials are generally considered nonpressure intermodal tanks. IM 101 tanks are built to withstand a working pressure of 25.4 to 100 psi (175 kPa to 700 kPa) with a typical capacity of 5,000 to 6,000 gallons (20 000 L to 24 000 L). Tanks for liquids and solids may have top loading or bottom loading valves. These tanks are equipped with emergency remote shutoff devices. Similar to highway tankers, they may also have fusible links and nuts and pressure/vacuum relief devices depending on the product in transport.

Intermodal Tanks for Non-Refrigerated Liquefied Compressed Gases

Pressurized intermodal tank containers are less commonly encountered in transport **(Figure 8.64)**. These tanks are designated T-50, and are formerly known as Spec 51 or IMO Type 5. Typically, they are designed for working pres-

Figure 8.64 Pressure intermodal tanks typically contain liquefied compressed gas. *Courtesy of Rich Mahaney.*

sures of 100 to 500 psi (700 kPa to 3 500 kPa) with a total capacity up to 5,500 gallons (22 000 L). Pressure-type intermodal containers usually transport liquefied gases under pressure such as LPG, chlorine, or anhydrous ammonia.

NOTE: Chlorine tank inlets and discharge outlets must meet standards established by the Chlorine Institute.

When the shells intended for the transportation of non-refrigerated liquefied compressed gases are equipped with thermal insulation, a device must be provided to prevent any dangerous pressure from developing in the insulating shell in the event of a leak. When the protective covering is closed, it must be gas tight. The insulation must be of adequate thickness and constructed to prevent the entry of moisture and damage to the insulation.

Regardless of the size or configuration of the container, data plates must be attached to the frame rail. These tanks may have fittings located on the top and bottom ends. Safety equipment includes:

- Safety relief devices
- Fusible links and nuts
- Excess flow valves
- Emergency remote shutoffs

Intermodal Tanks for Refrigerated Liquefied Gases

Tanks for refrigerated liquefied gases are used to transport cryogenic liquids **(Figure 8.65)**. These tanks are designated T-75, and are formerly known as IMO Type 7 containers. These tanks must be of seamless or welded steel construction and usually are manufactured in 10 ft to 40 ft (3 m to 12 m) configurations. Capacities will typically range around 4,400 gallons (17 600 liters) and will accommodate pressure around 250 psig (1 825 kPa). These tanks carry liquefied gases such as nitrogen, oxygen, hydrogen, and argon.

These tanks have a thermal insulation system that must include a complete covering of the shell with effective insulating materials. A jacket must protect external insulation to prevent the entry of moisture and other damage under normal transport conditions. When a jacket is gas-tight, a device must be provided to prevent any dangerous pressure from developing in the insulation space.

Figure 8.65 Cryogenic intermodal tanks typically contain refrigerated liquefied gas. *Courtesy of Rich Mahaney.*

Each filling and discharge opening in intermodal containers must be fitted with at least three mutually independent shut-off devices in series:

1. A stop-valve situated as close as possible to the jacket
2. A stop-valve
3. A blank flange or equivalent device

The shut-off device closest to the jacket must be a self-closing device, which is capable of being closed from an accessible position on the portable tank that is remote from the valve within 30 seconds of actuation. This device must actuate at a temperature of not more than 250°F (121°C).

Marine Tank Vessels

Hazardous materials incidents involving vessels can be minor, such as a small spill that occurs at a port during loading or unloading, or major, such as a spill contaminating miles (kilometers) of river or coastline waters or a large spill inside a ship. First responders need to be aware of vessel types and cargos that are likely to contain hazardous materials.

NOTE: The IFSTA manual, **Marine Fire Fighting for Land-Based Firefighters**, includes more information on marine hazards and resources.

Tankers

Tankers or tank vessels are ships that exclusively carry liquid products in bulk. Modern tankers are capable of transporting large quantities of liquid products. Tankers often carry a variety of products in segregated tanks. Tankers can be divided into three general categories:

- Petroleum carriers
- Chemical carriers
- Liquefied flammable gas carriers

Petroleum Carriers

Marine petroleum carriers transport crude oil or refined petroleum products. Marine carriers range in size from general purpose tankers with the capacity to carry 10,000 to 25,000 deadweight tonnes to ultra-large crude carriers with the capacity to carry 320,000 to 550,000 deadweight tonnes, all inclusive of fuel, stores, and crew (Hamilton). Because the overall weight of a type of crude or refined petroleum product will vary by its density, the cargo capacity may be variable in terms of total gallons (liters).

When entering the U.S. and Canada, the operator of any tank vessel carrying petroleum products is required by law to maintain vessel emergency response plans. These plans identify and ensure the availability of both a salvage company with expertise and equipment and a company with pollution incident response capabilities in the area(s) in which the vessel operates. Due to the size and scope of product release from these vessels, responders should not overlook the availability of preplanned resources during a marine fire fighting or emergency response.

Chemical Carriers

Chemical carriers transport multiple commodities and are sometimes nicknamed *floating drugstores*. They may carry oils, solvents, gasoline, sulfur, and other products (many classified as hazardous materials) in 30 to 58 separate tanks. Each tank usually has its own pump and piping, resulting in a maze of piping on the deck of a chemical carrier **(Figure 8.66)**. Thousands of chemical carriers operate worldwide.

Figure 8.66 Chemical carriers have a maze of piping on their deck.

Regulations do not require chemical carriers to carry placards. The only way to positively identify hazardous cargo is to ask the master or mate (captain or first officer) or obtain the cargo plan that identifies where each product is stowed on the vessel.

Liquefied Flammable Gas Carriers

Liquefied flammable gas carriers transport liquefied natural gas (LNG) and liquefied petroleum gas (LPG) and typically use large insulated spherical tanks for product storage. However, other configurations of gas carriers look very similar to ordinary tankers. **Coffer dams** (empty spaces between compartments) designed to contain low volume leakage from the tanks isolate the tanks within the vessel's hull. Cargo piping is located above the main deck so that any piping leaks vent to the atmosphere rather than inside the vessel. LPG carriers often transport a large number of pressure vessels at a time **(Figure 8.67)**.

Coffer Dam — Narrow, empty space (void) between compartments or tanks of a vessel that prevents leakage between them; used to isolate compartments or tanks.

Figure 8.67 LPG carriers may look similar to other tankers, but they may contain a large number of pressure vessels.

In U.S. ports that handle LNG and LPG carriers, regulations require the captain of the port to maintain LNG/LPG vessel management and emergency contingency plans. In Canada, regulations require each port handling hazardous shipments to conduct an evaluation that defines all threats to the port and environment and prepare contingency plans to manage emergencies. Responders consult these plans for area-specific guidance in handling emergencies involving these vessels.

Cargo Vessels

Cargo is shipped in the following vessel types:

- **Dry bulk carrier** — Carries products such as coal, wood chips, grain, iron ore, sand, gravel, salt, and fertilizers. The cargo is loaded directly into a hold without packaging, much like liquid in a tanker. Some of these cargoes generate dust (for example, grain), creating the possibility of an explosion.

- **Liquid bulk carrier** — Primary liquid bulk cargoes are chemicals (that may or may not be flammable) and liquid hydrocarbons. Liquid hydrocarbons carried in bulk include crude oils and refined oil products such as diesel fuel, gasoline, lubricating oils, and kerosene. These products vary widely in their characteristics and some can be highly volatile. The hazards presented are similar to those found at any petrochemical refinery or bulk storage facility.

- **Break bulk carrier** — Has large holds to accommodate a wide range of products such as vehicles, pallets of metal bars, liquids in drums, or items in bags, boxes, and crates.

- **Container vessel** — Carries cargo in standard containers that measure 8 feet (2.5 m) wide with varying heights and lengths. Container vessels may transport intermodal tanks (each enclosed in an open framework with standard container-size dimensions) **(Figure 8.68)**.

- **Roll On/Roll Off Vessel** — Has large stern and side ramp structures that are lowered to allow vehicles to be driven on and off the vessel. This vessel can be visualized as a floating, moving, multilevel parking garage.

Figure 8.68 Container vessels carry a large number of intermodal containers.

Barges

Barges are typically box-shaped, flat-decked vessels used for transporting cargo. In most instances, towing or pushing vessels are used to move barges because they are not self-propelled. A barge can transport virtually anything **(Figure 8.69)**. Barges may serve as floating warehouses with hazardous goods, vehicles, or railcars inside.

Figure 8.69 Some barges are designed for use with hazardous materials.

Some barges are configured as floating barracks for military or construction crews while others are used as bulk oil and chemical tankers. Some barges strictly transport intermodal containers. Some barges carry LNG in cylinders that may not be visible until a person is on board.

Air Freight Cargo

The DOT restricts the shipment of many hazardous materials aboard aircraft. In many cases these restrictions apply to passenger aircraft; but some may also apply to passenger air freight or cargo air freight. If a hazardous material is allowed to be shipped in the air, the hazard may be limited by the nature of the chemical and/or limited quantity allowed on board **(Figure 8.70)**.

Figure 8.70 Many restrictions apply to hazardous materials shipped via air. *Courtesy of John Demyan.*

The following are examples of materials that are prohibited on passenger aircraft but may be permitted on cargo aircraft with certain restrictions:

- Gasoline
- Aerosols containing flammable liquids
- Infectious substances
- Fireworks
- Petroleum oil

Aircraft will not display outer markings or placards indicating the type of cargo that may be carried. There will be little opportunity to evaluate any hazards from a safe distance. The DOT requires that shipping papers be carried within the aircraft to identify any hazardous materials on board. The shipping papers, known as an **air bill**, should be located on the flight deck. DOT labeling requirements that apply to products shipped by other modes of transportation are also required for air freight.

> **Air Bill** — Shipping document prepared from a bill of lading that accompanies each piece or each lot of air cargo. *Similar to* Waybill.

Pipelines

Pipelines primarily carry liquid petroleum products and natural and manufactured gases. The DOT requires that most pipelines be buried 30 to 36 inches (750 mm to 900 mm) below ground level. While this requirement is helpful in protecting the pipeline, it can be counterproductive when identifying the location of the pipeline.

Principles of Pipeline Operation

Pipelines are ideal because they allow the shipping of a product under pressure from one point to another without the need for off-loading. Product is introduced into the pipeline at an injection station located at the beginning of the pipeline system. Storage facilities equipped with pumps and compressors to aid in product movement can also be located at the injection station.

Compressor stations for gas pipelines and pump stations for liquid pipelines are located along the pipeline right-of-way to aid in product movement. These stations are located strategically along the pipeline based on the topography of the land and any operational considerations within the network. These stations work to maintain the pressure needed to move the product along the pipeline. The pipeline may end at a bulk storage facility which may be its final destination, or there may be partial delivery stations located within the pipeline's route.

Basic Identification

Many types of materials, particularly petroleum products, are transported across both the U.S. and Canada in an extensive network of pipelines. Most of these pipelines are buried in the ground. The U.S. DOT Pipeline and Hazardous Materials Safety Administration (PHMSA) regulates pipelines that carry hazardous materials across state borders, navigable waterways, and federal lands in the U.S. In Canada, the Canadian National Energy Board regulates oil and natural gas pipelines. PHMSA is introduced in Chapter 7 of this manual.

Where pipelines cross under (or over) roads, railroads, and waterways, pipeline companies must provide markers. These companies must also provide markers at sufficient intervals along the rest of the pipeline to identify the pipe's location. However, technicians should be aware that pipeline markers do not always mark the exact location of the pipeline. Technicians should not assume that the pipeline runs in a straight line between markers. Pipeline markers in the U.S. and Canada include the signal words *caution*, *warning*, or *danger* (representing an increasing level of hazard) and contain information describing the transported commodity and the name and emergency telephone number of the carrier **(Figure 8.71)**. Most pipelines have aerial markers for easy identification.

Figure 8.71 Pipeline markers are required to include specific information.

Construction Features

Pipeline construction can vary based on what products are intended to be shipped within the pipeline. Pipelines can range from 6 to 48 inches (150 mm to 1 200 mm) in diameter depending on their function. Distribution lines may be as small as a half-inch while transmission lines will usually be a larger diameter.

Pipeline construction is a multistep process that may include preconstruction surveys to clearing and grading of the pipeline right-of-way. Trenches are excavated to the specified depths, and the piping materials are aligned with the trench. The lengths of pipe are usually welded together. All piping must be inspected prior to the pipe being lowered to the ground and backfilled. Prior to use, the pipeline must be hydrostatically tested to federal regulations.

Valves are placed at regular intervals along the pipeline and act as a gateway **(Figure 8.72)**. When open, they allow the product to flow freely. The valves may be closed to either stop the flow or to isolate an area of pipeline that may require maintenance.

Figure 8.72 Pipeline valves are located at intervals to control the flow of the product. *Courtesy of Rich Mahaney.*

In addition to the compressor stations used to aid the movement of product, metering stations and valves can also be found along the pipeline's right-of-way. Metering stations allow pipeline companies to monitor the product in transit and measure the flow of a product.

The control station will monitor and manage all products within its pipeline. Most of the data received in the control center is provided by a **Supervisory Control and Data Acquisition (SCADA)** system. The SCADA system is a sophisticated communication system that will take measurements and collect data along the pipeline. It will transmit that data to a central point, usually located at the control station.

Using a sophisticated system like SCADA, control station operators can view the entire pipeline and act quickly should a leak or pressure loss occur. Remote closing of valves along the right-of-way can greatly reduce the impact of environmental damage.

The SCADA system can monitor:

- Flow rate
- Pressure
- Temperature
- Operational status

> **Supervisory Control and Data Acquisition (SCADA)** — System that monitors and controls coded signals from preset locations within an infrastructure (pipeline system), industry (manufacturing system), or facility (building system).

The U.S. DOT has documented that pipelines are the safest and most cost-effective means to transport large volumes of natural gas and hazardous liquids. The most common damages to pipelines occur from careless digging during third-party construction activities. Pipelines are well monitored; however, valves may leak.

Fixed Facility Containers

This section of the chapter focuses on storage tanks that hold bulk quantities of hazardous materials. Non-bulk packages that may be found at fixed facilities are discussed earlier in this chapter. Other containers at fixed facilities can include the following:

- Buildings
- Machinery
- Pipelines
- Open piles or bins
- Storage cabinets
- Aboveground storage tanks
- Underground storage tanks
- Reactors
- Vats
- Other fixed, on-site containers

Identifying the type of hazardous material present at a fixed facility can be much more difficult than simply recognizing the presence of a hazardous material. Most storage tanks are designed to meet the specific needs of both the facility and the commodity. Fixed facility tanks are manufactured to meet the needs of the products to be stored. Similar to highway cargo tanks and railway tanks, fixed facility tanks with rounded ends and pressure relief valves are designed to withstand higher pressures. Tanks with flatter ends may store liquids with low vapor pressures.

Fixed facilities with bulk liquids or gases may have features that will assist technicians if there is an incident:

- Fire protection systems can extinguish fires in their incipient stages.
- Monitoring and detection systems can immediately alert personnel to potential leaks or other problems.
- Pressure relief and vacuum relief protection devices can help release container pressure, thereby preventing catastrophic failures.
- Dikes and impoundments around fixed facility containers can help provide product spillage control in the event of a leak or release.
- Adequate tank spacing and the ability to transfer product between containers can minimize the likelihood that an incident involving one container will affect another.

Nonpressure Tanks

Nonpressure/atmospheric storage tanks are designed to hold contents under little pressure **(Table 8.14, p. 310)**. The maximum pressure contained in an atmospheric tank is 0.5 psi (3.5 kPa). Common types of atmospheric tanks include:

- Horizontal tanks
- Floating roof tanks
- Vapordome roof tanks
- Ordinary cone roof tanks
- Lifter roof tanks

Table 8.14
Atmospheric/Nonpressure Storage Tanks

Tank Type	Descriptions
	**Horizontal Tank** Cylindrical tanks sitting on legs, blocks, cement pads, or something similar; typically constructed of steel with flat ends. Horizontal tanks are commonly used for bulk storage in conjunction with fuel-dispensing operations. Old tanks (pre-1950s) have bolted seams, whereas new tanks are generally welded. A horizontal tank supported by unprotected steel supports or stilts (prohibited by most current fire codes) may fail quickly during fire conditions. **Contents:** Flammable and combustible liquids, corrosives, poisons, etc.
	**Cone Roof Tank** Have cone-shaped, pointed roofs with weak roof-to-shell seams that break when or if the container becomes overpressurized. When it is partially full, the remaining portion of the tank contains a potentially dangerous vapor space. **Contents:** Flammable, combustible, and corrosive liquids
	**Open Top Floating Roof Tank** Large-capacity, aboveground holding tanks. They are usually much wider than they are tall. As with all floating roof tanks, the roof actually floats on the surface of the liquid and moves up and down depending on the liquid's level. This roof eliminates the potentially dangerous vapor space found in cone roof tanks. A fabric or rubber seal around the circumference of the roof provides a weather-tight seal. **Contents:** Flammable and combustible liquids
 Vents around rim provide differentiation from Cone Roof Tanks	**Covered Top Floating Roof Tank** Have fixed cone roofs with either a pan or deck-type float inside that rides directly on the product surface. This tank is a combination of the open top floating roof tank and the ordinary cone roof tank. **Contents:** Flammable and combustible liquids

Continued

Table 8.14 (concluded)

Tank Type	Descriptions
	**Covered Top Floating Roof Tank with Geodesic Dome** Floating roof tanks covered by geodesic domes are used to store flammable liquids.
	**Lifter Roof Tank** Have roofs that float within a series of vertical guides that allow only a few feet (meters) of travel. The roof is designed so that when the vapor pressure exceeds a designated limit, the roof lifts slightly and relieves the excess pressure. **Contents:** Flammable and combustible liquids
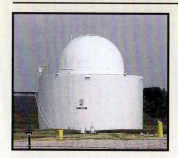	**Vapordome Roof Tank** Vertical storage tanks that have lightweight aluminum geodesic domes on their tops. Attached to the underside of the dome is a flexible diaphragm that moves in conjunction with changes in vapor pressure. **Contents:** Combustible liquids of medium volatility and other nonhazardous materials
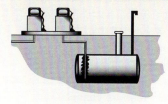 **Fill Connections Cover** 	**Atmospheric Underground Storage Tank** Constructed of steel, fiberglass, or steel with a fiberglass coating. Underground tanks will have more than 10 percent of their surface areas underground. They can be buried under a building or driveway or adjacent to the occupancy. This tank has fill and vent connections located near the tank. Vents, fill points, and occupancy type (gas/service stations, private garages, and fleet maintenance stations) provide visual clues. Many commercial and private tanks have been abandoned, some with product still in them. These tanks are presenting major problems to many communities. **Contents:** Petroleum products **NOTE:** First responders should be aware that some natural and manmade caverns are used to store natural gas. The locations of such caverns should be noted in local emergency response plans.

Pressure Storage Tank — Class of fixed facility storage tanks divided into two categories: low-pressure storage tanks and pressure vessels.

Low-Pressure Storage Tank — Class of fixed-facility storage tanks that are designed to have an operating pressure ranging from 0.5 to 15 psi (3.45 kPa to 103 kPa) {0.03 bar to 1.03 bar}.

Pressure Vessel — Fixed-facility storage tanks with operating pressures above 15 psi (103 kPa) {1.03 bar}.

Pressure Tanks

Pressure storage tanks are designed to hold contents under pressure **(Table 8.15)**. The NFPA uses the term *pressure tank* to cover both **low-pressure storage tanks** and **pressure vessels**. Low-pressure storage tanks have operating pressures from 0.5 to 15 psi (3.45 kPa to 103 kPa). Pressure vessels (including many large cryogenic liquid storage tanks) have pressures of 15 psi (103 kPa) or greater.

Table 8.15
Low-Pressure Storage Tanks and Pressure Vessels

Tank/Vessel Type	Descriptions
	**Dome Roof Tank** Generally classified as low-pressure tanks with operating pressures as high as 15 psi (103 kPa). They have domes on their tops. **Contents:** Flammable liquids, combustible liquids, fertilizers, solvents, etc.
	**Spheroid Tank** Low-pressure storage tanks. They can store 3,000,000 gallons (11 356 200 L) or more of liquid. **Contents:** Liquefied petroleum gas (LPG), methane, and some flammable liquids such as gasoline and crude oil
	**Noded Spheroid Tank** Low-pressure storage tanks. They are similar in use to spheroid tanks, but they can be substantially larger and flatter in shape. These tanks are held together by a series of internal ties and supports that reduce stresses on the external shells. **Contents:** LPG, methane, and some flammable liquids such as gasoline and crude oil
	Horizontal Pressure Vessel Have high pressures and capacities from 500 to over 40,000 gallons (1 893 L to over 151 416 L). They have rounded ends and are not usually insulated. They usually are painted white or some other highly reflective color. **Contents:** LPG, anhydrous ammonia, vinyl chloride, butane, ethane, compressed natural gas (CNG), chlorine, hydrogen chloride, and other similar products
	**Spherical Pressure Vessel** Have high pressures and capacities up to 600,000 gallons (2 271 240 L). They are often supported off the ground by a series of concrete or steel legs. They usually are painted white or some other highly reflective color. **Contents:** Liquefied petroleum gases and vinyl chloride

Pressure tanks may be found in different configurations based on the needs of the facility. Horizontal pressure tanks will be easy to distinguish because of the rounded ends. Other pressure tanks may be spherical.

Pressure tanks may also be stored below ground. While it will be impossible in these instances to see the tank's characteristics, the gauges, controls, and other identifiers can still be visible above ground.

Cryogenic Tanks

Cryogenic liquid storage tanks may come in many different shapes and will have round roofs **(Table 8.16)**. As with all cryogenic tanks, they will be heavily insulated and will rest on legs instead of being placed directly on the ground.

> **Cryogenic Liquid Storage Tank** — Heavily insulated, vacuum-jacketed tanks used to store cryogenic liquids; equipped with safety-relief valves and rupture disks.

Table 8.16
Cryogenic Liquid Storage Tank

Tank/Vessel Type	Description
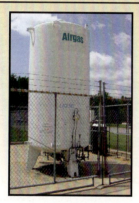	**Cryogenic Liquid Storage Tank** Insulated, vacuum-jacketed tanks with safety-relief valves and rupture disks. Capacities can range from 300 to 400,000 gallons (1 136 L to 1 514 160 L). Pressures vary according to the materials stored and their uses. **Contents:** Cryogenic carbon dioxide, liquid oxygen, liquid nitrogen, etc.

Other Storage Facility Considerations

Emergency responders should work diligently to preplan incidents at facilities that may store large and/or varied quantities of hazardous materials within their jurisdiction. In some cases, facilities may employ staff members who are well-trained in the use of the materials within the facility. In other cases, materials and their containers may not be handled correctly.

Laboratories

Laboratories are unique facilities that can store large quantities of hazardous materials. While the hazardous chemicals may be numerous and varied, they will likely be stored in small quantities and not in bulk. Laboratories can be found in just about any community and may range from a high school chemistry lab to a university, hospital, or private industry laboratory.

Each lab will hold different products based on its primary objective. A chemical inventory should assist the responding agencies should an incident occur. While it may be possible to refer to the Tier II reporting forms, many of the products in these facilities may not be in reportable quantities to be covered by this regulation.

Batch Plants

A batch plant is a manufacture and distribution facility that can produce materials such as concrete or asphalt. Batch plants can have a variety of tanks and storage bins based on the material being produced. It is not uncommon to find silos and nonpressure storage tanks where different materials and aggregates are stored. Emergency responders should be familiar with these types of facilities in their area and understand the hazards of each facility. Responders should contact facility personnel who can help identify dangers in the facility.

Non-Regulated and Illicit Container Use

Containers are manufactured to hold specific products and predetermined measurements including volumes, weights, and pressures. Containers that are not used correctly may put the emergency responder in harm's way. For example, when emergency responders encounter containers used in a manner inconsistently or incompatibly with its function, the container may not remain intact.

In this kind of situation, the emergency responder should attempt to mitigate the incident with a goal of preserving as much of the container's structural integrity as possible. Standard overpacking may not be a feasible option. Offloading the container may have to occur before it is overpacked and transported.

Radioactive Materials Packaging

All shipments of radioactive materials (RAM) must be packaged and transported according to strict regulations. These regulations protect the public, transportation workers, and the environment from potential exposure to **radiation**. The type of packaging used to transport radioactive materials is determined by the activity, type, and form of the material to be shipped. Depending upon these factors, radioactive material is shipped in one of five basic types of packaging listed in order of increasing level of radioactive hazard. Radioactive materials were introduced in Chapter 4 of this manual.

> **Radiation** — Energy from a radioactive source emitted in the form of waves or particles, as a result of the decay of an atomic nucleus; process known as *radioactivity*. Also called Nuclear Radiation.
>
> **Excepted Packaging** — Container used for transportation of materials that have very limited radioactivity.
>
> **Industrial Packaging** — Container used to ship radioactive materials that present limited hazard to the public and the environment, such as smoke detectors.

Excepted

Excepted packaging is used to transport materials that have a limited amount of radioactivity such as articles manufactured from natural or depleted uranium or natural thorium. Excepted packaging is only used to transport materials with extremely low levels of radioactivity that present no risk to the public or environment. Excepted packaging is not marked or labeled as such. Because of its low risk, excepted packaging is exempt from several labeling and documentation requirements. Empty radioactive materials packaging is treated as excepted packaging.

Industrial

Industrial packaging is a container that retains and protects its contents during normal transportation activities. Materials that present limited hazard to the public and the environment are shipped in these packages. Industrial packages are not identified as such on the packages or shipping papers. Items that can be shipped in industrial packaging include:

- Slightly contaminated clothing
- Laboratory samples
- Smoke detectors

Type A Packaging

Type A packages must demonstrate their ability to withstand a series of tests without releasing their contents **(Figure 8.73)**. The package and shipping papers will have the words Type A on them. Regulations require that the package protect its contents and maintain sufficient shielding under conditions normally encountered during transportation. Radioactive materials with relatively high specific activity levels are shipped in Type A packages. Items typically shipped in Type A packages include radiopharmaceuticals (radioactive materials for medical use) and certain qualified industrial products.

> **Type A Packaging** — Container used to ship radioactive materials with relatively high radiation levels.

Figure 8.73 Type A packages are built for durability under stress. *Courtesy of Tom Clawson.*

Type B Packaging

Type B packages must not only demonstrate their ability to withstand tests simulating normal shipping conditions, but must also withstand severe accident conditions without releasing their contents **(Figure 8.74)**. Type B packages are identified as such on the package as well as on shipping papers. The size of these packages can range from small containers to those weighing over 100 tons (100 tonnes). Type B packages are often very large and heavy and provide shielding against radiation. Radioactive materials that exceed the limits of Type A package requirements must be shipped in Type B packages. These include:

- Materials that would present a radiation hazard to the public or the environment if there were a major release.
- Materials with high levels of radioactivity such as spent fuel from nuclear power plants.

> **Type B Packaging** — Container used to ship radioactive materials that exceed the limits allowed by Type A packaging, such as materials that would present a radiation hazard to the public or the environment if there were a major release.

Figure 8.74 Type B packages are designed for durability under severe conditions. *Courtesy of the National Nuclear Security Administration, Nevada Site Office.*

Type C Packaging

Type C Packaging — Container used to ship highly reactive radioactive materials intended for transport via aircraft.

Type C packages are rarely encountered and are used for high-activity materials (including plutonium) transported by aircraft **(Figure 8.75 a-c)**. They are designed to withstand severe accident conditions associated with air transport without loss of containment or significant increase in external radiation levels. The Type C package performance requirements are significantly more stringent than those for Type B packages.

Figure 8.75 a-c Type C containers are built to withstand extreme stress and severe airplane accidents. *Courtesy of the National Nuclear Security Administration.*

316 Chapter 8 • Analyzing the Incident: Assessing Container Condition, Predicting Behavior, and Estimating Outcomes

Descriptions and Types of Radioactive Labels

Packages of radioactive materials must be labeled on opposite sides with the distinctive warning label. Each of the three label categories — RADIOACTIVE WHITE-I, RADIOACTIVE YELLOW-II, or RADIOACTIVE YELLOW-III — will bear the unique trefoil symbol for radiation. Class 7 Radioactive I, II, and III must always contain the isotope name and radioactive activity.

Radioactive II and III labels will also provide the **transport index (TI)** indicating the degree of control to be exercised by the carrier during transportation. The number in the transport index box indicates the maximum radiation level measured (mrem/hr) at one meter from the surface of the package. Packages with the Radioactive I label have a transport index of 0. When a package containing radioactive materials has a suspected breach, the current readings outside the container can be compared to the transport index to determine whether the readings have changed.

> **Transport Index (TI)** — Number placed on the label of a package expressing the maximum allowable radiation level in millirem per hour at 1 meter (3.3 feet) from the external surface of the package.

Chapter Notes

Code of Federal Regulations (CFR), Transportation, Title 49. 2011.

Hamilton, Mason T. "Oil tanker sizes range from general purpose to ultra-large crude carriers on AFRA scale." U. S. Energy Information Administration, 16 September 2016. Accessed: http://www.eia.gov/todayinenergy/detail.cfm?id=17991

Indian Springs Manufacturing. "Chlorine Institute Recovery Vessel." Accessed: http://www.indiansprings.com/chlorine-institute-recovery-vessel/

Soraka, Walter, CPP. 2008. Illustrated Glossary of Packaging Terminology, Second Edition. Naperville, IL: Marion Street Press, Inc.

Chapter Review

1. What are some common types of non-bulk containers and what kinds of hazardous materials do they carry?
2. What are the two types of IBCs and what types of materials do they transport?
3. Where are leaks typically found in ton containers?
4. What features can be used to identify nonpressure cargo tanks?
5. What leak hazards are associated with low-pressure cargo tanks?
6. What are some hazards associated with corrosive liquid tanks?
7. What are common leak points for high-pressure cargo tanks?
8. What are some common products carried in cryogenic tanks and what hazards might they present?
9. What products are commonly transported in tube trailers and what hazards are associated with them?
10. What are common products transported in dry bulk carriers and what hazards do they present?
11. What are three kinds of tank car markings and where are they located?
12. What safety features are found on railway tank cars?
13. What are the types of railway tank cars and what hazards are associated with them?
14. How can responders distinguish between the two major frame construction types of intermodal tank containers?
15. What safety equipment is present on intermodal tanks for non-refrigerated liquefied compressed gases?
16. What are the three mutually independent shut-off devices present on intermodal tanks that transport flammable refrigerated liquefied gases?
17. What hazards are associated with the three general categories of marine tank vessels?
18. What documentation can be used to determine what hazardous materials are aboard an aircraft?
19. How can information about the products carried in pipelines be identified?
20. What are some features of fixed facilities that will assist technicians at incidents involving bulk liquids or gases?
21. What other storage facilities may be harboring hazardous materials?
22. What are the five basic types of radioactive material packaging?

8-1
Evaluate the condition of a hazardous materials container. [NFPA 1072 7.2.3]

WARNING: If this skill involves the use of actual hazardous material samples, hazardous materials can cause serious injury or fatality. Appropriate personal protective equipment (PPE) must be worn and safety precautions must be followed. The following skill sheet demonstrates general steps; specific hazmat incidents may differ in procedure. Always follow the AHJ's procedures for specific incidents.

Step 1: Select and use appropriate PPE.
Step 2: Inspect container and its closures.

Step 3: Assess condition of container and closure, including types of damage and stress.
Step 4: Identify level of risk associated with container and closure damage and stress.
Step 5: Follow safety procedures.
Step 6: Avoid/minimize hazards.

Step 7: Decontaminate personnel, tools, and equipment.
Step 8: Communicate condition of container and closures and level of risk associated with this condition.

Planning the Response: Developing Response Objectives

Photo courtesy of Barry Lindley.

Chapter Contents

Risk-Based Response (RBR) 323	Response Modes 331
Risk versus Harm 326	Strategic Goals and Tactical Objectives 335
Response Models 326	PPE Selection and Decontamination Selection 341
APIE 326	**Developing a Site Safety Plan** 341
GEDAPER 327	Medical Surveillance Plan 343
Eight Step Process© 327	Pre-Entry Evaluation 344
HazMatIQ© 327	Post-Entry Evaluation 344
D.E.C.I.D.E. 327	Backup and Rapid Intervention 345
Emergency Response Guidebook 328	Emergency Procedures 345
Developing an Incident Action Plan 329	**Chapter Notes** 345
Preincident Surveys 330	**Chapter Review** 346
Incident Priorities 330	**Skill Sheets** 347

Chapter 9

Key Terms

Containment..339
Defensive Operations 333
Hazard and Risk Assessment 323
Hazard Class..325
Local Emergency Response Plan
(LERP)..332
Nonintervention Operations....................332
Offensive Operations..............................334
Postincident Analysis (PIA)....................340

Postincident Critique 340
Preincident Survey 323
Rapid Intervention Crew or Team
(RIC/RIT)...345
Response Model......................................326
Termination 339

Planning the Response: Developing Response Objectives

JPRs Addressed In This Chapter

7.3.1, 7.3.4

Learning Objectives

After reading this chapter, students will be able to:

1. Explain the concept of risk-based response at hazmat incidents. [NFPA 1072, 7.3.1, 7.3.4]
2. Describe the principles of risk management. [NFPA 1072, 7.3.1]
3. List types of response models used to manage a hazmat incident. [NFPA 1072, 7.3.1]
4. Describe processes for developing an incident action plan. [NFPA 1072, 7.3.4]
5. Describe processes for developing a site safety plan. [NFPA 1072, 7.3.4]
6. Skill Sheet 9-1: Develop response objectives and response options for a hazmat/WMD incident. [NFPA 1072, 7.3.1]
7. Skill Sheet 9-2: Develop a site safety and control plan. [NFPA 1072, 7.3.4]

Handwritten notes:

1. what are the risk of the incident, what will you need to do, what will protect your people and others. ~~life safety, properti~~ inci risk v. harm, APIE., risk a lot to save a lot, risk a little to save a little, risk nothing to save nothing
2. assess risks to make good decisions
3. APIE, GEDAPER, 8 step, HAZMATIO, DECIDE
4. APIE
5. figure out where it will spread. exposures. clean up

Chapter 9
Planning the Response: Developing Response Objectives

This chapter introduces the following topics related to incident action planning during a hazmat incident:

- Risk-based response (RBR)
- Risk versus harm
- Response models
- Developing an incident action plan
- Developing a site safety plan

Risk-Based Response (RBR)

Hazmat technicians must bring a risk-based response to hazmat incidents. This approach entails creating a planned and systematic approach to the incident. Risk-based response was introduced in Chapter 7 of this manual.

Planning for hazardous materials incidents includes:

- Assessing jurisdictional risks with **preincident surveys** and **hazard and risk assessments** (Figure 9.1)

> **Preincident Survey** — Assessment of a facility or location made before an emergency occurs, in order to prepare for an appropriate emergency response. *Also known as* Preplan.

> **Hazard and Risk Assessment** — Formal review of the hazards and risks that may be encountered by firefighters or emergency responders; used to determine the appropriate level and type of personal and respiratory protection that must be worn. *Also known as* Hazard Assessment.

Figure 9.1 Trains that run near populated areas must be evaluated for potential hazards and risks, including derailment. *Courtesy of Barry Lindley.*

- Developing a comprehensive resource management plan
- Determining response and staffing models
- Ensuring responders have the training, tools, and equipment necessary to respond to the hazards present in the community **(Figure 9.2)**

Figure 9.2 Jurisdictions that may not have high capacity trains will still likely have to consider the potential of a fuel leak from roadway transportation. *Courtesy of Barry Lindley.*

- Establishing mutual aid agreements

Many factors shape the development of the incident action plan and incident objectives. Although the Incident Commander is responsible for developing the initial incident action plan and identifying the response objectives (strategies) and response options (tactics), all responders must understand the process and know the tasks they may be asked to perform. Regardless of the depth and detail of the hazard and risk assessment, it must be based on circumstances, facts, and science. Steps for developing response objectives and response options for a hazmat/WMD incident are included in **Skill Sheet 9-1**.

Considerations include:

- Initial incident analysis, including the hazardous material(s) involved, the type(s) of container(s) present, and general incident circumstances.
- Pre-established incident priorities.
- Predetermined procedures such as SOPs.
- Results from size-up and hazard assessment results.

The response to hazardous material incidents should be fact-based. As technicians continually assess the incident, they can look to obtain more information on the products and assess these products in greater detail. An analysis of the hazard (severity) and risk (probability) will inform the technician of many important considerations.

Technicians should consider:

- Severity of the hazard via **hazard classes**. Hazardous materials will fall into one or more of these categories.
- Potential for fire, reactivity, or radioactivity.
- Probability of exposure or occurrence based on:
  — The chemical's physical state
  — The location of the incident (urban or rural)
  — The situation presented
  — Fixed facility or transportation

When assessing vulnerability, consider what may be vulnerable and how exposure could occur. Include people, property, and environment. Assess the likelihood of exposure by evaluating the following factors:

- The chemical and physical properties of the product or chemical
- The container and the different stresses that may affect it **(Figure 9.3)**

> **Hazard Class** — Group of materials designated by the Department of Transportation (DOT) that shares a major hazardous property.

Figure 9.3 A vessel may suddenly develop a heat-induced tear (HIT) when overpressurized. *Courtesy of Barry Lindley.*

- The environment, including weather conditions and topography
- Proximity to the hazard

Hazard and risk assessment can be broken down into two specific categories:

- The initial hazard and risk assessment
- The detailed hazard and risk assessment

The initial hazard and risk assessment occurs at the onset of the incident. Technicians can obtain valuable information based on dispatch information, incident location, and preplan information if available.

When conducting a detailed hazard and risk assessment, evaluate technical data from relevant sources and references in conjunction with air monitoring and metering results. As new information develops from this assessment, reevaluate and revise the plan of action if necessary.

Risk versus Harm

Some risks have a high reward benefit and some have little return. During risk-based response, technicians continually assess the risk. If the risks of an established mission outweigh the benefits, the responder should look for ways to reduce the risk. The IFSTA Principles of Risk Management state the following:

- Activities that present a significant risk to the safety of members shall be limited to situations where there is a potential to save endangered lives.
- Activities that are routinely employed to protect property shall be recognized as inherent risks to the safety of members, and actions shall be taken to avoid these risks.
- No risk to the safety of members shall be acceptable when there is no possibility to save lives or property.

A risk-benefit analysis means that members must carefully evaluate the incident including victim viability and available resources. If the proper resources are not present or if the rescue is extremely complicated with little chance of a successful outcome, technicians should carefully consider any actions.

Response Models

> **Response Model —** Framework for resolving problems or conflicts using logic, research, and analysis.

For an Incident Commander, the ability to make decisions and solve problems effectively is a necessary skill. Problem-solving and decision-making are fluid processes in hazmat incidents, so an IC's understanding of a problem (and consequent plans to address it) may change as more information becomes available and/or conditions change. However, technicians must look at the issues present and move in an orderly manner to a successful mitigation strategy. Using a **response model** can simplify the problem-solving process because most models incorporate an entire problem-solving process:

- An information gathering or input stage
- A processing or planning stage
- An implementation or output stage
- A review or evaluation stage

As outlined in the following sections, there are many models from which to choose. Most models move through each of the basic stages in problem solving and decision making, and departmental policy will often dictate the model used.

APIE

As discussed earlier in this manual, *APIE* is a simple response model used as the framework for NFPA 1072. This response model contains four basic problem-solving elements:

1. Analyze the incident.
2. Plan the initial response.
3. Implement the response.
4. Evaluate progress.

GEDAPER

The GEDAPER response model was developed by David Lesak and has been adopted and embedded into the curriculum of the National Fire Academy. The acronym GEDAPER stands for:

G - Gather information

E - Estimate potential course and harm

D - Determine strategic goals

A - Assess tactical options and resources

P - Plan of action implementation

E - Evaluate operations

R - Review the process

Eight Step Process©

Gregory G. Noll, Michael S. Hildebrand, and James G. Yvorra developed the Eight Step Incident Management Process©, which is a tactical decision-making model that focuses on hazmat/WMD incident safe operating practices. The eight steps are as follows:

1. Site management and control
2. Identify the problem
3. Hazard assessment and risk evaluation
4. Select protective clothing and equipment
5. Information management and resource coordination
6. Implement response objectives
7. Decontamination
8. Terminate the incident

HazMatIQ©

HazMatIQ© is a four-step decision-making response model that is used by major fire and law enforcement agencies across the United States. The HazMatIQ© system is a proprietary risk-based response system that will assist responders through a four-step process that includes:

- A quick chemical size-up using supplied charts
- A streamlined chemical hazard research process
- Meter prediction (reagent paper)
- Selection of mission-specific PPE

D.E.C.I.D.E.

Ludwig Benner developed the D.E.C.I.D.E. response model (Benner 1975). The D.E.C.I.D.E. mnemonic stands for:

D - Detect the presence of a hazardous material

E - Estimate likely harm without intervention

C - Choose response objectives

I - Identify response options

D - Do best option

E - Evaluate progress

Emergency Response Guidebook

The *Emergency Response Guidebook (ERG)* presents a response process approach to incident size-up and safety (see Information Box). This approach works well for lower levels of hazmat response but may not be practical for the Technician-Level responder. The ERG is a basic response resource that is often shared with technicians, with slight modifications to processes such as setting up command and evaluations.

ERG Safety Precautions

RESIST RUSHING IN!

APPROACH CAUTIOUSLY FROM UPWIND, UPHILL OR UPSTREAM:

- Stay clear of **Vapor**, **Fumes**, **Smoke** and **Spills**
- Keep vehicle at a safe distance from the scene

SECURE THE SCENE:

- Isolate the area and protect yourself and others **(Figure 9.4)**

Figure 9.4 In 2015, the response to an oil spill in Mount Carbon, WV, included evacuations of the nearby residential areas. *Courtesy of Barry Lindley.*

IDENTIFY THE HAZARDS USING ANY OF THE FOLLOWING:

- Placards
- Container labels
- Shipping documents
- Rail Car and Road Trailer Identification Chart
- Material Safety Data Sheets (MSDS)
- Knowledge of persons on scene
- Consult applicable guide page

ASSESS THE SITUATION:
- Is there a fire, a spill or a leak?
- What are the weather conditions?
- What is the terrain like?
- Who/what is at risk: people, property or the environment?
- What actions should be taken – evacuation, shelter-in-place or dike?
- What resources (human and equipment) are required?
- What can be done immediately?

OBTAIN HELP:
- Advise your headquarters to notify responsible agencies and call for assistance from qualified personnel

RESPOND:
- Enter only when wearing appropriate protective gear
- Rescue attempts and protecting property must be weighed against you becoming part of the problem
- Establish a command post and lines of communication
- Continually reassess the situation and modify response accordingly
- Consider safety of people in the immediate area first, including your own safety

ABOVE ALL: Do not assume that gases or vapors are harmless because of lack of a smell—odorless gases or vapors may be harmful. Use **CAUTION** when handling empty containers because they may still present hazards until they are cleaned and purged of all residues.

Developing an Incident Action Plan

An Incident Action Plan is an organized course of events that addresses all phases of an incident within a given timeframe. Based on the size of the incident, the AHJ may call for the plan to be either written or unwritten. As a hazmat incident escalates, having a written IAP becomes very important. The IAP contains the overall strategic goals and tactical objectives along with support requirements for a given operational period during an incident (Occupational Safety and Health).

An IAP is necessary to affect successful outcomes during emergency operations. An IAP may ultimately consist of several ICS forms bundled together that document the actions developed by the IC, general staff, or other members that are present at a planning meeting (Emergency Response Plan).

The following ICS forms will guide the collection of information needed in a formal written IAP:

- Incident briefing including site sketch (ICS Form 201)
- Incident objectives (ICS Form 202)
- Organizational assignment list (ICS Form 203)
- Division assignment list (ICS Form 204)
- Incident radio communications plan (ICS Form 205)

- Medical plan (ICS Form 206)
- Site safety plan (ICS Form 208 HM)

The IAP essentially ties together the entire problem-solving process by stating what the analysis has found, explaining what the plan is, and describing how it shall be implemented. These pieces, as a whole, will indicate the response options for the incident. Once the plan is established and resources are committed, it is necessary to assess its effectiveness. Gather and analyze information so that necessary modifications may be made to improve the plan if necessary. An IAP update must follow each formal planning meeting conducted by the Planning Section. The IC must approve the plan prior to its distribution. Responders must develop an IAP for each operational period after the initial action phase.

Preincident Surveys

Gathering information about a specific incident should begin before the potential becomes a reality, as much as possible. Staff should gather preincident information through district familiarization and building inspections. Emergency responders have a responsibility to know the location of fixed hazards, the probable location of transport hazards, the ways to locate preplanned information and procedures, and any applicable SOPs. Thorough preplanning can deliver the following information:

- Exposures (including people, property, and environment)
- Types, quantities, and locations of hazardous materials in a given area
- Dangers of any hazardous chemicals within the preplanned area
- Building features
- Site characteristics
- Possible access/egress difficulties
- Inherent limitations of an organization's control capabilities
- Contact telephone numbers of responsible parties and site experts

Incident Priorities

Technicians must make all plans with incident priorities in mind. However, incidents are dynamic, so these priorities may change according to the situation. Decisions during the problem-solving process of all hazmat incidents must be made with the following response priorities in mind:

- **Life safety** — The first priority is the safety of emergency responders and civilians. If responders do not protect themselves first, they cannot protect the public. Life safety must be a consideration from the moment an incident is reported until its termination — from the response to the scene until the ride back to the station. A dead, injured, or unexpectedly contaminated technician becomes part of the problem, not the solution. Decisions weighing the life safety of responders versus the life safety of the public must be based upon a careful risk/benefit analysis. Adopting a policy of cautious assessment before taking action is vital. A risk/benefit analysis should consider the following variables:
    — Risk to rescuers

- — Ability of rescuers to protect themselves
- — Probability of rescue
- — Difficulty of rescue
- — Capabilities and resources of on-scene resources
- — Possibilities of explosions or sudden material releases
- — Available escape routes and safe havens
- — Constraints of time and distance

- **Incident stabilization** — If there is no immediate threat to either responders or civilians, the next consideration is stabilizing the incident.
- **Protection of property and the environment** — When the first two priorities are satisfied, technicians can address conservation (or protection) of property and the environment. Stabilizing the incident can minimize environmental and property damage. If the situation calls for it, these priorities can be changed, but generally, emergency responders need to consider them in the order presented.
- **Societal restoration** — The IAP should include considerations to restore normal functions to the incident scene; for example, resuming transportation routes and allowing evacuees to return to their homes/businesses/traffic flow.

Response Modes

Strategies are divided into three strategic response (operation) modes (**Figure 9.5**):

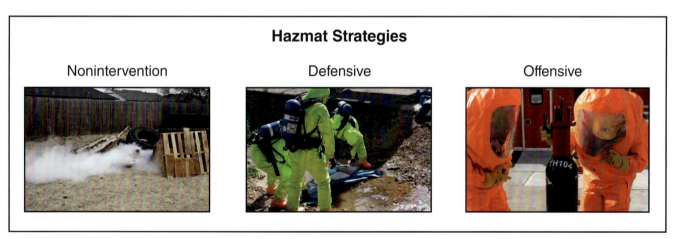

Figure 9.5 The IC may change strategies at a hazmat incident depending on the resources available and the conditions on the scene.

- **Nonintervention** — Allows the incident to run its course on its own
- **Defensive** — Provides confinement of the hazard to a given area by performing diking, damming, or diverting actions
- **Offensive** — Includes actions to control the incident such as plugging a leak

Similarly, three incident-based elements will affect the selection of strategic mode, and must always be considered as the IAP is developed:

- **Value** — Value relates directly to the incident priorities of life safety, incident stabilization, and property conservation. Value is stated in terms of yes or no — either there is value (yes) or there is no value (no). Once value has been determined with a yes, technicians can assess the degree of value. If a civilian life hazard exists (savable victim or victims), the value is high. If no civilian life hazard exists but environmental harm may be prevented or property may be saved, the value is somewhat less. If no civilian life hazard exists and responder actions will have little effect on environmental or property protection, there is no value. In the absence of value, technicians may choose either a nonintervention or defensive strategy.

- **Time** — There may be a limited window of opportunity to intervene before an incident escalates dramatically (such as cooling of a liquefied gas container exposed to direct flame impingement on its vapor space); estimated time span during which offensive operations may be initiated. In other cases, the reaction and response times of Technician-Level responders may be the driving factor in selecting the strategic mode for incident operations.

- **Size/Complexity** — Size and/or complexity of the incident most frequently drives the need to conduct protective action (evacuation or protection in place) concurrently with incident control operations. Resource requirements are driven by tactical requirements.

Selection of the strategic mode is based on the risk to responders, their level of training, and the balance between the resources required and those available. The safety of first responders is the foremost consideration in selecting a mode of operation.

The mode of operation may change during the course of an incident. For example, first-arriving responders may be restricted to nonintervention or defensive mode. After the arrival of the hazmat team, the IC may switch to offensive mode and initiate offensive tactics. The sections that follow describe response operations and the ways they may be applied to an incident.

Nonintervention Operations

Nonintervention operations are operations in which responders take no direct actions on the problem. Not taking any action is the only safe strategy in many types of incidents and the best strategy in certain types of incidents when mitigation is failing or otherwise impossible. An example of a situation for nonintervention is a vessel that cannot be adequately cooled because it is exposed to fire **(Figure 9.6)**. In such incidents, responders should evacuate personnel in the area and withdraw to a safe distance. A hazmat team may select the nonintervention mode when one or more of the following circumstances exist:

- The facility or **local emergency response plan (LERP)** calls for the mode based on a preincident evaluation of the hazards present at the site.
- The situation is clearly beyond the capabilities of responders.
- Explosions are imminent.
- Serious container damage threatens a massive release.

During nonintervention operations, technicians should:

- Withdraw to a safe distance.

Nonintervention Operations — Operations in which responders take no direct actions on the actual problem.

Local Emergency Response Plan (LERP) — Plan detailing how local emergency response agencies will respond to community emergencies; required by U.S. Environmental Protection Agency (EPA) and prepared by the Local Emergency Planning Committee (LEPC).

Figure 9.6 In 2009, a train derailment in Cherry Valley, IL, lead to fire exposure to a tank containing denatured fuel ethanol. *Courtesy of Barry Lindley.*

- Report scene conditions to the telecommunications center.
- Initiate an incident management system.
- Call for additional resources as needed.
- Isolate the hazard area and deny entry.
- Commence evacuation where needed.

Defensive Operations

Defensive operations are those in which responders seek to confine the emergency to a given area without directly contacting the hazardous materials involved **(Figure 9.7)**. Select the defensive mode when one of the following two circumstances exists:

Defensive Operations — Operations in which responders seek to confine the emergency to a given area without directly contacting the hazardous materials involved.

Figure 9.7 Defensive operations may include limiting the capability of a spill to spread to other areas, as is done around the periphery of this spill. *Courtesy of Barry Lindley.*

1. The facility or LERP calls for the mode based on a preincident evaluation of the hazards present at the site.
2. Responders have the training and equipment necessary to confine the incident to the area of origin.

In defensive operations, take the following actions:

- Report scene conditions to the telecommunications center.
- Initiate an incident management system.
- Call for additional resources as needed.
- Isolate the hazard area and deny entry.
- Establish and indicate zone boundaries.
- Commence evacuation where needed.
- Control ignition sources.
- Use appropriate defensive control tactics.
- Protect exposures.
- Perform rescues when safe and appropriate.
- Evaluate and report incident progress.
- Perform emergency decontamination procedures.

Offensive Operations

Offensive operations are those where responders take direct action on the material, container, or process equipment involved in the incident (**Figure 9.8**). These operations may result in contact with the material. Offensive operations are predominately conducted by hazardous materials technicians with personnel trained to lower levels of hazmat response serving in support roles.

> **Offensive Operations** — Operations in which responders take aggressive, direct action on the material, container, or process equipment involved in an incident.

Figure 9.8 In 2012, a train derailment in Farragut, TN, released oleum, also known as fuming sulfuric acid. Response included the deployment of unmanned master stream appliances for vapor dispersion. *Courtesy of Barry Lindley.*

Strategic Goals and Tactical Objectives

An IAP is an organized approach to planning for the mitigation of a serious situation. The premise of the IAP is to develop the necessary strategy and tactics to affect a positive and safe outcome. Developed strategies are broad in nature and define what needs to be accomplished. The National Fire Academy recommends using SMART modeling for the development of strategies. SMART is a mnemonic for the following:

- **S**pecific
- **M**easureable
- **A**ction oriented
- **R**ealistic
- **T**ime sensitive

Once the strategies are in place, technicians then develop tactics. Tactics are the operational tasks that are used to accomplish the strategies. Tactics should be measureable in both time and performance. They should be evaluated to ensure that they will meet the strategic goals that were put in place. Some typical hazmat strategies and tactics are presented in the sections following.

Isolation

The isolation perimeter may be comprised of an inner and outer perimeter, and it may be expanded or reduced in size as needed. In most cases, the outcomes of an on-site risk assessment determine the initial isolation perimeter established by first-arriving responders.

Once resources have been committed to an incident, it is easier to reduce the isolation perimeter in size than it is to extend it **(Figure 9.9)**. If resources have arrived and have been tasked at an incident, it may be difficult to then disengage and relocate those resources should the initial perimeter be inadequate. One example of this situation might be when the perimeter fails to ensure emergency responder safety, crowd control, crime scene preservation, or general safety for other hazards associated with toxic plumes or hazardous materials.

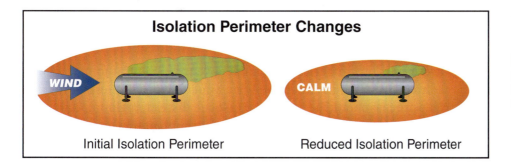

Figure 9.9 The initial isolation perimeter should include factors that may cause the incident to expand quickly.

A risk assessment or size-up of the incident is necessary in order to determine an appropriate size for the isolation perimeter. Where appropriate, the technician may consult with other on-site agency commanders regarding the perimeter size to ensure that the spatial requirements and tactical objectives of other agencies can be met.

NOTE: From a risk-management perspective, it is better to encompass a larger area that can be reduced in size once incident site conditions have been assessed for risks such as secondary devices, unidentified hazardous materials, and atmospheric monitoring.

The isolation perimeter is also used to control both access and egress from the incident site. Unauthorized personnel may be kept out, while responders may direct witnesses and persons with information about the incident to a safe location until they can be interviewed and released. Another important aspect of scene control at hazmat/WMD incidents is the establishment of hazard control zones and staging areas.

Identification

Positively identifying the involved product is the safest and most effective means of planning mitigation activities. During the planning process, product identification should be established as thoroughly as possible to allow the planning team to make critical decisions on personal protective equipment ensembles and decontamination procedures. In some instances, product identification is a relatively simple task. In other instances, hazmat technicians may encounter circumstances that prevent a full identification of a product. Technicians may gather product identification clues from the following sources:

- Information from occupancies and preincident surveys
- Containers
- Transportation placards, labels, or markings
- Technical markings
- Written resources such as shipping papers
- Monitoring and detection devices

Notification

Emergency response plans must ensure that responders understand their role in notification processes and predetermined procedures such as standard operating procedures (SOPs). Notification may include such items as incident-level identification and public emergency information/notification. It is better to dispatch more resources than necessary in an initial response to ensure sufficient resources are available to combat incident conditions. Responders should be familiar with the assets available in their jurisdictions.

Because hazardous materials incidents have the potential to overwhelm local resources, technicians must know the procedure to request additional assets. This process should be described in local, district, regional, state, and national emergency response plans as well as through mutual/automatic aid agreements.

Notification also involves contacting law enforcement whenever responders suspect a terrorist or criminal incident, as well as notifying other agencies such as public works or the local emergency operations center that an incident has occurred. Procedures will differ depending on the AHJ. Always follow SOPs and emergency plans for notification procedures.

The local emergency response plan (LERP) is the first resource responders should turn to if they need to request outside assistance for an incident

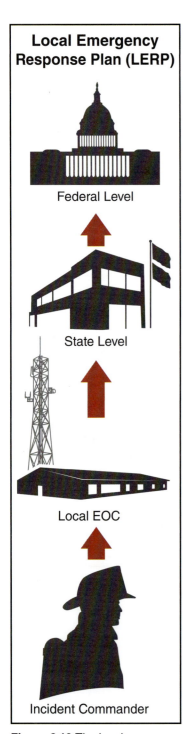

Figure 9.10 The local emergency response plan (LERP) should include details on how to request outside aid when needed.

(Figure 9.10). The local response agency should be closely tied to the community's Emergency Operations Center (EOC). If local assets are insufficient to manage the emergency, responders will make requests to the EOC for additional assistance. The EOC may then request federal assistance per AHJ. Even if additional assistance is not required for an incident, the proper authorities (local, state, and federal) must be informed that an incident has occurred.

Protection

Protection is the overall goal of ensuring safety of responders and the public. Protection goals also include measures to protect property and the environment. Protection goals are accomplished through such tactics as the following:

- Identifying and controlling materials and hazards
- Avoiding contact with hazardous materials
- Maximizing distance between people and hazardous areas (hazard control zones and incident perimeters)
- Using and wearing appropriate PPE (see Chapter 11, Personal Protective Equipment) **(Figure 9.11)**

Figure 9.11 Responders should wear the correct PPE for the incident.

- Using time, distance, and shielding, when appropriate
- Conducting rescues
- Providing decontamination (see Chapter 12, Decontamination)
- Providing emergency medical care and first aid
- Ensuring victims and responders stay upwind, upstream, and uphill of hazardous materials
- Taking any other measures to protect responders and the public, including conducting evacuations and sheltering in place

The protection and safety of emergency responders is the first priority at any incident. If emergency responders are injured or incapacitated, they will be unable to assist in any mitigation effort. Measures specifically intended to protect responders include:

- Ensuring accountability of all personnel

- Tracking and identifying all personnel working at an incident
- Working as part of a team or buddy system
- Assigning safety officers

Rescue

Due to the extremely complex nature of a hazardous materials incident, rescue may be one of the most difficult tasks to accomplish. Based on the nature of the incident, victims may be found in a variety of locations. They may be located out in the open, within a structure or confined space, or even under rubble or debris. Tools and equipment will differ for each scenario. An evaluation must also be made based on the viability of the victims.

Fire Control

Fire control is the strategy of minimizing the damage, harm, and effect of fire at a hazmat incident. If a fire is present in addition to a release (spill or leak), the incident is considerably more complicated. Based on risk and hazard assessment, technicians must make a decision whether to extinguish the fire and if so, how. The best course of action may be to protect exposures and let the fire burn until the fuel is consumed. Implement this strategy if the products of combustion are less of a hazard than the leaking chemical or if extinguishment efforts will place firefighters in undue risk. Withdrawal may be the safest (and best) tactical option if there is a threat of catastrophic container failure, boiling liquid expanding vapor explosion (BLEVE) or other explosion, or if the resources needed to control the incident are unavailable.

When containers or tanks of flammable liquids or gases are exposed to flame impingement, deploy water streams for maximum effective reach in order to prevent a BLEVE or other violent container failure. Commonly, responders can best achieve this cooling by directing a stream (or streams) at areas where there is direct flame impingement on the tank as well as along the tank's top so that water runs down both sides. This water stream cools the tank's vapor space. The piping and steel supports under tanks should also be cooled to prevent their collapse.

Do not extinguish gas-fed fires burning around relief valves or piping unless turning off the supply can stop the leaking product. An increase in the intensity of sounds or fire issuing from a relief valve indicates pressure within the container is increasing and container failure may be imminent.

Confinement

Diking, damming, diverting, and retaining are ways to confine a hazardous material. Take these actions to control the flow of liquid hazardous materials away from the point of release. Responders can use available earthen materials or materials carried on their response vehicles to construct curbs that direct or divert the flow away from:

- Gutters
- Storm sewers
- Outfalls
- Drains
- Flood-control channels

In some instances, it may be desirable to direct the flow into certain locations in order to capture and retain the material for later pickup and disposal **(Figure 9.12)**. Dams may be built that permit surface water or runoff to pass over (or under) the dam while holding back the hazardous material. Any construction materials that contact the spilled material must be disposed of in accordance with applicable regulations.

Figure 9.12 Dams can be designed to separate contaminants from water. *Courtesy of Rich Mahaney.*

Containment

A leak involves the physical breach in a container through which product is escaping. The goal of leak control is to stop or limit the escape or to contain the release either in its original container or by transferring it to a new one. This is often referred to as **containment**. The type of container involved, the type of breach, and properties of the material determine tactics and tasks relating to leak control. Leak control and containment are generally considered offensive actions. Containment tactics are discussed in depth in Chapter 13 of this manual.

Recovery and Termination

The last strategic goals for the proper management of a hazardous materials emergency are normally the recovery and **termination** efforts. There is a distinct difference between these functions. Recovery deals with returning the incident scene and responders to a preincident level of readiness. Termination involves returning the site to its owners, after evaluation. This evaluation leads to an improvement of future response capabilities based upon problems that were identified during the original incident.

Containment — The act of stopping the further release of a material from its container.

Termination — The phase of an incident in which emergency operations are completed and the scene is turned over to the property owner or other party for recovery operations.

The major goals of the recovery phase are as follows:

• Return the operational area to a safe condition.

• Debrief personnel before they leave the scene.

• Return the equipment and personnel of all involved agencies to the condition they were in before the incident.

On-scene recovery efforts are directed toward returning the scene to a safe condition **(Figure 9.13)**. These activities may require the coordinated effort

Figure 9.13 Recovery efforts may require significant work to restore a safe condition. *Courtesy of Barry Lindley.*

of numerous agencies, technical experts, and contractors. Generally, fire and emergency services organizations do not conduct remedial cleanup actions unless those actions are absolutely necessary to eliminate conditions that present an imminent threat to public health and safety. If such imminent threats do not exist, contracted remediation firms under the oversight of local, state/provincial and federal environmental regulators generally provide for these cleanup activities. In these situations, the fire and emergency services organization may also provide control and safety oversight according to local SOPs.

On-scene debriefing, conducted in the form of a group discussion, gathers information from all operating personnel, including law enforcement, public works, and EMS responders. This stage should include the collection of important observations, actions taken, and the timeline of those actions.

In addition, one important step in the debriefing process is to provide information to personnel concerning the signs and symptoms of overexposure to the hazardous materials, which is referred to as the hazard communication briefing (required by OSHA in the U.S.). This debriefing process must be thoroughly documented. The information provided to responders before they leave the scene should include:

- Identity of material involved
- Potential adverse effects of exposure to the material
- Actions to be taken for further decontamination
- Signs and symptoms of an exposure
- Mechanism by which a responder can obtain medical evaluation and treatment
- Exposure documentation procedures

Operational recovery involves those actions necessary to return the resource forces to a level of preincident readiness. These actions involve the following:

- Release of units
- Resupply of materials and equipment
- Decontamination of equipment and PPE
- Preliminary actions necessary for obtaining financial restitution

The financial effect of hazardous materials emergencies can be far greater than any other activity conducted by the fire and emergency services. Normally, a fire and emergency services organization's revenues obtained from taxes or subscriber fees are calculated based upon the equipment and personnel necessary to conduct fire suppression and other emergency activities. Communities should have in place the necessary ordinances to allow for the recovery of costs incurred from such emergencies. Another vital part of this process is the proper documentation of costs through the use of forms such as the Unit Log and other tracking mechanisms.

In order to conclude an incident, the technician must ensure that all strategic goals have been accomplished and the requirements of laws have been met. The team must complete documentation, analysis, and evaluation. The termination phase involves two procedural actions: **postincident critiques** and **analysis** (also known as an After Action Report or AAR). Analysis includes study of all postincident reports and critiques.

> **Postincident Critique** — Discussion of the incident during the Termination phase of response. Discussion includes responders, stakeholders, and command staff, to determine facets of the response that were successful and areas that can be improved upon.

> **Postincident Analysis (PIA)** — Overview and critique of an incident including feedback from members of all responding agencies. Typically takes place within two weeks of the incident. In the training environment it may be used to evaluate student and instructor performance during a training evolution.

PPE Selection and Decontamination Selection

PPE selection is an important part of the incident response plan. The AHJ should have a written management process regarding selection and use of PPE, and this process should be founded on a risk-based response model. PPE is addressed further in Chapter 11 of this manual.

In addition to PPE, decon selection is an important part of the incident response plan. The AHJ should have a written management process regarding selection and use of decon procedures, founded on a risk-based response model. Types of decon include: gross, emergency, mass, and technical. Decon is addressed further in Chapter 12 of this manual.

Developing a Site Safety Plan

A site specific health and safety plan is an important tool in many occupations. Industries, such as mining and construction, use site specific health and safety plans regularly for the safety of their employees (ICS 208 HM). In hazardous materials response, this plan is not only a vital document for the safety of emergency responders, it is also a requirement under OSHA 29 CFR 1910.120. Each hazardous materials incident presents unique conditions, and each site changes as dynamically as each chemical. An in-depth site specific health and safety plan is required for a Technician-Level hazmat response. Steps involved in developing a site safety and control plan are presented in **Skill Sheet 9-1**.

The depth and detail of a site specific health and safety plan may change based on the size and scope of the incident and the products involved. In many cases, only portions of the plan may be needed. However, for hazmat incidents of a larger magnitude, responders may need all aspects of the plan. The site specific health and safety plan (HASP) template developed by OSHA includes 15 chapters and should be considered for various emergency incidents (e-HASP).

WARNING!
If PPE fails in the hot zone, the technician must evacuate immediately.

Steps involved in developing response objectives and response options for a hazmat/WMD incident are presented in **Skill Sheet 9-2**. Specific elements of a health and safety plan should line up with the NIMS-ICS forms. The site specific health and safety plan ensures that all personnel are briefed on the following:

- Incident response objectives
- Emergency evacuation
- Event status
- PPE
- Communication methods
- Emergency procedures
- Product hazards

The following is a brief overview of the components of a site specific health and safety plan and the associated ICS forms **(Table 9.1, P. 342)**:

**Table 9.1
Health and Safety Plan and NIMS-ICS /
Local Forms**

Health and Safety Plan Component	NIMS-ICS / Local Forms
Site Security, Control and Communications	201
Roles, Responsibilities, and Authority	203
Incident Objectives	202
Communications	205 and 216
Medical Plan	206
Hazardous Materials Data Sheet	Local
Personal Protective Equipment Selection Sheet	Local
Plan Annex	Local
Decontamination Plan Outline	Local

- **Site security, control, and communications** — ICS Form 201 is used to develop a site map and/or sketch and should include the following information:

    — Orientation of the site with compass points indicated

    — Natural and man-made topographical features

    — Prevailing wind direction

    — Command post location

    — Staging area

    — Access control points (zones)

    — Decontamination corridor

    — Drainage points

- **Roles, responsibilities, and authority** — ICS Form 203 is used to list the person responsible for each job function.

- **Incident objectives** — ICS Form 202 is used to list the incident objectives and name assignments for each team that is assigned.

- **Communications** — ICS Form 205 (Incident Radio Communications) and ICS Form 216 (Radio Requirement Work Sheet) is used to ensure that all personnel involved in entry team activities remain in constant contact. While radios may be the obvious method, communication may also include visual or verbal as well.

- **Medical plan** — ICS Form 206 establishes the medical plan. The medical plan documents the name and location of the nearest hospital or medical facility.

- **Hazardous materials data sheet** — This sheet is completed for each hazardous substance and should list all known or suspected hazardous substances and concentrations.

- **Personal protective equipment selection sheet** — This sheet lists all PPE that will be used and the resource documents that suggest the recommendation.
- **Plan annex** — This component outlines the control zones and all monitoring information along with the outlined action levels based on the reading of the monitoring equipment.
- **Decontamination plan outline** — This outline includes setup, location, solutions to be used, and equipment needed for decontamination operations.

To finalize the site specific health and safety plan, a signature page is included so that all of the key Command Staff personnel can indicate their acceptance of the plan. The required signatures include (but are not limited to) the IC, Safety Officer, Operations Officer, and the Hazmat Branch Director/Group Supervisor. More information regarding medical surveillance is included in IFSTA's **Occupational Safety, Health, and Wellness** manual.

Medical Surveillance Plan

Each response can introduce health risks to a technician. Health precautions that will help to indicate the scope and severity of those problems will be established by the AHJ. This monitoring is directly to the type of incident and response, and independent of other medical surveillance as indicated in Chapter 10 of this manual. Incident-specific data is referred to as pre- and post-entry medical monitoring. Although not specifically required by 29 CFR 1910.120, the AHJ may specify a baseline set of vital signs as each team member joins a hazmat response team. This data will be helpful to recognize the possible signs and symptoms of exposure, contamination, or a heat-related illness **(Figure 9.14)**. Employment-related examinations are discussed further in Chapter 10 of this manual.

NOTE: The National Fallen Firefighters Foundation (NFFF) identified 16 Life Safety Initiatives to address the causes of preventable deaths (2004).

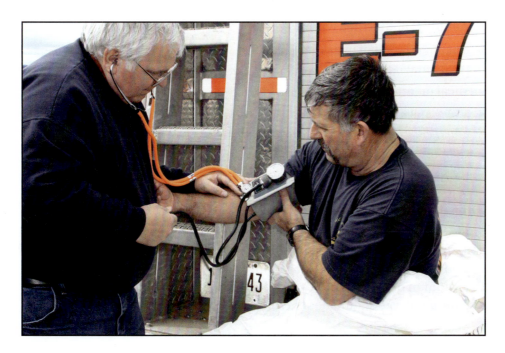

Figure 9.14 Medical surveillance includes collecting baseline data and comparing it to data after an exposure or activity.

Hazmat team members and emergency responders have a right to medical surveillance. Several laws, regulations, and national standards ensure that these rights are protected.

- 29 CFR 1910.120(f) *Hazardous Waste Operations and Emergency Response*, NFPA 1582 *Standard on Comprehensive Occupational Medical Program for Fire Department*, or the AHJ specify minimum requirements for a medical surveillance program.
- The Environmental Protection Agency (EPA) has also instituted policy for all response personnel that would not be covered by the OSHA standard.
- The NFPA has developed NFPA 1500 *Standard on Fire Department Occupational Safety and Health Program*.

Pre- and post-entry medical monitoring are vital for the protection of hazmat team members. This monitoring not only ensures that technicians are healthy before donning chemical protective clothing, but it also allows for a baseline for post-entry medical surveillance and gives the medical team comparison data to review while the emergency responder is in rehab.

Pre-Entry Evaluation

Develop any medical surveillance program under consultation with a physician. If possible and practical, a physician should be present during medical evaluations. However, the AHJ may allow for other levels of medical professionals to oversee the medical plan.

A detailed pre-entry medical surveillance report should include items as recommended by the organization's local healthcare professionals/doctors:

- A record of vital signs including blood pressure, pulse, respirations, body temperature, and body weight
- Evaluation of skin including rashes, wounds, and open sores
- An evaluation of mental status
- An outline of medical history including medications, any alcohol consumption, any medical treatment or diagnosis, and any symptoms of fever, nausea, vomiting, or diarrhea within 72 hours
- Documentation of hydration

Depending on the circumstances, the AHJ may also require some types of medical monitoring. Examples include:

- NFPA 1584, *Standard on the Rehabilitation Process for Members during Emergency Operations and Training Exercises*
- OSHA 29 CFR 1910.1000, Toxic and Hazardous Substances, including monitoring requirements for asbestos exposure testing

Post-Entry Evaluation

Post-entry evaluation includes many of the evaluation points on the pre-entry surveillance form. Monitor vital signs every 5 to 10 minutes until they return to a normal range; otherwise, additional medical testing may be necessary. Also monitor body weight; a drastic loss of body weight is a cause for concern and the employee should be sent for further testing.

Post-entry monitoring does not have to conclude at the end of the incident. Based on the products involved in the incident, it may be beneficial to continue post-exposure surveillance for a few days after the incident to ensure that response personnel are safe and cared for.

Backup and Rapid Intervention

NFPA and OSHA mandate the use of buddy systems and backup personnel at hazmat incidents. A buddy system is a system of organizing personnel into work groups in such a manner that each member has a buddy or partner. The purpose of the buddy system is to provide rapid help in the event of an emergency. In addition to the buddy system and back-up intervention, many AHJs also establish requirements for **rapid intervention crews or teams (RIC/RITs)** at hazmat incidents.

In addition to using the buddy system, backup personnel must be standing by with equipment to provide assistance or rescue if needed. Qualified basic life support personnel (as a minimum) must also be standing by with medical equipment and transportation capability.

The minimum number of personnel necessary for performing tasks in the hazardous area is four — two working in the area itself and two standing by as backup. Backup personnel must don the same level of PPE as entry personnel.

> **Rapid Intervention Crew or Team (RIC/RIT)** — Two or more firefighters designated to perform firefighter rescue; they are stationed outside the hazard and must be standing by throughout the incident. *Previously known as* Rapid Intervention Team (RIT).

Emergency Procedures

Emergency procedures should also include alternate communication methods should radio communications fail. The AHJ should adopt all signals for a variety of emergencies to communicate with personnel outside the hot zone. Practice alternate communication methods in advance of an emergency.

> **WARNING!**
> If one member of the entry team experiences a PPE issue, both members must exit together.

In case immediate evacuation is needed, establish an emergency evacuation signal. This signal should be a combination of radio communications, public address system, and air horn blasts. In many jurisdictions, three long blasts of apparatus air horns mean there should be an emergency evacuation into the safe refuge point.

Chapter Notes

"16 Firefighter Life Safety Initiatives." Everyone Goes Home, National Fallen Firefighters Foundation (NFFF), 2004. Accessed: http://www.everyonegoeshome.com/16-initiatives/

Benner, Ludwig, Jr. "D.E.C.I.D.E. In Hazardous Materials Emergencies." Fire Journal, 69:4, July 1975. Reprint: National Transportation Safety Board. Accessed: http://www.ludwigbenner.org/HMdocs/DECIDE_reprint.pdf

"Emergency Response Plan for Hazardous Materials." Environmental Health and Safety, University of Chicago, 2010. Accessed: http://safety.uchicago.edu/pp/emergency/hazmat.shtml

e-HASP (Health and Safety Plans) Software, Version 2.0. OSHA, March 2006. Accessed: https://www.osha.gov/dep/etools/ehasp/index.html

Occupational Safety and Health Guidance Manual for Hazardous Waste Site Activities, NIOSH/ OSHA/ USCG/ EPA, 1985. Accessed: https://www.osha.gov/Publications/complinks/OSHG-HazWaste/all-in-one.pdf

"Site Safety and Control Plan." ICS 208 HM fillable form. Accessed: https://www.pdffiller.com/en/project/51162660.htm?form_id=100116089

Chapter Review

1. What is the difference between an initial and a detailed hazard and risk assessment?
2. What are some basic principles of risk management?
3. What is the basic problem-solving process that most response models incorporate?
4. What three items should be included in the IAP in order to tie together the entire problem-solving process?
5. What are some of the components of site specific health and safety plans?

9-1

Develop response objectives and response options for a hazmat/WMD incident.
[NFPA 1072 7.3.1]

WARNING: If this skill involves the use of actual hazardous material samples, hazardous materials can cause serious injury or fatality. Appropriate personal protective equipment (PPE) must be worn and safety precautions must be followed. The following skill sheet demonstrates general steps; specific hazmat incidents may differ in procedure. Always follow the AHJ's procedures for specific incidents.

Step 1: Identify or develop response objectives for the incident.

Step 2: Identify or develop response options for each response objective.

Step 3: Make recommendations for response objectives and response options to the Incident Commander or hazardous materials officer.

9-2
Develop a site safety and control plan. [NFPA 1072 7.3.4]

WARNING: If this skill involves the use of actual hazardous material samples, hazardous materials can cause serious injury or fatality. Appropriate personal protective equipment (PPE) must be worn and safety precautions must be followed. The following skill sheet demonstrates general steps; specific hazmat incidents may differ in procedure. Always follow the AHJ's procedures for specific incidents.

Step 1: Identify tasks and resources required to meet response objectives.

Step 2: Address specified response objectives and options.

Step 3: Ensure plan is consistent with emergency response plan and policies and procedures.

Step 4: Ensure plan is within the capability of available resources, including personnel.

Step 5: Prepare an incident action plan.

Step 6: Identify site safety and control components.

Step 7: Identify points for a safety briefing.

Step 8: Identify pre-entry tasks.

Step 9: Identify atmospheric and physical safety hazards (when confined space is involved).

Step 10: Collect and preserve legal evidence.

Implementing and Evaluating the Action Plan: Incident Management

Chapter Contents

Incident Management Systems............ 353
    Incident Command 353
    Unified Command 355

Section Leaders and Supervisors in the Hazmat Branch 356
    Hazmat Branch Director/Group Supervisor 356
    Entry Team Leader 357
    Decontamination Leader 358
    Site Access Control Leader 358
    Hazmat Safety Officer 358
    Hazmat Logistics Officer 359
    Safe Refuge Area Manager 359
    Medical Officer 360
    Information and Research Officer 360

Hazmat Incident Levels and Evaluation ...361
    Level I 361
    Level II 361
    Level III 362
    NIMS Incident Typing System 362
    Evaluating Effectiveness 363

Forms and Logs 364
    Activity Logs 364
    Exposure Records 365
    Entry/Exit Logs 365

Medical Monitoring and Exposure Reporting 366
    Initial and Annual Hazmat Team Physical 366
    Immediate Care 367
    Post-Exposure Monitoring 367

Chapter Review 368
Skill Sheets 369

chapter 10

Key Terms

Incident Management System (IMS) 353
National Incident Management System - Incident Command System (NIMS-ICS) 353

Safety Officer 358
Unified Command (UC) 355

Implementing and Evaluating the Action Plan: Incident Management

JPRs Addressed In This Chapter

7.4.1

Learning Objectives

After reading this chapter, students will be able to:

1. Describe incident management systems. [NFPA 1072, 7.4.1]
2. Explain responsibilities of different positions within the Hazardous Materials Branch of the Incident Command Structure. [NFPA 1072, 7.4.1]
3. Recognize systems for identifying levels of hazmat incidents and evaluating effectiveness. [NFPA 1072, 7.4.1]
4. Identify forms, logs, and other documentation used at hazmat incidents. [NFPA 1072, 7.4.1]
5. Explain requirements for medical monitoring and exposure reporting of hazardous materials responders. [NFPA Standard 7.4.1]
6. Skill Sheet 10-1: Perform the duties of an assigned function within the incident command system. [NFPA 1072 7.4.1]
7. Skill Sheet 10-2: Assess and communicate progress of assigned tasks at a hazmat/WMD incident. [NFPA 1072, 7.4.1]

Chapter 10
Implementing and Evaluating the Action Plan: Incident Management

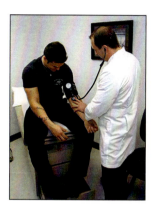

This chapter includes a detailed discussion of the following topics:
- Incident management systems
- Hazardous Materials Branch Director/Group Supervisor
- Hazmat incident levels and evaluation
- Forms and logs
- Medical monitoring and exposure reporting

Incident Management Systems

Hazardous materials incidents can be dynamic and tax the resources of any response agency. Not only will this type of incident be complex based on the hazards that may be present, but the nature of the incident may require personnel and resources that far exceed a simple Incident Command Structure. Because of this, it is important for hazmat technicians to understand the National Incident Management System (NIMS) and its functions.

Many U.S. federal regulations dictate how a hazardous materials incident should be managed. Both Occupational Safety and Health Administration (OSHA) and Environmental Protection Agency (EPA) regulations dictate that a hazardous materials incident must be operated with an **Incident Management System (IMS)** in place. In 2003, the U.S. implemented and mandated the use of NIMS. Most incidents are managed under the **National Incident Management System - Incident Command System (NIMS-ICS)**. Steps for performing the duties of an assigned function within the incident command system are provided in **Skill Sheet 10-1**.

NOTE: NIMS-ICS is not used universally outside of the United States. Follow the Incident Management System dictated by the authority having jurisdiction.

The specific organizational structure for a hazmat incident will be based on the incident's needs. An Incident Commander (IC) and a Safety Officer will be appointed at every hazmat incident. Other positions default to the IC until such a time as they require additional oversight. The following sections introduce positions that may be necessary at a hazmat incident.

Incident Command

Responders assigned to Command and Branch functions at hazmat incidents must be highly detail-oriented. At routine emergencies, those in charge may be able to develop a plan of action in their head and maneuver through it without much difficulty. At a hazmat incident, this cannot happen. Command must ensure written documentation of all incident details. Furthermore, the Inci-

> **Incident Management System (IMS)** — System described in NFPA 1561, *Standard on Emergency Services Incident Management System and Command Safety*, that defines the roles, responsibilities, and standard operating procedures used to manage emergency operations. Such systems may also be referred to as Incident Command Systems (ICS).

> **National Incident Management System - Incident Command System (NIMS-ICS)** — The U.S. mandated incident management system that defines the roles, responsibilities, and standard operating procedures used to manage emergency operations; creates a unified incident response structure for federal, state, and local governments.

dent Action Plan (IAP) must be in writing and must include specific elements and information such as:

- Preincident surveys
- Response modes
- Hazard assessments
- Goals and objectives
- Site safety plan

Not only is this information critical for safe operation at a hazardous materials incident, but it is also a requirement under many of the federal standards that dictate the response for these types of incidents.

The Incident Commander (IC) bears all the responsibility of managing the incident unless responsibility is transferred by assigning staff to other functional areas. The primary functional areas (sometimes referred to as General Staff functions) include the following **(Figure 10.1)**:

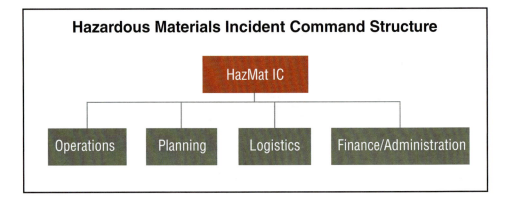

Figure 10.1 General staff functions include resources that must be accommodated at any incident of any size or complexity.

- Operations Section
- Planning Section
- Logistics Section
- Finance/Administration Section

General Responsibilities

The IC is required to perform the following functions at hazmat incidents:

- Establish the site safety (also called scene safety) plan.
- Implement a site security and control plan to limit the number of personnel operating in the control zones.
- Designate a Safety Officer.
- Identify the materials or conditions involved in the incident.
- Implement appropriate emergency operations.
- Ensure that all emergency responders (not just those from their own organizations) wear appropriate personal protective equipment (PPE) in restricted zones.
- Establish a decontamination plan and operation.
- Implement post incident emergency response procedures (incident termination).

Specific Responsibilities

The IC has three critical responsibilities:

1. Assume and announce Command and establish an effective operating position.
2. Rapidly evaluate the situation by conducting a thorough and complete size-up.
3. Initiate, maintain, and control the communication process.

The IC must also:

- Identify the incident objectives, develop an IAP, and assign resources that are consistent with local SOPs.
- Develop an effective Incident Command organization.
- Provide strategies and tactics.
- Review, evaluate, and revise the IAP.

Unified Command

A single agency within a solitary jurisdiction can manage most incidents. In these instances, the functions and Branches are typically filled by personnel from a single agency. However, based on the complexity or nature of a hazmat incident, a single agency may not be able to effectively mitigate the incident. Where multiple agencies and/or jurisdictions are expected to work together, a **Unified Command (UC)** structure is the most efficient and safest means of operation **(Figure 10.2)**.

> **Unified Command (UC)** — In the Incident Command System, a shared command role in which all agencies with geographical or functional responsibility establish a common set of incident objectives and strategies. In Unified Command there is a single Incident Command Post and a single operations chief at any given time.

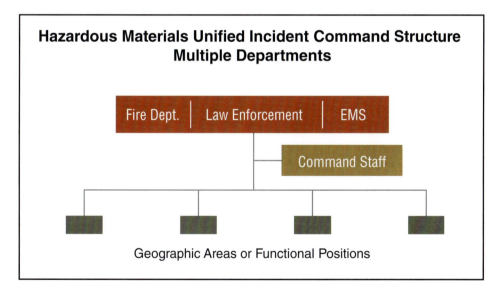

Figure 10.2 Unified Command puts decision makers from multiple organizations at the same level.

The concept of UC is simple: All agencies that are functionally involved in the incident contribute to the process. The representatives in the UC structure help to determine strategies and the overall incident objectives and help plan jointly for tactical objectives. To function well in this structure, responders must set common objectives and strategies. These strategies and tactics aid in the development of the IAP.

The UC structure should consist of key officials from each jurisdiction that have decision-making authority. As an option, a UC structure could include landowners, facility personnel, or other agencies or individuals who have functional expertise or capabilities.

The IAP falls under the authority of a single individual. In a UC structure, the Operations Section Chief directs the IAP. The jurisdiction having the greatest involvement may appoint the Operations Section Chief, but all agencies represented in the UC must agree on this designation.

Section Leaders and Supervisors in the Hazmat Branch

Based on the severity and complexity of the incident, all positions within the Branch may not have to be filled. However, it is highly advisable that these positions be staffed with personnel trained to the Technician Level **(Figure 10.3)**. The following text gives a brief overview of the duties of each Section Leader or Supervisor. Each of these Section Leaders or Supervisors should document his or her activities at the incident using a Unit/Activity Log (in the U.S., this will be ICS Form 214). The Incident Commander must be at least an Operations-Level hazmat responder.

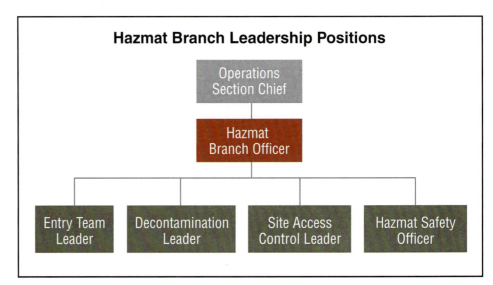

Figure 10.3 Responders trained to the hazmat Technician Level will understand the implications of a hazmat incident better than those who are not.

What This Means To You

OSHA Requirements

Further requirements for Incident Command are addressed in OSHA 1910.120(q). Even if you live in a state/province that has not adopted OSHA, you may be held responsible for this information. For example, the U.S. EPA has adopted some OSHA requirements that may be enforceable regardless of the state in which it is applied.

Hazmat Branch Director/Group Supervisor

The Hazmat Branch Director/Group Supervisor reports directly to the Operations Section Chief. The Branch Director/Group Supervisor is responsible for the Hazmat Branch and works closely with all other leaders who have been assigned to the Branch. According to NFPA 472, a Hazmat Branch Director/Group Supervisor, is certified to the Technician Level. NFPA 1072 does not include JPRs for certifications required of Hazmat Branch Directors/Group Supervisors.

NOTE: OSHA does not specify a Hazmat Branch.

A Hazmat Branch will be established at many hazmat incidents and will manage all tactical operations carried out within the control zones. The Hazmat Branch Director/Group Supervisor also coordinates all work conducted by a number of officers, including:

- Information and Research Officer
- Entry Team Leader
- Decontamination Leader
- Site Access Control Leader
- Hazmat Safety Officer

The Hazmat Branch Director/Group Supervisor manages all operations including:

- Site safety
- Rescue
- Entry operations
- Decontamination
- Containment or confinement activities
- Sampling
- Medical monitoring

The Hazmat Branch Director/Group Supervisor's duties may be far ranging. This officer should have input in the following areas:

- Developing control zones and access control points
- Evaluating and recommending public protection actions in conjunction with the Operations Section Chief
- Developing a site safety plan
- Developing the IAP

The Hazmat Branch Director/Group Supervisor should work with other key positions, such as the Incident Safety Officer, to develop public protective actions. Directly related to the incident, the Hazmat Branch Director/Group Supervisor may work with the Research Officer and other Information and Research Officers to ascertain relevant weather data and monitor the hazard site in order to develop the IAP. It is the Hazmat Branch Director/Group Supervisor's responsibility to conduct safety meetings with the entire Hazmat Branch. It is critical that the Hazmat Logistics Officer maintain thorough and appropriate documentation of all activities.

Entry Team Leader

The Entry Team Leader reports to the Hazmat Branch Director/Group Supervisor. The Entry Team Leader is responsible for all personnel assigned to entry operations and directs any rescue operation within the hot zone (exclusion zone). While the Entry Team Leader is responsible for all response personnel movement within the hot zone, he or she must also consider the movement of any contaminated victims. The Entry Team Leader must work closely and communicate with other members of the Hazmat Branch and other branches, especially the Decontamination Leader, the Research Officer, and the Safe Refuge Area Manager, if these functions have been activated.

Because of the intimate knowledge that the Entry Team Leader will hold based on information received from crews working within the hot zone, this officer must be allowed to recommend actions to mitigate any situation occurring in the hot zone. The Entry Team Leader is also responsible for the following:

- Communicating all actions to the Hazmat Branch Director/Group Supervisor
- Monitoring, recording, and maintaining entry and exit logs that outline air use, PPE, and time in the hot zone

Decontamination Leader

The Decontamination Leader reports directly to the Hazmat Branch Director/Group Supervisor and works closely with the Safe Refuge Area Manager, the Medical Branch, and the Entry Team Leader. The Decontamination Leader is responsible for all decon operations and personnel inside the warm zone (contamination reduction zone) and also holds the following responsibilities:

- Executing all decontamination requirements and guidelines outlined in the IAP
- Coordinating the transfer of contaminated patients requiring medical attention to the Medical Branch
- Maintaining documentation on any contamination (kept on a separate log)

Site Access Control Leader

The Site Access Control Leader works closely with the Entry and Decontamination Team Leaders and reports directly to the Hazmat Branch Director/Group Supervisor. The Site Access Control Leader is responsible for the following:

- Controlling all movement through the various routes and corridors of the hazmat incident site
- Controlling contamination
- Keeping appropriate records
- Helping establish control zones and creating a safe refuge area
- Appointing a Safe Refuge Area Manager
- Tracking the movement of all persons passing through contamination control lines
- Ensuring that all persons leaving restricted areas have received proper decontamination
- Coordinating with the Medical Branch to ensure that both responders and victims receive appropriate medical observation

Hazmat Safety Officer

Per NFPA 1500, *Standard on Fire Department Occupational Safety and Health Program*, a **Safety Officer** must be certified to the level of the incident response. During a Technician-Level response, the Safety Officer must be a trained technician, according to NFPA 472. Based on the complexity of the incident, there may be multiple Safety Officers assigned to oversee different tasks. The Incident Safety Officer may serve in the capacity of overseeing the health and safety of the personnel involved in the entire operation, whereas the Hazmat Safety Officer will serve as the safety officer overseeing the entry operation **(Figure 10.4)**.

> **Safety Officer** — Member of the IMS command staff responsible to the Incident Commander for monitoring and assessing hazardous and unsafe conditions and developing measures for assessing personnel safety on an incident. *Also known as* Incident Safety Officer.

Figure 10.4 The Hazmat Safety Officer must be trained to the level of the response, and duties may include controlling the access to the incident response area.

The Hazmat Safety Officer reports directly to the Incident Safety Officer and works in conjunction with the Hazmat Branch Director/Group Supervisor and the Entry Team Leader. The person in this position coordinates safety-related activities that have a direct effect on Hazmat Branch operations. The Hazmat Safety Officer advises the Hazmat Branch Director/Group Supervisor on all aspects of health and safety and has the authority to stop or prevent any action he or she deems unsafe.

The Hazmat Safety Officer should be involved in the preparation and implementation of the site safety plan and must advise the Hazmat Branch Director/Group Supervisor if a deviation from this plan must occur. The Hazmat Safety Officer's primary objective is to ensure the protection of all Hazmat Branch personnel from physical, environmental, or chemical hazards or exposures.

The Hazmat Safety Officer is responsible for maintaining the following documentation:

- Medical records for Hazmat Branch personnel
- Exposure records, PPE logs, and any other documentation that may be required by the AHJ

Hazmat Logistics Officer

The Hazmat Logistics Officer works in conjunction with the Logistics Section Chief and is responsible for ensuring that all materials needed to support the Hazmat Branch are in place. This position maintains an ongoing inventory of resources needed to control the incident, which may include self-contained breathing apparatus cylinders, foam concentrates, PPE, and any supplies that may be needed by the entry personnel to control or contain any chemicals involved.

The Hazmat Logistics Officer reports directly to the Hazmat Branch Director/Group Supervisor and works closely with all other leaders who have been assigned to the Hazmat Branch. Documentation for the Hazmat Logistics Officer is paramount.

Safe Refuge Area Manager

The Safe Refuge Area Manager works closely with the Entry and Decontamination Leaders along with the Medical Officers and reports directly to the Site Access Control Leader. The Safe Refuge Area Manager is responsible for:

- Evaluating and determining the priority of victims for treatment
- Collecting information from victims
- Helping to prevent victims from spreading contamination

Victims leaving the scene often have direct knowledge and information from inside the hot zone that may be helpful in determining mitigation activities. The Safe Refuge Area Manager may be the first person available to interview these people and must relay any pertinent information to the IC.

Medical Officer

This position reports directly to the Hazmat Branch Director/Group Supervisor. The Medical Officer works closely with the Decontamination Leader, Safe Refuge Area Manager, and Hazmat Safety Officer. More information about medical emergency procedures is included in the IFSTA manual, **Fire and Emergency Services Safety Officer**.

The Medical Officer is responsible for the following:

- Medical evaluation and monitoring of personnel operating at the site **(Figure 10.5)**

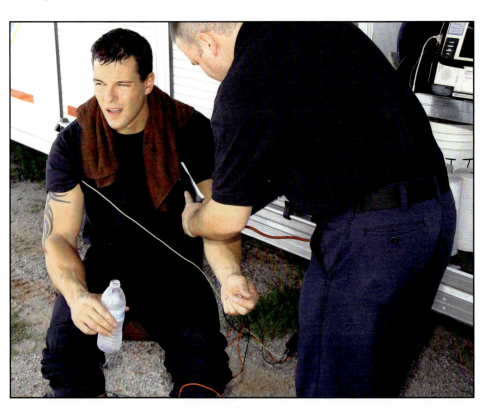

Figure 10.5 Rehabilitation at an incident is important to determine whether a responder was affected adversely by the experience.

- Treatment of personnel exposed to products at the incident
- Triage, as necessary
- Maintenance of medical records for both responders and victims

Information and Research Officer

The Information and Research Officer (or research specialist) reports directly to the Hazmat Branch Director/Group Supervisor. This position works in conjunction with the Planning Section Chief and the Incident Safety Officer.

The Information and Research Officer provides technical information assistance to the Hazmat Branch using a variety of reference sources. These sources may include:

- Facility representatives
- Technical references
- Computer databases
- Chemical Transportation Emergency Center (CHEMTREC) and Canadian Transport Emergency Centre (CANUTEC)

NOTE: These resources were explored in greater detail in Chapter 7 of this manual.

The Information and Research Officer's role goes beyond using reference materials. The Information and Research Officer may be required to perform the following tasks:

- Interpret environmental monitoring information
- Analyze samples obtained in the hot zone
- Determine PPE compatibility
- Assist in projecting the potential environmental effects of the release

Hazmat Incident Levels and Evaluation

After technicians have determined the initial scope of an incident, they can then identify the level of the incident in accordance with established guidelines. The U.S. National Response Team (a group of 15 federal agencies) uses a hazardous materials incident level system. These levels have been integrated into many local emergency plans and the levels of response range from Level I (least serious) to Level III (most serious) **(Table 10.1, p. 362)**. Based on the severity of the incident, technicians can identify the level of involvement and necessary resources.

Level I

Level I incidents are typically within the capabilities of the fire or emergency services organization or other first responders having jurisdiction. Level I incidents are the least serious and the easiest to handle. Evacuation (if required) is limited to the immediate area of the incident.

Level II

Level II incidents are beyond the capabilities of the first responders on the scene and may be beyond the capabilities of the first response agency/organization having jurisdiction. Some jurisdictions may require additional capabilities or resources. Level II incidents may require the services of a formal hazmat response team. A properly trained and equipped response team could be expected to perform the following tasks:

- Dike and confine within the contaminated areas
- Perform plugging, patching, and basic leak control activities
- Sample and test unknown substances
- Perform various levels of decontamination

**Table 10.1
Examples of Level I/II/III
Hazardous Materials Incidents
(U.S. National Response Team System)**

Level	Examples
I	• A natural gas leak that may have occurred from construction crews damaging a low-pressure gas line buried below the street • Broken containers of consumer commodities such as paint, thinners, bleach, swimming pool chemicals, and fertilizers
II	• Spill or leak requiring limited-scale evacuation • Any major accident, spillage, or overflow of flammable liquids • Spill or leak of unfamiliar or unknown chemicals • Accident involving extremely hazardous substances • Rupture of an underground pipeline • Fire that is posing a boiling liquid expanding vapor explosion (BLEVE) threat in a storage tank
III	• Those that require an evacuation extending across jurisdictional boundaries • Those beyond the capabilities of the local hazardous material response team • Those that activate (in part or in whole) the federal response plan

Level III

Level III incidents require resources from state/provincial agencies, federal agencies, and/or private industry. They also require Unified Command. A Level III incident is the most complex of all hazardous materials incidents and may require large-scale evacuation. Most likely, no single agency will manage the incident. Successful handling of the incident requires a collective effort from several of the following resources/procedures:

- Specialists from industry and governmental agencies
- Sophisticated sampling and monitoring equipment
- Specialized leak and spill control techniques
- Decontamination on a large scale

NIMS Incident Typing System

While the incident types introduced in the previous sections are used in many jurisdictions, NIMS has incorporated an incident typing system to assist in decisions concerning resource requirements. Five incident levels have been incorporated into the NIMS response protocols **(Table 10.2)**.

**Table 10.2
Characteristics of Level 5/4/3/2/1
Hazardous Materials Incidents
(NIMS Incident Typing System)**

Level	Characteristics
5	• The incident can be handled with one or two single resources with up to six personnel. • Only the IC position is activated. • A written Incident Action Plan is not required. • The incident is contained within the first operational period and often within an hour to a few hours after resources arrive on scene.
4	• Command staff and general staff functions are activated only if needed. • Several resources are required to mitigate the incident. • The incident is usually limited to one operational period in the control phase. • The agency administrator may have briefings and ensure the complexity analysis and delegation of authority is updated. • No written IAP is required but a documented operational briefing will be completed for all incoming resources. • The role of the Agency Administrator includes operational plans, which may include objectives and priorities.
3	• When capabilities exceed the initial attack, the appropriate positions should be added to match the complexity of the incident. • Some or all of the Command and General Staff positions may be activated, as well as Division/Branch and/or Unit Leader level positions. • A Type 3 Incident Management Team (IMT) or Incident Command organization manages initial action incidents with a significant number of resources, an extended attack incident until containment/control is achieved, or an expanding incident until transition to a Type 1 or 2 team. • The incident may extend into multiple operational periods. • A written IAP may be required for each operational period.
2	• Extends beyond the capabilities for local control and is expected to go into multiple operational periods. A Type 2 incident may require the response of resources out of area, including regional and/or national resources, to effectively manage the operations, Command, and general staffing. • Most or all of the Command and General Staff positions are filled. • A written IAP is required for each operational period. • Many of the functional units are needed and staffed. • Operations personnel normally do not exceed 200 per operational period and total incident personnel do not exceed 500 (guidelines only). • The Agency Administrator is responsible for the incident complexity analysis, agency administrator briefings, and the written delegation of authority.
1	• The most complex type of incident, requiring national resources to safely and effectively manage and operate. • All Command and General Staff positions are activated. • Operations personnel often exceed 500 per operational period and total personnel will usually exceed 1,000. • Branches need to be established. • The Agency Administrator will have briefings and ensure that the complexity analysis and delegation of authority are updated. • Use of resource advisors at the incident base is recommended. • There is a high impact on the local jurisdiction, requiring additional staff for office administrative and support functions.

Evaluating Effectiveness

Evaluating effectiveness is the natural ending place for any problem-solving process. If an IAP is effective, the IC should receive favorable progress reports from the Hazmat Branch Director/Group Supervisor and the incident should begin to stabilize. However, if mitigation efforts are failing or the situation is growing worse, the team must reevaluate and possibly revise the plan. The team must also reevaluate the plan as new information becomes available and circumstances change. If the initial plan is not working, the team must change it either by selecting new strategies or by changing the tactics used to achieve them. In accordance with predetermined communication procedures, responders must communicate the status of the planned response and the progress of their actions to the IC. Steps for assessing and communicating progress of assigned tasks at a hazmat/WMD incident are provided in **Skill Sheet 10-2**.

Forms and Logs

There may be many forms that should be filled out for both safety and liability reasons. The AHJ will dictate which forms need to be completed both during and after the incident.

NIMS has encouraged the standardization of many forms **(Table 10.3)**. However, some response agencies may have customized some forms in an effort to streamline the reporting process. Many NIMS forms are readily available by downloading them from the FEMA website.

Based on the scope and size of the incident, the IC will dictate which forms need to be completed during the incident. Many of these ICS forms are used for the development of the IAP. However, some forms are much more critical and the team must complete them to satisfy federal, state, or provincial requirements. These forms are discussed in the following sections.

Activity Logs

Activity logs are a running record of all the events that occurred during the incident. These logs should be filled out by many of the Section Chiefs and Leaders of those assigned within the Hazmat Branch structure. Activity logs should be a chronological list that includes the following information:

- Incident name
- Operational period
- Unit name/designation
- Unit leader
- Personnel assigned

The main body of information for this form is the log of all activity that has occurred. Each entry into the activity log should include the time of the event and a brief description of each significant occurrence that has taken place during the incident. This can include task assignments, task completions, injuries, difficulties encountered, and hazards identified. These logs become the timeline during the investigation phase and for all required reporting.

Table 10.3
NIMS-ICS Form Numbers and Titles

Form Number	Title
201	Incident Briefing
202	Incident Objectives
203	Organization Assignment List
204	Assignment List
205	Incident Radio Communications Plan
206	Medical Plan
207	Incident Organization Chart
208	Safety Message/Plan
208 HM	Site Safety Plan
211	Incident Check-in List
214	Activity Log
215	Operational Planning Worksheet
215A	Incident Action Plan Safety Analysis

Exposure Records

Exposure records are a vital record that should be completed by either the Decontamination Leader or the Hazmat Safety Officer. These exposure records should be kept in conjunction with any medical records for employees who have worked in proximity to the hazard. Exposure records should include the following information:

- Type of exposure
- Length of exposure
- Description of PPE used
- Type of decontamination used including any decontamination solutions
- On-scene and follow-up medical attention and/or assistance

Because exposures to hazardous chemicals may not present any signs or symptoms for many years, it is a legal requirement to retain medical records per the AHJ.

What This Means To You
OSHA Requirements for Medical Paperwork

OSHA and other legal requirements may include minimum numbers of years that these documents are kept. Those requirements don't necessarily include periodic reviews of that information.

Hazmat technicians should keep copies of their personal records for their own reference, and understand the information. Technicians should also make sure their primary care physician also receives these files and understands their implications.

Entry/Exit Logs

The Entry Team Leader should maintain the entry/exit logs. The entry/exit logs are an organized approach to managing not only the personnel entering the hot zone but also those within the decontamination line. Entry/exit logs should document the following information:

- Name of the Entry Team Leader
- Radio frequency that will be used for entry team activities
- Name of each individual who will enter the hot zone
- Name of each individual serving on the backup team
- Name of each individual serving on the decontamination team
- The time each person goes on air and the air cylinder gauge reading
- The time each person enters the decontamination line
- The time each person exits the decontamination line and the time they are off air

At the conclusion of the incident, the Hazmat Safety Officer should review the entry/exit logs. The Hazmat Safety Officer will then be required to submit these logs to the Command Staff for documentation retention.

Medical Monitoring and Exposure Reporting

Effective medical monitoring will help keep personnel safe on an emergency scene, and also may prevent many illnesses and injuries that can plague emergency responders. In the U.S., medical monitoring is a requirement of OSHA 29 CFR 1910.120. OSHA and NFPA requirements are separate and may overlap. More information on the topic of medical monitoring and response were included in Chapter 9 of this manual.

A medical monitoring plan should include:

- A baseline and annual physical that is used to update the medical history of the hazmat team member.

- An exposure physical that may be conducted any time a hazmat team member is exposed to a chemical. Exposure records shall be kept 30 years after employment.

- An exit physical when the employee's hazmat team tenure ends.

The baseline physical is a detailed and comprehensive exam that will review the employee's health history including any previous chemical exposure. The complete physical exam should include vision and hearing tests along with extensive laboratory blood work and urinalysis. A chest X-ray and EKG should also be included. These tests are used to obtain baseline numbers for future reference should there be a chemical exposure. Medical surveillance may go far beyond the basics. A qualified medical practitioner should review the medical surveillance program regularly.

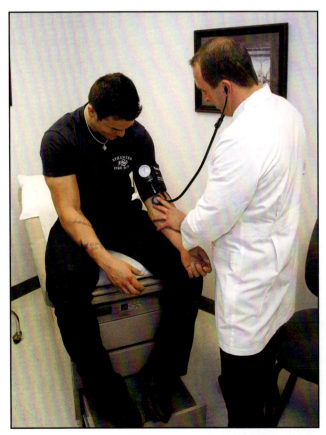

Figure 10.6 Baseline and annual evaluations are an important piece of a responder's medical history.

Initial and Annual Hazmat Team Physical

A health and safety program starts at acceptance onto a hazmat response team, or at the very least, ahead of allowing a team member to participate in a response. Baseline information is vital to accurately track the well-being of emergency response personnel **(Figure 10.6)**. To obtain an accurate snapshot of the health of a response team member, the acceptance screening should consider items recommended by the organization's local healthcare professionals/doctors such as:

- Total physical evaluation
- Body composition
- Laboratory tests and screenings
- Vision tests
- Hearing evaluation
- Lung function (spirometry)
- Electrocardiogram (ECG/EKG)
- Cancer screening
- Immunizations and infectious disease screening

The department's healthcare provider can work with team members and administrators to ensure that all employees are tracked for the possible hazards that may be associated with a response. Some tests may be required based on specific hazards that may be located in proximity to the response area. Physicians can perform blood tests to detect heavy metals such as lead, mercury, or arsenic. Physicians can also perform a liver function test if the responder may have been exposed to large quantities of petroleum-based solvents.

Medical requirements do not end after the initial physical evaluation. Periodic medical examinations should be a part of an employer's overall safety plan. These physicals should occur annually and new data should be compared to the original baseline data which allows for detection of any changes of the body's systems.

Initial and recurring medical exams provide information to the employer regarding an employee's fitness for duty. Per patient confidentiality laws, the employer may not receive quantified results from the examinations. Employees should coordinate with their primary care providers to receive copies of their results from the medical examinations.

Immediate Care

While some teams incorporate hazmat-trained paramedics, advanced life support personnel should be available to treat response team members should they require medical attention. As part of a team's preplanning process, personnel should identify hospitals that are equipped to handle hazmat, chemical, and toxicological emergencies. These facilities may have specially trained providers and equipment that may make a significant impact in an emergency.

Post-Exposure Monitoring

Post-exposure monitoring is necessary even if the technician is not feeling any symptoms of an exposure. If the technician was treated for an exposure to a chemical or toxic substance, additional testing may be required even after medical treatment is complete. These tests will help to confirm that the technician is healthy and will not suffer the adverse effects of an exposure in the future. Some tests may even quantify an exposure. These tests are helpful in determining if additional testing and treatment are necessary.

Documentation of an exposure is vitally important and should be completed in all cases if there is a confirmed or suspected exposure. All documentation should be in writing, and both the supervisor and the affected employee should complete and review all forms. Specific required forms may vary based on employer, local, state/provincial, or federal requirements. These forms may be used as evidence should there be a worker's compensation claim. OSHA 1910.1020, Access to Employee Exposure and Medical Records, requires that all exposure reports be maintained for the employee's length of employment plus an additional 30 years. However, employees should maintain their own copies of exposure reports that pertain to them.

Chapter Review

1. How does the responsibility for the Incident Action Plan change under a Unified Command structure?
2. List and define different positions and their responsibilities within the Hazmat Branch of the Incident Command System.
3. What types of resources will likely be required for the different levels of hazmat incidents?
4. What forms and logs will most likely be required at hazmat incidents?
5. What components should be included in a medical monitoring plan?

10-1

Perform the duties of an assigned function within the Incident Command System.
[NFPA 1072 7.4.1]

WARNING: If this skill involves the use of actual hazardous material samples, hazardous materials can cause serious injury or fatality. Appropriate personal protective equipment (PPE) must be worn and safety precautions must be followed. The following skill sheet demonstrates general steps; specific hazmat incidents may differ in procedure. Always follow the AHJ's procedures for specific incidents.

Step 1: Perform assigned duties in the Hazardous Materials Branch or group organization.

Step 2: Communicate observations to Hazardous Materials Branch Director/Supervisor, ICS operations section chief, or Incident Commander.

SKILL SHEETS

10-2

Assess and communicate progress of assigned tasks at a hazmat/WMD incident.
[NFPA 1072 7.5]

WARNING: If this skill involves the use of actual hazardous material samples, hazardous materials can cause serious injury or fatality. Appropriate personal protective equipment (PPE) must be worn and safety precautions must be followed. The following skill sheet demonstrates general steps; specific hazmat incidents may differ in procedure. Always follow the AHJ's procedures for specific incidents.

Step 1: Compare actual behavior of material and container to predicted behavior.

Step 2: Determine effectiveness of response options and actions in accomplishing response objectives.

Step 3: Modify response options and actions based on incident status review.

Step 4: Communicate status of response options and actions.

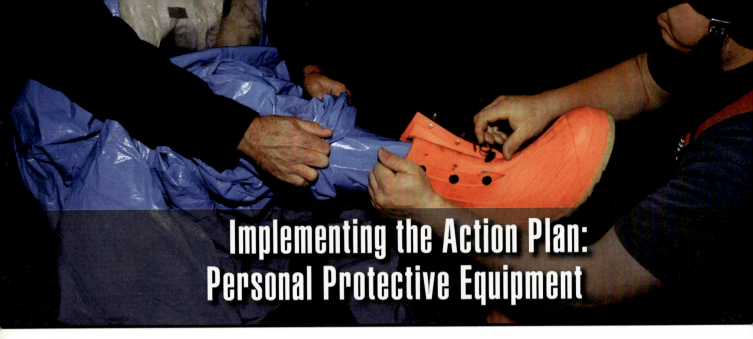

Implementing the Action Plan: Personal Protective Equipment

Chapter Contents

Respiratory Protection 375
   Standards for Respiratory Protection at Hazmat/WMD Incidents 377
   Self-Contained Breathing Apparatus (SCBA) 378
   Supplied Air Respirators 380
   Air-Purifying Respirators 380
   Respiratory Equipment Limitations 384

Protective Clothing Overview 385
   Standards for Protective Clothing and Equipment at Hazmat/WMD Incidents 386
   Structural Firefighters' Protective Clothing 388
   High Temperature-Protective Clothing 389
   Flame-Resistant Protective Clothing 390
   Chemical-Protective Clothing (CPC) 390

PPE Ensembles and Classification 395
   Levels of Protection 396
   Typical Ensembles of Response Personnel 402

PPE Selection Factors 404
   Selecting PPE for Unknown Environments 407
   Selecting Thermal Protective Clothing 408
   Selecting for Chemical Resistance/Compatibility 408

PPE-Related Stresses 410
   Heat Emergencies 411
   Heat-Exposure Prevention 411
   Cold Emergencies 413
   Psychological Issues 414

PPE Use 414
   Pre-Entry Inspection 415
   Safety and Emergency Procedures 416
   Donning and Doffing of PPE 423

PPE Inspection, Testing, Maintenance, Storage, and Documentation 425
   Inspection 425
   Testing 426
   Maintenance and Storage 426
   Documentation and Written PPE Program 427

Chapter Review 428
Skill Sheets 429

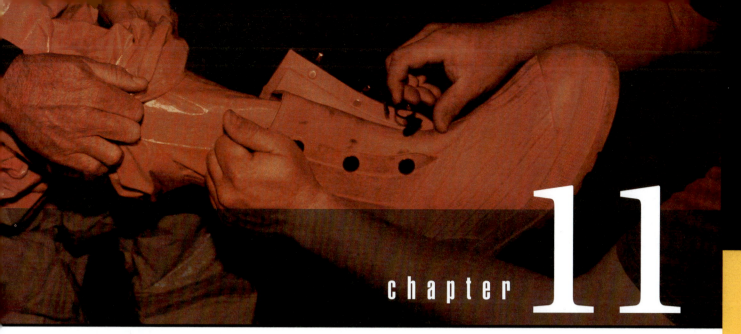

Chapter 11

Key Terms

Air-Purifying Respirator (APR) 380
Chemical Degradation 391
Chemical Protective Clothing (CPC) 390
Emergency Breathing Support System (EBSS) 380
Encapsulating 391
End-of-Service-Time Indicator (ESTI) 379
Flame-Resistant (FR) 390
Frostbite 413
Heat Cramp 411
Heat Exhaustion 411
Heat Rash 411
Heat Stroke 411

Hypothermia 413
Level A PPE 396
Level B PPE 396
Level C PPE 396
Level D PPE 396
Liquid Splash-Protective Clothing 391
Penetration 391
Permeation 391
Powered Air-Purifying Respirator (PAPR) 376
Self-Contained Breathing Apparatus (SCBA) 378
Supplied Air Respirator (SAR) 380
Trench Foot 413
Vapor-Protective Clothing 393

Implementing the Action Plan: Personal Protective Equipment

JPRs Addressed In This Chapter

7.3.2, 7.4.2, 7.4.3.1, 7.4.3.2, 7.4.3.3, 7.4.3.4, 7.4.4.2

Learning Objectives

After reading this chapter, students will be able to:

1. Describe types of respiratory protection used at hazmat/WMD incidents. [NFPA 1072, 7.3.2, 7.4.2]
2. Describe types of protective clothing used at hazmat/WMD incidents. [NFPA 1072, 7.3.2, 7.4.2, 7.4.3.1, 7.4.3.2, 7.4.3.3, 7.4.3.4, 7.4.4.2]
3. Explain types and classifications of PPE ensembles. [NFPA 1072, 7.3.2, 7.4.2]
4. Identify factors for selecting PPE at hazmat/WMD incidents. [NFPA 1072, 7.3.2]
5. Define stresses related to working in different types of PPE. [NFPA 1072, 7.3.2]
6. Explain procedures and considerations for using PPE at a hazmat/WMD incident. [NFPA 1072, 7.4.2]
7. Describe methods for inspection, testing, maintenance, storage, and documentation of PPE used at hazmat/WMD incidents. [NFPA 1072, 7.4.2]
8. Skill Sheet 11-1: Select PPE appropriate to a hazmat/WMD incident. [NFPA 1072, 7.3.2]
9. Skill Sheet 11-2: Don, work in, undergo technical decontamination while wearing, and doff structural fire fighting personal protective equipment. [NFPA 1072, 7.4.2]
10. Skill Sheet 11-3: Don, work in, undergo technical decontamination while wearing, and doff a Level C ensemble. [NFPA 1072, 7.4.2]
11. Skill Sheet 11-4: Don, work in, undergo technical decontamination while wearing, and doff liquid splash-protective clothing. [NFPA 1072, 7.4.2]
12. Skill Sheet 11-5: Don, work in, undergo technical decontamination while wearing, and doff vapor-protective clothing. [NFPA 1072, 7.4.2]
13. Skill Sheet 11-6: Inspect, test, maintain, and document chemical protective clothing serviceability. [NFPA 1072, 7.4.2]

Handwritten notes:
1. scba, air purifying respirator, supplied O2 respirator
2. Level A (vapor) B (splash) C (apr) d (bunker gear)
4. is it a vapor or gas, rote of exposure, mechanism of harm, hazards
5. heat stress
6. inspect, buddy system, follow guidelines, break through time
7.

Chapter 11
Implementing the Action Plan: Personal Protective Equipment

This chapter addresses PPE considerations for hazardous materials response at the Technician Level and covers the following topics:

- Respiratory protection
- Protective clothing
- PPE ensembles and classification
- PPE selection
- PPE-related stresses
- PPE use
- PPE inspection, testing, maintenance, and documentation

Respiratory Protection

Respiratory protection is a primary concern for first responders because inhalation is the most significant route of entry for hazardous materials. When correctly worn and used, protective breathing equipment protects the body from inhaling hazardous substances. Respiratory protection is, therefore, a vital part of any PPE ensemble used at hazmat/WMD incidents.

The basic types of protective breathing equipment used by responders at hazmat/WMD incidents are:

- Self-contained breathing apparatus (SCBA)
  — Closed-circuit SCBA **(Figure 11.1)**

Figure 11.1 Closed-circuit SCBA includes a system to purify exhaled air.

Figure 11.2 Open-circuit SCBA exhales into the atmosphere.

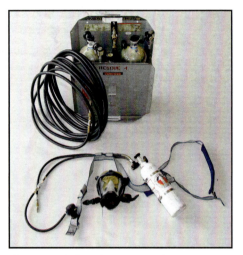

Figure 11.3 Supplied air respirators (SARs) are connected to an air compressor that is often too large to be worn by a responder.

Figure 11.4 Air-purifying respirators (APRs) rely on filters to remove contaminants from otherwise breathable air.

— Open-circuit SCBA **(Figure 11.2)**
- Supplied air respirators (SARs) **(Figure 11.3)**
- Air-purifying respirators (APR) **(Figure 11.4)**
  — Particulate-removing
  — Vapor-and-gas-removing
  — Combination particulate- and vapor-and-gas-removing
- **Powered air-purifying respirators (PAPR) (Figure 11.5)**

Handwritten annotation: N: ot resistant to oil / R: esistant to oil / P: oil or non oils or present

Powered Air-Purifying Respirator (PAPR) — Motorized respirator that uses a filter to clean surrounding air, then delivers it to the wearer to breathe; typically includes a headpiece, breathing tube, and a blower/battery box that is worn on the belt.

WARNING!
SCBA must be worn during emergency operations at terrorist/hazmat incidents until air monitoring and sampling determines other options are acceptable.

Each type of respiratory protection equipment has limits to its capabilities. For example, open-circuit self-contained breathing apparatus (SCBA) offers a limited working duration based upon the quantity of air contained within the apparatus' cylinder and the rate at which the air is used.

WARNING!
SCBA air consumption rates will vary.

Depending on the PPE you are issued, you may also need to be familiar with powered-air hoods, escape respirators, and combined respirators. The sections that follow discuss U.S. and international standards for respiratory protection, and respiratory equipment including each type's basic limitations. These features may affect the equipment's suitability for use in a particular response option.

Standards for Respiratory Protection at Hazmat/WMD Incidents

Because of the extreme hazards associated with chemical (such as military nerve agents), biological, radioactive, and nuclear materials that could be used in terrorist attacks, the U.S. Department of Homeland Security has adopted recommended standards developed by the National Institute for Occupational Safety and Health (NIOSH) and the NFPA for respiratory equipment to protect responders at hazmat/WMD incidents. NIOSH also certifies SCBA and recommends ways to select and use protective clothing and respirators at biological incidents. Depending on their location, responders may also need to be familiar with standards regarding respiratory equipment issued by the ISO (International Standards Organization), the European Union, or other authorities. OSHA 29 CFR 1910.134 is the mandatory respiratory standard in the U.S.

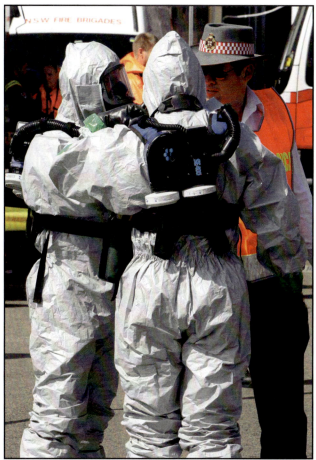

Figure 11.5 Powered air-purifying respirators (PAPRs) require less effort on the part of the responder. *Courtesy of Steven Baker, New South Wales Fire Brigades.*

The NIOSH and NFPA standards relating to respiratory equipment at hazmat/WMD incidents (including design, certification, and testing requirements) are as follows:

- **NIOSH *Chemical, Biological, Radiological and Nuclear (CBRN) Standard for Open-Circuit Self-Contained Breathing Apparatus (SCBA)*** — This standard establishes performance and design requirements to certify SCBA for use in CBRN exposures for use by first responders.

- **NIOSH *Standard for Chemical, Biological, Radiological, and Nuclear (CBRN) Full Facepiece Air-Purifying Respirator (APR)*** — This standard specifies minimum requirements to determine the effectiveness of full-facepiece APRs (commonly referred to as gas masks) used during entry into CBRN atmospheres that are not immediately dangerous to life and health (IDLH). Atmospheres that are above IDLH concentrations require the use of SCBA.

- **NIOSH** *Standard for Chemical, Biological, Radiological, and Nuclear (CBRN) Air-Purifying Escape Respirator and CBRN Self-Contained Escape Respirator* — This standard specifies minimum requirements to determine the effectiveness of escape respirators that address CBRN materials identified as inhalation hazards from possible terrorist events for use by the general working population.

- **NFPA 1852,** *Standard on Selection, Care, and Maintenance of Open-Circuit Self-Contained Breathing Apparatus (SCBA)* — This standard specifies the minimum requirements for the selection, care, and maintenance of open-circuit self-contained breathing apparatus (SCBA) and combination SCBA/SAR that are used for respiratory protection during fire fighting, rescue, and other hazardous operations.

- **NFPA 1981,** *Standard on Open-Circuit Self-Contained Breathing Apparatus (SCBA) for Emergency Services* — This standard specifies the minimum requirements for the design, performance, testing, and certification of open-circuit self-contained breathing apparatus (SCBA) and combination open-circuit self-contained breathing apparatus and supplied air respirators (SCBA/SAR) for the respiratory protection of fire and emergency responders where unknown, IDLH (immediately dangerous to life and health), or potentially IDLH atmospheres exist.

Responders in the U.S. should also be familiar with the following standards:

- **OSHA Regulation 29** *CFR* **1910.134,** *Respiratory Protection* — The major requirements of this OSHA Respiratory Protective Standard include: permissible practices; definitions; respiratory protection program; selection of respirators; medical evaluations; fit testing; use, maintenance, and care of respirators; identification of filters, cartridges, and canisters; training; program evaluation; and record keeping.

- **NIOSH Regulation 42** *CFR* **Part 84,** *Approval of Respiratory Protective Devices* — The purpose of this NIOSH regulation is to:

  — Establish procedures and prescribe requirements that must be met in filing applications for approval by NIOSH of respirators or changes or modifications of approved respirators

  — Provide for the issuance of certificates of approval or modifications of certificates of approval for respirators that have met the applicable construction, performance, and respiratory protection requirements set forth in this part

  — Specify minimum requirements and to prescribe methods to be employed by NIOSH and by the applicant in conducting inspections, examinations, and tests to determine the effectiveness of respirators used during entry into or escape from hazardous atmospheres

Self-Contained Breathing Apparatus (SCBA)

Self-contained breathing apparatus (SCBA) is an atmosphere-supplying respirator for which the user carries the breathing-air supply. SCBA is perhaps the most important piece of PPE a responder can wear at a hazmat incident in terms of preventing dangerous exposures to harmful substances. The unit consists of the following:

Self-Contained Breathing Apparatus (SCBA) — Respirator worn by the user that supplies a breathable atmosphere that is either carried in or generated by the apparatus and is independent of the ambient atmosphere. Respiratory protection is worn in all atmospheres that are considered to be immediately dangerous to life and health (IDLH). *Also known as* Air Mask *or* Air Pack.

- Facepiece
- Air hoses
- Harness assembly
- Pressure regulator
- Compressed air cylinder
- **End-of-service-time indicators (ESTIs)**

> **End-of-Service-Time Indicator (ESTI)** — Warning device that alerts the user that the respiratory protection equipment is about to reach its limit and that it is time to exit the contaminated atmosphere; its alarm may be audible, tactile, visual, or any combination thereof.

In the U.S., NIOSH and the Mine Safety and Health Administration (MSHA) must certify all SCBA for immediately dangerous to life and health (IDLH) atmospheres. SCBA that are not NIOSH/MSHA certified must not be used. The apparatus must also meet the design and testing criteria of NFPA 1981 in jurisdictions that have adopted that standard by law or ordinance. In addition, American National Standards Institute (ANSI) standards for eye protection apply to the facepiece lens design and testing.

NIOSH classifies SCBA as either closed-circuit or open-circuit. Two types of SCBA are currently being manufactured in closed- or open-circuit designs: pressure-demand, or positive-pressure. SCBA may also be either a high- or low-pressure type. However, use of only positive-pressure open-circuit or closed-circuit SCBA is allowed in incidents where personnel are exposed to hazardous materials.

NIOSH has entered into a Memorandum of Understanding with the National Institute of Standards and Technology (NIST), OSHA, and NFPA to jointly develop a certification program for SCBA used in emergency response to terrorist attacks. Working with the U.S. Army Soldier and Biological Chemical Command (SBCCOM), they developed a new set of respiratory protection standards and test procedures for SCBA used in situations involving WMD. Under this voluntary program, NIOSH issues a special approval and label identifying the SCBA as appropriate for use against chemical, biological, radiological, and nuclear agents. The SCBA certified under this program must meet the following minimum requirements:

- Approval under NIOSH 42 CFR 84, Subpart H
- Compliance with NFPA 1981
- Special tests under NIOSH 42 CFR 84.63(c):
  — *Chemical Agent Permeation and Penetration Resistance Against Distilled Sulfur Mustard (HD [military designation]) and Sarin (GB [military designation])*
  — *Laboratory Respirator Protection Level (LRPL)*

NIOSH maintains and disseminates a list of the SCBAs approved under this program. This list is entitled "CBRN SCBA" and contains the name of the approval holder, model, component parts, accessories, and rated duration. This list is maintained as a separate category within the NIOSH Certified Equipment List.

NIOSH authorizes the use of an additional approval label on apparatus that demonstrate compliance to the *CBRN SCBA* criteria. This label is placed in a visible location on the SCBA backplate (on the upper corner or in the area of the cylinder neck) **(Figure 11.6)**. The addition of this label provides visible and easy identification of equipment for its appropriate use.

Figure 11.6 NIOSH certification indicates the parameters that the apparatus will meet.

Supplied Air Respirators

The **supplied air respirator (SAR)** or airline respirator is an atmosphere-supplying respirator where the user does not carry the breathing air source. The apparatus usually consists of the following:

- Facepiece
- Belt- or facepiece-mounted regulator
- Voice communications system
- Up to 300 feet (90 m) of air supply hose
- Emergency escape pack or **emergency breathing support system (EBSS)** **(Figure 11.7)**
- Breathing air source (either cylinders mounted on a cart or a portable breathing-air compressor)

Because of the potential for damage to the air-supply hose, the EBSS provides enough air, usually 5, 10, or 15 minutes' worth, for the user to escape a hazardous atmosphere. SAR apparatus are not certified for fire fighting operations because of the potential damage to the airline from heat, fire, or debris.

NIOSH classifies SARs as Type C respirators. Type C respirators are further divided into two approved types. One type consists of a regulator and facepiece only. The second type consists of a regulator, facepiece, and EBSS, and may also be referred to as a SAR with escape (egress) capabilities. The second type is used in confined-space environments, IDLH environments, or potential IDLH environments. SARs used at hazmat or CBR incidents must provide positive pressure to the facepiece.

SAR apparatus have the advantage of reducing physical stress to the wearer by removing the weight of the SCBA. The air supply line is a limitation because of the potential for mechanical or heat damage. In addition, the length of the airline (no more than 300 feet [90 m] from the air source) restricts mobility. Problems with hose entanglement must also be addressed. Other limitations are the same as those for SCBA: restricted vision and communications.

Air-Purifying Respirators

Air-purifying respirators (APRs) contain an air-purifying filter, canister, or cartridge that removes specific contaminants found in ambient air as it passes through the air-purifying element. Based on which cartridge, canister, or filter is being used, these purifying elements are generally divided into the three following types:

- Particulate-removing APRs
- Vapor-and-gas-removing APRs
- Combination particulate-removing and vapor-and-gas-removing APRs

Figure 11.7 Emergency air is back-up for a system that may become disconnected or damaged.

Supplied Air Respirator (SAR) — Atmosphere-supplying respirator for which the source of breathing air is not designed to be carried by the user; not certified for fire fighting operations. *Also known as* Airline Respirator System.

Emergency Breathing Support System (EBSS) — Escape-only respirator that provides sufficient self-contained breathing air to permit the wearer to safely exit the hazardous area; usually integrated into an airline supplied air respirator system.

Air-Purifying Respirator (APR) — Respirator that removes contaminants by passing ambient air through a filter, cartridge, or canister; may have a full or partial facepiece.

APRs may be powered (PAPRs) or nonpowered. APRs do not supply oxygen or air from a separate source, and they protect only against specific contaminants at or below certain concentrations. Combination filters combine particulate-removing elements with vapor-and-gas-removing elements in the same cartridge or canister.

Respirators with air-purifying filters may have either full facepieces that provide a complete seal to the face and protect the eyes, nose, and mouth or half facepieces that provide a complete seal to the face and protect the nose and mouth **(Figure 11.8)**. Half-face respirators will NOT protect against CBR materials that can be absorbed through the skin or eyes and therefore are not recommended for use at hazmat/WMD incidents except for some explosive attacks where the primary hazard is dust or particulates.

Disposable filters, canisters, or cartridges are mounted on one or both sides of the facepiece **(Figure 11.9)**. Canister or cartridge respirators pass the air through a filter, sorbent, catalyst, or combination of these items to remove specific contaminants from the air. The air can enter the system either from the external atmosphere through the filter or sorbent or when the user's exhalation combines with a catalyst to provide breathable air.

Figure 11.8 The top two facepieces cover the whole face; the bottom two cover the nose and mouth.

No single canister, filter, or cartridge protects against all chemical hazards. Therefore, you must know the hazards present in the atmosphere in order to select the appropriate canister, filter, or cartridge. Responders should be able to answer the following questions before deciding to use APRs for protection at an incident:

- What is the hazard?
- What is the oxygen level?
- Is the hazard a vapor or a gas?
- Is the hazard a particle or dust?
- Is there some combination of dust and vapors present?
- What concentrations are present?
- Does the material have a taste or smell?

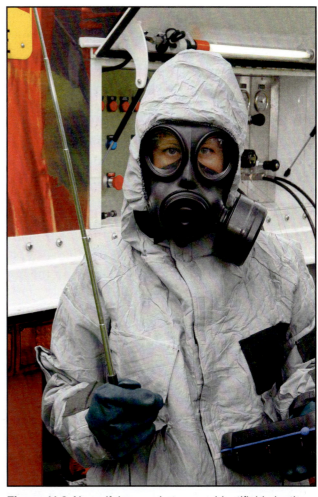

Figure 11.9 Air-purifying respirators are identifiable by the external filters, canisters, or cartridges mounted on the facepiece. *Courtesy of Steven Baker, New South Wales Fire Brigades.*

> **WARNING!**
> Do not wear APRs during emergency operations where unknown atmospheric conditions exist. Wear APRs only in controlled atmospheres where the hazards present are completely understood and at least 19.5 percent oxygen is present.

APRs do not protect against oxygen-deficient or oxygen-enriched atmospheres, and they must not be used in situations where the atmosphere is immediately dangerous to life and health (IDLH). APRs can only be used if the hazardous material has a taste or smell. The three primary limitations of an APR are as follows:

- Limited life of its filters and canisters
- Need for constant monitoring of the contaminated atmosphere
- Need for a normal oxygen content of the atmosphere before use

Take the following precautions before using APRs:

- Know what chemicals/air contaminants are in the air.
- Know how much of the chemicals/air contaminants are in the air.
- Ensure that the oxygen level is between 19.5 and 23.5 percent.
- Ensure that atmospheric hazards are below IDLH conditions.

At hazmat/WMD incidents, APRs may be used after the hazards at the scene have been properly identified. In some circumstances, APRs may also be used in other situations (law enforcement working perimeters of the scene or EMS/medical personnel) and escape situations. APRs used for these CBRN situations should utilize a combination organic vapor/high efficiency particulate air (OV/HEPA) cartridge (see the sections that follow).

Particulate-Removing Filters

Particulate filters protect the user from particulates, including biological hazards, in the air. These filters may be used with half facepiece masks or full facepiece masks. Eye protection must be provided when the full facepiece mask is not worn.

Particulate-removing filters are divided into nine classes, three levels of filtration (95, 99, and 99.97 percent), and three categories of filter degradation. The following three categories of filter degradation indicate the use limitations of the filter:

- **N** — **N**ot resistant to oil
- **R** — **R**esistant to oil
- **P** — Used when oil or nonoil lubricants are **P**resent

Particulate-removing filters may be used to protect against toxic dusts, mists, metal fumes, asbestos, and some biological hazards. High-efficiency particulate air (HEPA) filters used for medical emergencies must be 99.97 percent efficient, while 95 and 99 percent effective filters may be used depending on the health risk hazard.

Particle masks (also known as *dust masks*) are also classified as particulate-removing air-purifying filters. These disposable masks protect the respiratory system from large-sized particulates. Dust masks provide very limited protection and should not be used to protect against chemical hazards or small particles such as asbestos fibers.

Vapor-and-Gas-Removing Filters

As the name implies, vapor-and-gas-removing cartridges and canisters are designed to protect against specific vapors and gases. They typically use some kind of sorbent material to remove the targeted vapor or gas from the air. Individual cartridges and canisters are usually designed to protect against related groups of chemicals such as organic vapors or acid gases. Many manufacturers color-code their canisters and cartridges so it is easy to see what contaminant(s) the canister or cartridge is designed to protect against. Manufacturers also provide information about contaminant concentration limitations.

Powered Air-Purifying Respirators (PAPR)

The PAPR uses a blower to pass contaminated air through a canister or filter to remove the contaminants and supply the purified air to the full facepiece. Because the facepiece is supplied with air under a positive pressure, PAPRs offer a greater degree of safety than standard APRs in case of leaks or poor facial seals. For this reason, PAPRs may be of use at hazmat/WMD incidents for personnel conducting decontamination operations and long-term operations. Airflow also makes PAPRs more comfortable to wear for many people.

Several types of PAPR are available. Some units are supplied with a small blower and are battery operated. The small size allows users to wear one on their belts. Other units have a stationary blower (usually mounted on a vehicle) that is connected by a long, flexible tube to the respirator facepiece.

WARNING!
Do not use PAPRs in explosive or potentially explosive atmospheres.

As with all APRs, PAPRs should only be used in situations where the atmospheric hazards are understood and at least 19.5 percent oxygen is present. PAPRs are not safe to wear in atmospheres where potential respiratory hazards are unidentified, nor should they be used during initial emergency operations before the atmospheric hazards have been confirmed. Continuous atmospheric monitoring is needed to ensure the safety of the responder.

Combined Respirators

Combination respirators include SAR/SCBA, PAPR/SCBA, and SAR/APR. These respirators can provide flexibility and extend work duration times in hazardous areas. SAR/SCBAs will operate in either SAR or SCBA mode, for example, using SCBA mode for entry and exit while switching to SAR mode for extended work. PAPR/SCBA is a bulky combination. When using PAPR/SCBA combinations, it is necessary to know the composition of the atmosphere. The PAPR mode

Figure 11.10 An SCBA facepiece may fog with exertion and humidity.

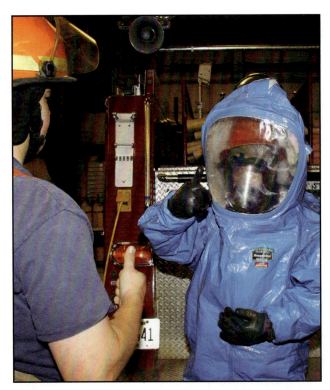

Figure 11.11 SCBA facepieces plus any additional gear may trigger claustrophobic reactions in some responders. Make sure the responders are comfortable after they are fully in their equipment.

allows a longer operational period if conditions are safe for use. SAR/APRs will also operate in either mode, but the same limitations that apply to regular APRs apply when operating in the APR mode. All combinations require specific training to use.

Supplied-Air Hoods

Powered- and supplied-air hoods provide loose fitting, lightweight respiratory protection that can be worn with glasses, facial hair, and beards. Hospitals, emergency rooms, and other organizations use these hoods as an alternative to other respirators, in part because they require no fit testing and are simple to use.

Respiratory Equipment Limitations

The advantages of using self-contained respiratory protection are independence, maneuverability, and protection from toxic and/or asphyxiating atmospheres.

Limitations imposed by equipment and air supply include:

- **Limited visibility** — Facepieces reduce peripheral vision, and facepiece fogging can reduce overall vision **(Figure 11.10)**.

- **Decreased ability to communicate** — Facepieces hinder voice communication if the facepiece is not equipped with a microphone or speaking diaphragm.

- **Increased weight** — Depending on the model, the protective breathing equipment can add 25 to 35 pounds (12.5 to 17.5 kg) of weight to the emergency responder.

- **Decreased mobility** — The increase in weight, change in profile, and splinting effect of the harness straps reduce the wearer's mobility.

- **Inadequate oxygen levels** — APRs cannot be worn in IDLH or oxygen-deficient atmospheres.

- **Chemical specific** — APRs can only be used to protect against certain chemicals. The specific type of cartridge depends on the chemical to which the wearer is exposed.

- **Psychological stress** — Facepieces may cause some users to feel confined or claustrophobic **(Figure 11.11)**.

- **Air supply limitations** — Open- and closed-circuit SCBA have maximum air-supply durations that limit the amount of time a first responder has to perform the tasks at hand.

- **Protection from finite conditions** — Non-NIOSH certified SCBAs may offer only limited protection in environments containing chemical warfare agents.

> **CAUTION**
> Personnel wearing respiratory equipment must have good physical conditioning, mental soundness, and emotional stability due to the physiological and psychological stresses of wearing PPE.

Protective Clothing Overview

Personal protective clothing and equipment chosen for hazmat incident responses must be based on specific chemical properties and a hazard analysis of the overall situation. The total ensemble will include outer garments, respiratory protection, gloves, and boots **(Figure 11.12)**. Careful consideration should be given to all aspects of the ensemble based on the given and expected hazards of the response option.

Figure 11.12 Most hazardous materials incidents will necessitate full protective ensembles including respiratory protection.

> **WARNING!**
> The wrong choice or use of CPC can be deadly.

Protective clothing must be worn whenever an emergency responder faces potential hazards arising from thermal hazards and chemical, biological, or radiological exposure. Skin contact with hazardous materials can cause a

Figure 11.13 Bomb suits protect against hazards such as fragmentation, overpressure, impact, and heat. *Courtesy of USMC, photo by Cpl Brian A. Tuthill.*

Figure 11.14 Body armor protects against ballistic threats. *Courtesy of U.S. Coast Guard, photo by PA1 Charles C. Reinhart.*

variety of problems, including chemical burns, allergic reactions and rashes, diseases, and absorption of toxic materials into the body. Protective clothing is designed to prevent these problems. Body armor and bomb suits can be worn to protect against ballistic hazards and shrapnel from explosives **(Figure 11.13)**.

No single combination or ensemble of protective equipment (even with respiratory protection), can protect against all hazards. For example, fumes and chemical vapors can penetrate fire fighting turnout coats and pants, so the protection they provide is not complete. Similarly, chemical-protective clothing (CPC) offers no protection from fires.

While technological advances are being made to improve the versatility of all types of PPE (for example, developing more chemical-resistant turnouts and more fire-resistant CPC), you must understand your PPE's limitations in order to stay safe. The sections that follow discuss the various standards that apply to protective clothing as well as the different clothing types that will commonly be used at hazmat/WMD incidents.

Standards for Protective Clothing and Equipment at Hazmat/WMD Incidents

As with respiratory protection, the U.S. Department of Homeland Security (DHS) has adopted NIOSH and NFPA standards for protective clothing used at hazmat/WMD incidents. Primarily, these apply to clothing worn at chemical and biological incidents in regards to chemical-protective clothing (CPC). However, responders should be familiar with any standards pertaining to design, certification, and testing requirements of any type of protective clothing they may need to use, including body armor, structural fire fighting gear, and bomb suits **(Figure 11.14)**. Depending on their location, responders may also need to be familiar with standards regarding respiratory equipment issued by the ISO, the European Union, or other authorities.

The primary NFPA standards addressing CPC are:

- **NFPA 1991, *Standard on Vapor-Protective Ensembles for Hazardous Materials Emergencies*** — The purpose of this standard is to establish a minimum level of protection for emergency response personnel against adverse vapor, liquid splash, and particulate environments during hazardous materials incidents and from specific chemical and biological terrorism agents in vapor, liquid splash, and particulate environments during CBRN terrorism incidents. The ensemble totally encapsulates the wearer and SCBA.

- **NFPA 1992,** *Standard on Liquid Splash-Protective Ensembles and Clothing for Hazardous Materials Emergencies* — This standard specifies minimum design, performance, certification, and documentation requirements; test methods for liquid splash-protective ensembles and liquid splash-protective clothing; and additional optional criteria for chemical flash fire protection.

- **NFPA 1994,** *Standard on Protective Ensembles for First Responders to CBRN Terrorism Incidents* — NFPA 1994 sets performance requirements for protective ensembles used in response to CBRN terrorism incidents. The standard defines three classes of ensembles (Class 2, 3, and 4) based on the protection required for different hazard types (vapors, liquids, and particulates) and airborne contaminant levels. Descriptions of the classes are:

  — Class 2 ensembles are intended for use at terrorism incidents involving vapor or liquid chemical or particulate hazards where the concentrations are at or above IDLH level requiring the use of CBRN compliant SCBA.

  — Class 3 ensembles are intended for use at terrorism incidents involving low levels of vapor or liquid chemical or particulate hazards where the concentrations are below IDLH, permitting the use of a CBRN compliant APR or PAPR.

  — Class 4 ensembles are intended for use at terrorism incidents involving biological or radiological particulate hazards where the concentrations are below IDLH levels permitting the use of CBRN compliant APR or PAPR. The ensembles are not tested for protection against chemical vapor or liquid permeability, gas-tightness, or liquid integrity.

In the U.S., the following guidance documents and related consensus standards also apply:

- **OSHA Regulation 29** *CFR* **1910.120,** *Hazardous Waste Operations and Emergency Response (HAZWOPER) Standard* — This Federal regulation applies to five distinct groups of employers and their employees. This includes any employees who are exposed, or potentially exposed to hazardous waste including emergency response operations for releases of, or substantial threat of the release of, hazardous substances regardless of the location.

- **OSHA Regulation 29** *CFR* **1910.132,** *Personal Protective Equipment* — This standard applies to personal protective equipment for eyes, face, head, and extremities and protective clothing, respiratory devices, and protective shields and barriers. The major requirements include: permissible practices; definitions; hazard assessment and equipment selection; training; and the proper care, maintenance, useful life, and disposal; program evaluation; and record keeping.

- **EPA Regulation 40 CFR Part 311,** *Worker Protection* — The EPA promulgated a standard identical to 29 CFR 1910.120 (OSHA's HAZWOPER Standard) to protect employees of State and local governments engaged in hazardous waste operations in States that do not have an OSHA-approved State plan.

- **OSHA Regulation 29 CFR 1910.156,** *Fire Brigades* — This standard identifies PPE requirements for industrial fire brigades. In many states, this also applies to fire departments.

Structural Firefighters' Protective Clothing

Structural fire fighting clothing is not a substitute for chemical-protective clothing; however, it does provide some protection against many hazardous materials. The atmospheres in burning buildings, after all, are filled with toxic gases, and modern structural firefighters' protective clothing with SCBA provides adequate protection against some of those hazards. The multiple layers of the coat and pants may provide short-term exposure protection from such materials as liquid chemicals; however, there are limitations to this protection. For example, structural fire fighting clothing is neither corrosive-resistant nor vapor-tight **(Figure 11.15)**. Liquids can soak through, acids and bases can dissolve or deteriorate the outer layers, and gases and vapors can penetrate the garment. Gaps in structural fire fighting clothing occur at the neck, wrists, waist, and the point where the pants and boots overlap.

Inadequate Vapor Protection

Figure 11.15 Structural fire fighting protective equipment is designed to protect against impact and high temperatures, but will not protect against corrosives or vapors.

Some hazardous materials can permeate (pass through at the molecular level) and remain in structural fire fighting clothing. Chemicals absorbed into the equipment can cause repeated exposure or a later reaction with another chemical. In addition, chemicals can permeate the rubber, leather or neoprene in boots, gloves, kneepads, and SCBA facepieces making them unsafe to use **(Figure 11.16)**. It may be necessary to discard equipment exposed to permeating chemicals.

While there is much debate among experts as to the degree of protection provided by structural fire fighting protective clothing (and SCBA) at hazmat/WMD incidents, there may be circumstances under which it will provide limited protection for short-term duration operations such as an immediate rescue. Agency emergency response plans and SOPs should specify the conditions and circumstances under which it is appropriate for emergency responders to rely on firefighter structural protective clothing and SCBA during operations at hazmat/WMD incidents.

Figure 11.16 Choose the correct type of material for a specific response. *Courtesy of Brent Cowx and Jonathan Gormick, Vancouver Fire & Rescue Services.*

Structural fire fighting protective clothing will provide protection against thermal damage in an explosive attack, but limited or no protection against projectiles, shrapnel, and other mechanical effects from a blast. It will provide adequate protection against some types of radiological materials, but not others. In cases where biological agents are strictly respiratory hazards, structural fire fighting protective clothing with SCBA may provide adequate protection. However, in any case where skin contact is potentially hazardous, it is not sufficient. Materials must be properly identified in order to make this determination, and any time a terrorist attack is suspected but not positively identified, it should be assumed that responders wearing only structural fire fighting protective clothing with SCBA are at some level of increased risk from potential hazards such as explosives, radiological materials, and chemical or biological weapons.

High Temperature-Protective Clothing

High temperature-protective clothing is designed to protect the wearer from short term exposures to high temperature in situations where heat levels exceed the capabilities of standard fire fighting protective clothing. This type of clothing is usually of limited use in dealing with chemical hazards. Two basic types of high-temperature clothing that are available are as follows:

1. **Proximity suits** — Permit close approach to fires for rescue, fire-suppression, and property-conservation activities such as in aircraft rescue and fire fighting or other fire fighting operations involving flammable liquids **(Figure 11.17)**. Such suits provide greater heat protection than standard structural fire fighting protective clothing.

Figure 11.17 High temperature-protective clothing is intended for use in temperatures higher than are commonly found in structural fire fighting. *Courtesy of William D. Stewart.*

2. **Fire-entry suits** — Allow a person to work in total flame environments for short periods of time; provide short-duration and close-proximity protection at radiant heat temperatures as high as 2,000°F (1 100°C). Each suit has a specific use and is not interchangeable.

> **WARNING!**
> High temperature-protective clothing is not designed to protect the wearer against chemical hazards.

Several limitations to high temperature-protective clothing are as follows:

- Contributes to heat stress by not allowing the body to release excess heat

Chapter 11 • Implementing the Action Plan: Personal Protective Equipment 389

> **Flame-Resistant (FR)** — Material that does not support combustion and is self-extinguishing after removal of an external source of ignition.

- Is bulky
- Limits wearer's vision
- Limits wearer's mobility
- Limits communication
- Requires frequent and extensive training for efficient and safe use
- Is expensive to purchase
- Integrity of suit is designed for limited exposure time

Flame-Resistant Protective Clothing

Many hazardous materials response personnel wear everyday **flame-resistant (FR)** work apparel **(Figure 11.18)**. This apparel is designed for continuous wear during work activities in designated areas in which there is minimal risk for exposure to the following:

- Hot or molten materials
- Hot surfaces
- Radiant heat
- Flash fires
- Flame
- Electrical arc discharge

This protective apparel will not ignite or melt under exposure to fire or radiant heat. Flame resistance in material can be achieved by using inherently flame resistant fibers or by treating the material with a flame retardant chemical:

- **Inherently Flame Resistant (IFR)** — Fibers that do not support combustion due to their chemical structure. They are flame resistant without chemical additives. The high-temperature-resistant polymers in IFR fibers provide an inert barrier between the wearer and the hazard. Protective properties of the fabric are permanent and cannot be washed out or removed.

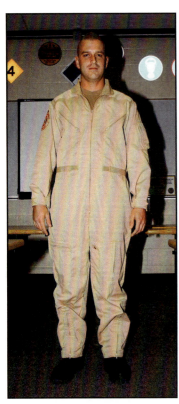

Figure 11.18 Flame-resistant apparel may be worn for some incidents, or as a base layer under other types of protective equipment.

- **Flame-Retardant** — A chemical compound that can be incorporated into a textile item during manufacture or applied to a fiber, fabric, or other textile item during processing to reduce its flammability. These fire retardants can be removed under some circumstances, such as washing.

Chemical Protective Clothing (CPC)

> **Chemical Protective Clothing (CPC)** — Clothing designed to shield or isolate individuals from the chemical, physical, and biological hazards that may be encountered during operations involving hazardous materials.

The purpose of **chemical protective clothing (CPC)** is to shield or isolate individuals from the chemical, physical, and biological hazards that may be encountered during hazardous materials operations. CPC is made from a variety of different materials, none of which protects against all types of chemicals. Each material provides protection against certain chemicals or products, but only limited or no protection against others. The manufacturer of a particular suit must provide a list of chemicals for which the suit is effective. Selection of appropriate CPC depends on the specific chemical and on the specific tasks to be performed by the wearer.

> **WARNING!**
> CPC is not intended for fire fighting activities, nor for protection from hot liquids, steam, molten metals, welding, electrical arc, flammable atmospheres, explosive environments or thermal radiation.

CPC is designed to afford the wearer a known degree of protection from a known type, concentration, and length of exposure to a hazardous material, but only if it is fitted properly and worn correctly. Improperly worn equipment can expose and endanger the wearer.

Most protective clothing is designed to be impermeable to moisture, thus limiting the transfer of heat from the body through natural evaporation. This can contribute to heat disorders in hot environments. Other factors include the garment's **degradation**, **permeation**, and **penetration** abilities and its service life. A written management program regarding selection and use of CPC is required. Regardless of the type of CPC worn at an incident, it must be decontaminated. Responders who may be called upon to wear CPC must be familiar with (and comfortable going through) their local procedures for technical decontamination. Decontamination is discussed further in Chapter 12 of this manual.

Design and testing standards generally recognize two types of CPC: liquid splash-protective clothing and vapor-protective clothing. The sections that follow describe these two types and, in addition, discuss operations where CPC is required, written management programs that specify CPC use, the ways in which CPC can be damaged, and considerations for the service life of CPC.

> **Chemical Degradation** — Process that occurs when the characteristics of a material are altered through contact with chemical substances.
>
> **Permeation** — Process in which a chemical passes through a protective material on a molecular level.
>
> **Penetration** — Process in which a hazardous material enters an opening or puncture in a protective material.

> **WARNING!**
> No single type of CPC protects against all chemical hazards.

> **WARNING!**
> You must have sufficient training to operate in conditions requiring the use of chemical-protective clothing.

Liquid Splash-Protective Clothing

Liquid splash-protective clothing is designed to protect users from chemical liquid splashes but not against chemical vapors or gases. NFPA 1992 sets the minimum design criteria for one type of liquid splash-protective clothing. Liquid splash-protective clothing can be **encapsulating** or nonencapsulating.

> **Liquid Splash-Protective Clothing** — Chemical-protective clothing designed to protect against liquid splashes per the requirements of NFPA 1992, *Standard on Liquid Splash-Protective Suits for Hazardous Chemical Emergencies*; part of an EPA Level B ensemble.
>
> **Encapsulating** — Completely enclosed or surrounded, as in a capsule.

Figure 11.19 An encapsulating suit covers all skin and equipment.

Figure 11.20 A nonencapsulating suit will have some potential gaps that may need to be sealed.

Figure 11.21 Any gaps in a nonencapsulating suit must be sealed with a tape compatible with the expected chemicals.

An encapsulating suit is a single, one-piece garment that protects against splashes or, in the case of vapor-protective encapsulating suits, also against vapors and gases **(Figure 11.19)**. Boots and gloves are sometimes separate, or attached and replaceable. Two primary limitations to fully encapsulating suits are:

- Impairs worker mobility, vision, and communication
- Traps body heat which might necessitate a cooling system, particularly when SCBA is worn

A nonencapsulating suit commonly consists of a one-piece coverall, but sometimes is composed of individual pieces such as a jacket, hood, pants, or bib overalls **(Figure 11.20)**. Gaps between pant cuffs and boots and between gloves and sleeves are usually taped closed **(Figure 11.21)**. Limitations to nonencapsulating suits include:

- Protects against splashes and dusts but not against gases and vapors
- Does not provide full body coverage: parts of head and neck are often exposed
- Traps body heat and contributes to heat stress

Neither encapsulating and nonencapsulating liquid splash-protective clothing are resistant to heat or flame exposure, nor do they protect against projectiles or shrapnel. The material of liquid splash-protective clothing is made from the same types of material used for vapor-protective suits (see following section).

When used as part of a protective ensemble, liquid splash-protective ensembles may include an SCBA, an airline (supplied air respirator [SAR]), or a full-face, air-purifying, canister-equipped respirator. Class 3 ensembles described in NFPA 1994 use liquid splash-protective clothing. This type of protective clothing is also a component of EPA Level B chemical protection ensembles.

Vapor-Protective Clothing

Vapor-protective clothing is designed to protect the wearer against chemical vapors or gases and offers a greater level of protection than liquid splash-protective clothing **(Figure 11.22)**. NFPA 1991 specifies requirements for a minimum level of protection for response personnel facing exposure to specified chemicals. This standard sets performance requirements for vapor-tight, totally encapsulating chemical-protective (TECP) suits and includes rigid chemical-resistance and flame-resistance tests and a permeation test against twenty-one challenge chemicals. NFPA 1991 also includes standards for performance tests in simulated conditions **(Figure 11.23)**.

> **Vapor-Protective Clothing** — Gas-tight chemical-protective clothing designed to meet NFPA 1991, *Standard on Vapor-Protective Suits for Hazardous Chemical Emergencies*; part of an EPA Level A ensemble.

Figure 11.22 Vapor-protective clothing is totally encapsulating.

Figure 11.23 Chemical protective equipment must meet some benchmarks for flame resistance.

Vapor-protective ensembles must be worn with positive-pressure SCBA or combination SCBA/SAR. Vapor-protective ensembles are components of ensembles to be used at chemical and biological hazmat/WMD incidents. These suits are also primarily used as part of a Level A protective ensemble, providing the greatest degree of protection against respiratory, eye, or skin damage from hazardous vapors, gases, particulates, sudden splash, immersion, or contact with hazardous materials.

Limitations to vapor-protective suits include:

- They melt and burn when exposed to fire, so they cannot be used in potentially flammable atmospheres
- Does not protect the user against all chemical hazards

- Impairs mobility, vision, and communication **(Figure 11.24)**
- Does not allow body heat to escape, so can contribute to heat stress, which may require the use of a cooling vest

Vapor-protective ensembles are made from a variety of special materials. No single combination of protective equipment and clothing is capable of protecting a person against all hazards.

Mission Specific Operations Requiring Use of Chemical-Protective Clothing

Chemical-protective clothing must be worn in certain circumstances. Without regard to the level of training required to perform them, these are operations that may require the use of CPC:

- Site survey
- Rescue
- Spill mitigation
- Emergency monitoring
- Decontamination
- Evacuation

If responders are involved in any of these activities, consideration must be given to what type of protective equipment is necessary given the known and/or unknown hazards present at the scene. Always follow AHJ SOP/Gs for operations requiring use of chemical-protective clothing **(Figure 11.25)**.

Figure 11.24 Responders wearing vapor-protective ensembles must move carefully to prevent snagging, tripping, or otherwise damaging the suit.

Written Management Programs

All emergency response organizations that routinely use CPC must establish a written Chemical-Protective Clothing Program and Respiratory Protection Management Program. A written management program includes policy statements, procedures, and guidelines. Copies must be made available to all personnel who may use CPC in the course of their duties or job.

The two basic objectives of any management program are protecting the user from safety and health hazards and preventing injury to the user from incorrect use or malfunction. To accomplish these goals, a comprehensive CPC management program includes the following elements:

- Hazard identification
- Medical monitoring
- Environmental surveillance
- Selection, care, testing, and maintenance
- Training

NOTE: Further information on medical monitoring is addressed in this manual in Chapter 5.

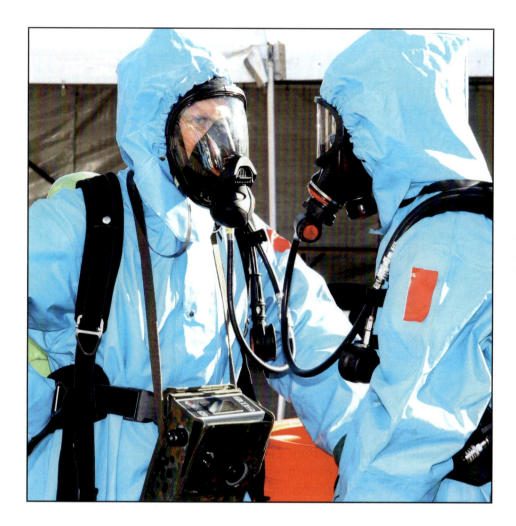

Figure 11.25 Work with at least one other responder in similar equipment when wearing chemical protective clothing.

Service Life
Each piece of CPC has a specific service life over which the clothing is able to adequately protect the wearer. For example, a Saranex/Tyvek® garment may be designed to be a coverall (covering the wearer's torso, arms, and legs) intended for liquid splash-protection and single use. If any garment is contaminated, it is best practice to remove it from service. Always follow AHJ SOP/Gs and manufacturer's specifications in regards to serviceability.

> **WARNING!**
> Never use CPC that is beyond its expiration date and/or has exceeded its service life.

All potentially contaminated CPC requires proper decontamination when the wearer leaves a potentially hazardous area. Chapter 12, Decontamination, provides more information about contamination and decontamination of CPC.

PPE Ensembles and Classification
To achieve adequate protection, an ensemble of respiratory equipment and clothing is typically used. When determining the appropriate ensemble,

Level A — Highest level of skin, respiratory, and eye protection that can be given by personal protective equipment (PPE), as specified by the U.S. Environmental Protection Agency (EPA); consists of positive-pressure self-contained breathing apparatus, totally encapsulating chemical-protective suit, inner and outer gloves, and chemical-resistant boots.

Level B — Personal protective equipment that affords the highest level of respiratory protection, but a lesser level of skin protection; consists of positive-pressure self-contained breathing apparatus, hooded chemical-protective suit, inner and outer gloves, and chemical-resistant boots.

Level C — Personal protective equipment that affords a lesser level of respiratory and skin protection than levels A or B; consists of full-face or half-mask APR, hooded chemical-resistant suit, inner and outer gloves, and chemical-resistant boots.

Level D — Personal protective equipment that affords the lowest level of respiratory and skin protection; consists of coveralls, gloves, and chemical-resistant boots or shoes.

consideration is given to the hazards present and the actions that need to be performed. For example, simple protective clothing such as gloves and a work uniform in combination with a face shield or safety goggles may be sufficient to prevent exposure to biological hazards such as bloodborne pathogens. At the other end of the spectrum, a vapor-protective, totally-encapsulating suit combined with positive-pressure SCBA may be needed when dealing with extremely hazardous, corrosive, and/or toxic vapors or gases, especially if the hazardous materials can damage other types of PPE and readily be absorbed through the skin.

WARNING!
No single type of PPE protects against all hazards.

While the EPA has established a set of chemical-protective PPE ensembles providing certain protection levels that are commonly used by fire and emergency service organizations, other organizations such as law enforcement, industrial responders, and the military may have their own standard operating procedures or equivalent procedures guiding the choice and use of appropriate combinations of PPE. Law enforcement personnel may be equipped with a far different PPE ensemble than a firefighter, hazmat technician, civil support response team, or environmental cleanup person working at the same hazmat/WMD incident. The sections that follow describe a variety of factors concerning PPE ensembles.

CAUTION
Always follow your agency's SOP/Gs in determining the level of PPE necessary to perform a task.

Levels of Protection

There are different levels of protective equipment used at incidents involving hazardous materials/WMD: **Level A**, **Level B**, **Level C**, and **Level D**. They can be used as the starting point for ensemble creation; however, each ensemble must be tailored to the specific situation in order to provide the most appropriate level of protection.

Selecting protective clothing and equipment by how they are designed or configured alone is not sufficient to ensure adequate protection at hazmat incidents. Just having the right components to form an ensemble is not enough. The EPA levels of protection do not define or specify what performance (for example, vapor protection or liquid splash protection) the selected clothing or equipment must offer, and they do not identically mirror the performance requirements of NFPA performance standards.

Level A

The Level A ensemble provides the highest level of protection against vapors, gases, mists, and particles for the respiratory tract, eyes, and skin **(Figure 11.26)**. Level A protection provides very little protection against fire. Level A PPE is primarily used at incidents including highly toxic and corrosive gases. Operations Level responders do not typically operate in situations requiring Level A protection. Responders who are required to wear Level A PPE must be appropriately trained.

The elements of Level A ensembles are:

- **Components** — Ensemble requirements are as follows:
  — Positive-pressure, full facepiece, SCBA, or positive-pressure airline respirator with escape SCBA approved by NIOSH **(Figure 11.27)**

Figure 11.26 A Level A ensemble protects skin and respiration against an IDLH atmosphere.

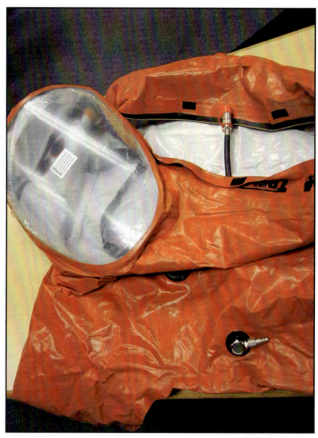

Figure 11.27 Level A ensembles may include airline respirators if the suit is equipped with an airline pass-through valve. *Courtesy of Brent Cowx and Jonathan Gormick, Vancouver Fire & Rescue Services.*

— Vapor-protective suits: TECP suits constructed of protective-clothing materials that meet the following criteria:

- Cover the wearer's torso, head, arms, and legs
- Include boots and gloves that may either be an integral part of the suit or separate and tightly attached **(Figure 11.28)**

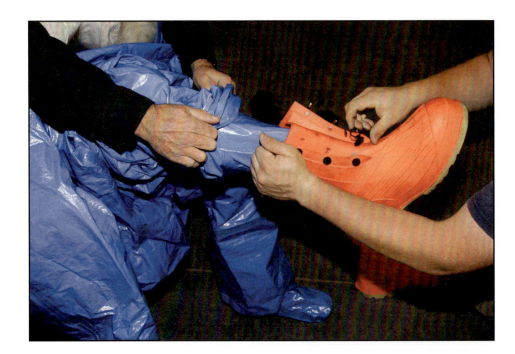

Figure 11.28 Boot liners may be an integrated part of an encapsulating suit.

- Enclose the wearer completely by itself or in combination with the wearer's respiratory equipment, gloves, and boots
- Provide equivalent chemical-resistance protection for all components of a TECP suit (such as relief valves, seams, and closure assemblies)
- May meet the requirements in NFPA 1991
  — Coveralls (optional)
  — Long underwear (optional)
  — Chemical-resistant inner gloves **(Figure 11.29)**

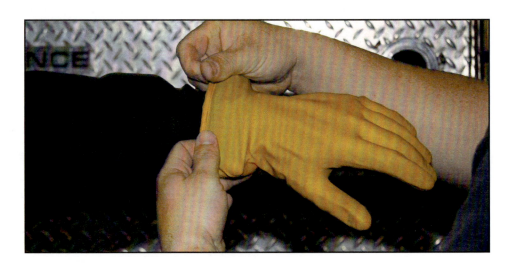

Figure 11.29 Responders don inner gloves before donning the rest of the suit to provide a chemical resistant base layer.

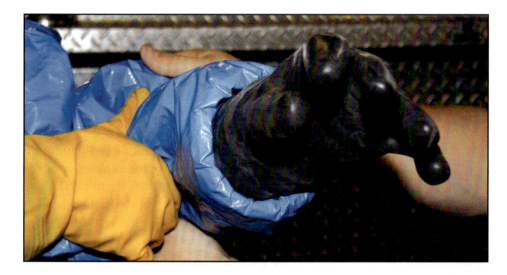

Figure 11.30 Outer gloves may be an integrated part of a fully encapsulating suit.

— Chemical-resistant outer gloves **(Figure 11.30)**
— Chemical-resistant boots with steel toe and shank
— Hardhat (under suit) (optional)
— Disposable protective suit, gloves, and boots (can be worn over totally-encapsulating suit, depending on suit construction)
— Two-way radios (worn inside encapsulating suit) **(Figure 11.31)**

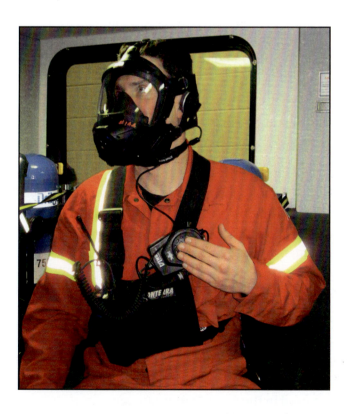

Figure 11.31 Portable radios and other communication systems may be worn inside an encapsulating suit. *Courtesy of Brent Cowx and Jonathan Gormick, Vancouver Fire & Rescue Services.*

- **Protection provided** — Highest available level of respiratory, skin, and eye protection from solid, liquid, and gaseous chemicals.
- **Use** — Level A ensembles are used when risk analysis indicates it is appropriate. For example, Level A protection may be appropriate when site

Chapter 11 • Implementing the Action Plan: Personal Protective Equipment operations and work functions involve a high potential for splash, immersion, or exposure to unexpected vapors, gases, or particulates of material that are harmful to skin or capable of damaging or being absorbed through the intact skin.

Level B

Level B protection requires a garment that includes an SCBA or a supplied air respirator and provides protection against splashes from a hazardous chemical (**Figure 11.32**). This ensemble is worn when the highest level of respiratory protection is necessary but a lesser level of skin protection is needed. Level B protection provides very little protection against fire. The Level B CPC ensemble may be encapsulating or nonencapsulating.

The elements of Level B ensembles are as follows:

- **Components** — Ensemble requirements are as follows:
  — Positive-pressure, full facepiece, SCBA, or positive-pressure airline respirator with escape SCBA approved by NIOSH
  — Hooded chemical-resistant clothing that may meet the requirements of NFPA 1992: overalls and long-sleeved jacket, coveralls, one- or two-piece (encapsulating or nonencapulating) chemical-splash suit, and disposable chemical-resistant overalls
  — Coveralls (optional)
  — Chemical-resistant inner gloves
  — Chemical-resistant outer gloves
  — Chemical-resistant boots with steel toe and shank
  — Disposable, chemical-resistant outer boot covers (optional)
  — Hardhat (outside or on top of nonencapsulating suits or under encapsulating suits)
  — Two-way radios (worn inside encapsulating suit or outside nonencapsulating suit)
  — Face shield (optional)

- **Protection provided** — Ensembles provide the same level of respiratory protection as Level A but have less skin protection. Ensembles provide liquid splash-protection, but no protection against chemical vapors or gases.

- **Use** — Ensembles may be used in the following situations:
  — Type and atmospheric concentration of substances have been identified and require a high level of respiratory protection but less skin protection.
  — Atmosphere contains less than 19.5 percent oxygen or more than 23.5 percent oxygen.

Figure 11.32 Level B ensembles protect skin against liquid chemical splashes and may be used where the atmosphere is IDLH for respiration.

— Presence of incompletely identified vapors or gases is indicated by a direct-reading organic vapor detection instrument, but the vapors and gases are known not to contain high levels of chemicals harmful to skin or capable of being absorbed through intact skin.

— Presence of liquids or particulates is indicated, but they are known not to contain high levels of chemicals harmful to skin or capable of being absorbed through intact skin.

Level C

Level C protection differs from Level B in the area of equipment needed for respiratory protection **(Figure 11.33)**. Level C is composed of a splash-protecting garment and an air-purifying device (APR or PAPR). Level C protection provides very little protection against fire. Level C protection includes any of the various types of APRs. Periodic air monitoring is required when using this level of PPE. Level C equipment is only used by emergency response personnel under the following conditions:

- The specific material is known
- The specific material has been measured
- This protection level is approved by the IC after all qualifying conditions for APRs and PAPRs have been met:
  — The product is known
  — An appropriate filter is available
  — The atmospheric oxygen concentration is between 19.5 to 23.5 percent
  — The atmosphere is not IDLH

The elements of Level C ensembles are as follows:

- **Components** — Ensemble requirements are as follows:
  — Full-face or half-mask APRs, NIOSH approved
  — Hooded chemical-resistant clothing (overalls, two-piece chemical-splash suit, and disposable chemical-resistant overalls)
  — Coveralls (optional)
  — Chemical-resistant inner gloves
  — Chemical-resistant outer gloves
  — Chemical-resistant boots with steel toe and shank
  — Disposable, chemical-resistant outer boot covers (optional)
  — Hardhat
  — Escape mask (optional)
  — Two-way radios (worn under outside protective clothing)
  — Face shield (optional)

- **Protection provided** — Ensembles provide the same level of skin protection as Level B but have a lower level of respiratory protection. Ensembles provide liquid splash-protection but no protection from chemical vapors or gases on the skin.

- **Use** — Ensembles may be used in the following situations:

Figure 11.33 Level C ensembles protect against liquid chemical splashes and may not be used where an atmosphere is IDLH.

— Atmospheric contaminants, liquid splashes, or other direct contact will not adversely affect exposed skin or be absorbed through any exposed skin.

— Types of air contaminants have been identified, concentrations have been measured, and an APR is available that can remove the contaminants.

— All criteria for the use of APRs are met.

— Atmospheric concentration of chemicals does not exceed IDLH levels. The atmosphere must contain between 19.5 and 23.5 percent oxygen.

Level D

Level D ensembles consist of typical work uniforms, street clothing, or coveralls **(Figure 11.34)**. Level D protection can be worn only when no atmospheric hazards exist.

The elements of Level D ensembles are as follows:

- **Components** — Ensemble requirements are as follows:
  - Coveralls
  - Gloves (optional)
  - Chemical-resistant boots/shoes with steel toe and shank
  - Disposable, chemical-resistant outer boot covers (optional)
  - Safety glasses or chemical-splash goggles
  - Hardhat
  - Escape device in case of accidental release and the need to immediately escape the area (optional)
  - Face shield (optional)

- **Protection provided** — Ensembles provide no respiratory protection and minimal skin protection.

- **Use** — Ensembles are typically not worn in the hot zone and are not acceptable for hazmat emergency response above the Awareness Level. Level D ensembles are used when both of the following conditions exist:
  - Atmosphere contains no hazard.
  - Work functions preclude splashes, immersion, or the potential for unexpected inhalation of or contact with hazardous levels of any chemicals.

Figure 11.34 Level D ensembles protect against some nonchemical hazards and may not be used exclusively where an atmosphere is IDLH.

Typical Ensembles of Response Personnel

The ensemble worn at an incident will vary depending on the mission of the responder. PPE for technical rescue personnel will differ from that of hazmat response teams, and so forth. However, responders of any discipline must be aware of what hazards are present at the incident, and what PPE is necessary to protect against the hazards to which they may be exposed **(Figure 11.35)**. For example, if respiratory hazards exist at the incident, all personnel who might be exposed to these hazards must wear respiratory protection regardless of their mission. It is important that personnel who may need to use such PPE be trained to do so. The sections that follow will outline some of the ensembles used by emergency responders at hazmat/WMD incidents, keeping in mind that the nature of the incident will dictate the PPE requirements.

Figure 11.35 Common hazards in high-angle technical rescue include abrasion and maneuvering in precarious positions, and may also include entering a low-oxygen atmosphere.

Fire Service Ensembles

Fire service personnel will wear ensembles appropriate for their mission at the incident, including typical fire fighting operations (such as fire extinguishment), hazardous materials response, and urban search and rescue (US&R).

The majority of responders will initially be wearing structural fire fighting protective clothing ensembles (turnout gear) that may offer limited protection against hazmat/WMD hazards **(Figure 11.36)**. These ensembles may be appropriate for conducting some operations (such as rescue) at hazmat/WMD incidents given appropriate protective measures such as limited exposure times.

Responders trained to use CPC at hazmat events may don EPA Level A or B ensembles as described in previous sections. Chemical-protective ensembles must be designed to protect the wearer's upper and lower torso, head, hands, and feet. Ensemble elements must include protective garments, protective gloves, and protective footwear. Ensembles must accommodate appropriate respiratory protection.

Figure 11.36 Structural fire fighting ensembles protect against thermal hazards as well as some chemical hazards.

Chapter 11 • Implementing the Action Plan: Personal Protective Equipment **403**

Law Enforcement Ensembles

Law enforcement personnel typically wear ballistic protection and no respiratory protection. PPE may also be assigned for emergency situations such as terrorist attacks. Law enforcement personnel must be trained to use the PPE they are assigned, whatever it may be.

Body armor is designed to protect against ballistic threats. Body armor is commonly used by law enforcement personnel, but some fire service and EMS agencies use it, particularly when operating in dangerous situations or areas where attacks might be likely. Body armor should always be replaced if it has been impacted or damaged.

Bomb disposal suits must provide full body protection against fragmentation, overpressure, impact, and heat. Normally designed to meet appropriate military specifications, they incorporate high-tech materials and ballistic plates in a head-to-toe ensemble. Helmets are usually designed with built-in communications capabilities and forced-air ventilation systems, some of which also provide protection/filtration against CBR materials. Bomb suits are very heavy and significantly impair dexterity and range of motion. New technology is being incorporated into next generation bomb suits to improve protection against CBR materials.

EMS Ensembles

EMS PPE must provide blood- and body-fluid pathogen barrier protection. PPE ensembles should include outer protective garments, gloves, footwear, and face protection. The items might be configured to cover only part of the upper or lower torso such as arms with sleeve protectors, torso front with apron-styled garments, and face with face shields **(Figure 11.37)**.

At hazmat incidents, EMS personnel not working in the hot zone may achieve some protection using a high-quality respirator, butyl rubber gloves, and a commercial chemical overgarment (elastic wrists and hood closures with built-in boots) that provides some liquid-droplet and vapor protection. This level of protection may or may not be adequate for personnel conducting triage and decontamination operations in the warm zone, depending on circumstances.

PPE Selection Factors

The risks and potential hazards present at an incident will determine the PPE needed. As with any emergency response, a hazard assessment will guide the responder to both the response objectives and the equipment that will be needed to mitigate the emergency. When conducting a hazard assessment, the responder should ask questions to help develop a response plan. Some questions that may help guide this process include the following:

- Has the chemical product been identified or classified?

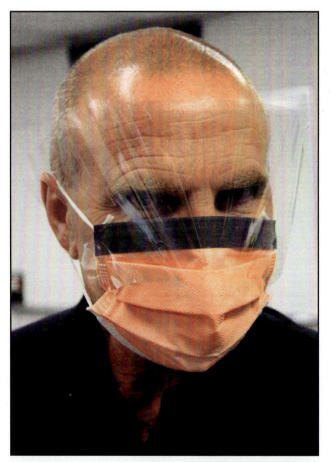

Figure 11.37 Infection control protective clothing may include disposable face shields, provided there is a safe atmosphere.

- How much product is involved?
- What harm can this product cause?
- What are the objectives of the mission?
- What equipment will be needed, and is it available?

The answers to these questions will allow the emergency responder to begin formulating a response plan. In the response plan, consideration for protective equipment and CPC must be given. **Skill Sheet 11-1** provides steps for selecting PPE.

General selection factors to be considered are as follows:

- **Chemical and physical hazards** — Both chemical and physical hazards must be considered and prioritized. Depending on what materials are present, any combination of hazards may need to be protected against. Flammability is an important hazard to consider since many types of PPE do not provide thermal protection.

- **Physical environment** — The ensemble components must be appropriate for whatever varied environmental conditions are present:
  - Industrial settings, highways, or residential areas
  - Indoors or outdoors
  - Extremely hot or cold environments
  - Uncluttered or rugged sites
  - Required activities involving entering confined spaces, lifting heavy items, climbing ladders, or crawling on the ground

- **Exposure duration** — The ensemble components' protective qualities may be limited by many factors including exposure levels, material chemical resistance, and air supply. Assume the worst-case exposure so that appropriate safety margins can be added to the ensemble wear time.

- **Available protective clothing or equipment** — An array of different clothing or equipment should be available to personnel to meet all intended applications. Reliance on one particular clothing type or equipment item may severely limit the ability to handle a broad range of hazardous materials or chemical exposures. In its acquisition of equipment and protective clothing, the responsible authority should attempt to provide a high degree of flexibility while choosing protective clothing and equipment that is easily integrated and provides protection against each conceivable hazard.

- **Compliance with regulations** — Agencies responsible for responding to CBR incidents should select equipment in accordance with regulatory standards for response to such incidents, such as NIOSH standards and NFPA 1994 **(Figure 11.38)**.

Protective clothing selection factors include the following:

- **Clothing design** — Manufacturers sell clothing in a variety of styles and configurations.

Figure 11.38 Mission-oriented protective posture (MOPP) ensembles are designed for safe response to incidents involving some chemical, biological, and radioactive materials.

Design considerations include the following:
— Clothing configuration
— Seam and closure construction
— Components and options
— Sizes
— Ease of donning and doffing
— Clothing construction
— Accommodation of other selected ensemble equipment
— Comfort
— Restriction of mobility

- **Material chemical resistance** — The chosen material(s) must resist permeation, degradation, and penetration by the respective chemicals. Mixtures of chemicals can be significantly more aggressive towards protective clothing materials than any single chemical alone. One permeating chemical may pull another with it through the material. Other situations may involve unidentified substances. Details:

Figure 11.39 An ensemble must be compatible with the conditions that the responder will meet. *Courtesy of the United States Air Force.*

— Very little test data are available for chemical mixtures. If clothing must be used without test data, clothing that demonstrates the best chemical resistance against the widest range of chemicals should be chosen **(Figure 11.39)**.

— In cases of chemical mixtures and unknowns, serious consideration must be given to selecting protective clothing.

- **Physical properties** — Clothing materials may offer wide ranges of physical qualities in terms of strength, resistance to physical hazards, and operation in extreme environmental conditions. Comprehensive performance standards (such as those from NFPA) set specific limits on these material properties, but only for limited applications such as emergency response. Users may also need to ask manufacturers the following questions:

— Does the material have sufficient strength to withstand the physical demands of the tasks at hand?

— Will the material resist tears, punctures, cuts, and abrasions?

— Will the material withstand repeated use after contamination and decontamination?

— Is the material flexible or pliable enough to allow needed tasks?

— Will the material maintain its protective integrity and flexibility under hot and cold extremes?

— Is the material subject to creation of a static electrical charge and discharge that could provide an ignition source?

- Is the material flame-resistant or self-extinguishing (if these hazards are present)?
- Are garment seams in the clothing constructed so they provide the same physical integrity as the garment material?

- **Ease of decontamination** — The degree of difficulty in decontaminating protective clothing may dictate whether disposable clothing, reusable clothing, or a combination of both is used.

- **Ease of maintenance and service** — The difficulty and expense of maintaining equipment should be considered before purchase.

- **Interoperability with other types of equipment** — Interoperability issues should be considered, for example, whether or not communications equipment can be integrated into the ensemble.

- **Cost** — Equipment needed to meet response requirements must be purchased within budget constraints.

Selecting PPE for Unknown Environments

In an environment that has an unidentified chemical (in response terms, an "unknown"), observations of the product and container, air monitoring instrumentation, and detection devices should be relied upon to identify the potential hazards. Such hazards include:

- Thermal/flammable
- Radiological
- Mechanical
- Asphyxiation
- Corrosive/chemical

A common industry tool used to determine when it is necessary to leave the work area is the lower explosive limit (LEL) sensor. The hazmat response community typically uses 10% of the LEL as a low level warning. However, if CPC is being worn, the action levels for flammability should be significantly lowered **(Figure 11.40)**. Some AHJs use a LEL of 1% in these instances.

Figure 11.40 Chemical protective clothing is not safe against thermal hazards, so most jurisdictions will set a low LEL action level when that equipment is in use.

The selection of protective clothing is driven by the mission. This decision is based on a risk analysis and AHJ SOPs. A common issue with all CPC is the lack of thermal protection. While flash protection is available, it does not provide thermal protection. Flash protection protects the ensemble and not the technician from the thermal hazards that are conducted through the material.

> **WARNING!**
> Air monitoring and instrumentation should reinforce the selection and use of chemical-protective clothing. The improper selection and use of chemical-protective clothing in a flammable atmosphere may be deadly.

Selecting Thermal Protective Clothing

The selection of thermal protective clothing should be based on the task or tasks that need to be performed and what other garments may be needed to complete the mission. For example, if the task will require the use of a Level A or B chemical-protective suit and the chemical has flammable properties, some type of thermal protection should also be used in conjunction with atmospheric monitoring **(Figure 11.41)**. Remember that the actions of ventilation and removing ignition sources can also greatly improve the flammable atmosphere. In some cases, structural PPE may be necessary.

Figure 11.41 Chemical suits with flash protection are not rated for long-term use in a high-heat environment. *Courtesy of the United States Air Force.*

Selecting for Chemical Resistance/Compatibility

The standard levels of protection refer to the ensemble style. The capabilities and limitations of each selected garment must be understood before a responder relies on the ensemble. For example, each garment fabric will have a specific resistance to a given chemical **(Table 11.1)**. No single garment can offer protection against all chemicals. To some degree, all materials will allow a chemical to pass through if given enough time. As covered earlier, chemical-protective clothing can be compromised by three principal methods:

- Permeation
- Degradation
- Penetration

Table 11.1
Sample Compatibility Chart

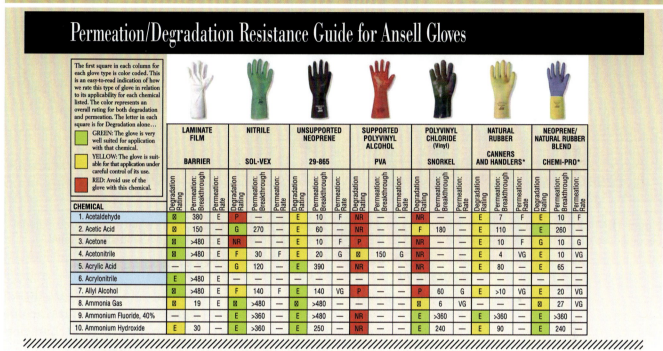

Courtesy of Ansell

Permeation

Permeation is a process that occurs when a chemical passes through a fabric at a molecular level **(Figure 11.42)**. As temperature increases, permeation increases.

In most cases, there is no visible evidence of chemicals permeating a material. The rate at which a compound permeates CPC depends on factors such as the chemical properties of the compound, nature of the protective barrier(s) in the CPC, and concentration of the chemical on the surface of the CPC. Most CPC manufacturers provide charts on *breakthrough time* (time it takes for a chemical to permeate the material of a protective suit) for a wide range of chemical compounds. Permeation

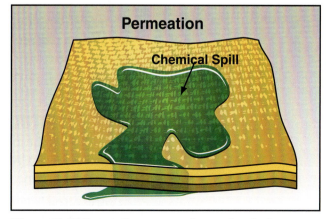

Figure 11.42 Permeation occurs when a chemical passes through a material at a molecular level.

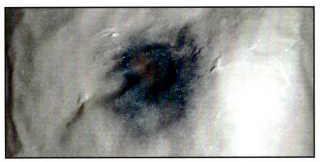

Figure 11.43 Permeation may not be apparent on the exterior of an ensemble, but may develop a visible indicator after several months. *Courtesy of Barry Lindley.*

data also includes information about the permeation rate (or the speed) at which the chemical moves through the CPC material after it breaks through **(Figure 11.43)**.

WARNING! Decontamination will not stop permeation from occurring.

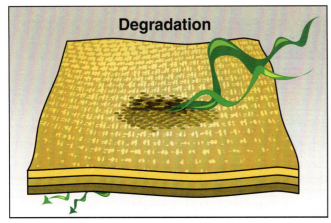

Figure 11.44 Chemical degradation occurs when a material decomposes on contact with a chemical.

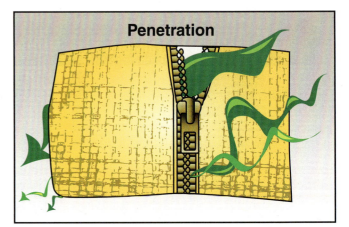

Figure 11.45 Penetration occurs when a hazardous material enters an opening or puncture in a protective material.

Degradation

Chemical degradation occurs when the characteristics of a material are altered through contact with chemical substances **(Figure 11.44)**. Examples include cracking, brittleness, blistering, and other changes in the structural characteristics of the garment. The most common observations of material degradation are discoloration, swelling, loss of physical strength, or deterioration.

Penetration

Penetration is a process that occurs when a hazardous material enters an opening or a puncture in a protective material **(Figure 11.45)**. Rips, tears, and cuts in protective materials, as well as unsealed seams, buttonholes, and zippers, are considered penetration failures. Such openings are often the result of faulty manufacture or problems with the inherent design of the suit. **Appendix B** explains protective clothing materials' manufacturing processes.

PPE-Related Stresses

Regardless of the job, all work may have related stresses. In a hazmat response, the stressors can be magnified exponentially. The responder must be aware of these stressors and be able to overcome them to effectively mitigate an incident.

Stress from wearing the personal protective clothing may come in various forms, such as:

- Heat-related due to the type of garments and the layers needed to protect responders.
- Cold-related because of either ambient temperature or the temperatures of the chemical products.
- Psychological in nature because of the constricting and confining equipment used in hazmat response.

Heat Emergencies

Wearing PPE or other special full-body protective clothing puts you at considerable risk of developing health effects ranging from transient heat fatigue to serious illness (heat stroke) or even death. Heat disorders include:

- **Heat stroke** (the most serious; see Safety Alert)
- **Heat exhaustion**
- **Heat cramps**
- **Heat rashes**

> **Heat Stroke** — Heat illness in which the body's heat regulating mechanism fails; symptoms include (a) high fever of 105° to 106° F (40.5° to 41.1° C), (b) dry, red, hot skin, (c) rapid, strong pulse, and (d) deep breaths or convulsions. May result in coma or even death. *Also known as* Sunstroke.

> **Heat Exhaustion** — Heat illness caused by exposure to excessive heat; symptoms include weakness, cold and clammy skin, heavy perspiration, rapid and shallow breathing, weak pulse, dizziness, and sometimes unconsciousness.

> **Heat Cramp** — Heat illness resulting from prolonged exposure to high temperatures; characterized by excessive sweating, muscle cramps in the abdomen and legs, faintness, dizziness, and exhaustion.

> **Heat Rash** — Condition that develops from continuous exposure to heat and humid air; aggravated by clothing that rubs the skin. Reduces the individual's tolerance to heat.

Heat Stroke

Heat stroke occurs when the body's system of temperature regulation fails and body temperature rises to critical levels. This condition is caused by a combination of highly variable factors, and its occurrence is difficult to predict. Heat stroke is a serious medical emergency and requires immediate medical treatment and transport to a medical care facility. The primary signs and symptoms of heat stroke are:

- Confusion
- Irrational behavior
- Loss of consciousness
- Convulsions
- Lack of sweating (usually)
- Hot, dry skin
- Abnormally high body temperature (for example, a rectal temperature of 105.8°F [41°C])

When the body's temperature becomes too high, it causes death. The elevated metabolic temperatures caused by a combination of workload and environmental heat load, both of which contribute to heat stroke, are also highly variable and difficult to predict. If you or any other responders show signs of possible heat stroke, obtain professional medical treatment immediately.

Heat-Exposure Prevention

Responders wearing protective clothing need to be monitored for the effects of heat exposure **(Figure 11.46, p. 412)**. Methods to prevent and/or reduce the effects of heat exposure include the following:

- **Fluid consumption** — Use water or commercial body-fluid-replenishment drink mixes to prevent dehydration. You should drink generous amounts

Figure 11.46 Heat-related stresses are a significant consideration during response while wearing chemical protective ensembles. *Courtesy of the United States Air Force.*

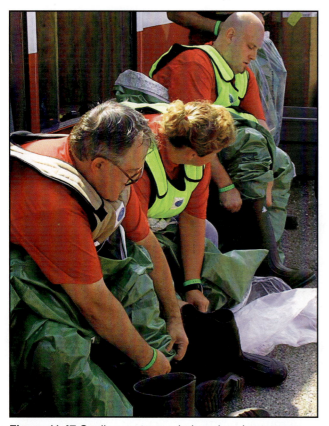

Figure 11.47 Cooling vests may help reduce heat stress.

of fluids both before and during operations. Drinking 7 ounces (200 ml) of fluid every 15 to 20 minutes is better than drinking large quantities once an hour. Balanced diets normally provide enough salts to avoid cramping problems. Details:

- Before working, drinking chilled water is good.
- After a work period in protective clothing and an increase in core temperature, drinking room-temperature water is better. It is not as severe a shock to the body.

- **Air cooling** — Wear long cotton undergarments, moisture-wicking modern fabrics, or similar clothing to provide natural body ventilation. Once PPE has been removed, blowing air can help to evaporate sweat, thereby cooling the skin. Wind, fans, blowers, and misters can provide air movement. However, when ambient air temperatures and humidity are high, air movement may provide only limited benefit.

- **Ice cooling** — Use ice to cool the body. Take care to not damage skin with direct contact with ice, as well as to not cool off an individual too quickly. Ice will also melt relatively quickly.

- **Water cooling** — Use water to cool the body. When water (even sweat) evaporates from skin, it cools. Provide mobile showers and misting facilities or evaporative cooling vests **(Figure 11.47)**. Water cooling becomes less effective as air humidity increases and water temperatures rise.

- **Cooling vests** — Wear cooling vests beneath PPE as allowed by the AHJ.

- **Rest/rehab areas** — Provide shade, humidity changes (misters), and air-conditioned areas for resting **(Table 11.2)**.

- **Work rotation** — Rotate responders exposed to extreme temperatures or those performing difficult tasks frequently.

- **Proper liquids** — Avoid liquids such as alcohol, coffee, and caffeinated drinks (or minimize their intake) before working. These beverages can contribute to dehydration and heat stress.

- **Physical fitness** — Encourage responders to maintain good physical fitness.

NOTE: NFPA 1584, *Standard on the Rehabilitation Process for Members during Emergency Operations and Training Exercises*, addresses many of these issues.

Table 11.2
NOAA's National Weather Service Heat Index

Temperature (°F)

Relative Humidity (%)	80	82	84	86	88	90	92	94	96	98	100	102	104	106	108	110
40	80	81	83	85	88	91	94	97	101	105	109	114	119	124	130	136
45	80	82	84	87	89	93	96	100	104	109	114	119	124	130	137	
50	81	83	85	88	91	95	99	103	108	113	118	124	131	137		
55	81	84	86	89	93	97	101	106	112	117	124	130	137			
60	82	84	88	91	95	100	105	110	116	123	129	137				
65	82	85	89	93	98	103	108	114	121	126	130					
70	83	86	90	95	100	105	112	119	126	134						
75	84	88	92	97	103	109	116	124	132							
80	84	89	94	100	106	113	121	129								
85	85	90	96	102	110	117	126	135								
90	86	91	98	105	113	122	131									
95	86	93	100	108	117	127										
100	87	95	103	112	121	132										

Likelihood of Heat Disorders with Prolonged Exposure or Strenuous Activity

☐ Caution ☐ Extreme Caution ☐ Danger ☐ Extreme Danger

Courtesy of NOAA.

Cold Emergencies

Cold temperatures may be caused by weather and/or other conditions such as exposure to cryogenic liquids. Prolonged exposure to freezing temperatures can result in health problems as serious as **trench foot**, **frostbite**, and **hypothermia**.

The primary environmental conditions that cause cold-related stress are low temperatures, high/cool winds, dampness, cold water, and standing/walking/working on cold, snowy and/or icy surfaces. Wind chill, a combination of temperature and velocity, is a crucial factor to evaluate when working outside **(Table 11.3, p. 414)**. For example, when the actual air temperature of the wind is 40°F (4.5°C) and its velocity is 35 mph (55 km/h), the exposed skin experiences conditions equivalent to the still-air temperature of 11°F (-12°C). Rapid heat loss may occur when exposed to high winds and cold temperatures.

You can prevent cold disorders with the following precautions:

- Being active
- Wearing warm clothing/layers
- Avoiding cold beverages
- Rehabbing in a warm area
- Dressing appropriately

Trench Foot — Foot condition resulting from prolonged exposure to damp conditions or immersion in water; symptoms include tingling and/or itching, pain, swelling, cold and blotchy skin, numbness, and a prickly or heavy feeling in the foot. In severe cases, blisters can form, after which skin and tissue die and fall off.

Frostbite — Local tissue damage caused by prolonged exposure to extreme cold.

Hypothermia — Abnormally low body temperature.

Table 11.3
Wind Chill Chart

Wind Speed (mph) \ Temperature (°F)	Calm	40	35	30	25	20	15	10	5	0	-5	-10	-15	-20	-25	-30	-35	-40	-45
5		36	31	25	19	13	7	1	-5	-11	-16	-22	-28	-34	-40	-46	-52	-57	-63
10		34	27	21	15	9	3	-4	-10	-16	-22	-28	-35	-41	-47	-53	-59	-66	-72
15		32	25	19	13	6	0	-7	-13	-19	-26	-32	-39	-45	-51	-58	-64	-71	-77
20		30	24	17	11	4	-2	-9	-15	-22	-29	-35	-42	-48	-55	-61	-68	-74	-81
25		29	23	16	9	3	-4	-11	-17	-24	-31	-37	-44	-51	-58	-64	-71	-78	-84
30		28	22	15	8	1	-5	-12	-19	-26	-33	-39	-46	-53	-60	-67	-73	-80	-87
35		28	21	14	7	0	-7	-14	-21	-27	-34	-41	-48	-55	-62	-69	-76	-82	-89
40		27	20	13	6	-1	-8	-15	-22	-29	-36	-43	-50	-57	-64	-71	-78	-84	-91
45		26	19	12	5	-2	-9	-16	-23	-30	-37	-44	-51	-58	-65	-72	-79	-86	-93
50		26	19	12	4	-3	-10	-17	-24	-31	-38	-45	-52	-60	-67	-74	-81	-88	-95
55		25	18	11	4	-3	-11	-18	-25	-32	-39	-46	-54	-61	-68	-75	-82	-89	-97
60		25	17	10	3	-4	-11	-19	-26	-33	-40	-48	-55	-62	-69	-76	-84	-91	-98

Frostbite occurs in 15 minutes or less

Courtesy of NOAA

Psychological Issues

The use of CPC can be a very confining experience. Whether working in a Level A fully encapsulated suit or even a lower level suit, CPC will be much more confining than structural fire fighting clothing and equipment. This confinement may cause claustrophobia in responders. In addition to the confinement of the protective clothing, just knowing the hazards of the chemicals involved may be disconcerting to the responder.

Psychological issues may be preventable through adequate training. As the responder works with the equipment and gains familiarity, confidence will build. Still, the mind is a very powerful organ and severe claustrophobia may be debilitating for a responder. If this is the case, emergency response in CPC may not be suitable for some responders.

PPE Use

There is much more to the use of PPE ensembles than just putting on the ensemble. While it is imperative that you be proficient in donning the equipment, you must also be able to safely function in the suit while performing both simple and challenging tasks. As familiarity increases, so will the comfort levels. Increased comfort levels will help reduce your stress. This reduced stress can help increase both your proficiency and work time.

Pre-Entry Inspection

It is necessary to check equipment prior to an entry into a hazardous atmosphere **(Figure 11.48)**. A thorough visual inspection should uncover any defects or deformities in the protective equipment. In addition to the visual inspection,

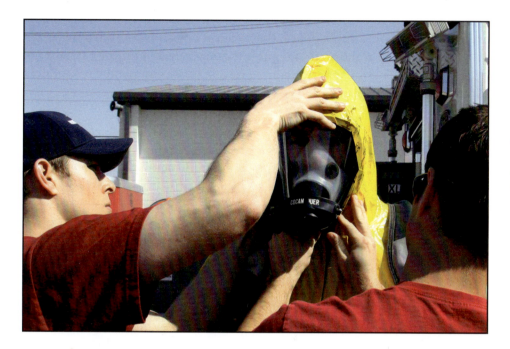

Figure 11.48 Proper fit, usage, and safety of equipment must be confirmed before a responder enters a hazardous atmosphere.

confirm all pressure test completion dates, and conduct an operational check of the following items:

- Breathing apparatus
- All zippers and closures
- Valves
- Communications equipment
- Any equipment that will be taken or used in the hot zone

READY

Components of the entry briefing can be summarized by the acronym, READY:

R Radio — Do you have a radio? Is it set on the correct channel? Perform a radio check.

E Equipment — What equipment is required, and do personnel know how to use it? What are the emergency signals?

A Air — Is your air cylinder full? What is the predetermined working time?

D Details — What is your team requested to do, typically no more than three items? The team should repeat this back to ensure accurate information transfer.

Y Yes — If all of the above steps are complete, entry team should finish donning PPE and proceed to do entry.

Safety and Emergency Procedures

In addition to issues such as cooling, preventing dehydration, and medical monitoring, there are other safety and emergency issues involved with wearing PPE. For example, responders using PPE at hazmat incidents must be familiar with their local procedures for going through the technical decontamination process (see Chapter 12, Decontamination). Anytime emergency responders are to enter an IDLH atmosphere, they should always work in teams of two or more, with a minimum of two equally trained and equipped personnel outside the IDLH atmosphere ready to rescue other emergency responders should the need arise.

Safety Briefing

A safety briefing will be conducted before responders enter the hot zone **(Figure 11.49)**. The safety briefing will cover relevant information including:

Figure 11.49 A safety briefing is designed to provide necessary information to responders before they begin their roles of responding to an incident or supporting other responders.

- Incident status (based on the preliminary evaluation and subsequent updates)
- Identified hazards
- Description of the site
- Tasks to be performed
- Expected duration of the tasks
- Escape route or area of refuge
- PPE and health monitoring requirements
- Incident monitoring requirements
- Notification of identified risks
- Communication procedures, including hand signals
- Any associated reports or documentation as required by the AHJ after using PPE at an incident.

Air Management

Anytime a limited air supply such as SCBA is worn, air management is an important consideration. Emergency procedures should be developed for responder loss of air supply. These may vary depending on the AHJ. To ensure adequate work time, calculate estimated times for the following tasks **(Figure 11.50)**:

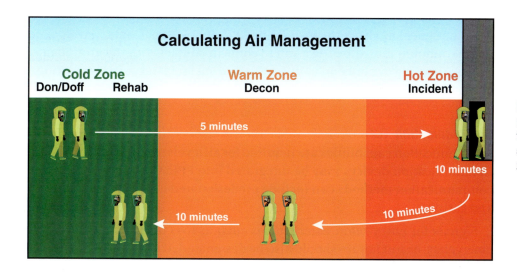

Figure 11.50 Air management must include estimates of the time needed to perform a task safely and return to the cold zone after decontamination.

- Walk to the incident
- Return from the incident
- Decon
- Work time
- Safety time (extra time allocated for emergency use)

Air must be allocated for these estimated times. Responders should have a plan in place for dealing with air emergencies.

Many organizations have SOP/Gs that explain calculations for doing this and/or designate maximum entry times (such as 20 minutes) based on the air supply available. It may be necessary to stock SCBA cylinders of different sizes and volumes in an agency's cache of equipment.

NOTE: A cylinder's service pressure and rating are not a true indication of the overall work time. The only constant is the amount of air the cylinder will contain when it is full.

Contamination Avoidance

The terms contamination and exposure are sometimes used interchangeably, but the concepts are actually very different. Contamination can be defined as a condition of impurity resulting from contact or mixture with a foreign substance. In other words, the hazardous material has to touch or be touched by another object. In contrast, exposure means that a hazardous material has entered or potentially entered your body via the routes of entry, for example, by swallowing, breathing, or contacting skin or mucous membranes.

Most hazmat responses will likely include contamination, which can increase the risk of exposure. Because of this, contamination should be avoided as best as possible. If avoidance is not possible and you need to protect the suit from damage, put something between your suit and the ground/contamination (options include: thick cardboard; rug; visqueen (plastic sheeting); absorbent pillows, pigs, booms, socks, and pads; knee pads).

A responder should consider the following best practices:

- Always try to reduce any contact with the product. Avoid walking through and touching the product whenever possible.
- Do not kneel or sit on the ground in CPC, if possible. Contact avoidance is paramount, but allowing a suit to come in contact with the ground may cause chafing or abrasion on the suit allowing for faster suit degradation.
- Protect monitoring instruments as best as possible.

Communication

Communication capabilities are required for all levels of personal protection. Communication devices may be integrated into PPE. Other non emergency communication methods can include predesignated hand signals, motions, and gestures.

Signals for entry-team emergencies such as loss of air supply, medical emergency, or suit failure should also be designated. If possible, entry teams, backup personnel, and appropriate safety personnel at the scene should have their own designated radio channel.

Should responders lose radio communications or operate in an atmosphere not allowing radio communications, a backup system must be part of the operational plan. Hand signals used as the backup plan should be simple, easy to remember, and distinguishable from a distance. Hand signals should be designated for the following situations **(Table 11.4)**:

**Table 11.4
Hand Signals**

Loss of air supply

Loss of suit integrity

Buddy down

Loss of radio communications

- Loss of air supply
- Loss of suit integrity
- Responder down from injury or illness
- Emergency
- Loss of radio communications
- Responder/situation is okay

CAUTION
Follow hand signals specified by the AHJ.

All responders should follow local protocols for evacuation situations. Typically, these protocols will involve notifying the appropriate personnel (such as the Entry Team Leader and/or Hazmat Safety Officer), and exiting the hot zone as quickly as possible.

The remaining capabilities of equipment should also be communicated during evacuation situations. For example, if air supply is lost while wearing a vapor-protective suit, there is a limited amount of air in the suit itself that can be breathed if the SCBA face piece or regulator is removed.

In addition to entry-team signals, an emergency evacuation signal for all responders should be included in the incident action plan. The emergency signal should indicate that an immediate exit from the hot zone is necessary. The signal should be audible (air horns) and also broadcast over the radio frequency.

Buddy System

As in any fire fighting operation, it is important to understand that the team concept is the safest mode of operation. The same is true for a hazmat response into a hazardous environment. Entry teams should consist of two or more equally protected responders **(Figure 11.51)**. The priority must be on the responder's safety over and above the incident itself. Any time emergency responders are to enter an IDLH atmosphere, they should always work in teams of two or more, with a minimum of two equally trained and equipped personnel outside the IDLH atmosphere ready to affect rescue of other emergency responders.

Figure 11.51 Responders in a hazardous environment must always work with a buddy.

U.S. Requirements
OSHA 1910.120 and 1910.134 require responders to operate in buddy systems. Backup teams are also required.

Suit Integrity

During a hazmat response, it may be inevitable that a responder will come in contact with a product **(Figure 11.52)**. Responders working in CPC must be aware of the limitations of their protective clothing and how the CPC can be compromised over the course of a mission work period. Penetration, permeation, and degradation are critical concerns when wearing these garments. If a suit is compromised by penetration or degradation, it may be possible that the wearer will be able to identify the reduction in integrity. If the suit is compromised, this should be treated as a true emergency and a rapid exit of the work zone is necessary. Decontamination will be the highest priority and the responder should seek a medical evaluation once the clothing has been removed.

NOTE: Care should be taken not to increase the chances of decreasing the garment's chemical-resistive capabilities by causing physical damage to the suit by careless work practices.

Figure 11.52 Responders must know the limitations of their protective equipment while contacting a hazardous material.

Mission Work Duration

The overall duration of the mission can affect all PPE, but it can have a profound effect on respiratory protection and air supply. As was discussed in the previous section, the volume of air in a cylinder is a given as long as the cylinder is full at the onset of its use. Use time is dynamic and can be improved through training.

The following factors must be considered when determining the work time of a specific mission:

- **Safety buffer** — The rated size of the cylinder will not be an indication of the overall work time **(Table 11.5)**. Because of this, it is important to build a safety buffer into the estimated work duration. As a rule of thumb, approximately one-third of the cylinder's working time should be allotted for a safety buffer **(Table 11.6)**. A safety buffer is a time that should never be compromised.

Table 11.5
Breathing Air Cylinder Capacities

Rated Duration	Pressure	Volume
30-minute	2,216 psi (15 290 kPa)	45 ft³ (1 270 L) cylinders
30-minute	4,500 psi (31 000 kPa)	45 ft³ (1 270 L) cylinders
45-minute	3,000 psi (21 000 kPa)	66 ft³ (1 870 L) cylinders
45-minute	4,500 psi (31 000 kPa)	66 ft³ (1 870 L) cylinders
60-minute	4,500 psi (31 000 kPa)	87 ft³ (2 460 L) cylinders

• Rated duration does not indicate the actual amount of time that the cylinder will provide air.

Table 11.6
Air Cylinder Safety Buffers

Cylinder Rating	Safety Buffer
30-minute	10 minutes
45-minute	15 minutes
60-minute	20 minutes

- **Travel time** — Account for travel time in the mission work duration (**Figure 11.53**). For example, if there is an expected travel time of 5 minutes to the work location, you must account for a total of 10 minutes or longer to accommodate fatigue on the return.

Figure 11.53 The time it takes to travel from the donning area to the hot zone must be included in calculations of energy expenditures.

Figure 11.54 Decontamination time must be accounted for in the air supply calculation. *Courtesy of Joan Hepler.*

- **Decon** — Decon may be a dynamic consideration based on the amount of people to be decontaminated. If there are two people leaving the work zone and it is estimated that the average decon time will be 5 minutes, then the last person to go through decon will need a total of 10 minutes **(Figure 11.54)**.

- **Responder workload** — The harder responders work, the heavier they will breathe, and the more air they will consume **(Table 11.7)**.

- **Environmental factors** — While ambient temperatures will differ based on locale, elevated temperatures in all areas can change the air consumption during the mission.

Table 11.7
Energy Expenditures During Common Sports and Fire Fighting Activities

Common Sports Activities	Average Expenditure (in METs)
Basketball	8.3
Cycling (10 mph [16 kph])	7.0
Football (touch)	8.0
Racquetball	9.0
Weight (circuit) Training	9.1
Common Fire Fighting Activities	**Average Expenditure (in METs)**
Climbing an Aerial Ladder	9.3
Chopping at Medium Speed	11.0
Dragging a Supply Hoseline	10.2
Raising a Ladder 40 - 50 feet (12 – 15 m)	9.2
Climbing Stairs in PPE/SCBA/Hotel Pack	11.0
Carrying Victim Down a Ladder	10.1

NOTE: One MET is defined as the energy expenditure for sitting quietly, which, for the average adult, approximates 3.5 ml of oxygen uptake per kilogram of body weight per minute (1.2 kcal/min for a 70-kg individual). For example, a 2-MET activity requires two times the metabolic energy expenditure of sitting quietly.

- **Work mission statement** — Helpful in determining the estimated work time allowed in a hot zone. This can be a valuable tool because it will let entry personnel visualize the timetables of the entire mission. Command personnel will be able to adjust the estimated times should there be a delay or change in any one of the categories.

Emergency procedures should also be developed for responder loss of air supply. These may vary depending on the AHJ.

Donning and Doffing of PPE

You should always train with the protective clothing that you will be using in the field **(Figure 11.55)**. The donning process can be quite time consuming and confusing for a user who is not totally familiar with the garments. Instructions should be included with all PPE for the total donning and doffing process. **Skill Sheets 11-2 through 11-5** provide steps for donning and doffing PPE.

Donning of PPE

While it is imperative that you follow the manufacturer and department recommendations for the donning of PPE, the following guidelines will outline generic donning procedures that may be included in an agency's procedures:

- Preselect the donning and doffing area in the cold zone as close to the entry point as possible. It should be clearly delineated.
- Ensure that the donning and doffing area is isolated from distractions and sheltered from the elements, if possible.
- Plan for as many people (including assistants) as needed to be involved in the donning procedures **(Figure 11.56)**.

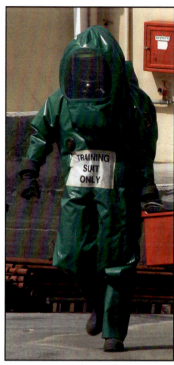

Figure 11.55 Responders must train with the equipment they will use, to develop competence and comfort with the resources.

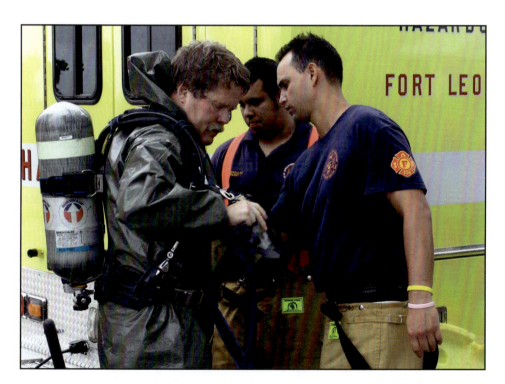

Figure 11.56 A good ratio is two assistants per one responder donning a chemical protective ensemble.

- Select an area that is large enough to accommodate all personnel involved in the donning and doffing procedures.
- Before starting the donning process, each entry and backup team member should be medically evaluated based on AHJ procedures.
- Continue hydration per AHJ procedures.
- Conduct a mission briefing before the donning process to ensure that all members are attentive and there are no distractions. The mission briefing should include the specifics of the mission such as IAP and site safety plan.
- Deploy chemical-protective clothing in an organized manner.

Chapter 11 • Implementing the Action Plan: Personal Protective Equipment   **423**

Figure 11.57 Chairs or benches with no back support may be used by responders with SCBA.

- Check all equipment visually and operationally prior to donning to ensure proper working order.
- Ensure that the entry team members have removed all their personal effects such as rings, wallets, badges, watches, and pins.
- Don appropriate undergarments at this time, if applicable.

The entry team should be seated so that breathing apparatus is accommodated, as necessary **(Figure 11.57)**. As described in Skill Sheets 2-5, the physical activity of the donning process should be conducted by assistants to allow the entry and backup personnel the opportunity to rest and reduce stress levels. Once the donning process has begun, the donning supervisor should prepare both the entry team and the backup team at the same rate. The teams should remain ready and off air until the entry order has been given.

Once the entry order has been given, the entry teams should be led to the entry access point. The safety officer should perform a final check of all equipment and closures before the teams are allowed to enter the hazard area. The backup team should be left off air and in a resting position until such a time that it may be called into service. Based on the hazards and chemicals involved, the backup team may be put on air and placed within the hot zone to reduce the travel time should the entry team need assistance with a rapid exit.

Doffing of PPE

Many times, the donning supervisor may also serve as the doffing supervisor. This will be very helpful based on the person's knowledge of the members and equipment that were utilized for the entry.

Upon exit, it can be assumed that the entry team has either been contaminated or potentially contaminated by the hazard, thus needing decontamination prior to doffing **(Figure 11.58)**. Based on the chemical hazards, it may be necessary to have the doffing personnel wear a lower level CPC. Any level of protection needed for doffing activities will be decided upon by the safety officer.

Figure 11.58 Plan to decontaminate an entry team even if they are wearing single use protective equipment.

The personnel assisting in the doffing procedures should watch for signs and symptoms of heat stress. The entry personnel who will be doffing their protective equipment will most likely be hot, tired, and anxious to remove the clothing.

All doffing procedures should follow the manufacturer's directions, but these generic procedures may be included in any department's policies and guidelines:

- Personnel who are doffing equipment should allow the assisting personnel to perform the work.
- Entry team members should only touch the inside of the garments and never the outside. Likewise, assisting personnel should only touch the outside of the garments. It is critical that cross contamination be avoided.
- Once the garments are removed, they should be zipped or stored so that the inside and outside surfaces cannot touch.
- All entry garments should be placed in a containment bag and appropriately marked.
- The breathing apparatus should be isolated and marked for appropriate decontamination **(Figure 11.59)**.
- The last item removed from the entry personnel should be the respirator facepiece. It should be removed by the user.
- All entry team and support team members must report immediately to rehab.

Figure 11.59 Remove the SCBA and isolate it for further decontamination.

PPE Inspection, Testing, Maintenance, Storage, and Documentation

The NFPA and OSHA require maintenance, testing, inspections, and record keeping of CPC. CPC manufacturers also require that extensive records be kept for garments purchased from them. It makes good sense to understand the history of a chemical-protective garment. Maintenance and testing of chemical-protective equipment is required. For more information on PPE inspection, testing, and maintenance, see **Skill Sheet 11-6**.

Inspection

Follow manufacturer's recommendations on inspection of PPE. Inspection and acceptance testing should be conducted when the garment is received into the inventory. If a garment is received and is not tested, the liability will fall to the end user. One cannot assume that all new equipment will be good. If a piece of protective clothing fails during the acceptance test, it is possible to return it to the manufacturer for a refund or exchange prior to payment.

Documentation will be required when placing a new protective garment into service. Some items that should be documented include:

- Specifications of the protective garment
- Suit identification number, if applicable
- Date it was placed in service
- Name and title of the person placing the suit into service
- Results of the acceptance test

Testing

A suit should be thoroughly examined if it has been used and decontaminated before it is allowed to be placed back into service per manufacturer's recommendations. After use, the following examinations should be performed:

- **Visual inspection** — Garments should be visually inspected for defects, deficiencies, and/or degradation.
- **Tactile inspection** — Degradation may not be visible. While wearing protective gloves, a tactile inspection by feel should be conducted to check for softness or stickiness.
- **Pressure test** — Level A CPC should be pressure tested to ensure they are gas-tight **(Figure 11.60)**. Other CPC should be pressure tested based on the manufacturer's recommendations.

Figure 11.60 Test a Level A ensemble to ensure that it forms a complete seal. *Courtesy of Brent Cowx and Jonathan Gormick, Vancouver Fire & Rescue Services.*

- **Soap bubble test** — If a suit does not pass a pressure test, the suit should be sprayed with a mild soapy solution while under pressure to help identify any leaks.
- **Light bar test** — CPC fabric and seams should be tested for mechanical damage. This is done by placing a fluorescent light inside the suit during inspection. The light will help to identify any deficiencies in the suit.
- **Documentation** — All results, both positive and negative should be recorded and documented.

NOTE: Perform all testing and inspections according to department protocols and manufacturer's directions.

Maintenance and Storage

All CPC should be stored and maintained according to the manufacturer's recommendations. CPC should be stored safely away from any contaminants and out of sunlight to avoid exposure to ultraviolet light **(Figure 11.61)**. A response agency may consider storing its ready-for-use CPC in a protective

Figure 11.61 Store all chemical protective equipment away from contaminants and sunlight.

container that may be sealed either by tab or lock. A sealed container will allow the next users of the equipment to know that the protective garments have been properly maintained, tested, inspected and can be used with confidence. If a container has a broken seal, then that container should be placed to the side and the equipment placed out-of-service until such a time that it can be retested, inspected, and deemed ready for use in accordance with manufacturer's recommendations.

Documentation and Written PPE Program

Record keeping is paramount for all hazmat equipment, but more so for CPC. Proper records should be kept on file that will allow a garment to be tracked over its life span. Documentation should follow manufacturer's recommendations and may include:

- Suit ID
- List of all repairs including who made the repair, how long the suit was out of service, what was done during the repair process, and when it was returned into service
- Suit use, including training and response
- Date, time, and duration of any person wearing the suit
- Results of all suit testing

NOTE: It is extremely important to document date, time, and duration of any chemical exposures and solutions used to decontaminate PPE. This may be done in safety, health, and wellness records or the PPE program, depending on the AHJ.

A written PPE plan is required and should be the cornerstone of any response agency's policies and procedures. The overall plan should be a part of the agency's employee health and safety program and may be site-specific if the situation allows. The written PPE plan should be specific to CPC and equipment and should address the following:

- The selection of CPC based on product hazards.
- Any limitations of the protective clothing during use, such as environmental considerations and contact with hazardous materials.
- Proper maintenance and storage of the protective equipment must be outlined.
- Training.
- Proper wear and fit of the protective garments along with specific donning and doffing procedures.
- In-depth inspection procedures must accompany this plan. Address inspection procedures for CPC prior to, during, and after use of the protective equipment. Documentation of these inspections is paramount.

As with any plan, its effectiveness must be evaluated. This may be done on a regular basis or after an incident or response. Any changes to the plan should be reviewed with agency personnel prior to implementation to ensure they are understood.

Chapter Review

1. What are the different types of respiratory protection available and what are their disadvantages?
2. What are the different types of protective clothing available and what are their disadvantages?
3. Give a brief description of the different levels of protection from A to D.
4. What are some common factors for selecting PPE for unknown environments, thermal protection, and chemical resistance/compatibility?
5. What are the main PPE-related stresses emergency responders will encounter?
6. What safety and emergency procedures should be in place when using PPE?
7. What is a written PPE plan and what should it address?

11-1
Select PPE appropriate to a hazardous materials/WMD incident. [NFPA 1072, 7.3.2]

WARNING: If this skill involves the use of actual hazardous material samples, hazardous materials can cause serious injury or fatality. Appropriate personal protective equipment (PPE) must be worn and safety precautions must be followed. The following skill sheet demonstrates general steps; specific hazmat incidents may differ in procedure. Always follow the AHJ's procedures for specific incidents.

Step 1: Select PPE ensemble for specified response option based on all identified hazards.

Step 2: Determine effectiveness of protective clothing based on its uses and limitations.

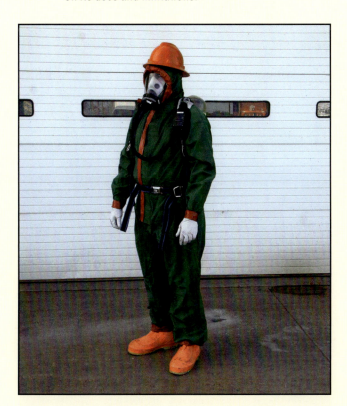

SKILL SHEETS 11-2

Don, work in, undergo technical decontamination while wearing, and doff structural fire fighting personal protective equipment. [NFPA 1072, 7.4.2]

WARNING: If this skill involves the use of actual hazardous material samples, hazardous materials can cause serious injury or fatality. Appropriate personal protective equipment (PPE) must be worn and safety precautions must be followed. The following skill sheet demonstrates general steps; specific hazmat incidents may differ in procedure. Always follow the AHJ's procedures for specific incidents.

Step 1: Perform a visual inspection of PPE and SCBA for damage or defects.
Step 2: Don protective trousers and boots.
Step 3: Don protective hood, pulling hood down around neck and exposing head.
Step 4: Don protective coat.
Step 5: Don SCBA. Ensure that the cylinder valve is fully open and that all straps are secured.
Step 6: Don SCBA facepiece and ensure a proper fit and seal.

Step 7: Pull hood up completely so that facepiece straps and skin are not exposed.
Step 8: Don helmet and secure.

Step 9: Don gloves.
Step 10: Ensure that all fasteners, straps, buckles, etc. are fastened.
Step 11: Ensure that no skin is exposed.
Step 12: Attach SCBA regulator to facepiece and make sure SCBA is functioning properly.
Step 13: Perform preentry checks according to AHJ's SOPs.

Step 14: Perform work assignment.
Step 15: Undergo technical decontamination per AHJ's SOPs.
Step 16: Doff PPE in reverse order according to AHJ's SOPs, avoiding contact with outer ensemble or surfaces that may be contaminated.

Step 17: Conduct a post-entry inspection of PPE for damage or defects according to AHJ's SOPs and document findings.

11-3

Don, work in, undergo technical decontamination while wearing, and doff a Level C ensemble. [NFPA 1072, 7.4.2]

SKILL SHEETS

WARNING: If this skill involves the use of actual hazardous material samples, hazardous materials can cause serious injury or fatality. Appropriate personal protective equipment (PPE) must be worn and safety precautions must be followed. The following skill sheet demonstrates general steps; specific hazmat incidents may differ in procedure. Always follow the AHJ's procedures for specific incidents.

NOTE: If using a PAPR, these steps will need to be modified according to AHJ's SOPs.

Step 1: Perform a visual inspection of PPE for damage or defects.

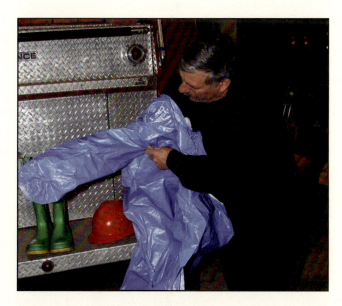

Step 2: Don Level C PPE and secure closures.

Step 3: Don work boots.
Step 4: Pull ensemble leg opening over the top of the work boots.

Step 5: Don respirator.
Step 6: Pull ensemble hood up completely so that facepiece straps and skin are not exposed.

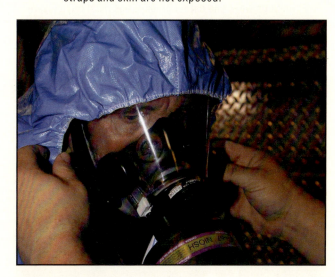

Step 7: Don inner protective gloves.
Step 8: Don outer protective gloves.

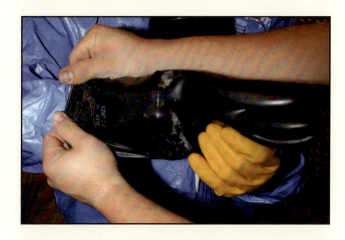

Step 9: Pull ensemble sleeves over the outside of the gloves.
Step 10: Breathe through respirator and ensure that respirator is functioning properly.
Step 11: Perform preentry checks as per AHJ's SOPs.
Step 12: Perform work assignment.
Step 13: After assignment has been performed, proceed to decontamination line.
Step 14: Undergo technical decontamination as per AHJ's SOPs.
Step 15: Doff ensemble according to AHJ's SOPs, avoiding contact with outer ensemble or surfaces that may be contaminated.
Step 16: Doff respirator according to AHJ's SOPs.
Step 17: Conduct a post-entry inspection of PPE for damage or defects according to AHJ's SOPs and document findings.
Step 18: Return to proper storage as per manufacturer's instructions.

SKILL SHEETS

11-4

Don, work in, undergo technical decontamination while wearing, and doff liquid splash-protective clothing. [NFPA 1072, 7.4.2]

WARNING: If this skill involves the use of actual hazardous material samples, hazardous materials can cause serious injury or fatality. Appropriate personal protective equipment (PPE) must be worn and safety precautions must be followed. The following skill sheet demonstrates general steps; specific hazmat incidents may differ in procedure. Always follow the AHJ's procedures for specific incidents.

NOTE: If using an encapsulating ensemble, these steps will need to be modified according to AHJ's SOPs.

Step 1: Perform a visual inspection of PPE and SCBA for damage or defects.

Step 2: Don liquid-splash PPE and secure closures.

Step 10: Attach SCBA regulator to facepiece and ensure proper operation.

Step 11: Perform preentry checks according to AHJ's SOPs.

Step 12: Perform work assignment.

Step 3: Don work boots according to AHJ's SOPs.
Step 4: Don the SCBA according to AHJ's SOPs.
Step 5: Don SCBA face peice and ensure a proper fit and seal.

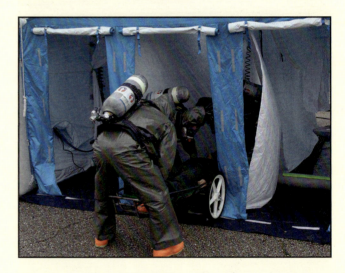

Step 13: Undergo technical decontamination as per AHJ's SOPs.
Step 14: To doff, remove PPE in reverse order of donning.
Step 15: Ensure medical monitoring is performed as per AHJ's SOPs.
Step 16: Conduct a post-entry inspection of PPE for damage or defects according to AHJ's SOPs and document findings.
Step 17: Return to proper storage as per manufacturer's instructions.

Step 6: Pull ensemble hood up completely so that facepiece straps and skin are not exposed.
Step 7: Don protective head gear (if required by AHJ).
Step 8: Don inner protective gloves.
Step 9: Don outer protective gloves.

11-5

Don, work in, undergo technical decontamination while wearing, and doff vapor-protective clothing. [NFPA 1072, 7.4.2]

WARNING: If this skill involves the use of actual hazardous material samples, hazardous materials can cause serious injury or fatality. Appropriate personal protective equipment (PPE) must be worn and safety precautions must be followed. The following skill sheet demonstrates general steps; specific hazmat incidents may differ in procedure. Always follow the AHJ's procedures for specific incidents.

Step 1: Perform a visual inspection of PPE and SCBA for damage or defects.

Step 2: Ensure the ensemble is the correct size.

Step 3: Ensure zipper is in good working order.

Step 4: Remove shoes, belts, and any objects that could damage ensemble.

Step 5: Don ensemble according to AHJ's SOPs.

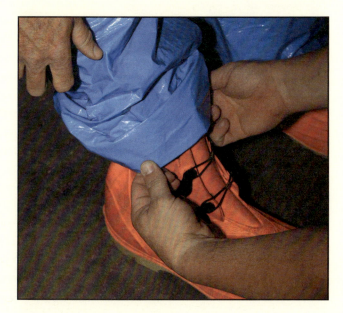

Step 6: Don the SCBA according to AHJ's SOPs.

Step 7: Turn on air supply, don SCBA facepiece, check seal and breathe normally to ensure SCBA operates properly.

Step 8: Don protective head gear (if required by AHJ).

Step 9: Perform preentry checks according to AHJ's SOPs.

Step 10: Perform checks per AHJ's SOP/Gs.

Step 11: Perform work assignment.

Step 12: Undergo technical decontamination as per AHJ's SOPs.

Step 13: To doff, remove PPE in reverse order of donning.

Step 14: Ensure medical monitoring is performed as per AHJ's SOPs.

Step 15: Conduct a post-entry inspection of PPE for damage or defects according to AHJ's SOPs and document findings.

Step 16: Return to proper storage as per manufacturer's instructions.

SKILL SHEETS

11-6

Inspect, test, maintain, and document chemical protective clothing serviceability.
[NFPA 1072, 7.4.2]

WARNING: If this skill involves the use of actual hazardous material samples, hazardous materials can cause serious injury or fatality. Appropriate personal protective equipment (PPE) must be worn and safety precautions must be followed. The following skill sheet demonstrates general steps; specific hazmat incidents may differ in procedure. Always follow the AHJ's procedures for specific incidents.

Inspection
Step 1: Ensure the suit's serviceability as per AHJ's SOPs.
Step 2: Visually inspect both the interior and exterior of the suit looking for damage or defects. *Photo courtesy of Barry Lindley.*

Step 3: Visually inspect the suit for any changes to the suit material.
Step 4: Check to ensure the zipper functions correctly.
Step 5: Check the function of all valves and passthroughs if applicable.

Testing (Pressure Test)
Note: These steps are applicable to vapor-protective suits.
Step 6: Test suit according to manufacturer's recommendations. *Photo courtesy of Brent Cowx.*

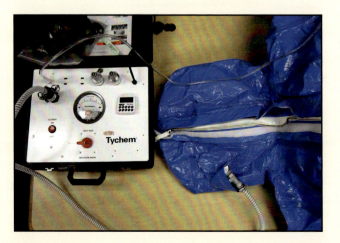

Step 7: Check and record pressure as per manufacturer's recommendations. *Photo courtesy of Brent Cowx.*
Note: Air loss should be observed over time. Follow all manufacturer recommendations.

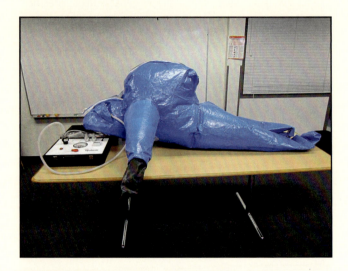

Maintenance
Note: These steps are applicable to vapor-protective suits.
Step 8: Maintain suit according to manufacturer's recommendations.

Documentation
Step 9: Document all findings during the inspection as per AHJ's requirements and remove any suits with defects or malfunctions from service.
Step 10: Return to proper storage as per manufacturer's instructions.

Implementing the Action Plan: Decontamination

Chapter Contents

Introduction to Decontamination 439	**Decontamination Implementation** 457
Decontamination Methods 440	Site Selection 457
Mass Decontamination 443	Factors Affecting Decontamination 458
Technical Decontamination 446	Decontamination Corridor Layout 459
Absorption 448	Other Implementation Considerations 463
Adsorption 449	Cold Weather Decontamination 464
Chemical/Biological Degradation 449	Medical Monitoring 465
Dilution 450	Evidence Collection and Decontamination 465
Disinfection 451	Termination 466
Evaporation 451	**Special Considerations** 466
Isolation and Disposal 452	Radiation 467
Neutralization 452	Pesticides 467
Solidification 453	Infectious Agents 467
Sterilization 453	Decontamination during a Terrorist Event 468
Vacuuming 454	**Chapter Review** 468
Washing 455	**Skill Sheets** 469
Assessing Decontamination Effectiveness 455	

436 Chapter 12 • Implementing the Action Plan: Decontamination

chapter 12

Key Terms

Absorption 442	Mass Decontamination 443
Adsorption 442	Neutralization..452
Ambulatory..445	Positive-Pressure Ventilation (PPV) 452
Berm 459	Sterilization 453
Biodegradable 450	Surfactant...455
Dilution 450	Technical Decontamination 446
Disinfection...451	

Implementing the Action Plan: Decontamination

JPRs Addressed In This Chapter

7.3.3, 7.4.4.1, 7.4.4.2

Learning Objectives

After reading this chapter, students will be able to:

1. Recognize general considerations for decontamination operations. [NFPA 1072, 7.3.3, 7.4.4.1, 7.4.4.2]
2. Describe decontamination methods. [NFPA 1072, 7.3.3, 7.4.4.1, 7.4.4.2]
3. Explain methods for assessing the effectiveness of decontamination. [NFPA 1072, 7.3.3, 7.4.4.1, 7.4.4.2]
4. Describe considerations for implementing decontamination. [NFPA 1072, 7.3.3, 7.4.4.1, 7.4.4.2]
5. Describe special considerations for decontamination involving uncommon hazards. [NFPA 1072, 7.3.3, 7.4.4.1, 7.4.4.2]
6. Skill Sheet 12-1: Perform mass decontamination on ambulatory people. [NFPA 1072, 7.4.4.1]
7. Skill Sheet 12-2: Perform mass decontamination on nonambulatory victims. [NFPA 1072, 7.4.4.1]
8. Skill Sheet 12-3: Perform technical decontamination on ambulatory people. [NFPA 1072, 7.4.4.2]
9. Skill Sheet 12-4: Perform technical decontamination on nonambulatory victims. [NFPA 1072, 7.4.4.2]
10. Skills Sheet 12-5: Select a decontamination method appropriate to a hazmat/WMD incident. [NFPA 1072, 7.3.3]

Chapter 12
Implementing the Action Plan: Decontamination

This chapter explains decontamination (decon) from a Technician-Level standpoint and builds on concepts learned at the Operations Levels:

- Introduction to decontamination
- Decontamination methods
- Assessing decontamination effectiveness
- Decontamination implementation
- Special considerations

Introduction to Decontamination

Responders at the hazardous materials Technician Level are trained to enter an area that may contain a hazardous chemical or substance. Hazmat technicians may become contaminated because their responsibilities often require them to work in close proximity to hazardous materials. For a review of basic concepts of decontamination, refer to IFSTA's **Hazardous Materials for First Responders**.

A decontamination method should be selected early in the decision-making process. This decision should not wait until hazmat technicians are dressed and ready to enter. Decontamination processes must be available and ready for implementation before responders enter the hot zone.

Per risk-based response, the chemical product will determine the decontamination selection method. Product-driven selection methods must consider properties of the involved chemicals when responders are determining decontamination needs. Other considerations include the availability of equipment and materials needed to adequately decontaminate themselves and civilians.

While washing and dilution in conjunction with isolation and disposal are the most cost-effective and common means of decontamination, other options may be available to the hazmat technician. Evaluate all decontamination options and consult with experts or specialists familiar with the chemicals that the response teams will encounter.

Ensure that the decontamination methods that will be used are compatible with the product, protective clothing, and equipment. Hazardous materials activities performed on scene will also influence the selection of the method.

The following resources can assist the hazmat technician in selecting a decontamination method and are also available to assist in many aspects of the hazmat response:

- Technical references such as Safety Data Sheets (SDSs) supplied from the chemical's manufacturer
- Chemical assistance agencies such as CHEMTREC in the United States, CANUTEC in Canada, and SETIQ in Mexico
- Local or regional poison control centers

Decontamination Methods

Decontamination methods can be divided into four broad categories: wet or dry methods and physical or chemical methods **(Figure 12.1)**. These methods vary in their effectiveness for removing different substances, and many factors may play a part in the selection decision such as weather conditions and the chemical and physical properties of the hazardous material(s). The response options may also affect which methods are used.

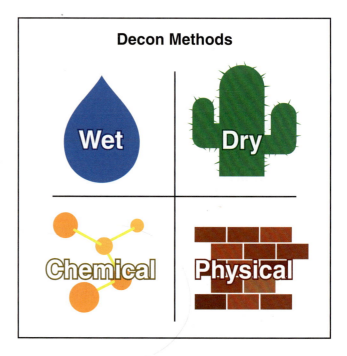

Figure 12.1 Four decontamination methods may be used to remove and contain hazardous materials.

As their names imply, wet and dry methods are categorized by whether they use water or other resources as part of the decon process. Wet methods usually involve washing the contaminated surface with solutions or flushing with a hose stream or safety shower **(Figure 12.2)**. Dry methods include scraping, brushing, and absorption **(Figure 12.3)**.

NOTE: The IFSTA manual, **Hazardous Materials for First Responders**, includes significant information regarding wet decontamination and dry decontamination methods.

Physical methods of decontamination remove the contaminant from a contaminated person without changing the material chemically (although wet methods may dilute the chemical). The contaminant is then contained (when practical) for disposal **(Figure 12.4)**. Examples of physical decontamination methods include:

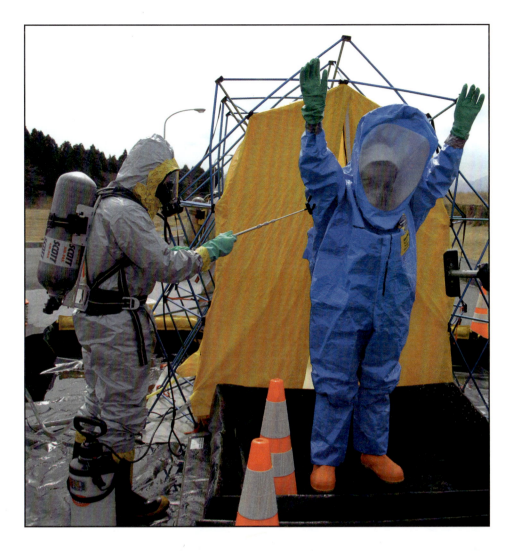

Figure 12.2 Wet decon uses water and water-based solutions to remove or neutralize contaminants. *Courtesy of U.S. Marine Corps, photo by Warren Peace.*

Figure 12.3 (above) Dry decon uses resources such as brushes and dry chemicals to remove or neutralize contaminants. *Courtesy of U.S. Army, photo by SSG Fredrick P. Varney, 133rd Mobile Public Affairs Detachment.*

Figure 12.4 (right) Contaminated items are often sorted for disposal and salvage.

Absorption — Penetration of one substance into the structure of another, such as the process of picking up a liquid contaminant with an absorbent.

Adsorption — Adherence of a substance in a liquid or gas to a solid. This process occurs on the surface of the adsorbent material.

- **Absorption**
- Brushing and scraping
- Evaporation
- Washing

- **Adsorption**
- Dilution
- Isolation and disposal
- Vacuuming

Chemical methods are used to make the contaminant less harmful by changing it through some kind of chemical process. For example, using bleach to sanitize tools and equipment that have been exposed to potentially harmful etiological agents is a form of chemical decontamination because the organisms are actually killed by the bleach. Examples of chemical decontamination methods include:

- Chemical degradation
- Disinfection
- Neutralization
- Sanitization
- Sterilization
- Solidification

The most effective means of decontamination is often as simple as removing the outer clothing or PPE that has been contaminated by the hazardous material. Removing clothes can remove a high percentage of the contaminant.

Additionally, flushing the contaminated surface with water is effective at removing the harmful substance or sufficiently diluting it to a safe level. For this reason, removal of contaminated clothing/PPE and flushing with water is usually sufficient for emergency and mass decon.

Technical decon requires additional effort to thoroughly remove all contaminants and involves washing with water and some type of soap, detergent, or chemical solution. The decision whether to perform emergency or technical decon is based on the hazardous material involved and the urgency in removing the victim from the contaminated environment.

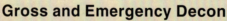

Gross and Emergency Decon

Gross and emergency decon do not require Technician-Level training to be effective. They are included here for reference. The IFSTA manual, **Hazardous Materials for First Responders**, includes more information on these decontamination tactics.

Gross Decon

Gross decontamination is a phase of decontamination where significant reduction of the amount of surface contamination takes place as quickly as possible. Traditionally, gross decon was accomplished by mechanical removal of the contaminant or initial rinsing from handheld hoselines, emergency showers, or other nearby sources of water at hazmat incidents.

Because of increased awareness of firefighters' cancer risk, gross decon is now recommended at all emergency incidents involving exposure to potentially hazardous substances, including the toxic products of combustion. This may be accomplished by doffing PPE at the scene and using wipes or other decon methods to remove soot from the face, head, and neck. PPE, tools, and equipment should be isolated, cleaned, and decontaminated according to SOPs, before reuse. It is recommended that structural firefighter protective clothing be machine washed in designated machines back at the station. Personnel should shower with soap and water thoroughly as soon as possible, even if wet methods of decon are used at the emergency incident scene.

Emergency Decon

The goal of emergency decontamination is to remove the threatening contaminant from the victim as quickly as possible — there is no regard for the environment or property protection. Emergency decon may be necessary for both victims and rescuers. If either is contaminated, individuals must remove their clothing (or PPE) and wash quickly. Victims may need immediate medical treatment, and they cannot wait for the establishment of a formal decontamination corridor. The following situations are examples of instances where emergency decontamination is needed:

- Failure of protective clothing
- Accidental contamination of emergency responders
- Immediate medical attention is required by emergency workers or victims in the hot zone

Emergency decontamination has the following advantages:

- Fast to implement
- Requires minimal equipment (usually just a water source such as a hoseline)
- Reduces contamination quickly
- Does not require a formal contamination reduction corridor or decon process

However, emergency decontamination has definite limitations. Removal of all contaminants may not occur, and a more thorough decontamination must follow. Emergency decontamination can harm the environment. If possible, measures must be taken to protect the environment, but such measures should not delay lifesaving actions. The advantage of eradicating a life-threatening situation far outweighs any negative effects that may result.

Mass Decontamination — Process of decontaminating large numbers of people in the fastest possible time to reduce surface contamination to a safe level. It is typically a gross decon process utilizing water or soap and water solutions to reduce the level of contamination, with or without a formal decontamination corridor or line.

Mass Decontamination

Mass decontamination is the process of rapidly reducing or removing contaminants from multiple persons (victims and responders) in potentially life-threatening situations **(Figure 12.5)**. Mass decon is initiated when the number of victims and time constraints do not allow the establishment of an in-depth decontamination process (such as technical decon). All agencies should have a mass decon plan as part of their overall emergency response plan (ERP). Responders must be familiar with the AHJ's procedures for implementing mass decon within the Incident Command system, including decontamination team positions, roles, and responsibilities. To determine the correct mass decon procedure, technicians must be familiar with established SOPs, ERPs, training, skills learned during drills/exercises, and preplans.

Figure 12.5 Mass decon refers to a large number of people being decontaminated at the same time.

NOTE: Technician-Level responders may not have learned the skills needed for mass decon with supervision of a Mission-Specific responder. These operations must include a responder with the appropriate skill set.

Chapter 12 • Implementing the Action Plan: Decontamination

Figure 12.6 Communication in mass decon areas may include hand signals.

Figure 12.7 Decon corridor paths should be clearly marked and instantly recognizable.

Figure 12.8 Water flow should be calibrated to soak everyone passing underneath.

The scene of an incident that requires the use of mass decon may be quite chaotic and difficult to control. To combat the chaos of the incident, responders should take the following actions:

- Communicate with victims by using hand signals, signs with pictures, apparatus public address systems, megaphones, or other methods to direct them to decon gathering areas and through the decon process itself **(Figure 12.6)**.

- Provide directions that are clear, easily understood, short, and specific since victims may be traumatized and/or suffering from exposure.

- Use barrier tape, traffic cones, or other highly visible means to mark decon corridors **(Figure 12.7)**.

Mass decon methods include dilution, isolation, and washing. Washing with a soap-and-water solution or universal decontamination solution will remove many hazardous chemicals and WMD agents. However, availability of such solutions in sufficient quantities cannot always be ensured. Therefore, technicians can most readily and effectively accomplish mass decon with a simple water shower system that merely dilutes the hazardous product and physically washes it away.

Mass decon showers should use a high volume, low pressure of water delivered in a fog pattern to ensure the showering process removes the hazardous material **(Figure 12.8)**. The actual showering time is an incident-specific decision but may be as long as two to three minutes per individual under ideal situations. When large numbers of potential victims queue for decon, showering time may be significantly shortened. Showering time may also depend upon the volume of water available. Decontamination is a circular process; use post-decon monitoring to ensure that all contaminants are removed before allowing victims to progress to the next area.

Emergency responders should not overlook existing facilities when identifying means for rapid decontamination methods. For example, the necessity of saving victims' lives would justify the activation of overhead fire sprinklers for use as showers, even if property damage were to result. Similarly, having victims wade and wash in water sources such as public fountains, chlorinated swimming pools, or swimming areas provides an effective, high-volume decontamination technique.

It is recommended that all victims undergoing mass decon remove clothing at least down to their undergarments before showering. Removal of clothing can remove significant amounts of the contaminant materials **(Figure 12.9)**. Technicians should encourage victims to remove as much clothing as possible, proceeding from head to toe. Contaminated clothing should be isolated in drums, appropriate bags, or other containers for later disposal. For more information about performing mass decontamination on ambulatory and nonambulatory victims, see **Skill Sheets 12-1** and **12-2**.

Ambulatory People

People, often responders, who are able to understand directions, talk, and walk unassisted are considered **ambulatory** and should be directed to an area of safe refuge within the isolation perimeter to await prioritization for decontamination.

Several of the following factors may influence the priority for ambulatory patients:

- Victims with serious medical symptoms (such as shortness of breath or chest tightness)
- Victims closest to the point of release
- Victims reporting exposure to the hazardous material
- Victims with evidence of contamination on their clothing or skin
- Victims with conventional injuries (e.g., broken bones, open wounds)

Figure 12.9 Contaminants are often the most prevalent on outer layers of clothing.

Nonambulatory Victims

Nonambulatory victims are civilians or responders who are unconscious, unresponsive, or unable to move unassisted **(Figure 12.10)**. These patients may be more seriously injured than ambulatory patients. They may have to remain in place if sufficient personnel are not available to remove them from the hot zone.

Ambulatory — People, often responders, who are able to understand directions, talk, and walk unassisted.

Figure 12.10 Victims who cannot move while unassisted must be aided through decon areas. *Courtesy of New South Wales Fire Brigades.*

Chapter 12 • Implementing the Action Plan: Decontamination **445**

Technical Decontamination

Technical decontamination is the planned and systematic removal of contaminants from personnel and/or equipment **(Figure 12.11, p. 448)**. Technical decontamination uses chemical or physical methods to thoroughly remove or neutralize contaminants primarily from entry team personnel's PPE and equipment. It may also be used on incident victims in nonlife-threatening situations. Technicians must be familiar with their organization's procedures for implementing technical decon within the ICS, and the advantages and disadvantages of different technical decon methods **(Table 12.1)**.

> **Technical Decontamination** — Using chemical or physical methods to thoroughly remove contaminants from responders (primarily entry team personnel) and their equipment; usually conducted within a formal decontamination line or corridor following gross decontamination. *Also known as* Formal Decontamination.

Table 12.1
Advantages and Disadvantages of Technical Decon Methods

Method	Advantages	Disadvantages
Absorption	• Many absorbent materials are inexpensive and readily available • Can be used as part of dry decon operations • Effective on flat surfaces	• Does not alter the hazardous material • Ineffective for decontaminating protective clothing and vertical surfaces • Disposal of contaminated absorbent materials may be problematic and expensive • Absorbent materials may increase in weight and/or volume as they absorb the hazardous material • Absorbent materials must be compatible with the hazardous material
Adsorption	• Contains the hazardous material better than absorbent materials • Transportation of materials to disposal is simplified • Off-gassing (release of vapors/gases) is effectively reduced • Adsorptive materials do not swell	• Process can generate heat • Application typically limited to remediation of shallow liquid spills • Adsorptive materials are expensive • Adsorptive material must be compatible with the hazardous material (they are product specific)
Chemical Degradation	• Can reduce cleanup costs • Reduces risk posed to the first responder when dealing with biological agents • Often utilizes commonly available, inexpensive materials such as bleach, isopropyl alcohol, or baking soda • Utilizes products that are readily available	• Takes time to determine the right chemical to use (which should be approved by a chemist) and set up the decon process • Can cause violent reactions if done incorrectly and may create heat and toxic vapors • Rarely used to decontaminate people • Can damage PPE
Dilution	• Lessens the degree of hazard present by reducing the concentration of the hazardous material • Easy to implement (water is usually available) • Is very effective in many circumstances requiring decon • Can be used to decon large pieces of equipment/apparatus	• Can't be used on materials that react adversely to water • May be problematic in cold weather • May create large amounts of contaminated run-off • May be impractical because of the amount of water required for effective dilution
Disinfection	• Kills most of the biological organisms present • Can be used on site • Can be accomplished using a variety of chemical or antiseptic products • Disinfecting agent may be as simple as antibacterial soap or detergent	• Limited to biological decon only • May be difficult to decon large pieces of equipment/apparatus • Disinfecting agent may be toxic or harmful

Continued

Table 12.1 (concluded)

Method	Advantages	Disadvantages
Evaporation	• No additional materials necessary • No runoff collection necessary • No (or very limited) expense incurred	• Applicable for a very limited number of chemicals • Generally limited to decon of tools and equipment, not people • May be dramatically affected by weather conditions (including wind, temperature, humidity, and rain) • Hazardous vapors may travel and cause problems • May require a long time to complete • May not be acceptable method to use depending on applicable laws and regulations
Isolation and Disposal	• Isolation can be quick and effective • Easily achieved with containers such as isolation drums, heavy plastic bags, and other means of containment	• Disposal and transport costs may be extremely high • May require replacement of equipment and PPE that cannot be decontaminated and placed back in service
Neutralization	• Chemically alters the hazardous material to reduce the degree of hazard present • Effective on most corrosives and some poisons • Neutralizing agents are readily available (soda ash, vinegar)	• May be very difficult to successfully implement • Rarely done on living tissue • May require large quantities of neutralizing agents • May create violent chemical reaction including the release of heat and hazardous vapors • Preplanning is usually necessary
Solidification	• Solids are easier to contain than liquids and gases • Reduces the amount of vapor production and off-gassing • Easier to clean up	• Requires specialized materials to implement
Sterilization	• Kills all microorganisms present	• Difficult or impossible to do onsite
Vacuuming	• Effective at removing dust and particulates • Effective indoors • Dry method, useful for cold weather operations in some situations	• Requires specialized vacuums equipped with hepa filters • May require high risk, negative air containment for decon area • Removing liquid chemical contamination requires special equipment • May require additional decon procedures to ensure complete decontamination (for example, washing) • Can't be used to decontaminate materials that react adversely to contact with water • May be problematic in cold weather • May create large amounts of contaminated run-off
Washing	• Quick and easy to implement (water is usually readily available) • Soap is readily available and inexpensive • Typically more effective than dilution alone • Is very effective in many circumstances requiring decon • Can be used to decon large pieces of equipment/apparatus	• Can't be used to decontaminate materials that react adversely to contact with water • May be problematic in cold weather • May create large amounts of contaminated run-off

Figure 12.11 Technical decon refers to the systematic removal or neutralization of contaminants. *Courtesy of USAF, photo by Chiakia Iramina.*

Hazmat technicians usually conduct technical decon within a formal decon line or corridor. The contaminants involved at the incident will determine the type and scope of technical decon. Resources for determining the correct procedures may include books, reference sources, computer programs, and databases such as:

- Emergency response centers (CHEMTREC, CANUTEC, SETIQ, etc.)
- NIOSH Pocket Guide
- Poison control centers
- Preincident plans
- Safety data sheets (SDSs)
- Technical experts
- WISER

Absorption

Absorption is the process of soaking up a chemical spill **(Figure 12.12)**. In the simplest of terms, think of a liquid spill on your counter. When you wipe up the spill with a paper towel, you are absorbing the liquid into the towel. The same theory can be applied to the fluid leak at a motor vehicle accident. Sand or other absorbent material compatible with the leak will soak the material into the medium.

The process of absorption does not change the chemical properties of a compound or its hazards. Instead, the compound is simply moved from one medium to another. Absorption can be helpful when attempting to decontaminate equipment or machinery that may not be able to withstand other decon processes such as washing or dilution. Absorption also creates much less waste than many other processes. Absorption is not a process that is generally used for the decontamination of responders or their protective equipment.

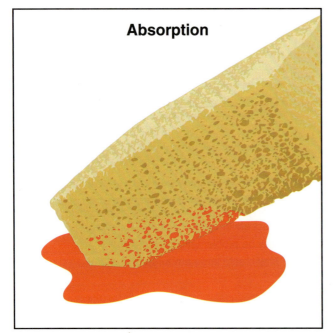

Figure 12.12 Absorption is a physical decon method often used to move a hazardous material from a solid surface.

448 Chapter 12 • Implementing the Action Plan: Decontamination

Because the process of absorption does not change the chemical properties of a product, the hazard will still exist and the end result will be a hazardous waste. The one exception to this rule would be if emergency responders absorbed an acid spill with an absorbent material such as soda ash. The end result would be a neutralized product that has been absorbed in the soda ash making it easier to move.

Hazmat technicians can accomplish the process of absorption by using a variety of resources ranging in complexity from absorbent pads and booms to particulate material such as cat litter. Other commercial products are available.

CAUTION
Take care to ensure that products used for absorption will be compatible with the material to be absorbed.

Adsorption

Adsorption is the adherence of a chemical to the surface of another medium **(Figure 12.13)**. This chemical method of decontamination allows for a liquid to adhere to a solid adsorbent material. Adsorption is useful for removing a liquid product floating on another liquid such as a hydrocarbon floating on water. Adsorbents tend not to swell like absorbents. Ensure that the adsorbent used is compatible with the hazardous material in order to avoid potentially dangerous reactions.

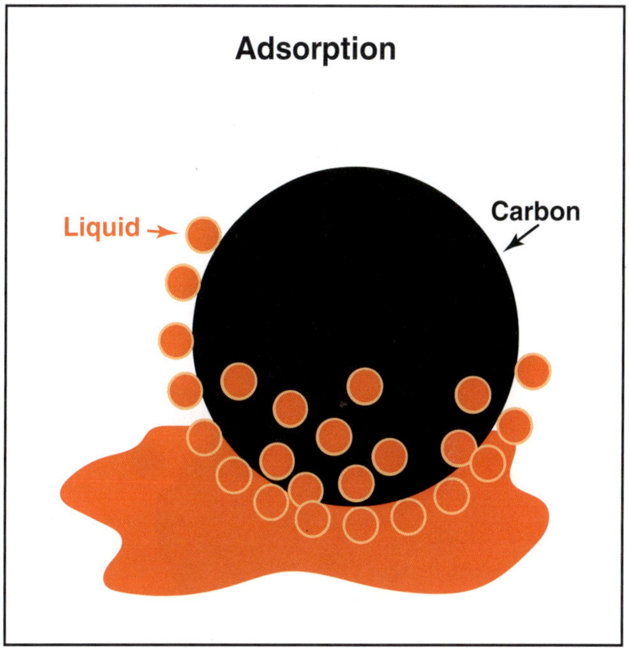

The adsorption process may produce high heat and spontaneous combustion as a by-product. Technicians must properly dispose of adsorbed materials.

A number of adsorbent materials are commercially available. Adsorbent materials are often found in round or pelletized form and will have high heat stability and small pores resulting in a higher surface area. Some adsorbent materials may include:

- Activated carbon or charcoal
- Silica
- Aluminum gel
- Earth clay (fuller's clay)

Figure 12.13 Adsorption is a chemical decon method often used to remove a liquid floating on another liquid.

Chemical/Biological Degradation

Chemical degradation and biological degradation are the processes in which a product is altered or degraded via action from a nonreactive substance into a less hazardous material. This may be a natural process such as the degradation of a product as it ages, or it may be a physical process such as the agitation that occurs when a product is exposed to weather. For chemical degradation to occur, a substance may be used to attack or alter the properties of another

Figure 12.14 Chemical and biological degradation are physical and chemical decon processes that occur when a material is broken down into less hazardous materials based on an interaction with other agents.

Biodegradable — Capable of being broken down into innocuous products by the actions of living things, such as microorganisms.

chemical. For **biodegradable** materials to decompose, enzymes or microbial agents may be used to reduce the hazards of a biological agent. In the simplest of terms, think of a pile of leaves left in place over time. As weather, insects, and microbes act on the leaves, they biodegrade into soil **(Figure 12.14)**.

Chemical/biological degradation is a complicated process, and it may be difficult to find specific information for the different agents that may be needed. In addition, it may be difficult to determine the correct concentrations to achieve decontamination safely. During the degradation process, heat may be a by-product of the reaction.

Because responders are intentionally exposing personnel and equipment to a chemical, they must take care to adequately research the chemicals and reactions that may occur. Do not use this type of decontamination for protective clothing, response equipment, or people.

> **WARNING!**
> Never use chemical or biological degradation processes to decontaminate skin.

Chemicals such as isopropyl alcohol, sodium hydroxide, sodium carbonate, and calcium oxide, can be used as an additive in washing solutions to aid in degrading a harmful product. These items used in the decontamination process will react with the host product and will change it into a less hazardous substance.

Dilution

Dilution is the process in which a solvent, such as water, lessens the concentration and hazards of a contaminant **(Figure 12.15)**. Dilution works best on products that may be water soluble or miscible. The use of large quantities of water may carry away or dilute a product to a concentration at which it will be less harmful. Dilution, in conjunction with washing, is an effective means for decontaminating emergency responders, victims, and equipment for many types of contamination.

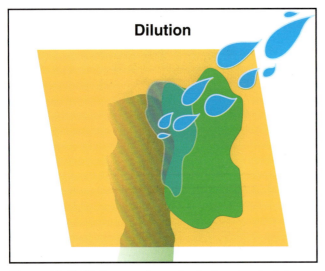

Figure 12.15 Dilution may be a physical or chemical decon process, depending on the solvent used to reduce the concentration of a target material.

Dilution — Application of water to a water-soluble material to reduce the hazard.

Viscous substances such as fuel oils, antifreeze, and different adhesives may be difficult to dilute. Even in cases where dilution is effective, the process may require a significant cleanup effort from contaminated runoff.

Dilution most commonly uses water, which is usually available in the large quantities needed to effectively dilute a product. Types of equipment needed for the dilution process include pumps, hoses, or other means of moving water.

> **CAUTION**
> Water may react violently with some materials.

> **Disinfection** – Any process that eliminates most biological agents; disinfection techniques may target specific entities. Often uses chemicals.

Disinfection

Disinfection is a process in which a significant percentage of pathogenic organisms are killed or controlled. Disinfectants are substances that are applied to nonliving objects in order to kill or eliminate microorganisms **(Figure 12.16)**. The process of disinfection can be performed in many ways and may range from household disinfectants such as floor cleaners to disinfectants for water treatment and filtration.

Disinfectants are effective when they can contact a surface and remain wet for an extended period of time. A disinfectant may have to remain on a surface for a few seconds to as long as ten minutes to be effective. Based on the orientation of the surface being disinfected, this requirement may make disinfection an impractical option.

The effectiveness of disinfectants vary based on the situation. Some disinfectants may be wide-spectrum in nature and can be used to destroy a wide range of microorganisms. Others may be more specific and affect only an isolated assortment of bacteria or viruses. Some types of disinfectants may include:

- **Air disinfectants** — Eliminate microorganisms suspended in air
- **Alcohols** — Disinfect skin
- **Aldehydes** — Eliminate spores or fungi
- **Oxidizing agents** — Attack the cellular structure of microorganisms
- **Phenolics** — Used in household disinfectant soaps and hand washes

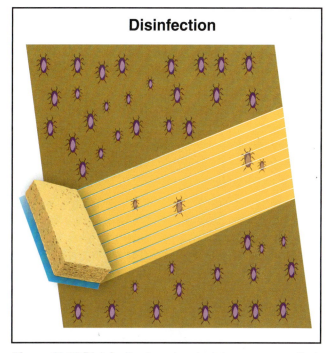

Figure 12.16 Disinfection is a chemical decon process that kills target microorganisms.

Evaporation

Evaporation allows the liquid portion of a chemical product to dry out **(Figure 12.17)**. Another term for this process is "off-gassing." Once the process is complete, the end result may be the total elimination of a chemical or a solid waste that may be dissolved or suspended in a solution.

Because evaporation happens naturally, there is often little need for responder intervention. Evaporation may be a tactic used for volatile chemical spills if the conditions allow for the product's safe evaporation, and if the vapors will not cause a safety issue.

Figure 12.17 Evaporation is a physical decon process that allows a chemical product to disseminate into the atmosphere.

Chapter 12 • Implementing the Action Plan: Decontamination

Positive-Pressure Ventilation (PPV) — Method of ventilating a room or structure by mechanically blowing fresh air through an inlet opening into the space in sufficient volume to create a slight positive pressure within and thereby forcing the contaminated atmosphere out the exit opening.

Products with low vapor pressure such as alcohols begin evaporating almost immediately. One major advantage of allowing a product to evaporate is that it will reduce the amount of hazardous waste that will need to be removed.

Technicians may use **positive-pressure ventilation (PPV)** fans to aid evaporation. PPV fans may also be used during decon of high vapor pressure products.

While evaporation may be a useful tactic for decontamination, vapors may be dangerous and not easily controlled. Due to the vapor pressure of a chemical, evaporation may not be a feasible method due to the length of time it may take for a product to evaporate.

Isolation and Disposal

Isolation and disposal are a two-part decontamination process that may work in tandem with other decon methods. Isolation is the act of segregating contaminated equipment in an effort to reduce exposure. Disposal is the final step in the decontamination process. Any material that cannot be decontaminated effectively must be disposed of safely and legally.

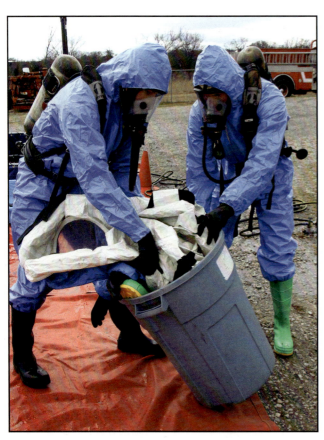

Figure 12.18 Isolation and disposal are physical decon processes used to remove contaminated items from proximity with noncontaminated items.

Isolation is advantageous in that contaminants do not spread to other items, and the hazards are isolated from personnel. Isolation and disposal are the simplest form of decontamination. Contaminated equipment and clothing are removed carefully to prevent contact with contaminated surfaces and contained in bags or drums to reduce the possibility of cross contamination **(Figure 12.18)**.

Technicians must always consider the risk of cross contamination when using isolation as a decontamination method. This method may also create a sizable amount of hazardous waste that will have to be removed or neutralized.

In order to effectively isolate contaminated PPE and equipment, response organizations will need to stock a variety of bags and drums. Technicians will use these to contain the contaminated items.

Neutralization

Neutralization — Chemical reaction in water in which an acid and base react quantitatively with each other until there are no excess hydrogen or hydroxide ions remaining in the solution.

Neutralization is a form of chemical degradation and is used in association with acids and bases. The main objective with neutralization is to bring the pH of the chemical in question as close to neutral as possible. Neutralization is typically a technique used to decontaminate equipment **(Figure 12.19)**.

Sodium carbonate is an example of a weak base that can be used to neutralize an acid. Examples of weak acids that can be used to neutralize a basic (alkaline) material include citric acid, ascorbic acid, or acetic acid.

Neutralization is entirely product specific and no single chemical can neutralize any chemical. This is generally a difficult procedure to perform because the correct chemicals must be used in the correct amounts. In addition, wastewater from the process must also be collected and disposed. The hazmat technician must have specific product information and should rely on the help of product experts when attempting this process. Neutralization may produce an exothermic reaction and should never be used on living tissue.

Required chemicals needed for neutralization must be specific to the contaminants in question. Use extreme care when attempting to neutralize a product because the chemical needed to accomplish this will present hazards of its own. Use pH paper to monitor pH levels.

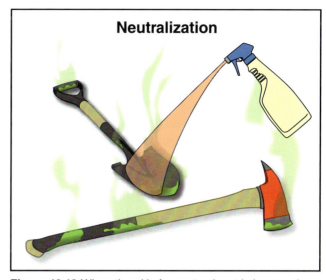

Figure 12.19 When the pH of a contaminant is known, the contaminated material can be treated to bring the pH closer to neutral.

Solidification

In solidification, an agent applied to a liquid spill combines with the liquid either chemically or physically to form a solid. Solidification limits the spread of a contaminant and allows for easier handling. Contractors usually conduct solidification to prepare a hazardous substance for disposal. The end result is typically a rubber-like substance that allows for easy removal and disposal.

While the solidification process is helpful, it may not alter the hazards of the target chemical. The reaction of the two chemicals may also produce heat or other dangerous by-products. This method of decontamination may also produce more hazardous waste than the original target chemical.

CAUTION
Solidification should not be used for personnel decontamination.

Technicians may use commercially available products for solidification tactics. The U.S. Environmental Protection Agency (EPA) has approved products such as Oil Bond® for use with static spills and in waterways **(Figure 12.20)**. These products can be mixtures of hydrocarbon polymers and can be either swept onto a spill or placed in booms or pillows for use in waterway spills. Solidification materials are most helpful for hydrocarbon spills such as gasoline, fuel oils, and hydraulic fluids. They will also work with aromatics and chlorinated solvents.

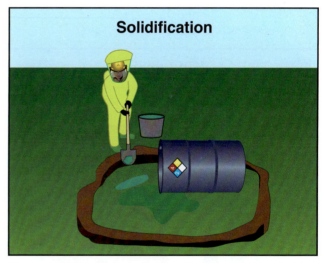

Figure 12.20 Some hazardous materials can be treated to turn into a solid that can be more easily contained.

Sterilization

In contrast to disinfection, **sterilization** results in destruction of all forms of microbial life. Sterilization is often necessary to ensure a complete decontamination.

Because of the equipment needed to accomplish sterilization, technicians may find this process impossible or impractical to complete on the scene of an emergency and in on-site decon lines. This process is best completed at facilities that can ensure that the sterilization process is both complete and successful.

Sterilization — Any process that destroys biological agents and other life forms. Often uses heat.

Technicians normally accomplish sterilization by using chemicals, steam, heat, or radiation **(Figure 12.21)**. Sterilization can also be accomplished in association with disinfection. Often, responders will disinfect tools and equipment on-scene and wait to sterilize resources at a later time and place.

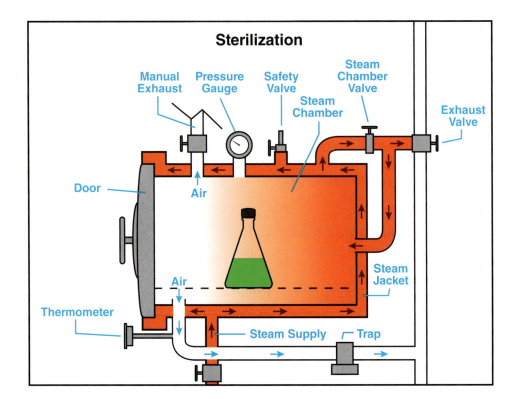

Figure 12.21 Sterilization is a chemical decontamination process that uses steam and heat to kill microbes on a surface or in a liquid.

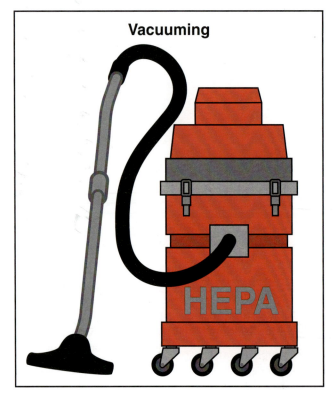

Figure 12.22 Industrial vacuums use HEPA filters to collect fine particles.

Vacuuming

Vacuuming is the process of using a negative pressure intake and high efficiency particulate air (HEPA) filter to remove dust, particulate, or other solid hazards from personnel and equipment **(Figure 12.22)**. This process has two advantages:

1. It will remove particulate down to three microns.

2. By isolating only the particulate being removed, it may reduce the overall waste that must be disposed of.

Vacuuming can be a time-consuming activity, especially when associated with decontamination. The vacuuming equipment needed may be costly and the filters must be changed frequently. Exhaust air from the vacuum must also be monitored to ensure that no product is bypassing the filters.

Vacuuming requires the use of specialized vacuums. The equipment being used must be compatible with the materials being vacuumed. Household vacuums cannot be used because the filtration on this style of equipment is not adequate to handle fine particulates.

Washing

Washing is similar to dilution in that they are both wet methods of decontamination **(Figure 12.23)**. Washing includes the use of solutions, solvents, and other **surfactants** that will help to reduce the adhesion between the contaminant and the material being cleaned. Surfactants work to lift the contaminant from the material and allow it to be removed when it is rinsed with a copious amount of water.

When using washing as the preferred method of decontamination, contain and properly dispose of the wastewater. The materials being removed may still have hazardous properties. Consider that washing may not be an adequate method of decontamination for viscous substances nor may it have a positive effect if the products are not miscible.

Technicians will need a supply of water along with a variety of cleaning solutions based on both the material being removed and the contaminated surface. Use the manufacturer's recommendations when applying decontamination solutions to chemical protective clothing.

Based on the products being removed, washing may include household laundry detergent that can act as an emulsifying agent. Other items, such as trisodium phosphate, can be used, but it may be too harsh and could possibly damage protective clothing.

Figure 12.23 The process of washing uses a soap or similar material to remove a substance from a surface.

Surfactant — Chemical that lowers the surface tension of a liquid; allows water to spread more rapidly over the surface of Class A fuels and penetrate organic fuels.

Assessing Decontamination Effectiveness

In the United States, OSHA requires that the effectiveness of decon operations be assessed. Because decon methods will vary with their effectiveness at removing different substances, responders should conduct an assessment of decon operations at the start of the decontamination process and as an ongoing assessment. Do this through the use of monitoring and detection devices or other equipment as well as visually. The hazardous materials involved will generally determine the technology or device needed to perform the assessment operations.

Technicians should perform post-decon checks as the individual exits the decon corridor **(Figure 12.24)**. If contamination is detected, redirect individuals through the decon process.

Check (or recheck) victims still complaining of symptoms or effects. If the effectiveness of decon is questioned, victims should go through decon again prior to transport.

Technicians will normally need to store tools and equipment in the decon area until the emergency phase

Figure 12.24 The effectiveness of the decontamination process must be confirmed before letting anyone leave the scene. *Courtesy of Steven Baker, New South Wales Fire Brigades.*

Chapter 12 • Implementing the Action Plan: Decontamination 455

of the operation is completed. After being decontaminated, check these tools to ensure that all contamination has been removed before placing them back in service.

Apparatus will also need to undergo decon if it has been exposed or potentially exposed to hazardous materials **(Figure 12.25)**. The same monitoring and detection equipment used to determine effectiveness of decon on victims and responders may be used on equipment, tools, and apparatus.

Figure 12.25 Apparatus may need to be decontaminated.

If technicians determine the decon process to be ineffective, they may repeat the process while introducing variables that may be more effective based on the results of the completed steps. A few methods for testing the effectiveness of the decontamination process can include the following:

- **Visual inspection** — Technicians can sometimes visually observe the effectiveness of decontamination using natural or ultraviolet light. Some contaminants may leave visible stains or residue while others may not. Some hydrocarbon-based products may become visible under ultraviolet light.

- **Wipe sampling** — The subject or item is wiped with a cloth swab, which is then sent out for analysis. However, results may not be immediately available.

- **Cleaning solution analysis** — This method evaluates the cleaning solution effectiveness during the final rinse. If contaminants are found in the final rinse water pool, it may suggest that additional cleaning and rinsing may be necessary.

- **Detection and monitoring** — As decontamination efforts are undertaken, technicians should use available and approved devices to monitor the results to determine the effectiveness of the decontamination process.

Decontamination Implementation

As indicated in the following sections, technicians should consider a number of factors while implementing decon at a hazmat incident. These, and other, factors may influence the timeliness and effectiveness of implementation.

Site Selection

Preplans should include predesignated areas for mass decon at locations likely to be targeted by terrorists, such as government buildings and stadiums. Hospitals must also have plans to decon potentially large numbers of victims who self-present (arrive via private transportation) at emergency rooms.

Consider the following factors when choosing a decontamination site:

- **Wind direction** — The decontamination site needs to be upwind of the hot zone to help prevent the spread of airborne contaminants into clean areas. If the decontamination site is improperly located downwind, wind currents will blow mists, vapors, powders, and dusts toward responders and victims. During long-term operations, the local weather service can provide assistance in predicting changes in the wind speed and direction.

- **Weather** — During cold weather, protect the site from blowing winds, especially near the end of the corridor. Shield victims from cold winds when they are removing protective clothing.

- **Accessibility** — The site must be away from the hazards, but adjacent to the hot zone so that persons exiting the hot zone can step directly into the decontamination corridor. An adjacent site eliminates the chance of contaminating clean areas. It also puts the decontamination site as close as possible to the actual incident.

- **Terrain and surface material** — The decontamination site should ideally be flat or sloped toward the hot zone. This terrain allows anything that may accidentally get released in the decontamination corridor to drain toward or into the contaminated hot zone. It also allows for persons leaving the decon corridor to enter into a clean area. If the site slopes away from the hot zone, contaminants could flow into a clean area and spread contamination.

- **Lighting and electrical supply** — The decontamination corridor should have adequate lighting to help reduce the potential for injury to personnel **(Figure 12.26)**. Selecting a decontamination site illuminated by streetlights, floodlights, or other types of permanent lighting reduces the need for portable lighting. If permanent lighting is unavailable or inadequate, technicians must use portable lighting. Ideally, the decontamination site will have a ready source of electricity for portable lighting, space heaters, water heaters, and other needs. However, if such a source is not available, portable generators will be needed.

- **Drains and waterways** — Avoid locating a decontamination site near storm and sewer drains, creeks,

Figure 12.26 Supplemental lighting and electrical generators should be available in case they are needed.

ponds, ditches, and other waterways unless the sewer system is approved for use as a contained system that can be managed and neutralized. If avoiding these openings is not possible, construct a dike to protect the storm drain opening or nearby waterway. Protect all environmentally sensitive areas if possible but never delay decon to protect the environment if the delay will increase injury to those affected by the event.

- **Water supply** — If technicians intend to use wet decon, sufficient water supply must be available at the decontamination site.

> **CAUTION**
> Protect the environment when possible, but prioritize helping people.

In addition to the above, technicians should consider other time and terrain factors. For example, the less time it takes personnel to get to and from the hot zone, the longer personnel can work in the hot zone. As indicated in the Air Management section of Chapter 11, crucial time periods should be considered:

- Travel time to and within the hot zone
- Time allotted to work in the hot zone
- Travel time back to the decontamination site
- Decontamination time

Finding ideal topography is not always possible, and responders may have to place some type of barrier to ensure confinement of an unintentional release. Considerations include:

- Employ diking around the site to prevent accidental release of contamination.
- Seek a hard, nonporous surface at the site to prevent ground contamination. Areas may include a driveway, parking lot, or street.
- Regardless whether the surface is porous, use salvage covers or plastic sheeting to form the technical decontamination corridor **(Figure 12.27)**. Covers or sheeting will help control the direction of any run-off and prevent contaminated water from soaking into the earth.

Figure 12.27 Protect the floor of the decon corridor from runoff or contamination.

Factors Affecting Decontamination

Several factors can affect how technicians perform decontamination. Based on the products being decontaminated, the amount of wash or rinse stations may vary. For example, if a person is contaminated with a viscous product such as a fuel oil, the product may not be overly hazardous, but it will also not be removed easily. It may take several wash/rinse stations to accomplish an adequate decontamination. However, a person contaminated with a volatile liquid may be easier decontaminated because most of the hazardous properties of the liquid may have already evaporated by the time the hazmat technician arrives at the decon corridor.

Chemical factors that affect decontamination can include the following:

- **Toxicity** — Chemicals pose a significant health risk. Thoroughly remove any chemicals that may pose a skin absorption hazard. Make all efforts to completely decontaminate all surfaces.

- **Water-soluble chemicals** — Washing or diluting is the best process for removing these chemicals, or others. However, technicians may need a surfactant to lift the product from the surface being decontaminated. The surfactant and contaminants will remain in the wastewater.

- **Water-reactive chemicals** — These products will be difficult to remove from garments or equipment because of the relative humidity in the air. There is a chance that water-reactive chemicals will already be undergoing a chemical reaction before arrival at the decontamination corridor. The most effective technique may be brushing, blowing, or vacuuming. Water dilution may be required after the initial product is removed from the surface. Because of the nature of water-reactive chemicals, use copious amounts of water to ensure the safety of all involved.

Decontamination Corridor Layout

The principle of decontamination is to reduce the amount of contaminants at each wash station and eliminate any possibility of cross contamination. Decon corridors should separate humans (responders and victims) and equipment/vehicles. The configuration of the decontamination corridor will vary and be specific to the products in question. Configure a decon corridor to bridge the space between the hot and warm zone. The corridor should be wide enough to accommodate the operation and long enough to accommodate all of the necessary wash stations **(Figure 12.28)**.

Figure 12.28 A decon corridor should be capable of containing all necessary stations.

Place the decon corridor on a base of water resistive sheet plastic. The plastic should be one continuous sheet to securely contain the decon waste. Build a **berm** (dike) into the decon base to assist with containment. A berm may be created quickly with a fire hose inflated with air or charged with water. For more information on performing technical decontamination involving ambulatory and nonambulatory victims, see **Skill Sheets 12-3** and **12-4**.

> **Berm** — Temporary or permanent barrier intended to control the flow of water. *Similar to* Dike *and* Dam.

An example of a simple wet decon corridor may include (**Figure 12.29**):

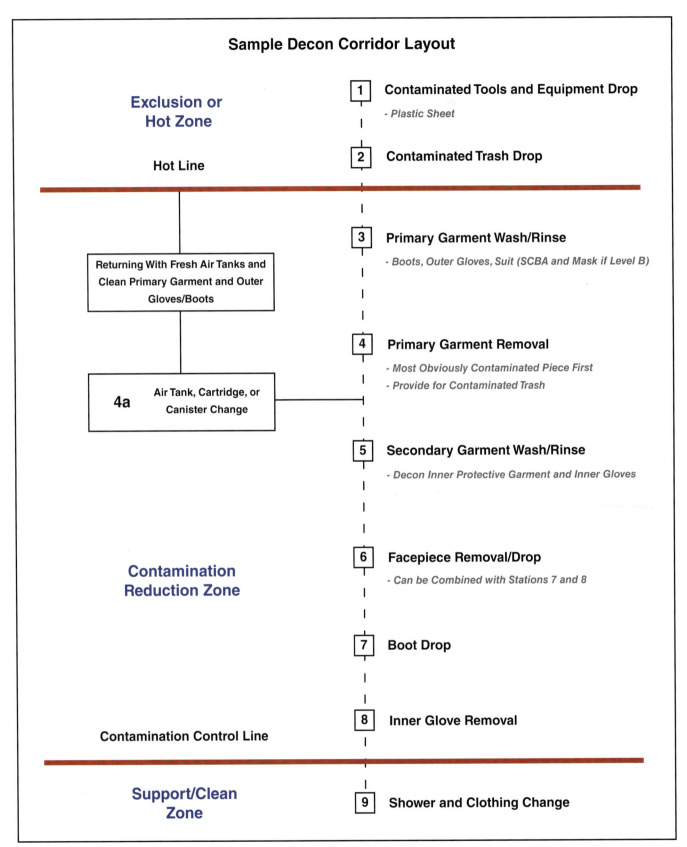

Figure 12.29 A decon corridor may be expanded or contracted based on the needs of the incident. *Original source courtesy of the U.S. Agency for Toxic Substances and Disease Registry (ATSDR).*

- **Gross tool drop** — Provides containment for the temporary storage of contaminated tools and equipment
- **Gross rinse** — Provides a liquid retention pool along with a showering nozzle or wand
- **Wash station** — Includes liquid retention pools, buckets with approved cleaning solutions, brushes for washing and scrubbing, and nozzles or wands for rinsing. Establish multiple wash stations as needed.
- **Undressing area** — Includes chairs and containers for removed PPE and clothing
- **Personal decon shower** — Based on the chemical hazards, it may be at the site or offsite
- **Medical monitoring** — Area where staff can monitor vitals and conditions of all persons exiting the decon corridor

Segregation of the Decontamination Line

Technicians should separate the physical area that will contain the decontamination line from the surrounding warm zone. The decon line should be visible and should stand out. Identify the decon line by cones, barrier tape, barricades, or brightly colored tarps. The visibility of the decon line ensures that responders in need of decontamination do not waste time looking for the decon line if their air supply is diminishing.

Containment

If washing or dilution is the primary decontamination method, make every effort to contain the contaminated wastewater so that it can be disposed of properly. Each wash/rinse station should include some kind of containment device. Options include pools with an elevated platform to stand on, or plastic sheeting with a berm.

To reduce the chances of cross contamination, contain loose items such as tools and clothing articles. Use separate containers, such as buckets, drums, and tarps, for items designated for disposal or reuse.

Personnel

Decontamination can be a personnel-intensive operation. Adequate personnel must be on-scene to perform the work needed. One staff member should be assigned to each station in the decon line. If staffing allows, each wash station should include a minimum of two personnel.

A supervisor should be assigned to oversee operations at the decontamination line. The supervisor should not be a working part of the line and should only oversee the operation. Having this supervisor will also allow for a safety margin if one of the members working the decon line should suffer a crisis or suit breach.

Equipment

The hazmat team must properly equip all entry-level personnel, as well as adequately equip the decon line. CPC in the proper size and style must be available for all decon line members. An active decon line may have as many

Figure 12.30 Showers should include water flow from multiple directions, if possible.

Figure 12.31 Long-handled brushes may be used to scrub CPC. *Courtesy of Joan Hepler.*

Figure 12.32 Patient transportation devices, such as this wheeled Stokes basket, are easily decontaminated for reuse at an incident.

as six to eight people, and the response agency must offer a relatively large number of suitable garments.

Technicians must also consider respiratory equipment for decon line members. For fire-based response organizations, availability of SCBA is usually not an issue. However, some agencies may not have access to sufficient sets of breathing apparatus.

Over and above CPC, much of the equipment used in the decontamination line may be available at hardware, department, or building supply stores. Hazmat teams often manufacture their own equipment and devices from readily available plumbing supplies. Essential equipment for a decon line includes the following:

- **Tarps, plastic sheeting, or containers** — Use these tools for creation of the equipment drop.

- **Base cover** — Comprised of tarps or plastic sheeting, this cover should be large enough to contain the entire decontamination effort.

- **Containment pools** — Portable pools, such as a children's wading pool, may be used to contain water runoff from washing and rinsing operations. Portable pools tend to be sturdy and will hold a large amount of liquid. Equip any washing or rinsing station with a pool.

- **Isolation steps** — Responders stand on steps provided in the containment pool with their feet out of water runoff.

- **Showers** — Commercially available or homemade showers should supply an adequate flow of water **(Figure 12.30)**. If possible, lateral flows should work in conjunction with vertical flows.

- **Brushes** — Soft bristled, synthetic car wash brushes can serve the purpose for scrubbing CPC. A long-handled brush is recommended **(Figure 12.31)**.

- **Solution containers** — Technicians can use commercially available buckets to hold a variety of decontamination solutions. Pump-type sprayers may also be used to deliver the solution.

- **Containers for discarded items** — Use trash cans, half barrels, or heavy bag-lined cardboard boxes to hold discarded or removed items during the decon process.

- **Rescue devices** — Patient-carrying devices used to carry nonambulatory victims may include wire-type Stokes baskets, or SKED® devices **(Figure 12.32)**. These work well for decontamination of victims and can be easily cleaned and reused. Other equipment that technicians can adapt for this purpose includes deer carriers intended for transporting hunted game across wildland environments.

- **Water transport** — Most decon operations can be supported by a standard 500 gallon (2 000 L) capacity tank as found in most fire apparatus. Hoselines for water transport should be of smaller diameter such as a garden hose and nozzle. Use manifold systems to split larger diameter hoselines into small garden hose-sized lines.

Procedures

Specific procedures should guide response personnel involved in decontamination. Technicians must conduct decon in an orderly manner and procedures should address normal operations as well as any emergency situations that may arise. The AHJ may establish procedures with regard to prioritizing entry into the decon corridor. Some of these criteria may include:

- Air levels
- Level of contamination (cleanest to dirtiest)
- Medical emergencies
- Compromise of PPE integrity

Decontamination procedures should contain provisions for medical evaluation along with proper hydration. In many instances, the performance of decontamination procedures proves to be more strenuous than the activities of entry personnel. Decon personnel are subject to the same stresses as any other response personnel wearing CPC. Because of this, decon personnel should be treated the same regarding work time and rehabilitation procedures for pre- and post-use of CPC.

Other Implementation Considerations

Performing decon operations on law enforcement and military personnel offers a unique challenge. These personnel often carry weapons and will not release them to civilian personnel during decon operations. Therefore, hazmat teams may need to include a hazmat-trained law enforcement officer involved in the decon operations with the sole responsibility of decontaminating weapons and ensuring their security as the operation progresses **(Figure 12.33)**. Give special consideration to the decontamination of weapons, ammunition, and other equipment that could be damaged by exposure to liquid decon solutions or water. Decon plans must take this equipment into consideration in accordance with local policies and procedures.

Figure 12.33 Law enforcement personnel who enter a decontamination corridor may need special security on their weapons and equipment.

In some instances, it may be necessary to establish a separate decontamination corridor for armed emergency services personnel leaving the hot zone. Armed responders should place their weapons in a hazmat recovery bin under the supervision of a suitably protected law enforcement officer as officers disarm and go through decon.

Before taking canines into the hot zone of hazmat/WMD incidents, emergency response personnel should take precautions (such as putting protective booties on dogs' feet) and should have their own procedures to decontaminate their animals. However, these animals will be processed through the decontamination corridor and personnel may have to assist **(Figure 12.34)**.

Figure 12.34 Animals may need to be decontaminated. *Courtesy of FEMA News Photos, photo by Jocelyn Augustino.*

Criminal suspects may also need to be decontaminated. Law enforcement must supervise the suspect throughout this process. If conducting technical decon, the AHJ or law enforcement may determine whether a suspect will go through the same decon steps established for responders and other victims, including whether handcuffs must be removed and decontaminated. Follow AHJ procedures for decontaminating criminal suspects.

Cold Weather Decontamination

Conducting wet decon operations in cold weather can be difficult to execute safely. Even if showers utilize warm water, runoff water can quickly turn to ice, creating a serious hazard for both victims and responders. If warm water is not available, susceptible individuals (elderly, very young, individuals with chemical injuries or pre-existing health conditions such as diabetes) can suffer cold shock or hypothermia.

If water is involved, hypothermia can occur at temperatures higher than expected in dry conditions. Give consideration to protecting victims from the cold. Answering the following questions will provide information on how best to protect victims:

- Are wet methods necessary, or can disrobing and dry methods accomplish effective decon?
- Is wind chill a factor?
- Is shelter available for victims during and after decon **(Figure 12.35)**?

- Is it possible to conduct decon indoors (e.g., sprinkler systems, indoor swimming pools, locker room showers)?
- If decon will be conducted indoors (at preplanned facilities, for example), how will victims be transported?
- If decon must be conducted outside in freezing temperatures, how will icy conditions be managed (e.g., sand, sawdust, kitty litter, salt)?

Figure 12.35 Private or sheltered areas may provide more security for decontamination procedures or other incident functions.

> **WARNING!**
> Individuals who have been exposed to chemical agents should undergo emergency decon immediately, regardless of ambient temperatures.

Individuals exposed to chemical agents should disrobe and thoroughly shower (**Figure 12.36**). Responders should provide dry clothing and warm shelter as soon as possible after the individuals shower.

Figure 12.36 Decontamination corridors should use warm water and privacy enclosures when possible.

Medical Monitoring

Technicians must give medical consideration to any personnel who have been working in the hot zone and those who have been decontaminated. Evaluate these individuals for both heat stress and potential exposure.

A physical examination of each person should include the assessment of vital signs at several intervals after the person has been through decon. Compare these vital signs to the vitals taken during the reentry examination. Thoroughly document all findings. The examination may include:

- Body temperature
- Pulse
- Respiratory rate
- Blood pressure
- Auscultation of lung sounds
- Evaluation of body weight
- Mental status
- Other tests or evaluations as required by the AHJ

Evidence Collection and Decontamination

Technicians will perform collection, preservation, and sampling of evidence under the direction of law enforcement per established procedures. Decontamination issues associated with these activities will also be determined in conjunction with law enforcement.

Technicians must appropriately package evidence collected on the scene by responders (for example, in approved bags or other evidence containers). Only the exterior of the packaging will be decontaminated as it passes from the hot zone to the cold zone **(Figure 12.37)**. When evidence passes through the decon corridor, responders must document chain of custody in writing.

Termination

After concluding decon activities, the hazmat team must hold a debriefing as soon as possible for those involved in the incident. In some cases, return of personal items may be a law enforcement function because of evidentiary issues. There may be circumstances in which responders immediately return personal effects to the persons undergoing decon. Provide exposed persons with as much information as possible about the delayed health effects of the hazardous materials involved.

Emergency response plans and/or SOPs may require additional reports and supporting technical documentation such as incident reports, after-action reports, and regulatory citations. Personnel may also need to complete and file exposure records. All responders who have been exposed or potentially exposed to hazardous materials are required to fill out these records according to agency SOPs.

Figure 12.37 Decontaminate only the exterior of packaged evidence.

Information recorded on the exposure report might include:

- Activities performed
- Product involved
- Reason for being there
- Equipment failures or malfunction of PPE
- Hazards associated with the product
- Symptoms experienced
- Monitoring levels in use
- Circumstances of exposure
- Exposure suspected/known
- Decon type undergone

Schedule examinations with medical personnel when an exposure has occurred. The individual, the individual's personal physician, and the individual's employer need to keep copies of these exposure records for future reference.

NOTE: Further information about termination activities can be found in Chapter 14.

Special Considerations

Throughout the course of a hazmat incident, the hazmat technician may encounter some uncommon hazards. Personnel must take special care to properly decontaminate personnel and equipment if confronted with any of these special hazards. For more information on selecting a decontamination method appropriate to a hazmat/WMD incident, see **Skill Sheet 12-5**.

Radiation

If called upon to respond to an incident involving radiation, the hazmat technician must take special precautions during mitigation and decontamination. Avoiding any potential radiation exposure is always the best policy, but if that is not possible, then practicing time, distance, and shielding is the next best tactic. Utilize the assistance of technical experts, hospitals, and universities. The U.S. Department of Energy may also offer assistance. They can be contacted through CHEMTREC.

Decontamination for radiation exposure should be a standalone decon line separate from other decon lines. Equip this line with its own containment devices for both liquids and solids. Monitor victims and responders with a radiation detector. Isolate clothes and equipment, and inform individuals that they should wash and be rinsed thoroughly in the decontamination line. A personal shower should follow soon after. In these instances, the exposed individual must wash methodically. Finally, monitor all persons again in areas where direct contact may have occurred such as the feet and hands and other sensitive areas such as the mouth, nose, and ears.

In the event that a victim cannot be fully decontaminated and still shows signs of exposure, the medical team and first responders may still need to protect against potential exposure. Personnel should provide some kind of shielding to protect against contaminating equipment and first responders. One simple way to provide this shielding is to lay double layers of blankets above the carrier, and then another double layer of blankets over the victim.

Pesticides

Pesticide manufacturers may be able to offer decontamination recommendations if contacted directly. CHEMTREC can provide valuable information for pesticide decontamination.

> **National Pesticide Information Center**
> The National Pesticide Information Center (NPIC) is another resource that can provide information on pesticide documentation. The NPIC phone number is 1-800-858-7378. The NPIC web page is: http://npic.orst.edu/.

Infectious Agents

Infectious agents may offer special challenges for emergency responders. In many cases, these agents may only be destroyed by sterilization. If responders face this type of decontamination process, most state health departments may be able to offer assistance. In addition, laboratories that may offer significant technical assistance include the Centers for Disease Control and Prevention, local hospitals, and university laboratories.

Decontamination during a Terrorist Event

The past decades have seen the threat of terrorism heighten worldwide. From the sarin attacks in Japan to the anthrax incidents in the U.S. that occurred after 9/11, responders must respond seriously to incidents that could potentially include terror attack elements.

Product or chemical identification is paramount in these situations, but quite often emergency responders do not have adequate field detection devices available at the onset of an incident. In these cases, outside help from technical specialists will be necessary. Technicians must know what agencies can assist in what ways before the incident takes place.

Another issue may be the sheer number of victims at this type of event. In these cases, technical decontamination may be impractical. If there are a large number of victims, mass emergency decon may be the most practical alternative **(Figure 12.38)**. Given that a large amount of contamination may be lost once clothing is removed, removal of clothing should be the first priority. Fog streams from an aerial device or multiple pumping apparatus set up as a corridor may be the safest and quickest alternative for decontaminating a large number of ambulatory victims.

Figure 12.38 Technical decon is impractical at incidents where a large number of people must be decontaminated. *Courtesy of David Lewis.*

Chapter Review

1. When should the decontamination processes be available and ready for implementation during a hazmat incident?

2. What are the four types of decontamination methods and in what situations might they be used?

3. What methods are used for determining the effectiveness of decontamination?

4. What factors must be considered when choosing a decontamination site?

5. Which types of hazards require special considerations for decontamination operations?

12-1
Perform mass decontamination on ambulatory people. [NFPA 1072, 7.4.4.1]

WARNING: If this skill involves the use of actual hazardous material samples, hazardous materials can cause serious injury or fatality. Appropriate personal protective equipment (PPE) must be worn and safety precautions must be followed. The following skill sheet demonstrates general steps; specific hazmat incidents may differ in procedure. Always follow the AHJ's procedures for specific incidents.

Step 1: Ensure proper decontamination method has been chosen to minimize hazards.

Step 2: Ensure that all responders are wearing appropriate PPE for performing mass decontamination operations.

Step 3: Ensure decontamination operations are set up in a safe area.

Step 4: Ensure the decontamination corridor provides privacy for victims.

Step 5: Prepare fire apparatus for use during mass decontamination.

Step 6: Set fire nozzle to fog pattern.

Step 7: Instruct all people to go through mass decontamination.

Step 8: Instruct each person to remove contaminated clothing, ensuring that victims do not come into further contact with any contaminants.

Step 9: Instruct each person to keep arms raised as they proceed slowly through the wash area.

Chapter 12 • Implementing the Action Plan: Decontamination   **469**

SKILL SHEETS

12-1 cont.
Perform mass decontamination on ambulatory people. [NFPA 1072, 7.4.4.1]

Step 10: Monitor for additional contamination using the appropriate detection device.

NOTE: If contamination is found, instruct each person to go through wash again, as appropriate.

Step 11: Instruct each person to move to a clean area to dry off.
Step 12: Send each person for medical treatment.

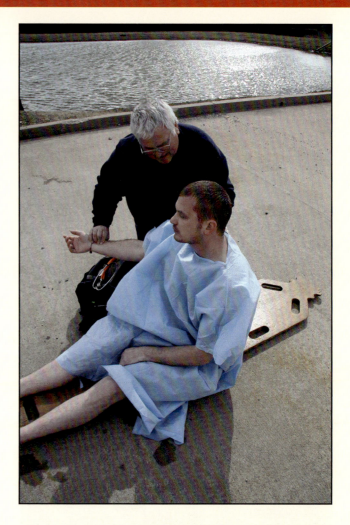

Step 13: Inform EMS personnel of contaminant involved and its hazards, if known.
Step 14: Ensure personnel, tools, and equipment are decontaminated.
Step 15: Terminate decontamination operations according to AHJ's policies and procedures.
Step 16: Complete required reports and supporting documentation.

12-2
Perform mass decontamination on nonambulatory victims. [NFPA 1072, 7.4.4.1]

WARNING: If this skill involves the use of actual hazardous material samples, hazardous materials can cause serious injury or fatality. Appropriate personal protective equipment (PPE) must be worn and safety precautions must be followed. The following skill sheet demonstrates general steps; specific hazmat incidents may differ in procedure. Always follow the AHJ's procedures for specific incidents.

Step 1: Ensure proper decontamination method has been chosen to minimize hazards.

Step 2: Ensure that all responders are wearing appropriate PPE for performing mass decontamination operations.

Step 3: Establish mass decontamination corridor for nonambulatory decontamination in a safe location and according to the AHJ's SOPs.

Step 4: Ensure the decontamination corridor provides privacy for victims.

Step 5: Establish an initial triage point to evaluate and direct victims.

Step 6: Perform lifesaving intervention if necessary.

Step 7: Transfer the victims to the nonambulatory wash area of the decontamination station on an appropriate backboard/litter device.

Step 8: Remove all clothing, jewelry, and personal belongings, and place in appropriate containers. Decontaminate as required, and safeguard personal belongings and items. Use plastic bags with labels for identification.

Step 9: Carefully undress nonambulatory victims, and avoid spreading the contamination when undressing. Do not touch the outside of the clothing to the skin.

NOTE: If biological agents are suspected, a fine water mist can be applied to trap the agent in the clothing and prevent the spread of contamination.

Step 10: Completely wash each victim's entire body using handheld hoses, sponges, and/or brushes and then rinse following AHJ's SOPs for safety.

NOTE: Clean the victim's genital area, armpits, folds in the skin, and nails with special attention. If conscious, instruct the victim to close his/her mouth and eyes during wash and rinse procedures.

Step 11: Transfer the victims from the wash and rinse stations to a drying station after completing the decontamination process. Ensure that the victims are completely dry.

Step 12: Monitor for additional contamination using the appropriate detection device.

NOTE: If contamination is detected, repeat decontamination wash and/or change decontamination method as appropriate.

Step 13: Have on-scene medical personnel reevaluate the victims' injuries.

Step 14: Ensure personnel, tools, and equipment are decontaminated.

Step 15: Terminate decontamination operations according to AHJ's policies and procedures.

Step 16: Complete required reports and supporting documentation.

SKILL SHEETS

12-3

Perform technical decontamination on ambulatory people. [NFPA 1072, 7.4.4.2]

WARNING: If this skill involves the use of actual hazardous material samples, hazardous materials can cause serious injury or fatality. Appropriate personal protective equipment (PPE) must be worn and safety precautions must be followed. The following skill sheet demonstrates general steps; specific hazmat incidents may differ in procedure. Always follow the AHJ's procedures for specific incidents.

Step 1: Ensure proper decontamination method has been chosen to minimize hazards.

Step 2: Ensure that all responders are wearing appropriate PPE for performing technical decontamination operations.

Step 3: Establish technical decontamination corridor for ambulatory decontamination according to the AHJ's SOPs.

Step 8: Instruct the person to undergo gross decontamination.

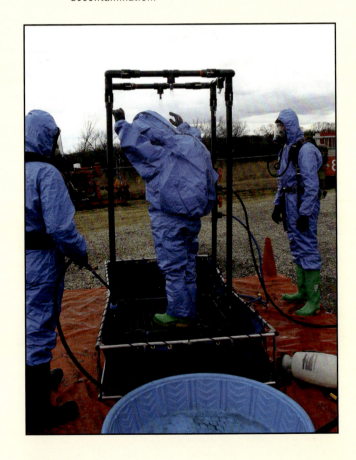

Step 4: Ensure the decontamination corridor provides privacy for victims.

Step 5: Establish an initial triage point to evacuate and direct persons.

Step 6: Perform lifesaving intervention if needed.

Step 7: If not an emergency provider, instruct person to remove potentially contaminated clothing and jewelry, ensuring he/she does not come in further contact with contaminants. Emergency responders may drop tools or equipment.

472   Chapter 12 • Implementing the Action Plan: Decontamination

12-3 cont.
Perform technical decontamination on ambulatory people. [NFPA 1072, 7.4.4.2]

Step 9: Instruct the person to undergo secondary decontamination wash.

NOTE: Emergency responders undergoing technical decon will doff PPE per SOPs after the secondary decontamination wash.

Step 10: Instruct the person to enter the privacy station, remove undergarments, and shower and wash thoroughly from the top down.

NOTE: Do NOT ask members of the public to remove their clothes to shower unless complete privacy is provided.

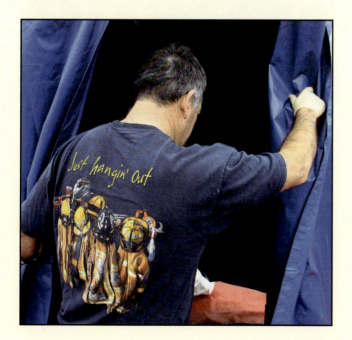

Step 11: Provide a clean garment for the person to wear after showering.

Step 12: Monitor for additional contamination using the appropriate detection device.

NOTE: If contamination is detected, repeat the decontamination sequence and/or change the decontamination method, as appropriate.

Step 13: Direct the person to the medical evaluation station.

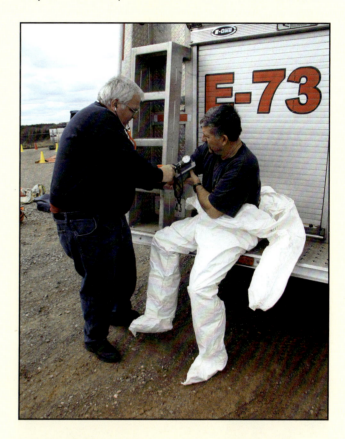

Step 14: Ensure personnel, tools, and equipment are decontaminated.

Step 15: Terminate decontamination operations according to AHJ's policies and procedures.

Step 16: Complete required reports and supporting documentation.

SKILL SHEETS

12-4

Perform technical decontamination on nonambulatory victims. [NFPA 1072, 7.4.4.2]

WARNING: If this skill involves the use of actual hazardous material samples, hazardous materials can cause serious injury or fatality. Appropriate personal protective equipment (PPE) must be worn and safety precautions must be followed. The following skill sheet demonstrates general steps; specific hazmat incidents may differ in procedure. Always follow the AHJ's procedures for specific incidents.

Step 1: Ensure proper decontamination method has been chosen to minimize hazards.

Step 2: Ensure that all responders are wearing appropriate PPE for performing technical decontamination operations.

Step 3: Establish technical decontamination corridor for nonambulatory decontamination according to the AHJ's SOPs.

Step 4: Ensure the decontamination corridor provides privacy for the victims.

Step 5: Establish an initial triage point to evaluate and direct victims.

Step 6: Perform lifesaving intervention if needed.

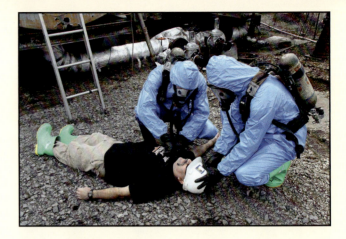

Step 7: Transfer the victim to the nonambulatory wash area of the decontamination station on an appropriate backboard/litter device.

Step 8: Remove all clothing, jewelry, and personal belongings, and place in appropriate containers. Decontaminate items as required, and safeguard. Use plastic bags with labels for identification.

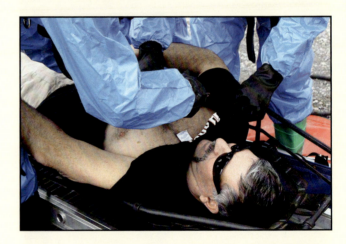

Step 9: Carefully undress nonambulatory victims, and avoid spreading the contamination when undressing. Do not touch the outside of the clothing to the skin.

NOTE: If biological agents are suspected, a fine water mist can be applied to trap the agent in the clothing and prevent the spread of contamination.

Step 10: Completely wash the victim's entire body using handheld hoses, sponges, and/or brushes, and then rinse.

NOTE: Clean the victim's genital area, armpits, folds in the skin, and nails with special attention. If conscious, instruct the victim to close his/her mouth and eyes during wash and rinse procedures.

Step 11: Transfer the victim from the wash and rinse stations to a drying station after completing the decontamination process. Ensure that the victim is completely dry.

12-4 cont.
Perform technical decontamination on nonambulatory victim. [NFPA 1072, 7.4.4.2]

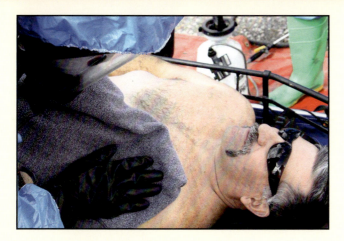

Step 12: Monitor for additional contamination using the appropriate detection device.

NOTE: If contamination is detected, repeat decontamination wash and/or change decontamination method, as appropriate.

Step 13: Have on-scene medical personnel reevaluate the vicitm's injuries.

Step 14: Ensure personnel, tools, and equipment are decontaminated.

Step 15: Terminate decontamination operations according to AHJ's policies and procedures.

Step 16: Complete required reports and supporting documentation.

12-5
Select a decontamination method appropriate to a hazmat/WMD incident.
[NFPA 1072 7.3.3]

WARNING: If this skill involves the use of actual hazardous material samples, hazardous materials can cause serious injury or fatality. Appropriate personal protective equipment (PPE) must be worn and safety precautions must be followed. The following skill sheet demonstrates general steps; specific hazmat incidents may differ in procedure. Always follow the AHJ's procedures for specific incidents.

Step 1: Identify decontamination method to minimize hazards for each response option (operations and methods).

Step 2: Identify equipment required to implement decontamination methods (operations and methods).

Implementing the Action Plan: Product Control

Chapter Contents

Introduction to Product Control **481**
- Nonintervention 483
- Defensive Operations 483
- Offensive Operations 485

Damage Assessment and Predicting Behavior **486**
- Early Size-Up 487
- Container Information 487
- Potential Container Stress 488
- Types of Container Damage 488
- Predicting Likely Behavior 492

Product Containment: Plugging and Patching **494**
- Plugging 495
- Patching 496
- Specialized Plugging Equipment 499

Cargo Tanks **501**
- Methodology of Leak Control 502
- Methods and Precautions for Fire Control ... 504
- Product Removal and Transfer Considerations ... 505

Pressurized Containers **507**
- Understanding Fittings on Pressure Containers ... 508
- Working with Fittings on Pressure Containers ... 509

Drums **513**
- Bung Leaks 513
- Chime Leaks 513
- Sidewall Punctures 513

Overpacking **514**
- Tasks 515
- Lab Packs 516

Other Basic Product Control Techniques .. **517**
- Vapor Suppression and Dispersion 517
- Blanketing/Covering 517

Specialized Product Control Techniques .. **518**
- Hot and Cold Tapping 518
- Product Transfer 519
- Flaring 519
- Venting 520
- Applying Large Plugging and Patching Devices ... 520
- Vent and Burn 520

Chapter Review **521**
Skill Sheets **522**

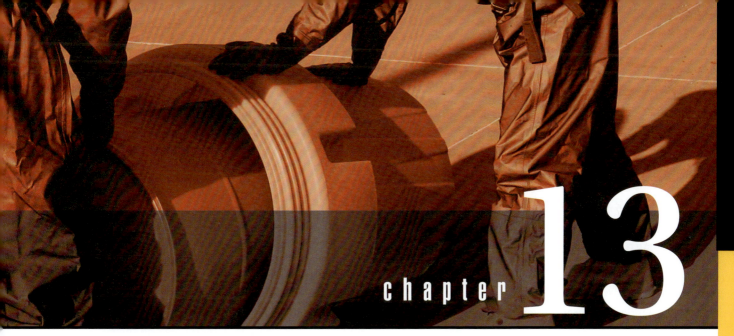

chapter 13

Key Terms

Autorefrigeration	508
Base Metal	490
Bonding	507
Bung	513
Chime	513
Cold Tapping	519
Ferrous Metal	487
Flaring	519
Fusible Plug	509
Grounding	507
Heat Sink	504
Hot Tapping	518
Jubilee Pipe Patch	498
Mild Steel	487
National Pipe Thread Taper (NPT)	506
Soft Patch	498
Static Electricity	506
Tapping	518
Vapor Dispersion	517
Vapor Suppression	517

Implementing the Action Plan: Product Control

JPRs Addressed In This Chapter

7.2.4, 7.2.5, 7.4.3, 7.4.3.1, 7.4.3.2, 7.4.3.3 , 7.4.3.4

Learning Objectives

After reading this chapter, students will be able to:

1. Describe the three basic types of product control operations. [NFPA 1072, 7.4.3.1, 7.4.3.2, 7.4.3.4]
2. Explain processes for assessing damage and predicting behavior of hazardous materials and their containers. [NFPA 1072, 7.2.4, 7.2.5, 7.4.3.1, 7.4.3.2]
3. Describe plugging and patching operations. [NFPA 1072, 7.4.3.2]
4. Describe product control techniques for cargo tanks. [NFPA 1072, 7.4.3.1, 7.4.3.2, 7.4.3.4]
5. Describe product control techniques for pressurized containers. [NFPA 1072, 7.4.3.1, 7.4.3.2, 7.4.3.4]
6. Describe product control techniques for drums. [NFPA 1072, 7.4.3.1, 7.4.3.2]
7. Describe overpacking techniques. [NFPA 1072, 7.4.3.3]
8. Describe other basic product control techniques. [NFPA 1072, 7.4.3.1, 7.4.3.2, 7.4.3.4]
9. Describe specialized product control techniques. [NFPA 1072, 7.4.3.1, 7.4.3.2, 7.4.3.4]
10. Skill Sheet 13-1: Perform absorption/adsorption. [NFPA 1072, 7.4.3.2]
11. Skill Sheet 13-2: Perform damming. [NFPA 1072, 7.4.3.1]
12. Skill Sheet 13-3: Perform diking operations. [NFPA 1072, 7.4.3.1]
13. Skill Sheet 13-4: Perform diversion. [NFPA 1072, 7.4.3.1]
14. Skill Sheet 13-5: Perform retention. [NFPA 1072, 7.4.3.1]
15. Skill Sheet 13-6: Perform vapor suppression. [NFPA 1072, 7.4.3.1]
16. Skill Sheet 13-7: Perform vapor dispersion. [NFPA 1072 7.4.3.1]
17. Skill Sheet 13-8: Perform dilution. [NFPA 1072, 7.4.3.1]
18. Skill Sheet 13-9: Predict the likely behavior of known hazardous materials at an incident. [NFPA 1072, 7.2.4]
19. Skill Sheet 13-10: Predict the likely effect of a hazardous materials incident within the endangered area. [NFPA 1072, 7.2.5]
20. Skill Sheet 13-11: Select appropriate product control techniques to mitigate a hazmat/WMD incident. [NFPA 1072, 7.4.3.1]
21. Skill Sheet 13-12: Plug a leaking container. [NFPA 1072, 7.4.3.2]
22. Skill Sheet 13-13: Patch a leaking container. [NFPA 1072, 7.4.3.2]
23. Skill Sheet 13-14: Apply a sleeve, jacket, clamp, or wrap to a leak. [NFPA 1072, 7.4.3.2]
24. Skill Sheet 13-15: Conduct liquid transfer operations involving a leaking nonpressure container. [NFPA 1072, 7.4.3.4]
25. Skill Sheet 13-16: Bond and ground a container. [NFPA 1072, 7.4.3.4]
26. Skill Sheet 13-17: Perform remote valve shutoff or activate emergency shutoff device. [NFPA 1072, 7.4.3.2]
27. Skill Sheet 13-18: Tighten or close leaking valves, closures, packing glands, and/or fittings. [NFPA 1072, 7.4.3.2]
28. Skill Sheet 13-19: Cap a leak. [NFPA 1072, 7.4.3.2]
29. Skill Sheet 13-20: Overpack a nonbulk container and/or radioactive materials package. [NFPA 1072, 7.4.3.3]

Chapter 13
Implementing the Action Plan: Product Control

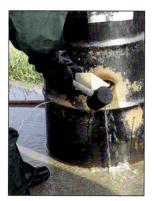

Regardless of the location of a hazardous materials incident, product control decisions start with an understanding of both the product and type of container involved. Based on this information, the Incident Commander must decide the type of operation to use while weighing the impact of each operation with response personnel, the public, and the environment.

This chapter provides information about the following topics:

- Introduction to Product Control
- Damage Assessment and Predicting Behavior
- Product Containment: Plugging and Patching
- Cargo Tanks
- Pressurized Containers
- Drums
- Overpacking
- Other Basic Product Control Techniques
- Specialized Product Control Techniques

Introduction to Product Control

Effectively controlling any released product(s) is essential for mitigation of hazmat incidents; therefore, technicians must use a risk-based response. Proper identification of the chemical will allow emergency response crews to make educated decisions on PPE selection and response tactics, but the use of monitoring equipment will help confirm the product's properties. If the product has not been identified, monitoring and direct reading instruments will give hazmat technicians quality, validated information for initial decisions and as the incident progresses.

CAUTION
Technicians must perform all procedures at hazardous materials incidents with tools, equipment, and PPE that are appropriate for the task and the products involved.

While an Incident Commander may opt to use nonintervention tactics at hazmat incidents, product control tactics are often defensive or offensive. While defensive operations typically do not involve direct contact with the product or container, offensive operations may. In the defensive stance, typically no work is performed on the container and the product is isolated while it is outside the container (commonly referred to as *spill control* or *confinement*) **(Figure 13.1)**. In the offensive stance, the hazmat technician will work to contain the product to its container (commonly referred to as *leak control* or *containment*) **(Figure 13.2)**. Regardless of the tactic, the safety of the responders is paramount. Hazmat technicians must use, inspect, and maintain

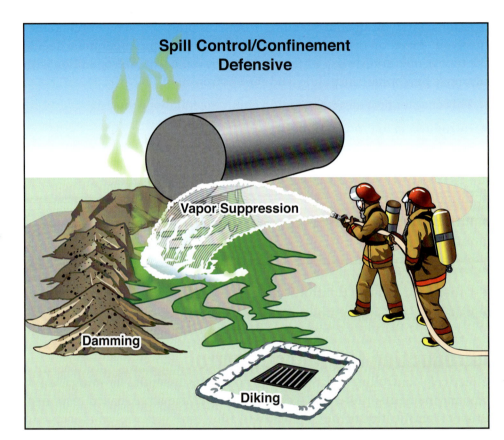

Figure 13.1 Defensive maneuvers isolate a spill outside of a container.

Figure 13.2 Offensive maneuvers contain a spill within its container.

product control tools and equipment per the AHJ's SOPs and manufacturer's recommendations and specifications.

Depending on the circumstances and available personnel, confinement and containment operations may occur simultaneously. This combination should not be confused with an offensive operation working simultaneously with a defensive operation. Hazmat technicians may need to contain the leak offensively, while confinement operations are occurring to limit the overall footprint of a hazardous materials incident.

Nonintervention

Using a risk-based response philosophy will help the IC determine if non-intervention is the correct option. Based on the conditions on scene, it may not be possible to approach a container to stop a leak or get close enough to a product to attempt containment. Safety must always be the first priority in any response operation, including:

- Responder safety and effectiveness
- Public safety
- Position and vulnerability of communities upwind and downwind
- Environmental health
- Long-term and short-term effects of material release

Defensive Operations

Defensive operations mean that emergency response personnel will work to confine the product's release and limit the physical size of the release area. Environmental protection is a special consideration because the hazardous product can be released directly or indirectly into the air, surface water, groundwater, or onto soil. Depending on the area affected, various confinement methods may be available to help restrict the spread of the hazardous product. Steps for adsorption/absorption, damming, diking, diversion, retention, vapor suppression, vapor dispersion, and dilution are provided in **Skill Sheets 13-1 through 13-8**, respectively.

NOTE: Some tactics may be considered offensive rather than defensive since they are typically performed by hazmat technicians in the hot zone (for example, neutralization and application of gels). See IFSTA's **Hazardous Materials for First Responders**, for more information on defensive product control tactics.

Figure 13.3 Vapor dispersion and cooling can safely mitigate an incident. *Courtesy of Barry Lindley.*

Defensive operation efforts may include any means available to limit the size of the release area. Based upon the type of release, some confinement techniques may include the following:

- Gas or vapor leaks
  — Ventilation
  — Dispersion **(Figure 13.3)**
  — Dissolution
  — Blanketing

Figure 13.4 Neutralization is one technique that can be used to mitigate liquid spills on a solid surface.

Figure 13.5 Floating booms can contain materials that float on water. *Courtesy of U.S. E.P.A.*

- Liquid leaks on a surface
  — Diking
  — Diverting
  — Absorbing
  — Applying gels
  — Neutralization **(Figure 13.4)**
  — Solidification
  — Emulsification
  — Retention
  — Adsorption
  — Dilution
- Liquid leaks in water
  — Booming **(Figure 13.5)**
  — Diverting
  — Damming
  — Adsorption
  — Dispersion
- Solid spills on a surface **(Figure 13.6)**
  — Blanketing/Covering

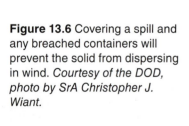

Figure 13.6 Covering a spill and any breached containers will prevent the solid from dispersing in wind. *Courtesy of the DOD, photo by SrA Christopher J. Wiant.*

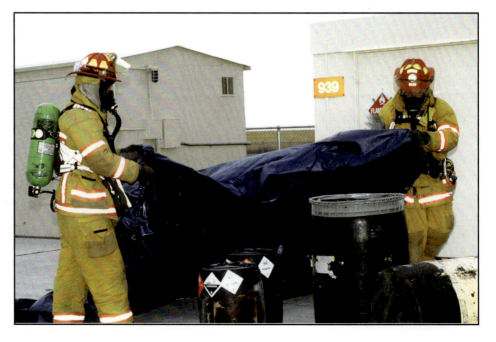

484 Chapter 13 • Implementing the Action Plan: Product Control

Hazardous Liquid Release

In an area of highway construction, an MC 312 corrosive liquid tank was involved in a rollover accident after hitting a concrete barrier. Product was leaking from the cargo tank, and a vapor cloud was blowing away from the highway into an unoccupied field. Liquid was draining towards a drainage ditch beside the highway. The driver was able to escape the cab of the vehicle without assistance, and no other vehicles were involved in the rollover. Shipping papers that the driver provided to emergency responders confirmed that the liquid was hydrochloric acid.

While highway patrol personnel managed scene control, the fire department response included a hazardous materials response team. The hazmat personnel identified two product control objectives to stop the growth of the hazardous materials incident: prevent released liquid from traveling down the drainage system into a nearby waterway, and stop the leak itself, which appeared to be from a damaged rupture disc and air pad valve.

Using earthmoving equipment from the construction site, a series of dikes were built across the drainage ditch to contain released product. After discussion and information sharing with the shipping company, hazmat crews dressed in Level A PPE, approached the vehicle, and drove wooden blocks into the holes to plug the leaks. The shipping company hired a contractor to perform additional cleanup at the incident, and the contractor's crew covered pools of liquid with soda ash. The contractor then pumped the ash/liquid mixture into recovery vessels. Finally, the contractor off-loaded the remaining product from the damaged vehicle into another MC 312.

Offensive Operations

Offensive operations are usually hands-on operations that place an emergency responder in close proximity to the released product. Personnel need to be properly trained to perform offensive operations and must take extreme care in all these activities. When determining if an offensive operation is practical, technicians must use a risk-based response philosophy that includes definitive product identification, research and response information, and response evidence such as meter readings. The first question that technicians must ask is, "Will an offensive tactic make a difference?" The safety of the emergency responder must remain the top priority, and an offensive tactic that will not have a positive outcome for either the public or the environment must not happen.

If the hazmat team decides that the emergency responders have the training and resources necessary to undertake an offensive operation, technicians must continuously answer the following questions:

- What is the material? What do we know about it?
- Are the correct resources and equipment available, based on the chemical and physical properties, to help mitigate the situation?
- Why is the material leaking?
- What is the risk versus benefit to the entry team for an offensive operation (**Figure 13.7**)?

Figure 13.7 Use risk-based response to assess the environment and the risks that the entry team will face.

- What hazards can be realized if offensive actions are not taken?
- What stresses caused the breach in the container and is the stress likely to continue? Can we change or remove the stressors?
- Is the material stable and can it be continuously monitored?
- Will the offensive operations have a positive outcome?

Though some departments may train technicians with the skills needed for more complicated responses, they are advised to keep mitigation and control tactics as simple as possible. The rest of this chapter includes references to types of mitigation strategies that are effective on particular types of containers, particularly as relevant to specific types of breaches in those containers.

Damage Assessment and Predicting Behavior

For a product to harm a person or the environment, it must leave its container. Hazmat technicians must be prepared to act when a container's construction and safety features have been compromised. When the construction specifications of a container are altered, the inherent safety features may be greatly reduced or totally eliminated, in some cases even releasing the product. Many types of containers were discussed briefly in Chapter 8.

The goal of a damage assessment is to determine the status of three primary concerns (**Figure 13.8**):

- The container construction material
- The type of stress to which the container has been subjected
- The internal pressure of the container

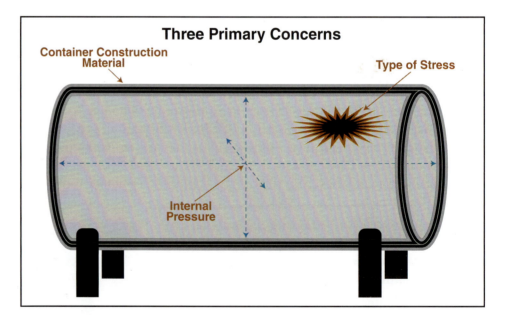

Figure 13.8 Three primary concerns during assessment are the container's construction material, stresses, and pressure.

Hazmat technicians will need additional training to perform damage assessment, and in many instances, a specialist may be required. However, being familiar with these concerns will give response personnel in-depth information on any type of container, from the smallest pail to large railcars. While the scope and magnitude may change based on the size of the incident and what types of products and containers are involved, the general principles are similar — assess the situation to minimize the risk. Steps for predicting a likely behavior of a hazardous material, and predicting the effects of those materials at an incident are provided in **Skill Sheets 13-9** and **13-10**, respectively.

Early Size-Up

Early size-up gives the response team an initial view of the problem and serves as a baseline to determine if conditions are improving or worsening. The information gathered is critical to successful incident mitigation. Only personnel wearing appropriate PPE should conduct the size-up, in accordance with available information.

Items to consider and identify during size-up may include:

- Number and type of containers involved
- Condition of containers and any visible stressors
- Container markings for product identification
- Orientation of the containers to determine stability for offensive operations
- Number of civilians injured or in harm's way

> **Ferrous Metal** — Metal in which iron is the main constituent element; carbon and other elements are added to the iron to create a variety of metals with various magnetic properties and tensile strengths; varieties include cast and wrought iron, steel and steel alloys; stainless steel, and high-carbon steel.

> **Mild Steel** — Class of steel in which a low-level of carbon is the primary alloying agent; available in a variety of formable grades. *Also called* Carbon Steel.

Container Information

Understanding the components and materials of a container is critical for a proper damage assessment. While some container materials are designed to withstand the daily stresses of use, they may fail more quickly when subjected to extreme stresses during a critical event. Some old containers that are still in use may withstand far less stress than those that have been manufactured to more recent standards.

Figure 13.9 Aluminum is a relatively soft metal that can deform easily when abraded or impacted.

Aluminum Containers

Aluminum containers tend to be relatively light and can withstand impact stress well. Aluminum does not react with hydrocarbons. However, these containers cannot be subjected to substantial internal pressures. They also do not plug well because the metal is relatively soft. Aluminum can be easily gouged and weakened and will not withstand friction such as "road burn" that can occur during highway accidents **(Figure 13.9)**.

Steel Containers

Steel is considered a **ferrous metal**. When a ferrous metal is strong, it will not be ductile; if it is strengthened, the ductility will be reduced. While some ductility is necessary, too much may allow the container to elongate, reducing the thickness of the wall.

Steel is easy to examine for metal elongation, heat stress, and fractures **(Figure 13.10)**. It is the easiest of all metals to plug. **Mild steel** can often withstand dents but does bend and distort easily and will chemically react with many materials.

Figure 13.10 Steel with enough ductility will deform instead of breaching, as shown in this drum that contained a polymerization reaction. *Courtesy of Barry Lindley.*

Chapter 13 • Implementing the Action Plan: Product Control   **487**

When dealing with ferrous metals, technicians must understand that when welded, the weakest point will be at either side of the welded seam. The heat incurred in welding can disrupt the original annealing process and affect the steel at the molecular level.

When inspecting a steel container for damage, the smaller the radius of the bend or the dent, the more likely the metal has been weakened and may fail. Such bends will reduce the overall thickness of the metal and increase the likelihood of failure. When small diameter bends or dents happen, fractures occur on the inside of the container opposite the damaged portion which will not be visible on inspection.

High Strength Low Alloy Containers

High strength low alloy steel has high carbon content. This material is extremely strong and abrasion resistive and is used in the construction of many pressure vessels. However, this metal reacts with corrosives, fractures easily, and is difficult to plug.

Austenitic Stainless Containers

This alloy is iron-based with a content of either chromium or nickel. This material is corrosion and abrasion resistive, and does not fracture.

Figure 13.11 Physical damage indicators can include wheel burn and product ignition. *Courtesy of Barry Lindley.*

Potential Container Stress

Potential stress on a container can include the following:

- Changes in temperature that may affect the product or container
- Physical damage that would disfigure or weaken the container **(Figure 13.11)**
- Chemical reactions such as polymerization or mismatched chemicals or containers
- Excessive product weight
- Increased product vapor pressure in a damaged tank

Unusual stresses may cause disintegration of the container. Some unusual stresses may include acid reactions and BLEVE due to increased vapor pressure. Stress may cause the following:

- Runaway cracking
- Punctures
- Tears
- Opening of closures
- Splits

Types of Container Damage

When a cargo container is involved in an accident it must first be inspected for damage, and the emergency responder must be aware of the different types of damage a container can sustain. What seems to be an insignificant blemish can be critical based on the container's date of manufacture and type of material used for construction.

In addition, measuring temperature and pressure is a critical aspect of damage assessment and behavior prediction. Even if a tank is undamaged and not releasing product, a catastrophic release may occur due to temperature and pressure.

When evaluating the condition of tanks and containers during size-up, use consistent terminology, such as the following notations:

- Undamaged, no product release
- Damaged, no product release
- Damaged, product release
- Undamaged, product release

The following sections detail the assessment criteria to consider during the inspection of the damaged cargo tank:

- Cracks
- Scores and gouges
- Dents
- Heat-affected areas

U.S. Federal Railroad Administration Research

The U.S. Federal Railroad Administration (FRA) issued a Damage Assessment of Railroad Tank Cars Involved in Accidents (2005). In this document, the agency outlined concerns that are specific to railcars. However, the theories presented can be applied to other types of tanks and containers as well.

NOTE: The type of railcars referenced in this report are commonly used in the U.S., Canada, and Mexico.

The following bullets are presented in the Damage Assessment document for use when evaluating the damage to a rail tank car:

- If the maximum depth of a wheel burn exceeds 1/8 inch (3.2 mm), unload the tank as soon as possible. If the depth of the wheel burn is less than 1/8 inch (3.2 mm), empty the tank at the closest loading facility, provided it is moved with care and not in ordinary train service.

- Sharp dents in the shell of the tank (cylindrical section), which are parallel to the long access, are the most serious as these dents drop the rating of the tank by 50 percent.

- For dents in the shell of tank cars built before 1967, unload the tank without moving it under the following conditions:
  — A minimum radius of curvature of 4 inches (100 mm) or less
  — Have a crack anywhere
  — Cross a weld
  — Include a score or a gouge

- For dents in the shell of tank cars built since 1967, unload the tank without moving it under the following conditions:
  — A minimum radius of curvature of 2 inches (50 mm) or less
  — Have a crack anywhere
  — Cross a weld
  — Include a score or a gouge
  — Show evidence of cold work

Rail tank cars built after 1988 use normalized steel. These have greater strength than those built between 1967 and 1988. Approximately 50 percent of tank cars in service were built prior to 1988.

SOURCE: https://www.fra.dot.gov/Elib/Document/1200

Cracks

Cracks in a **base metal**, no matter how small, are a reason for concern **(Figure 13.12)**. If the base metal of a container is cracked, off-load a container as soon as possible. When a dent, score, or gouge accompanies a crack, emergency responders should stabilize the tank without moving the container. Technicians should also consider a cracked pressurized container critical and likely to fail if the depth of the fracture cannot be determined.

Fillet welds are often used to attach ancillary items to a tank such as ladders and hose beds. A cracked fillet weld is not a critical issue unless the crack extends to the container itself. Cracks in tank car welds used to attach brackets or reinforcement plates are not critical unless the crack extends into the base metal.

Since 1995, engineers have been working with the DOT and the tank car industry to apply damage tolerance analysis (DTA) methods to analyze metal fatigue crack growth in the welded stub sills (underframes) of tank cars. The problem of fatigue crack growth in welded structures is certainly not unique to the tank car industry and has always been a concern to the offshore structures and maritime industries, among others.

Figure 13.12 Cracks in a container's structure may cause a container to fail.

Dents

Dents that run parallel to the long axis of a container can be critical based on the container's manufacture date **(Figure 13.13)**. Pressurized tanks constructed before 1966 of either 212-B or 515-B grade steel are considered to be in critical condition and likely to fail if they have a dent of less than 4 inches (100 mm). The same is true for a pressurized container built during or after 1966 that uses TC-128 steel and has a dent with an inside radius of less than 2 inches (50 mm).

Massive dents in heads of the tank are generally not serious unless gouges or cracks are present with the dents. Small dents in the head of the tank not exceeding 12 inches (300 mm) in diameter, in conjunction with cold work in the bottom of the dent, are marginal if they show a radius of curvature less than 4 inches (100 mm) for tanks built before 1967. Small dents in places other than the head of the tank may not be critical unless they are also associated with gouges and cracks.

Figure 13.13 Dents in a container may lead to failure depending on their orientation, association with other damage, and the container's construction material.

> **Base Metal** — In hazardous materials containers, the structural material of a containment vessel itself, independent of welding materials and external supports.

Scores and Gouges

Scores and gouges may not be as critical as a crack in the base metal, but they still must be monitored. If a score or gouge crosses a welded seam and the metal removed is no more than the weld reinforcement (the portion that protrudes above the container), and no base metal is removed, the vessel is not perceived to be in critical condition. If the score or gouge removes the base metal on the welded seam, it will be considered critical **(Figure 13.14)**. In addition, longitudinal scores or gouges that cross a weld and affect the heat-sensitive zones are critical. If scores or gouges cross the welded bead of a pressurized container and come in contact with the base metal, consider the container to be in critical condition.

When assessing scores, consider a longitudinal score that runs the length of the container to be the most dangerous. However, do not ignore circumferential scores, for such scores also constitute a longitudinal notch at any given section.

Technicians should unload tanks having scores or gouges in place when the internal pressure exceeds half of the allowable internal pressure allowed for the tank. **Tables 13.1** and **13.2**, for example, show the allowable pressure for 340W and 400W tanks respectively.

Figure 13.14 Scores and gouges may cause a container to fail if they weaken the area around a weld. *Courtesy of Barry Lindley.*

Table 13.1 Limiting Score Depths for 340W Tanks	
Depth of Score	**Maximum Safe Internal Pressure**
¹⁄₁₆ in. (1.59 mm)	191 PSIG (1 316.9 kPa) (89°F [32°C] for commercial propane)
⅛ in. (3.18 mm)	170 PSIG (1 172.1 kPa) (85°F [29°C] for commercial propane)
³⁄₁₆ in. (4.76 mm)	149 PSIG (1 027.3 kPa) (76°F [24°C] for commercial propane)
¼ in. (6.35 mm)	127 PSIG (875.6 kPa) (65°F [18°C] for commercial propane)

Note: In no case should a tank containing a score in excess of ¹⁄₁₆ in. (1.59 mm) for 340W tanks be shipped by rail, although the tank could be uprighted and even moved short distances for product transfer.

Table 13.2 Limiting Score Depths for 400W Tanks	
Depth of Score	**Maximum Safe Internal Pressure**
¹⁄₁₆ in. (1.59 mm)	228 PSIG (1 572 kPa) (108°F [42°C] for commercial propane)
⅛ in. (3.18 mm)	205 PSIG (1 413.4 kPa) (99°F [37°C] for commercial propane)
³⁄₁₆ in. (4.76 mm)	188 PSIG (1 296.2 kPa) (93°F [34°C] for commercial propane)
¼ in. (6.35 mm)	162 PSIG (1 116.9 kPa) (82°F [28°C] for commercial propane)

Note: In no case should a tank containing a score in excess of ⅛ in. (3.18 mm) in for 400W tanks be shipped by rail, although the tank could be uprighted and even moved short distances for product transfer.

Heat-Affected Areas: Welds, Rail Burn, Wheel Burn, and Road Burn

The heat-affected zone of any metal container can be described as the area of the base metal container which has its microstructure altered by welding

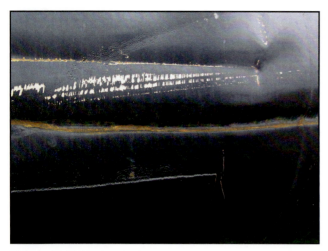

Figure 13.15 Mechanical stresses can generate enough heat to weaken a container.

or other heat-intensive operations. The heating then recooling of a base metal will change the properties immediately adjacent to the heated area. The extent of the property change will be different with each base metal, but the heat-affected areas will always be less ductile than the original base metal.

NOTE: Heat-induced tears, discussed in Chapter 8 of this manual, are a similar phenomenon.

Mechanical stresses, such as rail burn, can also lead to heat-affected areas **(Figure 13.15)**. Burn damage is similar to a gouge. If the depth of the burn or gouge exceeds 1/8 inch (3.2 mm), consider off-loading the container **(Figure 13.16)**. If burns cross the welded bead of a pressurized container and come in contact with the base metal, consider the container to be in critical condition.

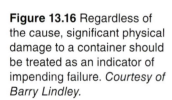

Figure 13.16 Regardless of the cause, significant physical damage to a container should be treated as an indicator of impending failure. *Courtesy of Barry Lindley.*

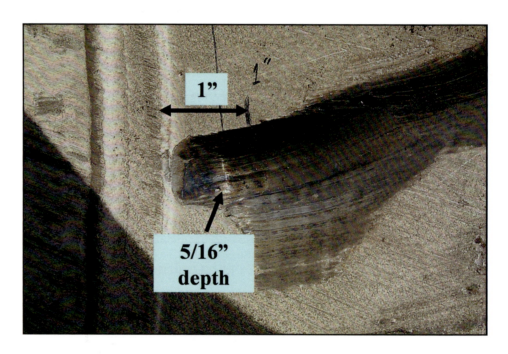

Predicting Likely Behavior

Before the start of any offensive operation, hazmat technicians must have an understanding of what is involved and what could happen during the operation. This process starts during size-up and should continue throughout the operation. Product identification and reconnaissance of the area and containers involved can help determine the likely behavior. Asking and answering the following questions can also prove helpful:

- Where is the hazardous material or container likely to go if released during the emergency?
- Why is the hazardous material likely to go there?
- When is the hazardous material likely to go there?
- How will the hazardous material get there?
- What harm will occur when the hazardous material gets there?

In order to predict the likely behavior of a product and its associated container, technicians must obtain as much information as possible including definitive identification of the contents of the container. Knowing the contents ensures that technicians can base their decisions on the chemical and physical properties of the product or products involved and the quantities involved. Determining the following properties can help emergency responders in predicting likely behavior:

- Does the hazardous material have energetic properties (such as being explosive) or can it create energy as a condition?
- Is the product flammable or combustible **(Figure 13.17)**?

Figure 13.17 Response to a container failure should tailor a response to the material involved. *Courtesy of Barry Lindley.*

- What is the state of matter and what is its vapor pressure?
- Is the product an aggressive chemical (corrosive)?
- What effects will this product have on human life?
- If the product leaves its container, can it mix with water? Can it be diluted?
- Where will the material go when it leaves its container?

An understanding of the container and its features is helpful as well. To determine the condition of the container, technicians must know what it was like in its original state. Understanding how the container was constructed can also help in determining how it may fail.

After the involved products are identified, determine the internal pressure of the tank. When a tank has been placed under thermal or chemical stress or has been damaged, the internal pressure of the tank may be enough to cause it to fail. If the tank fails while an offensive operation is underway, the results may be devastating. The internal pressure can also be a deciding factor on whether the tank can be plugged or patched or if other means are necessary to mitigate the incident. In some instances, the internal pressure of a container may have to be reduced before any offensive operation can begin.

Determining the internal pressure of containers can be difficult. The type of container will provide the first clue, though technicians must consider environmental conditions that can cause pressure changes such as temperature and altitude. Technicians can also read pressure gauges, if provided on the container.

An advanced method of determining pressure is to attach a pressure gauge to the vapor valve of a damaged container. Technicians can then read the pressure directly from that gauge.

Another advanced method involves determining the product temperature and estimating the vapor pressure. Do this using an infrared thermometer or thermal imager that provides a direct reading of the product's temperature. Once technicians determine this information, they can consult a reference source such as the Mathieson Gas Data book to provide temperature/pressure curves. They can then make an estimate of the tank pressure.

> **CAUTION**
> Insulated or jacketed tanks will prevent an accurate measurement of tank temperature and pressure.

Technicians may also need to determine the amount of product in a container. Use a thermal imager to check noninsulated tanks. Percussing the side of noninsulated tanks may yield a different sound at the liquid level and vapor space. Technicians can also check the container if it has a built-in level device. Most bulk fixed facility tanks and many pressure tanks will have level devices.

Technicians must also consider the following exposures and response conditions:

- What effect on life and the environment will the product have if no offensive actions are taken?
- What will the effects be if there is nonintervention?
- Is the container located indoors or outdoors?
- What are the weather conditions and what effects will they have on the product?
- Is it daytime or nighttime?
- Is visibility an issue?

Technicians must evaluate all these factors before any operation can begin. Taking these factors into account on a risk versus benefit basis will help the response team decide if an offensive mode of operation will be acceptable with a nominal risk, or if other response modes need to be considered.

Product Containment: Plugging and Patching

If an offensive operation is necessary, the hazmat technician must then identify the different containment techniques that can be used to restrain the product within its container. Containment techniques and procedures vary in their complexity. Technicians may achieve containment by simply uprighting a drum or using a golf tee to plug a small hole. However, more complex systems may be needed for large holes on large containers or systems. More complex containment methods may include using high-pressure pneumatic air bags or orchestrating the shutdown of an entire pumping system. Whatever the method used, plan the techniques and procedures carefully with consideration for operating times, personnel requirements, product hazards, and resource compatibility and quantity requirements. Steps for selecting appropriate product control techniques are provided in **Skill Sheet 13-11**.

WARNING!
The leak control system and equipment must be compatible with the quantity and type of material to be contained.

Figure 13.18 Wooden wedges and golf tees can be used to seal a hole in a container.

The following sections will describe different leak control and containment methods that may be available to the hazmat technician. All hazardous materials response personnel should be familiar with what containment equipment is available within the department's cache and must be trained in the proper use and techniques of that equipment.

Plugging

While there are many commercially available plug and patch kits on the market, effective plugging equipment can be as simple as homemade wedges, golf tees, and rubber gasket material (**Figure 13.18**). Steps for plugging a leaking container are provided in **Skill Sheet 13-12**.

CAUTION
Ensure that the plug is larger than the opening in the container and that it is compatible with both the container and its contents.

The simplest method of plugging a small hole is to insert an inert material inside the hole (**Figure 13.19**). For example, soft woods such as pine and Douglas fir are commonly used plugging materials, but they are incompatible with corrosives. Plastic plugs should be used for corrosives.

Plugs may work best if they are wrapped with a lightweight material before being inserted in the hole. Lightweight cloth or light gauge rubber gasket material are often helpful to keep the plug in place and fill some of the gap between the plug and the container. In many

Figure 13.19 Be careful to select a plug material that is compatible with the contained material and somewhat larger than the hole being plugged.

Chapter 13 • Implementing the Action Plan: Product Control **495**

cases, the use of a plug will not make a watertight fit. However, using a compatible caulk or putty in addition to the plug may reduce or even stop the leak.

Patching

Patching is often required when a breach in a container is too large to be plugged **(Figure 13.20)**. Patching procedures may be more complex based on the damage to the container and whether the protruding edges of the breach are on the side that is to be patched. Good reconnaissance of the container's affected area will assist hazmat technicians in determining the type of patch and additional equipment needed. Technicians may also use straps or pneumatic equipment in association with the patching material based on the size of the hole. Using a two- or three-part epoxy that is compatible with both the chemical and the material being patched will help secure the patch and reduce or eliminate any leakage. In addition to epoxy patches, some jurisdictions and applications may use magnetic or freeze patches.

NOTE: Even if a plug or patch cannot completely stop a leak, it can slow the leak until hazmat personnel can develop other response plans.

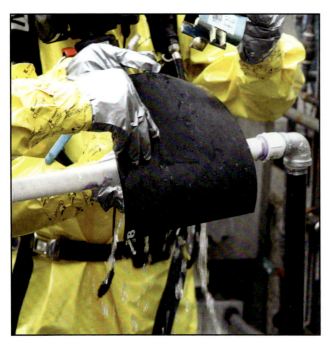

Figure 13.20 Patches can be used to seal a large or irregular breach.

Technicians may need to use whatever materials are available to improvise patches. While skill is necessary, patching often calls for a certain amount of creativity **(Figure 13.21)**. Prefabricated patches are usually the easiest to use and are in many cases effective. However, conditions may not allow for the use of these patches and alternate means may be necessary. The following sections introduce different types of patches that are commonly used. Steps for patching a leaking container are provided in **Skill Sheet 13-13**.

Figure 13.21 Effectively patching a breach may require some creativity. *Courtesy of Barry Lindley.*

Box Patch

A box patch is used when the protruding edges of the breach are located on the side of the container where the patch is to be attached **(Figure 13.22)**. The box must be made of a material that is compatible with both the container and the product and should be lined with a gasket material such as canvas, cloth, or rubber. Once the box is in place over the breach, use shoring straps or some other means to hold the box in place. The methods used depend upon the container's size, location, and orientation.

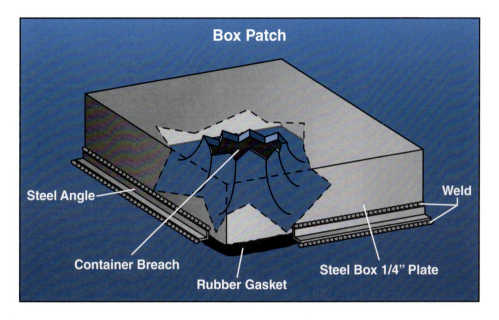

Figure 13.22 Box patches are used to elevate patch material from the surface of the container when the breach has caused an uneven surface.

Hook Bolts and Patch

Hook bolts are usually fabricated from round stock such as steel or aluminum. They come in a variety of diameters and shapes. Hook bolts are manufactured in such an orientation that the head of the bolt will pass through the hole into the container and hook to the inside of the container as it is fastened by a bolt or a nut on the outside. Hook bolts may be manufactured in the shape of a T, L, or J **(Figure 13.23)**. Wood or steel strongbacks may be used in association with the bolt to provide additional strength.

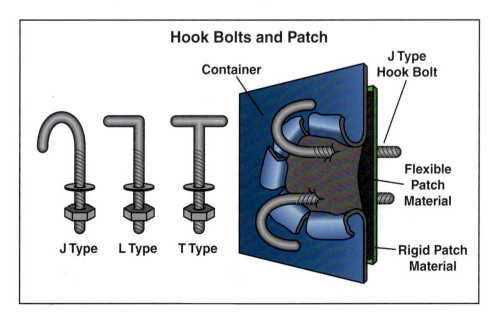

Figure 13.23 Hook bolts are secured to the internal surface of the container and are fastened outside of the patch material.

Chapter 13 • Implementing the Action Plan: Product Control

Soft Patch — Combination of gasket material and either wooden plugs or wedges inserted into a hole to patch a leak.

When using a hook bolt, fully insert the head end of the bolt through the container and rotate so that it cannot be pulled back through the hole. Insert a pad or gasket through the stock of the bolt along with a plank or other type of strongback. Tighten a nut on the threaded end of the bolt to secure the patch. Hook bolts may be used with a variety of patches and in various combinations.

Pipe Patches

Pipes may be patched using different methods in a similar manner as containers. Technicians may need different types of patches depending on conditions at the scene. A variety of items can assist emergency responders when repairing a pipe. Based on the size of the pipe and its content and pressure, items such as pipe wraps, bar clamps, and other clamps will aid in the patching of a leaking pipe. As with containers, emergency responders will always have to be mindful of compatibility with the product. Steps for applying a sleeve, jacket, clamp, or wrap are provided in **Skill Sheet 13-14**.

When patching pipes, technicians must also consider the pressure in the pipe **(Figure 13.24)**. If pressure is of slight concern, using a soft patch can control a split or a tear in a pipe. A **soft patch** is a combination of gasket material and wooden plugs or wedges that are inserted into the hole. Insert the plug or wedge into the pipe only far enough to control the leak. Pressure in the pipe can increase if the plug or wedge is inserted too far. Once inserted, cut or trim the excess soft patch material outside the pipe, and cover the damaged area with another piece of gasket material that can be secured in place using string, wire, and tape, or other means.

Figure 13.24 The choice of a pipe patch must take into consideration any pressure behind the leak.

Soft patches can be modified or improved to suit any use. To increase the integrity of the patch, use a piece of light gauge metal on the outside of the rubber gasket to act as a strongback.

Jubilee Pipe Patch — Modification of a commercial hose clamp; consists of a cylindrical sheet of metal with flanges at each end. The sheet metal can be wrapped over packing around a pipe leak. The flanges can be attached to one another using screws and nuts to form a seal.

If pressure in the pipe is more of a concern, a **jubilee pipe patch** may be in order. Jubilee patches (also called saddle patches) are a modification of a commercial hose clamp. This type of patch is usually manufactured from commercial sheet metal and is rolled into the needed diameter. Tabs are bent on each edge to form a flange. Holes are then drilled into the flange to allow the patch to be secured to the pipe.

To install a jubilee pipe patch, place a piece of rubber or other gasket material over the breach, and install and secure the patch over the rubber. The patch should be large enough to cover and overlap the damaged area by at least 2 inches (50 mm) on all sides. The bolts are then tightened to secure the patch in place.

An elbow patch can be used to secure a hole on a curved portion of pipe. Elbow patches are usually constructed of a woven-type cloth that incorporates a resin hardener. The cloth is shaped to contour the pipe and is then covered with multiple layers of PVC. It may be possible to repressurize the pipe once the resin has hardened and the patch is secure.

Specialized Plugging Equipment

Technicians may use a variety of equipment to control or contain a leaking container. This equipment may range from dome cover clamps for cargo tankers to chlorine kits for different sized chlorine containers. This equipment requires specialized training and knowledge of the product being dealt with. Departments must always take care to ensure that emergency responders have the background, knowledge, and training to deal with these types of materials. This equipment may require personnel trained to the specialist level to be safely used. A few examples of this equipment are included in the following sections. These resources were introduced in Chapter 8 of this manual.

NOTE: Technicians can modify some kits to contain a range of materials by changing the gasket material to match the intended spill.

A-Kit

The Chlorine Institute's Emergency Kit "A" (known as an "A-Kit") is designed for use with compressed gas cylinders (specifically, 100 and 150 pound [50 and 75 kg] capacity cylinders). Technicians can use the A-Kit's tools and equipment to contain and control leaks in and around the cylinder valve and in the side wall of chlorine cylinders **(Figure 13.25)**.

Figure 13.25 The "A-Kit" can control leaks from the valves and side walls of chlorine compressed gas cylinders.

B-Kit

The Chlorine Institute's Emergency Kit "B" (known as a "B-Kit") is designed for use with chlorine ton containers. Technicians can use the B-Kit's tools and devices to contain and control leaks in and around ton container valves. These kits can also control leaks in the side walls of the container **(Figure 13.26)**.

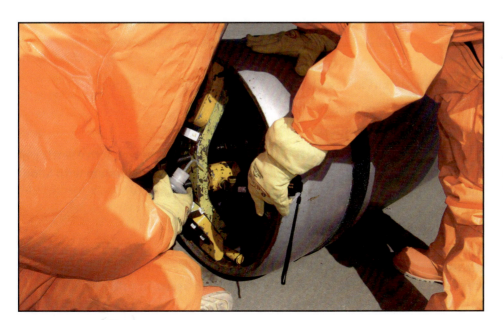

Figure 13.26 The "B-Kit" can control leaks from the valves and side walls of chlorine ton containers.

C-Kit
The Chlorine Institute's Emergency Kit "C" (known as a "C-Kit") is manufactured to the design specifications of the Chlorine Institute. The C-Kit contains devices to stop leaks at the safety valve or angle valves of standard DOT 105J500W chlorine tank cars, DOT MC331 cargo tanks, DOT 51 portable tanks in chlorine service, and barges.

The chlorine C-Kit is equipped with a variety of hand tools, but it also contains specific tools to help contain leaks in and around the angle valves and pressure relief devices **(Figure 13.27)**. Because of the highly specialized nature of this operation, only trained personnel should be allowed to work with the C-Kit.

Figure 13.27 The "C-Kit" can control leaks from the valves of tank cars and cargo tanks in chlorine service.

> **CAUTION**
> Other kits are available for specific materials. Only people who are trained in their safe and proper use should operate these kits.

Propane A- and B-Kits
Propane containers are common in industrial and household applications, and propane is a leading cause of line of duty death in the hot zone. When technicians determine a propane container or its valves, fittings, or attachments to be too unsafe to leave the product in it, the responders may select a kit to remove the propane via flaring. Each kit has fittings that match a common type of container.

> **WARNING!**
> Propane is the leading cause of LODD in the hot zone.

Two types of kits are available for flaring propane containers:

- A-Kit for DOT containers
- B-Kit for SME containers

Flaring is considered an advanced skill because it involves active setting of fires. At the same time, a technician can safely operate this system with about the same level of hazard as a backyard barbecue grill.

Midland Emergency Response Kit
Midland Manufacturing offers a detailed response kit specifically for use with pressurized tank cars. The Midland Emergency Response Kit is packaged in three individual containers and is designed to assist in mitigating leaks in valves from chlorine, liquid petroleum gas, or anhydrous ammonia tanks.

Because of the critical nature of this type of work, only personnel specially trained with this equipment and on large tank cars should be involved with this type of operation and type of recovery vessel.

Recovery Vessels

Some hazmat teams may have access to a chlorine and/or ammonia recovery vessel which is used to contain a leak from a 100-150 pound (50-75 kg) cylinder **(Figure 13.28)**. A recovery vessel may be used in applications when an A-Kit is ineffective or not available. A recovery team such as a "chlor rep" team or the chlorine vendor may have these vessels available. Some facilities may have their own equipment on site. In addition, some vendors may be willing to bring resources to a response. These vessels are also discussed in Chapter 8 of this manual.

Figure 13.28 Recovery vessels are used to temporarily contain chlorine or ammonia, depending on the type of recovery vessel.

Cargo Tanks

Transportation containers vary in size and type based on the product being transported. The following sections will introduce the concerns and issues that emergency responders encounter when these large containers become involved in emergency incidents. The construction and specifications of many transportation containers are included in Chapter 10 of this manual.

Discussions regarding transportation accidents involving hazardous materials often gravitate towards major incidents where the container has overturned and released its contents. While these incidents are indeed serious, they occur rarely when compared to hazardous materials incidents as a whole. In fact, the majority of problems associated with cargo tanks involve issues with valves and fittings. Tightening a fitting or closing a valve can mitigate many container releases. Because most of the tanks encountered are equipped with internal valves for product discharge, technicians may perform little (if any) field repair. If field repairs are possible, hazmat teams usually need specialists to complete the repair.

If a transportation container is involved in an incident, it is critical that technicians thoroughly inspect the damaged cargo tank to determine the extent of the damage. Inspect all accessible surfaces, and note the type, location, direction, and extent of the damage. It may be necessary to reinspect a cargo tank after any debris covering the damaged area of the tank has been removed or when the tank has been lifted or righted.

Technicians may find jacketed cargo tanks extremely difficult to inspect without removing the jacket. A lack of damage to the tank's jacket can be used as an indicator that the tank may only have sustained minimal damage or no damage at all.

The following sections explain product control in leaking containers and refer to specific classifications of containers including:

- MC 306/DOT 406 (the Fire Control section later in this chapter addresses aluminum varieties of these tanks)
- MC 307/ DOT 407
- MC 312/DOT 412

Methodology of Leak Control

MC 306/DOT 406, MC 307/DOT 407, and MC 312/DOT 412 cargo tanks can pose specific hazards because they may be under pressure. Based on the area of the leak, responders may encounter a liquid leak or a vapor leak.

Tank truck leaks may occur in multiple locations. If the tank has overturned, leaks may occur through the dome cover even if the mechanical stressors did not breach the tank. Breaches normally result from stresses caused by impact such as during the accident. The stress involved and the material of the container may be far ranging. Splits, tears, or punctures may be present in multiple locations on the tank.

Tanks may be breached in several locations; therefore, technicians should inspect the tank on as many sides as possible. Generally speaking, the lower the location of the leak on the tank, the more serious the problem. Leaks located below the level of the product will be relatively obvious, and should be controlled first. Leaks above the product level can also create hazards. For example, harmful vapors may be released from these leaks into the surrounding area, or fresh air may be drawn into the tank creating an explosive atmosphere in the vapor space if flammable or combustible liquids are involved. The following sections will introduce different types of leaks and breaches that responders may encounter when responding to incidents involving over the road tank trucks.

Dome Cover Leaks

Several resources and methods are available to stop dome leaks. These may range from wooden wedges to commercially available devices. If the container involved is carrying a flammable or combustible liquid, conduct monitoring operations continuously. As with all hazmat operations, response personnel should avoid direct contact with any materials involved.

If a 306/406 tank dome is leaking, technicians may use a dome cover clamp to secure the manway or fill cap to stop the leak **(Figure 13.29)**. A number of

Figure 13.29 Dome cover clamps may effectively stop a leak from a tank dome.

dome clamps are available and technicians must be trained to use the dome clamp provided by their jurisdiction. Be aware that most gaskets in this area will leak even when new. Prior to the application of the dome cover clamp, observe any conditions that would indicate reactivity with the container or any other products or materials. If conditions such as rising temperatures are detected indicating a possible reaction taking place, follow SOPs accordingly.

Reactivity with Container or Other Products

Reactivity with the container usually involves some type of corrosive action. If there is not an active leak it may be difficult to tell that a reaction is occurring with the container. One indication that a reaction may be happening is unexplained noise from the container such as creaking or growling. Additional indicators are smoking/fuming from the container or unexplained heating or vibration. Use extreme caution when these indicators are present because the container could come apart suddenly. If an internal reaction with the container is suspected, consider pulling back for further evaluation.

If there is an active leak, technicians may observe shiny new metal or bubbles/foaming around the leak area. This could indicate hydrogen formation leading to a flammable or explosive atmosphere. Take caution around this potential flammable area and use appropriate control measures to prevent ignition.

Most chemical reactions happen fairly fast, so if products have already mixed, it is possible the reaction may have already occurred. However, some reactions like oxidation or polymerization make take longer due to the nature of the materials. Indicators include heating, smoking/fume production, or unexplained fires. As the reaction mixture increases in temperature, the reaction will typically go faster and may become explosive. If observing this type of reaction, withdraw for further evaluation. Verify if the chemicals present will potentially react with themselves or with other chemicals.

If more than one dome is leaking, response personnel should evaluate the situation and determine the order that the dome cover clamps will be installed. These are general procedures and response personnel must be familiar with the specific equipment and procedures used in their department.

Addressing Breaches in Cargo Tanks

Any cargo tank involved in a motor vehicle accident or other type of incident may be subject to holes or punctures from various types of stress. Whether it is from road burn, wheel burn, or other dynamic influences, the responder may face the following:

- Irregular shaped holes
- Punctures
- Splits or tears
- Leaking valves and caps
- Leaking manway gaskets
- Leaking piping

While the volumes of product may be more substantial, the theory of patching cargo tanks is not any different from plugging or patching a drum or other type of pressure cylinder. To control leaks in these containers, use different-sized wooden plugs in association with lead wool or putty. All products used to control leaks must be compatible with both the product and the container. The use of cargo straps in conjunction with pneumatically operated air bags may also assist the responder in securing the leak.

Ensure that the tank is stable and properly shored before beginning any plugging or patching operation. Movement of the tank cannot only cause a collapse or roll of the rubble, but shifting load inside the container may also dislodge plugs that have been put in place. As always, wear proper PPE and maintain constant air monitoring.

NOTE: Responders must be prepared to control contents should container failure occur during plugging and patching operations.

Methods and Precautions for Fire Control

As previously stated, aluminum is a common material used in MC 306/DOT 406 tank construction. Aluminum has inherently good traits when it comes to tank construction and tends to be light compared to other materials such as steel.

Aluminum tanks will begin to melt at approximately 1,200°F [650°C]. If an aluminum tank is breached and the product ignites, the liquid will act as a **heat sink** and the fire will burn at the liquid level. As the liquid and vapors are consumed, the exposed aluminum shell can be expected to melt and eventually fail.

Heat Sink — In thermodynamics, any material or environment that absorbs heat without changing its physical state or appreciably changing temperature.

There is the possibility of a BLEVE whenever a flammable liquid is burning in a container. Responders should take precautions to prevent BLEVE. If no product is spilled from the tanker, it could take several hours for this type of tank to burn completely. For flammable liquid fires in aluminum containers, it may be wise to allow the fire to burn with limited intervention on the tank. While allowing the product to burn off is an option, consider exposures. In roadway incidents this includes bridges and overpasses. Response priorities should concentrate on the protection of life and exposures and confining any product that may have escaped the tank but has not ignited.

WARNING!
Depending on the materials and conditions, adding water may make a situation worse.

In many instances applying a foam blanket to the affected area is necessary. Refer to IFSTA's **Essentials of Fire Fighting**, **Hazardous Materials for First Responders**, and **Principles of Foam Fire Fighting** manuals for a review on foam operations. When using foam, ensure that the foam is compatible with the product that is burning. In addition, consider the available water supply and any foam runoff that might cause an environmental hazard.

Product Removal and Transfer Considerations

If an MC 306/DOT 406, MC 307/DOT 407, or MC 312/DOT 412 cargo tank is damaged, it may be necessary to remove the product into a viable container. Remove as much product as possible before righting a cargo tank. Options for product removal depend upon:

- The type of product
- Containers involved
- Container stress
- The orientation of the container
- Any incident-specific variables

CAUTION
It is dangerous to attempt to upright a damaged cargo tank that still has product in it.

Specialized contractors often perform the actual product transfer, but responders still maintain the Command structure of the incident and will be responsible for overall safety. Responders performing product removal and liquid transfer must be properly trained in the operation and authorized to perform it. Steps for conducting liquid transfer operations are provided in **Skill Sheet 13-15**.

When major cargo carriers are to be off-loaded, technicians should survey the tank and area to determine a safe method for transferring the product. Off-loading a product from a damaged tank can help to reduce stress on the tank and lessen the chance of total failure of the container when it is eventually moved.

When transferring a product, observe the following:

- Location of additional leak points, if any
- Size and compatibility of the transfer container
- Recirculation of the product
- Cooling and heating of products
- Availability of the proper tools and equipment to complete the job
- Increases to the scene size and control area

Different methods are available for the transfer of a product based on the needs of the incident including:

- **Portable pumps** — Come in a variety of sizes and are used to move liquids. These pumps may be electric, gasoline, diesel, power take-off (PTO), water, or air-powered. Select the type and size based on the type of product being transferred, the capacity of the receiving container, and the lift and flow capacities needed at the scene.
- **Pressure differential pumps** — Used to create a pressure differential between two tanks to transfer vapors and gases. A vapor compressor creates a positive pressure differential to move the product by pulling vapors from

the receiving tank, compressing them, and forcing them into the damaged tank car. The increased pressure in the damaged tank pushes the product into the receiving tank. Responders must consider how much pressure the damaged container can take. Pressure differential pumps may be used in tandem with portable pumps to increase the speed and efficiency of the transfer.

- **Vacuum trucks** — Have mounted vacuums and can handle a variety of products including flammables and corrosives.

- **Betts valves** — Emergency unloading fixture equipped with an internal self-closing stop valve. A Betts valve is designed to allow responders to unload an overturned tank through a capped 3 inch (75 mm) **National Pipe Thread Taper (NPT)** clean-out without spillage of the product.

The fixture permits the removal of the clean-out cap and product discharge through proper hoses to the level of the clean-out cap. This valve can be used in off-loading operations of MC 307/DOT 407 highway cargo tanks.

> **National Pipe Thread Taper (NPT)** — US standard for pipe threading developed with the intent to create a fluid-tight seal.

Safety Considerations

Safety is the primary concern for any transfer operation. Safety considerations for any transfer operation include:

- Only personnel involved should be in the general area of the transfer operation.
- Personnel should wear appropriate PPE.
- Equipment used in transfer must be compatible with materials involved.

Container Stability

Emergency response crews should shore and stabilize the container before the start of any transfer operation. Bracing must also be in place. Responders must be aware that once the transfer operation starts, it is possible that the container may shift.

Air Monitoring

Air monitoring should be an ongoing activity during transfer operations. The use of portable and gas-powered equipment may create an ignition source if the products that are being off-loaded and transferred have flammable properties. Constantly monitor LEL readings. While a reading of 0% LEL is ideal, the Incident Safety Officer will have to determine the acceptable LEL during the operation. The elimination of all ignition sources or the use of vapor suppression should be a priority for a safe operation. If vacuum trucks are being used, vent the vacuum exhaust downwind of the site.

Bonding and Grouping

The movement of products can generate **static electricity**. When two materials are in contact, electrons may move from one material to the other, which leaves an excess of positive charges on one material and negative charges on the other. When the materials are separated, the balance is disrupted creating what is known as static electricity. If technicians do not address static electricity in an off-loading or transfer operation, the result may be a flash fire or explosion. Steps for bonding and grounding a container are provided in **Skill Sheet 13-16**.

> **Static Electricity** — Accumulation of electrical charges on opposing surfaces, created by the separation of unlike materials or by the movement of surfaces.

Bonding and **grounding** multiple objects can reduce the risk of static electricity to an operation. When objects are bonded, a conductor is used to connect the objects together. Bonding the objects together can minimize or eliminate static electricity.

Grounding is the means of connecting the objects to the ground to minimize the buildup of static electricity **(Figure 13.30)**. Using a grounding rod or another object that is grounded can eliminate the charge differences between the objects and the ground. For a safe transfer operation, bond the two containers to each other and then use a grounding rod and cable to bring them to ground.

NOTE: Refer to Chapter 6 for more information on monitoring for grounding and bonding. Megger kits are a common tool for checking resistance as a part of grounding and bonding.

> **Bonding** — Connection of two objects with a metal chain or strap in order to neutralize the static electrical charge between the two. *Similar to* Grounding.

> **Grounding** — Reducing the difference in electrical potential between an object and the ground by the use of various conductors. *Similar to* Bonding.

To safely bond and ground two containers, begin with the damaged vehicle. The tank and all appliances (hoses and recovery basins) should be bonded and then connected to the receiving tank. Connect all bonding cables to clean surfaces that are free of paint or grease. Do not use cables with alligator clips; instead, use a clip that will offer a more secure connection. Use the following steps for bonding and grounding operations; however, also be familiar with the procedures that your organization uses:

Step 1: Verify safety of the atmosphere.

Step 2: Establish a ground field per departmental SOP/Gs.

Step 3: Connect the ground wire from the damaged tank to the grounding post.

Step 4: Connect the bonding cable from the damaged container to the receiving container.

Step 5: Connect the ground wires from the receiving container to a grounding post. This connection may be made before the grounding and bonding of the damaged container.

Step 6: Verify continuity and resistance throughout the grounding and bonding process.

Figure 13.30 Reduce the risk of a fire by grounding items that may generate a static electric charge.

Pressurized Containers

As described in Chapter 10, compressed gas cylinders, pressure vessels, ton containers, and other pressurized containers are manufactured with materials that can withstand high pressures. Based on the stresses that affect these containers, a pressurized container may leak before it ruptures. Leaks may occur in a variety of locations on pressurized containers including the valve assembly or cracks within the container's sidewalls.

Because pressurized containers are designed to hold high pressures (sometimes in excess of 6,000 psig [41 400 kPa]), care must be taken. While being mindful of the pressure issues, hazmat technicians must also definitively identify the product within the container to ensure that all PPE and equipment will be appropriate for the products involved and hazards present.

Whenever technicians are working with leaking pressurized vessels, they must consider:

- Pressure and pressure changes
- Availability and proper use of appropriate PPE
- Configuration and condition of the container
- Internal and external temperature of the container
- Size of the container
- Presence of **autorefrigeration**
- Possibility of BLEVE
- Expansion ratios of the gases involved
- Availability and use of a thermal imaging camera to assist in locating leak points
- Invisible fires that some gaseous products create
- Conditions that would indicate reactivity with the container or other products in proximity, for example, rising temperatures

Additionally, any time a pressure cylinder or ton container is involved, technicians should do the following:

- Locate the leaking container (if the container is in a group) by looking for escaping vapor.
- Ask for approval from the plant operator to close the connecting valves if the container is connected to additional piping or processing unit.

> **Autorefrigeration** — Rapid chilling of a liquefied compressed gas as it transitions from a liquid state to a vapor state.

Understanding Fittings on Pressure Containers

One of the simplest ways to contain a leak in pressure containers may be to close the valve. However, if the container is part of a pumping system, closing valves without direct knowledge of the entire pumping system can have a devastating effect **(Figure 13.31)**. Pumping systems can be extremely complex. Hazmat technicians must know exactly what valve or shutoff is being closed and what system it will affect. Pumping valve systems are often not marked so knowing exactly what the valve controls becomes extremely important. Technicians must perform any valve operation in concert with facility personnel. For example, at ammonia refrigeration systems, ammonia can be trapped in piping between valves. Pressure builds in these pipes which can cause them to fail. Steps for performing remote valve shutoff or activating an emergency shutoff device are provided in **Skill Sheet 13-17**.

Technicians must also understand how a valve operates. Some valves such as gas mains have stems. If the stem runs in line with the pipe, the valve is open and product is able to flow. If the stem runs across the pipe,

Figure 13.31 Some leaks may be controlled simply by closing a valve.

the valve is closed and product cannot flow. While this is a common orientation for valves, each valve can be different. Post indicator valves (PIVs) use a sign or marking when the valve is open or closed, but some types of valves, such as butterfly valves, have no marking **(Figure 13.32)**. Valves are discussed in detail in the IFSTA manual, **Fire Protection, Detection, and Suppression Systems**. Steps for tightening or closing leaking valves, closures, packing glands, and/or fittings are provided in **Skill Sheet 13-18**.

The hazmat technician may be able to make minor repairs to different systems. Because pressurized containers in facility systems may be complex, response personnel should not attempt these operations without the guidance of facility operators.

Some simple tasks that may be beneficial to product control may be tightening loose plugs or replacing missing plugs. These simple assignments may be crucial in preventing further spillage. The hazmat technician must be able to perform actions such as closing valves and tightening or replacing plugs while wearing appropriate PPE. Air monitoring must be conducted continuously especially if working with products that may be combustible or flammable. If working inside, give consideration to a release in a confined space and the possibility of an oxygen-deficient atmosphere.

Working with Fittings on Pressure Containers

The techniques used to control a specific type of leak from a pressurized container will be similar regardless of the container involved, even though the equipment used to control the leak may be different **(Figure 13.33, p. 510)**. Capping kits are designed for a variety of pressurized containers. Steps for capping a leak are provided in **Skill Sheet 13-19**.

The procedures and steps provided in the sections that follow are generalized for commercially available repair kits used by emergency response organizations. Hazmat technicians should be familiar with the cache of equipment that is available to their agency, and they must be trained specifically on such equipment. Always follow manufacturer recommendations and specific department operating procedures.

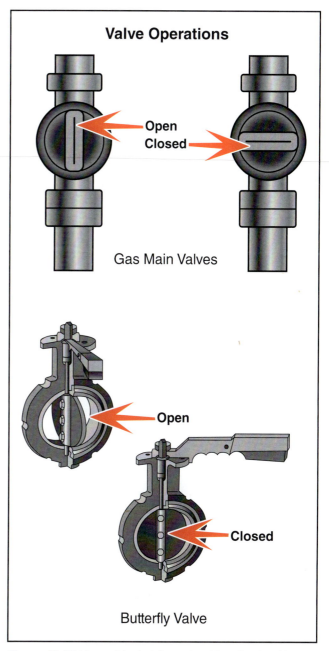

Figure 13.32 Hazmat technicians should understand how valves work.

Fusible Plug

Fusible plugs are common safety devices found in pressurized vessels. Depending on the plug's design, these devices are components that commonly leak and require attention. Tools needed when containing a fusible plug leak may include wrenches, clamps, set screws, blocks, and gasket material.

Once the equipment is in place, responders must be sure that the leak has been controlled. If the fusible plug is being replaced on a chlorine cylinder,

Fusible Plug — Safety device in pressurized vessels that consists of a threaded metal cylinder with a tapered hole drilled completely through its length; the hole is filled with a metal that has a low, predetermined melting point.

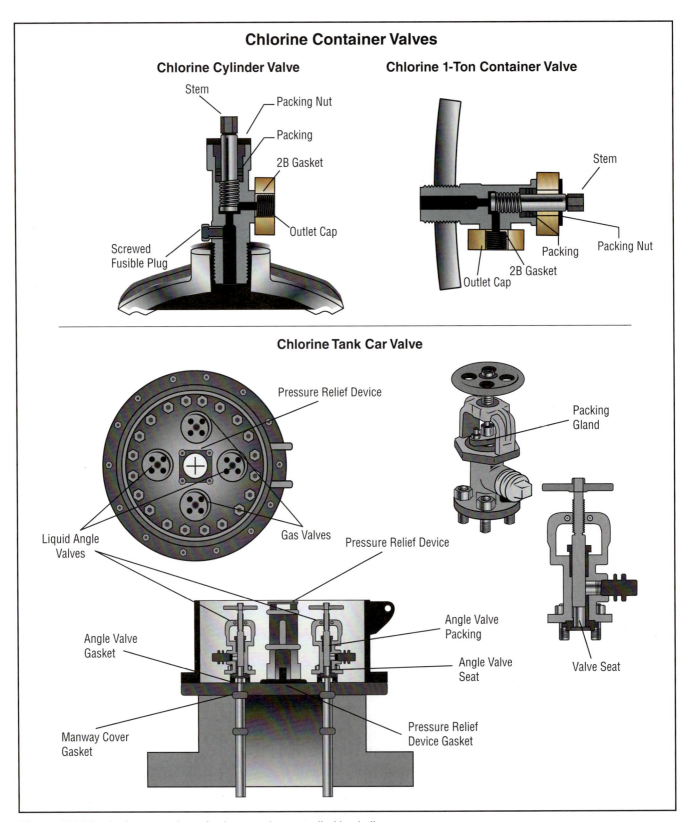

Figure 13.33 Leaks from a variety of valves can be controlled in similar ways.

use a compatible product such as liquid ammonia vapor to check for a leak **(Figure 13.34)**. Technicians must use products that are compatible with the materials involved. If a fusible plug cannot be replaced or the leak stopped, use an alternate means such as a hood assembly.

Figure 13.34 Always test a seal to ensure that a leak has been controlled.

Fusible Plug Threads

Sometimes it is not possible to tighten or replace a fusible plug. If the plug still leaks, responders may have to clamp off the fusible plug in order to control the leak. Tools that may be needed include:

- Hacksaw
- Wrenches
- Set screws
- Gasket material
- Files
- Clamps
- Blocks

Cylinder Wall

If the sidewall of a pressure vessel has been breached, technicians may need to apply a patch around the cylinder. Tools that may be needed include:

- Screws
- Yokes
- Patching equipment
- Scrapers
- Wrenches
- Chains
- Gaskets

Valve Blowout

There are times when the valve assembly will fail in the open position or completely blowout. If this happens, secure the valve opening by driving a drift pin into the opening and then sealing the valve unit with a hood assembly.

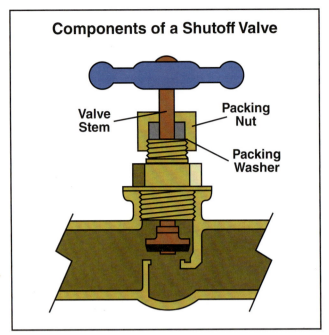

Figure 13.35 A packing nut is located between the valve stem and the container body.

NOTE: Unless responders are on the scene when the valve blowout occurs, the tank will be empty before a drift pin can be inserted.

Once the drift pin has been driven into the valve opening, stay clear of the valve opening in case it dislodges. The hood device assembly should then be assembled and applied.

Valve Gland

When a valve gland is leaking and in need of repair, it may be possible to tighten the packing nut to prevent further leakage **(Figure 13.35)**. Before tightening the nut, use the assembly on the cylinder or the appropriate wrench from a commercially available tool kit to ensure that the valve stem has been closed. Once the valve stem is closed, attempt to tighten the packing nut **(Figure 13.36)**. The packing nut should be tightened firmly but not overtightened. If the leak persists, attach the hood assembly. Follow the directions found in the valve blowout section to assemble the hood assembly.

Valve Inlet Thread

If the valve assembly is leaking around the inlet threads, it may be possible to tighten the valve assembly onto the cylinder in order to stop the leak. Attach a wrench to the valve assembly and using even pressure, tighten the valve slowly until the leak stops. If the leak persists, attach a hood assembly. Follow the directions found in the valve blowout section to assemble the hood assembly.

Valve Seat

Foreign matter may occasionally become lodged in the valve assembly. When this happens, the valve will not seat properly when closed. In these instances, simply opening and closing the valve may be enough to dislodge the foreign matter and allow the valve to seat correctly. If the cylinder is disconnected from its process and can be safely reconnected, responders should reconnect and attempt to open and close the valve stem in an effort to dislodge the foreign matter. If the cylinder cannot be safely reconnected to its process, the hazmat technician should apply an outlet cap and gasket. If the leak persists, attach a hood assembly. Follow the directions found in the valve blowout section to assemble the hood assembly.

Figure 13.36 Tighten the packing nut securely, but do not overtighten.

Valve Stem Blowout

If the response team experiences a valve stem assembly blowout, the repair will be similar to the steps used for a valve blowout. Hazmat technicians should immediately drive the drift pin into the valve body. Once this is accomplished, apply the hood assembly. Follow the directions found in the valve blowout section to assemble the hood assembly.

NOTE: In most instances the contents will have already been released.

> **Chime** — Reinforcement ring at the top (head) of a barrel or drum.

> **Bung** — Cork, plug, or other type of stopper used in a barrel, cask, drum, or keg.

Drums

Leaking drums are a fairly common type of hazmat incident. Typical low-pressure drums are flat pieces of metal rolled into tubes with two capped ends. A **chime** runs around the outer edge of each head. Various access holes are found in different types of drums, although openings are typically at the top. These openings or access holes are each closed with a right-handed screw plug known as a **bung**. Responders should pay attention to bulges in a surface of a drum because that may be an indicator of overpressure which could indicate the potential for a catastrophic failure.

Bung Leaks

Bung leaks are a common occurrence on drums, especially when they are lying on their side. Correcting a bung leak may be as simple as righting the drum so that the bung opening is in the vapor space **(Figure 13.37)**. Whenever responders are working with drums, product identification is critical. If the product is flammable, responders must ensure that vapor suppression activities are in place and that they are controlling vapors. When possible, an LEL of 0% is recommended. The AHJ will set action levels.

When attempting a bung-leak repair, the drum should always be upright if possible or the leak positioned at the highest point. Use a bung wrench to tighten the bung and stop the leak **(Figure 13.38)**. If necessary, overpack the drum. Overpacking will be discussed in a later section.

Figure 13.37 Position a drum so the vapor space is aligned with the breached area.

Chime Leaks

Similar to bung leaks, righting a barrel or drum so that the leak is in the vapor space is a simple way to stop a chime leak. Drum wrap or epoxies may also assist in stopping or controlling the leak. Make sure that the leak has stopped. Then overpack, seal, and properly mark the barrel.

Sidewall Punctures

Sidewall punctures may be more challenging to remedy than a bung or chime leak. These punctures can be as small as nail holes or as large as splits caused by forklift or tow motor impacts. As with any leaking container, the hazmat technician should look for any conditions that would indicate reactivity with the container or other products. If conditions such as rising temperatures are

Figure 13.38 Customized wrenches may be used to tighten a loose bung.

detected, follow SOPs accordingly. If the product is flammable, maintain acceptable LEL levels while working on the barrel and perform continuous monitoring.

Once the location of the leak is determined, orient the barrel so that the leak is at the highest point. Based on the size and orientation of the hole, use plugging and/or patching equipment to stop the leak. For small holes, epoxies are often the best. If possible, lead wool or other similar products can assist with the plug to help seal the barrel **(Figure 13.39)**. Once the leak has been stopped, trim the plugging material so that it is as flush to the barrel as possible. Cover the patched plugged area with tape or other appropriate material to help keep the plugging material in place. Then overpack, seal, and properly mark the barrel.

Overpacking

Overpacking is the process by which a damaged drum can be placed inside another container. Overpack drums are usually constructed of steel or plastic such as a polyethylene and may come in sizes ranging from 5 gallons to 119 gallons (20 L to 475 L) **(Figure 13.40)**. Overpacking is commonly used for liquids and solids; however, cylinder overpacks are becoming more commonly available. Drum liners can also be used with overpack drums. Capping kits are designed for a variety of pressurized containers. Steps for overpacking a nonbulk container and/or radioactive materials package are provided in **Skill Sheet 13-20**.

Technicians must address the following considerations when handling drums, especially regarding the drum's contents:

- Drums that have been involved in a transportation accident or that have been buried or abandoned for a long period of time may be compromised.

- Weakened drums may breach when overpacked or handled.

- Depending on the product contained within the drum, weights may be over 800 lbs (400 kg). Mechanical machinery may be needed to safely move these drums.

- Drums that contain liquid and have deteriorated to the point of being unmovable may need to have their contents off-loaded. Use hand pumps for transferring the contents if mechanical means are not available or may further damage the drum and cause a breach.

- Leaking barrels should be temporarily repaired before overpacking.

- Drums may be overpressurized.

Figure 13.39 A filler material will help create a tight seal in an irregularly shaped breach.

Figure 13.40 Overpack drums are manufactured in a variety of sizes and materials.

In addition, hazmat technicians should take the following personal precautions when handling or overpacking drums:

- Wear appropriate PPE.
- Keep fingers outside of drums.
- Use a drum upender or dolly to move drums whenever possible **(Figure 13.41)**.

Figure 13.41 A drum upender is a useful tool when maneuvering a drum.

- Maintain safe body positioning when moving a drum.
- Stay uphill of a drum.
- Chock drums in place to prevent their movement.

Tasks

When overpacking drums, responders must first consider the possible weight that they will have to handle. With this in mind, several tactical options are available. The main objective when handling drums is safety, both when dealing with the material and avoiding injuries when overpacking. Use the following methods to overpack a drum:

- **Slide-in Method** — Accomplished by sliding a drum on its side into an overpack drum. Use items such as 2x4 planks or pipes to assist the process. Lay the drum to be overpacked on its side with the pipes perpendicular to the barrel. Roll the drum onto the pipes. The pipes will assist with rolling the drum into the overpack **(Figure 13.42)**.

- **Rolling Slide-in Method (also called the V-Method)** — Position the drum and the overpack end-to-end at

Figure 13.42 Pipes can be strategically arranged to aid during overpacking procedures.

Chapter 13 • Implementing the Action Plan: Product Control **515**

Figure 13.43 Rolling the drum and overpack at an angle to each other will position part of the barrel inside the overpack with minimal effort.

a 30° angle (**Figure 13.43**). Roll both together in the same direction. A portion of the drum will enter the overpack and then responders will have to finish sliding it into the overpack.

- **Drum Lifting Method** — Completed with the assistance of mechanical equipment such as a forklift or crane. Lift the drum into the upright position. Place the overpack under the drum, and then lower it into the overpack. Ensure that personnel are not below the barrel when it is lifted.

- **Slip-Over Method (also called Inverted Method)** — The recovery drum is placed over the top of the drum (**Figure 13.44**). Tilt both drums onto the ground and right them with the drum inside the recovery drum. Consider product recovery by ensuring the drum bung openings are oriented towards the lid when righted.

Figure 13.44 An overpack drum can be placed over a leaking drum.

Lab Packs

Laboratory packs are used for the disposal of chemicals in process samples from universities, hospitals, or similar institutions. Lab packs often include multiple containers of compatible product surrounded by absorbent material. Lab packs may contain radioisotopes, shock-sensitive products and highly volatile, highly corrosive, or highly toxic chemicals.

In some instances, particularly in illicit labs, technicians may need to adulterate or contaminate the product. The process involves pouring the product directly into the absorbent material of the lab pack to ensure that the product cannot be reused in the future. Consider compatibility between the product and the absorbent material.

Other Basic Product Control Techniques

In addition to the techniques described in the earlier sections of this manual, some techniques are taught to all technicians. These techniques may be combined with others during scene control at a hazmat incident.

Vapor Suppression and Dispersion

Vapor suppression is the action taken to reduce the emission of vapors at a hazmat incident. Fire fighting foams are effective at suppressing vapors at spills of flammable and combustible liquids if the foam concentrate is compatible with the material. When suppressing vapors associated with hazardous materials, Class B foams are most commonly used **(Figure 13.45)**. Further information on vapor suppression and fire fighting foams may be found in IFSTA's **Hazardous Materials for First Responders** manual.

Vapor Suppression — Action taken to reduce the emission of vapors at a hazardous materials spill.

Vapor Dispersion — Action taken to direct or influence the course of airborne hazardous materials.

Figure 13.45 Class B foams are typically used for vapor suppression operations.

There are significant differences in Class B foams. For example, concentrates designed solely for hydrocarbon fires (such as regular fluoroprotein and regular Aqueous Film Forming Foam [AFFF]) will not extinguish polar solvent (alcohol-type fuel) fires regardless of the concentration at which they are used. Water-miscible materials such as alcohols, esters, and ketones destroy regular fire fighting foams and require an alcohol-resistant foam agent.

When vapors cannot be suppressed, dispersion may be a viable option. **Vapor dispersion** is the action taken to direct or influence the course of airborne hazardous materials. Use pressurized streams of water from hoselines or unattended master streams in a fog pattern to help disperse vapors. These streams create turbulence, which increases the rate of mixing with air and reduces the concentration of the hazardous material. After using water streams for vapor dispersion, responders must confine and analyze runoff water for possible contamination.

Figure 13.46 Covering a liquefied gas spill, in this case ammonia, with a blanket or tarp will suppress its rate of vaporization and cause it to autorefrigerate. *Courtesy of Rich Mahaney.*

NOTE: An AHJ may choose to use positive pressure ventilation fans to disperse vapors.

Blanketing/Covering

Personnel often perform blanketing or covering to prevent dispersion of hazardous materials. Responders must consider the compatibility between the material being covered and the material covering it.

Tools for blanketing or covering solids such as powders and dusts include tarps, plastic sheeting, salvage covers, or other materials (including foam). Blanketing/covering may also be done as a form of temporary mitigation for radioactive and biological substances, for example, to reduce alpha or beta radiation or prevent the spread of biological materials. Personnel can blanket/cover liquefied gas leaks to cause the released material to auto refrigerate beneath the tarp or covering **(Figure 13.46)**. As a temporary option, responders can cover openings of some liquid containers with plastic sheets or tarps to confine vapors. Blanketing of liquids is essentially the same as vapor suppression (see Vapor Suppression section) because it typically uses an appropriate aqueous (water) foam agent to cover the surface of a spill.

Specialized Product Control Techniques

Hazmat technicians must be aware of specialized techniques that may be necessary at some response operations. Technicians must also be aware of each technique's response limitations. Specialized response options specified in this manual are offered as an introduction to available options but are not typically considered a part of the hazmat technician response protocol.

Response personnel at the specialist level may conduct many of the following techniques, which are often complex and dangerous. Only personnel trained specifically in each of these respective techniques should attempt to perform the operations.

Hot and Cold Tapping

Tapping is the process of attaching a nozzle or outlet onto a container's tank or piping to assist with the removal or transfer of a product. Only trained personnel should conduct tapping operations.

Hot tapping is a technique in which a specialist conducts welding operations in order to attach the nozzle or outlet while the container or pipe still contains product **(Figure 13.47)**. This technique requires special training and practice. Use this technique only in emergency situations when the removal of a chemical from its container cannot be accomplished through normal means due to lack of container integrity or damage to the valves or piping.

Tapping — Process to attach a nozzle or outlet onto a container's tank or piping to assist with the removal or transfer of a product.

Hot Tapping — Using welding or cutting to attach a nozzle or outlet onto a container's tank or piping to assist with the removal or transfer of a product.

Figure 13.47 Hot tapping is the act of welding a nozzle or outlet to a container that still contains a hazardous material. *Courtesy of Barry Lindley.*

Cold tapping is similar to hot tapping. However, the specialist fastens a nozzle or outlet to the container or piping rather than welding it.

Cold Tapping — Fastening a nozzle or outlet onto a container's tank or piping to assist with the removal or transfer of a product.

Product Transfer

Product transfer is a technique that personnel can use to off-load a chemical or product from a damaged container into a receiving container. This process may also be conducted for overloaded containers. While this procedure may be used for any sized container, personnel must exercise special care when transferring products from rail cars. When working with railway containers, consider product transfer if the following conditions exist:

- The container is damaged and cannot be safely moved or rerailed.
- The rail tank is sound, but damage to the bolsters will not allow the car to be rerailed.
- Topography will preclude the container from being rerailed.
- The container is overloaded.
- Damaged valves cannot be repaired.

Based on the state of matter and the type of product, responders have a variety of options when considering product transfer. The tools and equipment needed to accomplish the operation will vary based on the product, its associated hazards, and the containers being used. Take care during any emergency product transfer, and only involve personnel specifically trained in this type of operation.

Flaring

Flaring is a technique that personnel can use to reduce the amount of product in a container **(Figure 13.48)**. This controlled burning process may be effective at reducing the volume of either gases or liquids through a flare pipe or other outlet. Flaring offers the following beneficial results:

Flaring — Controlled release and disposal of flammable gases or liquids through a burning process.

Figure 13.48 Flaring is a controlled burn of flammable gases and liquids. *Courtesy of Barry Lindley.*

- It will reduce the quantity of the chemical while decreasing the internal pressure of the container.
- The flammable gas is now a controlled burn, and the variables, including the rate of burn, are easier to control.

Take care to ensure that the flaring operation will not ignite other materials. Flaring can be performed on various types of containers, including tank cars.

There are two types of flaring operations: vapor flaring and liquid flaring. Vapor flaring is the burning of vapor of a compressed flammable gas at the outlet of a flare stand. Liquid flaring is the burning of liquefied flammable gas or flammable liquid (which vaporizes) at the end of a flare pipe in a pool.

Venting

Venting is a process in which trained responders release the product into the atmosphere in order to reduce the pressure within a container. This process is typically associated with nonflammable products such as liquefied compressed gases. The vented release may be directly into the atmosphere. If the product is toxic, it may be released after it has received a treatment process to reduce its toxicity. Conduct these operations only after consulting with the shipper, environmental specialists, and other container specialists. Because of the hazards involved, exercise extreme care when performing a venting operation.

Applying Large Plugging and Patching Devices

Many commercially available devices can be used for patching holes in larger containers. Manufacturers such as Vetter® and Paratech® make pneumatically inflatable bags that can be used on various-sized containers. Used in association with ratchet or other types of strapping devices, these bags can be inflated to control or even stop a leak on a variety of container sizes and materials. Flow-through plugs (Vetter®) can be used in association with the air bag for controlled off-loading operations.

Pneumatically inflatable bags are usually made of a nylon-type material and may have reinforcement on the edges. These are available in a variety of sizes. Covers are available for the bags that can protect them from products such as acids. Consider the compatibility of the product, the container, and the inflatable bag being used.

Vent and Burn

In some cases, venting and burning a hazardous material may be considered as a product control technique. This process is highly specialized and must only be considered when all other options are unavailable or will not work based on the circumstances.

The vent and burn process uses explosives to create controlled openings in a container. In this process, shaped charges are placed strategically around the container to open the container and burn off liquefied flammable compressed gas or flammable liquids. Use the vent and burn technique solely as a last resort.

Chapter Review

1. What is the difference between defensive and offensive operations?
2. What is the goal of damage assessment?
3. List causes of potential container stress.
4. List types of container damage and how they might be controlled.
5. What are some questions that responders should be asking when predicting likely behavior of damaged containers?
6. Describe three different types of patches used to repair hazardous materials containers.
7. Describe different types of leaks and breaches commonly encountered when responding to tank trucks and how they might be mitigated.
8. What variables must be considered when planning for product transfer and removal?
9. What are some ways that pressurized containers can fail?
10. What type of leaks occur on drums and what are some ways they may be corrected?
11. What considerations must be addressed when performing overpacking operations?
12. What are the similarities and differences between blanketing/covering and vapor suppression?
13. What are some of the specialized product control techniques that only specialized responders should conduct?

SKILL SHEETS

13-1
Perform absorption/adsorption. [NFPA 1072, 7.4.3.1]

WARNING: If this skill involves the use of actual hazardous material samples, hazardous materials can cause serious injury or fatality. Appropriate personal protective equipment (PPE) must be worn and safety precautions must be followed. The following skill sheet demonstrates general steps; specific hazmat incidents may differ in procedure. Always follow the AHJ's procedures for specific incidents.

Step 1: Ensure proper product control technique is chosen.

Step 2: Ensure that all responders involved in the control function are wearing appropriate PPE for performing absorption/adsorption operations and that appropriate hand tools have been selected.

Step 3: Select a location to efficiently and safely perform the absorption/adsorption operation.

Step 4: Select the most appropriate sorbent/adsorbent.

Step 5: Deploy the sorbent/adsorbent in a manner that most efficiently controls the spill.

Step 6: Upon mitigation of the incident, place any contaminated material, such as clothing, in an approved container for transportation to a disposal location.

Step 7: Seal and label the container and document appropriate information for department records.

Step 8: Decontaminate tools.

Step 9: Advance to decontamination line for decontamination.

Step 10: Inspect and maintain tools and equipment as per local SOPs and manufacturer's recommendations.

Step 11: Complete required reports and supporting documentation.

13-2
Perform damming. [NFPA 1072, 7.4.3.1]

WARNING: If this skill involves the use of actual hazardous material samples, hazardous materials can cause serious injury or fatality. Appropriate personal protective equipment (PPE) must be worn and safety precautions must be followed. The following skill sheet demonstrates general steps; specific hazmat incidents may differ in procedure. Always follow the AHJ's procedures for specific incidents.

NOTE: Instructor directions will need to determine if construction is overflow, underflow, or containment dam.

Step 1: Ensure proper product control technique is chosen.

Step 2: Ensure that all responders involved in the control function are wearing appropriate PPE for performing damming operations and that appropriate hand tools have been selected.

Step 3: Select a location to efficiently and safely perform damming operation.

Step 4: Construct the dam in a manner and location that most effectively controls the spill.

Step 5: Upon mitigation of the incident, place any contaminated material, such as clothing, in an approved container for transportation to a disposal location.

Step 6: Seal and label the container and document appropriate information for department records.

Step 7: Decontaminate tools.

Step 8: Advance to decontamination line for decontamination.

Step 9: Inspect and maintain tools and equipment as per local SOPs and manufacturer's recommendations.

Step 10: Complete required reports and supporting documentation.

SKILL SHEETS

13-3
Perform diking operations. [NFPA 1072, 7.4.3.1]

WARNING: If this skill involves the use of actual hazardous material samples, hazardous materials can cause serious injury or fatality. Appropriate personal protective equipment (PPE) must be worn and safety precautions must be followed. The following skill sheet demonstrates general steps; specific hazmat incidents may differ in procedure. Always follow the AHJ's procedures for specific incidents.

Step 1: Ensure proper product control technique is chosen.

Step 2: Ensure that all responders involved in the control function are wearing appropriate PPE for performing diking operations and that appropriate hand tools have been selected.

Step 3: Select a location to efficiently and safely perform the diking operation.

Step 4: Construct the dike in a manner and location that most efficiently controls and directs the spill to a desired location.

Step 5: Upon mitigation of the incident, place any contaminated material, such as clothing, in an approved container for transportation to a disposal location.

Step 6: Seal and label the container and document appropriate information for department records.

Step 7: Decontaminate tools.

Step 8: Advance to decontamination line for decontamination.

Step 9: Inspect and maintain tools and equipment as per local SOPs and manufacturer's recommendations.

Step 10: Complete required reports and supporting documentation.

13-4
Perform diversion. [NFPA 1072, 7.4.3.1]

WARNING: If this skill involves the use of actual hazardous material samples, hazardous materials can cause serious injury or fatality. Appropriate personal protective equipment (PPE) must be worn and safety precautions must be followed. The following skill sheet demonstrates general steps; specific hazmat incidents may differ in procedure. Always follow the AHJ's procedures for specific incidents.

Step 1: Ensure proper product control technique is chosen.

Step 2: Ensure that all responders involved in the control function are wearing appropriate PPE for performing diversion operations and that appropriate hand tools have been selected.

Step 3: Select a location to efficiently and safely perform the diversion operation.

Step 4: Construct the diversion in a manner and location that most effectively controls and directs the spill to a desired location.

Step 5: Working as a team, use hand tools to break the soil, remove the soil, pile the soil, and pack the soil tightly.

Step 6: Upon mitigation of the incident, place any contaminated material, such as clothing, in an approved container for transportation to a disposal location.

Step 7: Seal and label the container and document appropriate information for department records.

Step 8: Decontaminate tools.

Step 9: Advance to decontamination line for decontamination.

Step 10: Inspect and maintain tools and equipment as per local SOPs and manufacturer's recommendations.

Step 11: Complete required reports and supporting documentation.

SKILL SHEETS

13-5

Perform retention. [NFPA 1072, 7.4.3.1]

WARNING: If this skill involves the use of actual hazardous material samples, hazardous materials can cause serious injury or fatality. Appropriate personal protective equipment (PPE) must be worn and safety precautions must be followed. The following skill sheet demonstrates general steps; specific hazmat incidents may differ in procedure. Always follow the AHJ's procedures for specific incidents.

Step 1: Ensure proper product control technique is chosen.

Step 2: Ensure that all responders involved in the control function are wearing appropriate PPE for performing retention operations and that appropriate hand tools have been selected.

Step 3: Select a location to efficiently and safely perform the retention operation.

Step 4: Evaluate the rate of flow of the leak to determine the required capacity of the retention vessel.

Step 5: Working as a team, retain the hazardous liquid so that it can no longer flow.

Step 6: Upon mitigation of the incident, place any contaminated material, such as clothing, in an approved container for transportation to a disposal location.

Step 7: Seal and label the container and document appropriate information for department records.

Step 8: Decontaminate tools.

Step 9: Advance to decontamination line for decontamination.

Step 10: Inspect and maintain tools and equipment as per local SOPs and manufacturer's recommendations.

Step 11: Complete required reports and supporting documentation.

13-6
Perform vapor suppression. [NFPA 1072, 7.4.3.1]

WARNING: If this skill involves the use of actual hazardous material samples, hazardous materials can cause serious injury or fatality. Appropriate personal protective equipment (PPE) must be worn and safety precautions must be followed. The following skill sheet demonstrates general steps; specific hazmat incidents may differ in procedure. Always follow the AHJ's procedures for specific incidents.

Step 1: Ensure proper product control technique is chosen.

Step 2: Ensure that all responders involved in the control function are wearing appropriate PPE for performing vapor suppression operations.

Step 3: Select a location to efficiently and safely perform the vapor suppression operation.

Step 4: Evaluate the quantity and surface area of the hazardous material that has leaked.

Step 5: Determine the appropriate type of foam for the type of hazardous material present.

Step 6: Working as a team, deploy the foam eductor and foam, and advance the hoseline and foam nozzle to a position from which to apply the foam.

Step 7: Flow hoseline until finished foam is produced at the nozzle.

Step 8: Apply finished foam in an even layer covering the entire hazardous material spill area.

Step 9: Upon mitigation of the incident, place any contaminated material, such as clothing, in an approved container for transportation to a disposal location.

Step 10: Seal and label the container and document appropriate information for department records.

Step 11: Decontaminate tools.

Step 12: Advance to decontamination line for decontamination.

Step 13: Inspect and maintain tools and equipment as per local SOPs and manufacturer's recommendations.

Step 14: Complete required reports and supporting documentation.

SKILL SHEETS

13-7
Perform vapor dispersion. [NFPA 1072, 7.4.3.1]

WARNING: If this skill involves the use of actual hazardous material samples, hazardous materials can cause serious injury or fatality. Appropriate personal protective equipment (PPE) must be worn and safety precautions must be followed. The following skill sheet demonstrates general steps; specific hazmat incidents may differ in procedure. Always follow the AHJ's procedures for specific incidents.

Step 1: Ensure proper product control technique is chosen.

Step 2: Ensure that all responders involved in the control function are wearing appropriate PPE for vapor dispersion operations.

Step 3: Select a location to efficiently and safely perform the vapor dispersion operation.

Step 4: Working as a team, advance the hoseline to a position to apply agent through vapor cloud to disperse vapors.

Step 5: Constantly monitor the leak concentration, wind direction, exposed personnel, environmental impact, and water stream effectiveness.

Step 6: Upon mitigation of the incident, place any contaminated material, such as clothing, in an approved container for transportation to a disposal location.

Step 7: Seal and label the container and document appropriate information for department records.

Step 8: Decontaminate tools.

Step 9: Advance to decontamination line for decontamination.

Step 10: Inspect and maintain tools and equipment as per local SOPs and manufacturer's recommendations.

Step 11: Complete required reports and supporting documentation.

13-8
Perform dilution. [NFPA 1072, 7.4.3.1]

WARNING: If this skill involves the use of actual hazardous material samples, hazardous materials can cause serious injury or fatality. Appropriate personal protective equipment (PPE) must be worn and safety precautions must be followed. The following skill sheet demonstrates general steps; specific hazmat incidents may differ in procedure. Always follow the AHJ's procedures for specific incidents.

Step 1: Ensure proper product control technique is chosen.

Step 2: Ensure that all responders involved in the control function are wearing appropriate PPE for performing dilution operations.

Step 3: Select a location to efficiently and safely perform dilution operations.

Step 4: Evaluate the rate of flow of the leak to determine the required capacity of the retention area and the quantity of water required to dilute the material.

Step 5: Working as a team, monitor and assess the leak, and advance hoselines and tools to retention area.

Step 6: Flow water to dilute spilled material.

Step 7: Monitor any diking or dams to ensure integrity of retention area.

Step 8: Upon mitigation of the incident, place any contaminated material, such as clothing, in an approved container for transportation to a disposal location.

Step 9: Seal and label the container and document appropriate information for department records.

Step 10: Decontaminate tools.

Step 11: Advance to decontamination line for decontamination.

Step 12: Inspect and maintain tools and equipment as per local SOPs and manufacturer's recommendations.

Step 13: Complete required reports and supporting documentation.

SKILL SHEETS

13-9

Predict the likely behavior of known hazardous materials at an incident. [NFPA 1072, 7.2.4]

WARNING: If this skill involves the use of actual hazardous material samples, hazardous materials can cause serious injury or fatality. Appropriate personal protective equipment (PPE) must be worn and safety precautions must be followed. The following skill sheet demonstrates general steps; specific hazmat incidents may differ in procedure. Always follow the AHJ's procedures for specific incidents.

Step 1: Identify behavior of each hazardous material's container and contents.

Step 2: Identify any reactivity issues and hazards resulting from potential mixing of hazardous materials.

Step 3: Use approved process to predict likely behavior of materials and their containers when multiple materials are involved.

Step 4: Communicate description of likely behavior of the hazards.

13-10

Predict the likely effect of a hazardous materials incident within the endangered area. [NFPA 1072, 7.2.5]

WARNING: If this skill involves the use of actual hazardous material samples, hazardous materials can cause serious injury or fatality. Appropriate personal protective equipment (PPE) must be worn and safety precautions must be followed. The following skill sheet demonstrates general steps; specific hazmat incidents may differ in procedure. Always follow the AHJ's procedures for specific incidents.

Step 1: Determine or predict concentrations of materials within the endangered area.

Step 2: Identify physical, health, and safety hazards within the endangered area.
Step 3: Identify areas of potential harm in the endangered area.
Step 4: Identify potential outcomes within the endangered area.
Step 5: Communicate potential outcomes.

SKILL SHEETS

13-11

Select appropriate product control techniques to mitigate a hazmat/WMD incident.
[NFPA 1072, 7.4.3.1]

WARNING: If this skill involves the use of actual hazardous material samples, hazardous materials can cause serious injury or fatality. Appropriate personal protective equipment (PPE) must be worn and safety precautions must be followed. The following skill sheet demonstrates general steps; specific hazmat incidents may differ in procedure. Always follow the AHJ's procedures for specific incidents.

Step 1: Select approved PPE.
Step 2: Select approved product control technique.

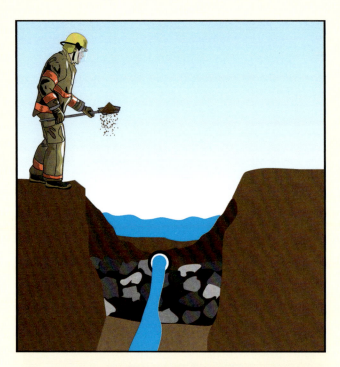
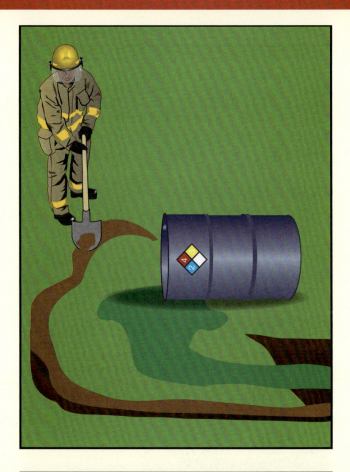
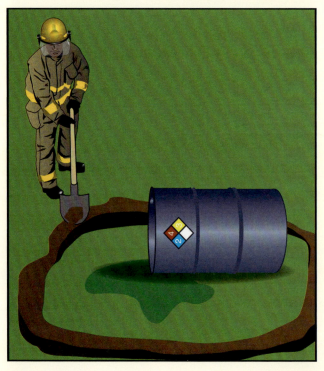
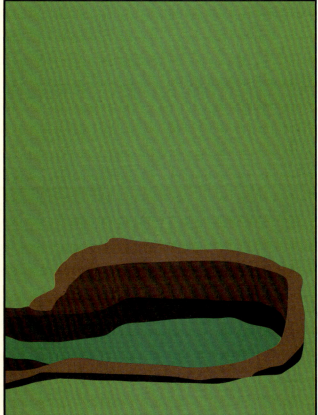

532   Chapter 13 • Implementing the Action Plan: Product Control

13-12
Plug a leaking container. [NFPA 1072, 7.4.3.2]

WARNING: If this skill involves the use of actual hazardous material samples, hazardous materials can cause serious injury or fatality. Appropriate personal protective equipment (PPE) must be worn and safety precautions must be followed. The following skill sheet demonstrates general steps; specific hazmat incidents may differ in procedure. Always follow the AHJ's procedures for specific incidents.

Step 1: Ensure proper product control technique is chosen.

Step 2: Ensure that all responders involved in the control function are wearing appropriate PPE for performing plugging operations and that appropriate hand tools have been selected.

Step 3: Select a location to efficiently and safely perform the plugging operation.

Step 4: Avoid direct contact with the hazardous material to the extent possible.

Step 5: Determine the location of the leak.

Step 6: Attempt to position container so that the location of leak is in the uppermost position.

Step 7: Use the proper tools and equipment to control and plug the leak.

Step 8: Decontaminate tools.

Step 9: Advance to decontamination line for decontamination.

Step 10: Inspect and maintain tools and equipment as per local SOPs and manufacturer's recommendations.

Step 11: Complete required reports and supporting documentation.

13-13
Patch a leaking container. [NFPA 1072, 7.4.3.2]

WARNING: If this skill involves the use of actual hazardous material samples, hazardous materials can cause serious injury or fatality. Appropriate personal protective equipment (PPE) must be worn and safety precautions must be followed. The following skill sheet demonstrates general steps; specific hazmat incidents may differ in procedure. Always follow the AHJ's procedures for specific incidents.

Step 1: Ensure proper product control technique is chosen.

Step 2: Ensure that all responders involved in the control function are wearing appropriate PPE for performing patching operations and that appropriate hand tools have been selected.

Step 3: Select a location to efficiently and safely perform the patching operation.

Step 4: Avoid direct contact with the hazardous material to the extent possible.

Step 5: Determine the location of the leak.

Step 6: Attempt to position container so that the location of leak is in the uppermost position.

Step 7: Use the proper tools and equipment to control and patch the leak.

Step 8: Decontaminate tools.

Step 9: Advance to decontamination line for decontamination.

Step 10: Inspect and maintain tools and equipment as per local SOPs and manufacturer's recommendations.

Step 11: Complete required reports and supporting documentation.

13-14
Apply a sleeve, jacket, clamp, or wrap to a leak. [NFPA 1072, 7.4.3.2]

SKILL SHEETS

WARNING: If this skill involves the use of actual hazardous material samples, hazardous materials can cause serious injury or fatality. Appropriate personal protective equipment (PPE) must be worn and safety precautions must be followed. The following skill sheet demonstrates general steps; specific hazmat incidents may differ in procedure. Always follow the AHJ's procedures for specific incidents.

Step 1: Ensure proper product control technique is chosen.

Step 2: Ensure that all responders involved in the control function are wearing appropriate PPE for applying a sleeve, jacket, clamp, or wrap to a leak and that appropriate hand tools have been selected.

Step 3: Select a location to efficiently and safely apply a sleeve, jacket, clamp or wrap to a leak.

Step 4: Avoid direct contact with the hazardous material to the extent possible.

Step 5: Determine the location of the leak.

Step 6: Attempt to position container so that the location of leak is in the uppermost position.

Step 7: Apply the sleeve, jacket, clamp or other wrap to control the leak.

Step 8: Decontaminate tools.

Step 9: Advance to decontamination line for decontamination.

Step 10: Inspect and maintain tools and equipment as per local SOPs and manufacturer's recommendations.

Step 11: Complete required reports and supporting documentation.

SKILL SHEETS

13-15

Conduct liquid transfer operations involving a leaking nonpressure container.
[NFPA 1072, 7.4.3.4]

WARNING: If this skill involves the use of actual hazardous material samples, hazardous materials can cause serious injury or fatality. Appropriate personal protective equipment (PPE) must be worn and safety precautions must be followed. The following skill sheet demonstrates general steps; specific hazmat incidents may differ in procedure. Always follow the AHJ's procedures for specific incidents.

Step 1: Ensure proper product control technique is chosen.

Step 2: Ensure that all responders involved in the control function are wearing appropriate PPE for performing liquid transfer operations and that appropriate hand tools have been selected.

Step 3: Select a location to efficiently and safely perform the liquid transfer operation.

Step 4: Protect exposures and personnel.

Step 5: Follow safety procedures.

Step 6: Minimize/avoid hazards.

Step 7: Complete hazards monitoring requirements.

Step 8: Ensure recovery container is compatible with liquid product.

Step 12: Decontaminate tools.

Step 13: Advance to decontamination line for decontamination.

Step 14: Inspect and maintain tools and equipment as per local SOPs and manufacturer's recommendations.

Step 15: Complete required reports and supporting documentation.

Step 9: Recognize the need to suppress vapors.

Step 10: Bond and ground containers following steps in **Skill Sheet 13-16**.

Step 11: Use approved liquid transfer method to transfer product to recovery container.

13-16
Bond and ground a container. [NFPA 1072, 7.4.3.4]

WARNING: If this skill involves the use of actual hazardous material samples, hazardous materials can cause serious injury or fatality. Appropriate personal protective equipment (PPE) must be worn and safety precautions must be followed. The following skill sheet demonstrates general steps; specific hazmat incidents may differ in procedure. Always follow the AHJ's procedures for specific incidents.

Caution: If product is flammable, ensure appropriate safety precautions are in place.

Step 1: Ensure proper product control technique is chosen.

Step 2: Ensure that all responders involved in the control function are wearing appropriate PPE for performing bonding and grounding operations.

Step 3: Select a location to efficiently and safely perform the bonding and grounding operation.

Step 4: Establish ground field.

Note: If applicable, check resistance between grounding rod and earth.

Step 5: Bond the containers.

Note: If applicable, check resistance between bonding clamp and container.

Step 7: If necessary, decontaminate tools and advance to decontamination line for decontamination.

Step 8: Notify the Incident Commander of the completed objective.

Step 9: Inspect and maintain tools and equipment as per local SOPs and manufacturer's recommendations.

Step 10: Complete required reports and supporting documentation.

Step 6: Ground the containers.

SKILL SHEETS

13-17

Perform remote valve shutoff or activate emergency shutoff device. [NFPA 1072, 7.4.3.2]

WARNING: If this skill involves the use of actual hazardous material samples, hazardous materials can cause serious injury or fatality. Appropriate personal protective equipment (PPE) must be worn and safety precautions must be followed. The following skill sheet demonstrates general steps; specific hazmat incidents may differ in procedure. Always follow the AHJ's procedures for specific incidents.

Step 1: Ensure proper product control technique is chosen.

Step 2: Ensure that all responders involved in the control function are wearing appropriate PPE for performing remote valve shutoff operations.

Step 3: Identify and locate the emergency remote control valve and/or emergency shutoff device.

Step 4: Operate the remote control valve and/or emergency shutoff device properly.

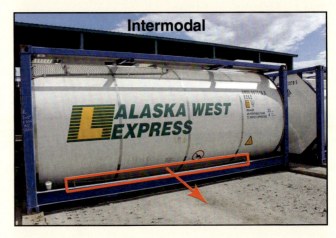

Step 5: If necessary, decontaminate tools and advance to decontamination line for decontamination.

Step 6: Notify the Incident Commander of the completed objective.

Step 7: Inspect and maintain tools and equipment as per local SOPs and manufacturer's recommendations.

Step 8: Complete required reports and supporting documentation.

13-18

Tighten or close leaking valves, closures, packing glands, and/or fittings.
[NFPA 1072, 7.4.3.2]

SKILL SHEETS

WARNING: If this skill involves the use of actual hazardous material samples, hazardous materials can cause serious injury or fatality. Appropriate personal protective equipment (PPE) must be worn and safety precautions must be followed. The following skill sheet demonstrates general steps; specific hazmat incidents may differ in procedure. Always follow the AHJ's procedures for specific incidents.

Note: If more than one container is present, attempt to locate the leaking container by looking for vapors or by using appropriate leak detection material or equipment.

If the container is connected to piping/process unit, close the valves that connect the container to the process and turn off the process after consulting with the process owner/operator.

Step 1: Ensure proper product control technique is chosen.

Step 2: Ensure that all responders involved in the control function are wearing appropriate PPE for performing tightening or closing of leaking valves, closures, packing glands, and/or fittings and that appropriate hand tools have been selected.

Step 3: Select a location to efficiently and safely perform tightening or closing operations.

Step 4: Remove obstructions from container if possible.

Step 5: Attempt to position container so that valve is in the uppermost position.

Step 6: If the container is disconnected from a process and can be reconnected, reconnect and gently open and close the valve stem to dislodge foreign matter from the seat using the valve handle or a wrench.

Step 7: Tighten or close leaking valves, closures, packing glands, and/or fittings as appropriate.

Step 8: Decontaminate tools.

Step 9: Advance to decontamination line for decontamination.

Step 10: Inspect and maintain tools and equipment as per local SOPs and manufacturer's recommendations.

Step 11: Complete required reports and supporting documentation.

SKILL SHEETS

13-19
Cap a leak. [NFPA 1072, 7.4.3.2]

WARNING: If this skill involves the use of actual hazardous material samples, hazardous materials can cause serious injury or fatality. Appropriate personal protective equipment (PPE) must be worn and safety precautions must be followed. The following skill sheet demonstrates general steps; specific hazmat incidents may differ in procedure. Always follow the AHJ's procedures for specific incidents.

Note: If more than one container is present, attempt to locate the leaking container by looking for vapors or by using appropriate leak detection material or equipment.

If the container is connected to piping/process unit, close the valves that connect the container to the process and turn off the process after consulting with the process owner/operator.

Step 1: Ensure proper product control technique is chosen.

Step 2: Ensure that all responders involved in the control function are wearing appropriate PPE for capping operations and that appropriate hand tools have been selected.

Step 3: Select a location to efficiently and safely perform capping operations.

Step 4: Remove obstructions from container if possible.

Step 5: Attempt to position container so that valve is in the uppermost position.

Step 6: If the container is disconnected from a process and can be reconnected, reconnect and gently open and close the valve stem to dislodge foreign matter from the seat using the valve handle or a wrench.

Step 7: Tighten or close leaking valves, closures, packing glands, and/or fittings as detailed in **Skill Sheet 13-18**.

Step 8: If leak persists, apply capping kit as per manufacturer's instructions.

Step 9: Decontaminate tools.

Step 10: Advance to decontamination line for decontamination.

Step 11: Inspect and maintain tools and equipment as per local SOPs and manufacturer's recommendations.

Step 12: Complete required reports and supporting documentation.

13-20

Overpack a nonbulk container and/or radioactive materials package.
[NFPA 1072, 7.4.3.3]

WARNING: If this skill involves the use of actual hazardous material samples, hazardous materials can cause serious injury or fatality. Appropriate personal protective equipment (PPE) must be worn and safety precautions must be followed. The following skill sheet demonstrates general steps; specific hazmat incidents may differ in procedure. Always follow the AHJ's procedures for specific incidents.

Slide-in Method

Step 1: Ensure proper product control technique is chosen.

Step 2: Ensure that all responders involved in the control function are wearing appropriate PPE for performing overpacking operations and that appropriate hand tools have been selected.

Step 3: Select a location to efficiently and safely perform the overpacking operation.

Step 4: Avoid direct contact with the hazardous material to the extent possible, including residue on the patched container or package.

Caution: Avoid placing hands in pinch-point areas while maneuvering the container or package.

Step 5: Place the damaged container or package in appropriate position using safe lifting techniques.

Step 6: Using a wedge, place the container or package onto a board or piece of pipe using safe lifting techniques.

Step 7: With the damaged container or package resting on the board or pipe, position the overpack container so its open top slides under the damaged container or package by several inches.

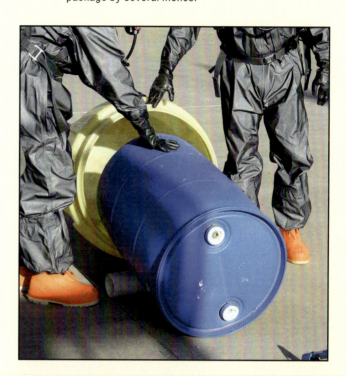

Step 8: As overpack container is held in place, team member(s) grasps the top of the damaged container or package and rolls/slides it on to the pipe into the overpack container.

Chapter 13 • Implementing the Action Plan: Product Control   541

13-20 cont.
Overpack a nonbulk container and/or radioactive materials package.
[NFPA 1072, 7.4.3.3]

Step 9: Lift the overpack container into an upright position using safe lifting techniques and attach the overpack container lid.

Step 10: Close the overpack container per manufacturer's instructions.
Step 11: Mark and label the overpack container according to appropriate regulations and AHJ's SOPs.
Step 12: Decontaminate tools.
Step 13: Advance to decontamination line for decontamination.
Step 14: Inspect and maintain tools and equipment as per local SOPs and manufacturer's recommendations.
Step 15: Complete required reports and supporting documentation.

Rolling Slide-in Method
Step 1: Ensure proper product control technique is chosen.
Step 2: Ensure that all responders involved in the control function are wearing appropriate PPE for performing overpacking operations and that appropriate hand tools have been selected.
Step 3: Select a location to efficiently and safely perform the overpacking operation.
Step 4: Avoid direct contact with the hazardous material to the extent possible, including residue on the patched container or package.

Caution: Avoid placing hands in pinch-point areas while maneuvering the container or package.
Step 5: Place the damaged container or package on its side using safe lifting techniques.
Step 6: With the damaged container or package on its side, position the overpack container at approximately 30 degrees to the base of the damaged container or package.
Step 7: Ensure that the area in front of the container or package is clear of obstructions.
Step 8: Roll both containers or packages and overpack container so that the damaged container or package is inserted into the overpack container.

Step 9: Repeat Steps 5-8 to get majority of container or package into the overpack container.

542 Chapter 13 • Implementing the Action Plan: Product Control

13-20 cont.
Overpack a nonbulk container and/or radioactive materials package.
[NFPA 1072, 7.4.3.3]

Step 10: Lift the overpack container into an upright position using safe lifting techniques and attach the overpack container lid.

Step 11: Close the overpack container as per manufacturer's instructions.
Step 12: Mark and label the overpack container according to appropriate regulations and AHJ's SOPs.
Step 13: Decontaminate tools.
Step 14: Advance to decontamination line for decontamination.
Step 15: Inspect and maintain tools and equipment as per local SOPs and manufacturer's recommendations.
Step 16: Complete required reports and supporting documentation.

Slip-Over Method
Caution: This method presents a high risk for pinching or crushing injuries to hands and feet, as well as strains and sprains. Use caution when moving the containers.
Step 1: Ensure proper product control technique is chosen.
Step 2: Ensure that all responders involved in the control function are wearing appropriate PPE for performing overpacking operations and that appropriate hand tools have been selected.
Step 3: Select a location to efficiently and safely perform the overpacking operation.
Step 4: Avoid direct contact with the hazardous material to the extent possible, including residue on the patched container or package.
Caution: Avoid placing hands in pinch-point areas while maneuvering the container or package.
Step 5: Stand the damaged container or package using safe

lifting techniques.
Step 6: Turn overpack container upside down and place over the damaged container or package.
Note: Step 6 may differ according to the type of overpack container used. Follow all AHJ's SOPs.
Step 7: With the damaged container or package inside the

Chapter 13 • Implementing the Action Plan: Product Control   **543**

13-20 cont.
Overpack a nonbulk container and/or radioactive materials package.
[NFPA 1072, 7.4.3.3]

overpack container, carefully lay the overpack container on its side.

Step 8: Lift the overpack container into an upright position

using safe lifting techniques and attach the overpack container lid.

Step 9: Close the overpack container as per manufacturer's instructions.

Step 10: Mark and label the overpack container according to appropriate regulations and AHJ's SOPs.

Step 11: Decontaminate tools.

Step 12: Advance to decontamination line for decontamination.

Step 13: Inspect and maintain tools and equipment as per local SOPs and manufacturer's recommendations.

Step 14: Complete required reports and supporting documentation.

Implementing the Action Plan: Incident Demobilization and Termination

Chapter Contents

Demobilization **549**	After Action Report 554
Termination **550**	Documentation 555
Debriefing 550	**Chapter Review** **556**
Postincident Critique 552	**Skill Sheets** **557**

chapter 14

Key Terms

After Action Report (AAR) 554
Debriefing 550
Demobilization 549

Implementing the Action Plan: Incident Demobilization and Termination

JPRs Addressed In This Chapter

7.4.3.4, 7.5.1, 7.6.1

Learning Objectives

After reading this chapter, students will be able to:

1. Explain demobilization.
2. Describe processes for incident termination. [NFPA 1072, 7.4.3.4, 7.5.1, 7.6.1]
3. Skill Sheet 14-1: Participate in a postincident critique. [NFPA 1072, 7.5.1, 7.6.1]

Chapter 14
Implementing the Action Plan: Incident Demobilization and Termination

Just as incident operations must be well coordinated, resource demobilization and incident termination must also be carefully planned. In hazardous materials incidents, the transition into demobilization and termination modes typically occurs as the incident priorities change from response to recovery. This chapter addresses incident demobilization and termination considerations. Termination was introduced in Chapter 9 of this manual.

Demobilization

As operations at an incident near completion, resources and equipment on scene may no longer be needed and may be better served elsewhere **(Figure 14.1)**. This is especially true for mutual aid units and other departments' equipment. **Demobilization** should be conducted in an organized manner, beginning with resources that are no longer needed. Demobilizing resources allows personnel to be rehabbed and units to be restocked and put back in service.

Demobilization — The process of identifying assets on the scene that are no longer needed and returning them to service.

Figure 14.1 The availability and necessity of resources and equipment should be evaluated throughout an incident.

Traditionally, demobilization has been thought of as taking place at the incident's conclusion. However, the Incident Commander (IC) should begin planning for demobilization at the incident's start. Careful demobilization planning can help eliminate waste and may eliminate potential fiscal and legal impacts. It ensures a controlled, safe, efficient, and cost-effective release of equipment and resources.

Demobilization should be conducted according to department policies and procedures and may vary based on the incident's size. Agency policies, procedures, and agreements must be considered prior to releasing assets. For example, on a small incident, resources may be released to finish work shifts.

At larger incidents, resources may have to be demobilized based on agreed-upon work schedules or other jurisdictional rules.

Demobilization must be conducted in concert with the IC and Incident Management team. Primary roles of the IC and Section leaders may include the following:

- **Incident Commander** — Approves resource orders and demobilization.
- **Operations Section** — Identifies operational resources that are, or may become, excess to the incident.
- **Planning Section** — Develops and implements the demobilization plan.
- **Logistics Section** — Implements transportation inspection program and handles special transportation needs.
- **Finance/Administration Section** — Processes claims, time records, incident costs, and assists in release priorities.

Using pre-developed checklists during demobilization will ensure that everything is completed. Such checklists are developed before the incident using many different resources.

Termination

Termination is the incident phase where emergency response activities near completion and recovery operations start to take priority. The responsibility for clean-up must be determined. The incident's administration may be passed on to local, state, or federal authorities, and cleanup contractors may be responsible for additional work.

Even though the incident may be winding down, critical steps must still be followed. The following sections cover the items that are required when terminating a hazardous materials incident. Three primary components must be incorporated during the termination of hazmat incidents:

- An incident debriefing
- A postincident critique
- After action procedures that include complete documentation of the incident and activities

NOTE: These activities are included in 29 *CFR* 1910.120; they are therefore mandated in the U.S.

Debriefing

In the simplest terms, a **debriefing** (also called a "hot wash" or "tailboard review") gathers information from all personnel who were involved in incident operations. The most effective form of debriefing may be a group discussion. The main debriefing objectives and discussion topics should include the following:

- Who responded to the incident?
- What operations took place?
- When did operations take place?
- Were operations successful?
- Were there any injuries and what treatment was provided?

> **Debriefing** — A gathering of information from all personnel that were involved in incident operations.

During the debriefing, it is necessary to evaluate the Site Safety Plan's accuracy. The plan should be evaluated to ensure that all aspects were in place at the appropriate times and that the plan covered all the expected issues with the anticipated outcomes.

Concepts of an Effective Debriefing

To be effective, a debriefing should start as soon as the incident's emergency phase has concluded. It is important to gather the information needed while it is still fresh in responders' minds. This discussion also provides an opportunity to deliver information to those who responded. Information gathered and delivered in a debriefing should include:

- A list of all personnel and agencies involved
- A listing of all substances involved
- Any exposures including people, equipment, or environmental factors
- Signs and symptoms from the exposure
- Postincident medical contact
- Secondary decontamination procedures
- Status of equipment
- Problems or issues with the equipment and/or personnel
- Incident organization
- Timeline of the incident
- Any other pertinent information

Delayed Symptoms

Some symptoms of exposure may not manifest until well after the incident. The debriefing must include information about potential delayed signs and symptoms. For example, some pesticide effects may not be noticed until 72 hours after initial exposure. Without this information, someone who has been exposed to potentially harmful materials may not seek medical care until it is too late.

The debriefing should be conducted in a distraction-free area. While this is not always possible on the emergency scene, a quiet area should be sought. Environmental conditions should be considered as well. Debriefing response personnel in extreme cold or heat may not have the best results.

To be effective, a hazmat team leader, company officer, or chief officer should lead an organized debriefing. While the IC is not required to lead the debriefing, someone who is able to facilitate good dialogue between participants should be in charge. Based on the chemical information and specific hazards, a responder with a more specialized background may be needed to effectively deliver the debriefing information.

It is best to limit the debriefing's overall time. After a long and strenuous work period, the emergency responders' attention span may be limited. A concise, organized approach that will deliver necessary information in a limited time frame will have the best results.

When to Conduct a Debriefing

Conduct the debriefing as soon as reasonably possible at the incident's conclusion so that information is fresh and accurate. While it may be very beneficial to conduct the debriefing with everyone who was involved at the same time, it is often not practical. However, it is imperative that each emergency responder be interviewed. The debriefing leader should meet and interview the participants and determine what actions they took during the incident.

At larger incidents, it may not be possible to interview and debrief everybody who was on the scene. In these instances, it may be necessary to debrief Section leaders **(Figure 14.2)**. These personnel will gather the necessary information and pass it on to the personnel who were in their charge **(Figure 14.3)**. This information will be used in the postincident critique/analysis.

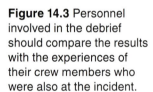

Figure 14.2 When an incident is very large, it may be most practical to gather the Section leaders for the on-scene debrief.

Figure 14.3 Personnel involved in the debrief should compare the results with the experiences of their crew members who were also at the incident.

Postincident Critique

A postincident critique (postincident analysis) is a more formal means of evaluating all the events that took place during the incident. This is a critical function of Incident Command and should be conducted in order to identify successes or difficulties that were encountered during the response. Once identified, these experiences can provide future direction, guide planning and training activities, and identify needed equipment and any additional SOPs.

The postincident critique/analysis should answer the following questions:

- What did we do right?
- What did we do wrong?
- What can we do differently in the future?

The postincident critique can aid in appraising concerns such as:

- Personnel and public safety
- Information and data management
- Interagency cooperation
- Overall function of the Incident Management System (IMS)
- Effectiveness of any operational techniques
- Public notification

Components of a Postincident Critique

The postincident critique should concentrate on all aspects of the incident including successes, failures, and needed improvements that can be made to ensure future success. An effective critique requires three key components to be successful:

- Direction
- Participation
- Solutions

A single individual should lead the critique **(Figure 14.4)**. While input is sought from all persons involved in the critique, the person conducting the session should act as a facilitator in order to stay on task and not allow the critique to stray from its primary mission. It is imperative that the facilitator remain neutral during the session, especially if there are sensitive subjects that must be brought forward concerning the incident. Typical elements of a postincident critique agenda include:

- A review of the response
- Input from key personnel
- Lessons learned

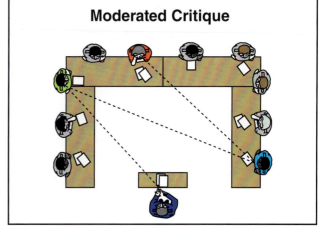

Figure 14.4 One individual should lead a postincident critique to moderate and guide the conversation through potentially heated topics.

The purpose of a critique is not to assign blame and criticism to individuals or groups. The critique must be seen as a positive event and allow for a free-flow of information. It is important to take the information that is extracted from the critique and apply it to operational considerations for future incidents.

Postincident Critique Participation

While it is necessary to have input from all incident participants at the postincident critique, doing so in an organized manner allows for better information flow. Dividing the critique into several levels can assist the facilitator in maintaining control and being able to record and document the information for future use. Levels of critique can include the following:

- **Participant level critique** — Key personnel will be able to make individual statements relevant to their activities at the incident. Participants should concentrate on major issues they feel warrant a discussion.

- **Section level critique** — This is a structured review of emergency operations. Through a group spokesperson, each operational section presents a summary of its assigned roles during the incident. The section spokesperson should discuss challenges and any unanticipated events that may have occurred. The section spokesperson may also discuss any recommendations on how the section could improve performance.

- **Group level critique** — This critique allows for a wider and more open forum. An open dialogue should be encouraged from group participants. The facilitator should reinforce any constructive comments, and all important points should be recorded and documented.

During the critique proceedings, meeting times should be limited to no more than ninety minutes. This is to ensure that those involved remain active and engaged throughout the process. Steps for participating in a postincident critique are shown in **Skill Sheet 14-1**.

After Action Report

The critique's proceedings should be thoroughly documented. To accomplish this, an individual can be designated to take notes during the critique sessions **(Figure 14.5)**. At the conclusion, generate and file a report that will deliver the lessons learned from the incident and any recommendations for improvement that may be appropriate for future responses. This document is called the **After Action Report (AAR)**.

> **After Action Report (AAR)** — A concise report that details and analyzes incident operations, provides lessons learned from the incident, and makes recommendations for improvement in future responses.

Figure 14.5 After the postincident critique, the After Action Report offers the critique's findings and is maintained as part of the incident's documentation.

AARs serve a vital role and should be shared with other organizations so that lessons learned at the incident may benefit others. Many injuries and fatalities have been prevented because departments shared their critique records and were not afraid to outline their deficiencies and what lessons they learned from each incident. Critique documents are an essential learning tool for not only the department involved, but others as well.

Documentation

Each emergency response organization will have its own unique requirements for reporting activities and actions at hazardous materials incidents. Some documentation requirements may be specified by department policy while others are mandatory under state or federal law.

National Fire Incident Reporting System (NFIRS) Reports

Most fire departments in the United States participate in the National Fire Incident Reporting System (NFIRS). NFIRS is a national system sponsored by the United States Fire Administration. This report will be the primary response record for a fire-based response. NFIRS also supplies a supplemental record for reporting hazardous materials incidents. The local report should be forwarded to the lead state fire agency and then on to the U.S. Fire Administration. Other incident reports may be required depending on the AHJ.

Debriefing Records

Debriefing records serve as the incident's chronological account. These should include any follow-up records on response personnel if problems arise at a later date. Debriefing records should include:

- All written documentation
- Photographs
- Video and audio tapes
- Computer-generated reports

NOTE: Debriefing records may be used in legal proceedings. Therefore, these documents must be accurate and secure.

Critique Records

Accurate critique records can be used in several ways including increasing the safety of on-scene operations. These records can also be useful for planning, training, and prioritizing hazards. Critique records should be reviewed after the incident to ensure that any specific issues have been rectified and operational procedures have been adjusted to allow for an organized and safe operation.

Reimbursement Logs

Reimbursement logs should be maintained in case any equipment used during the incident is eligible for reimbursement. FEMA has established a reimbursement schedule for a variety of equipment that may be used at an emergency scene. If a response agency plans to apply for reimbursement, accurate record keeping will be a vital component for eligibility.

Records Management

Because hazmat incidents are so complex, accurate and organized records must be kept. Records must maintain a permanent record for the health and safety of responders. In addition, the incident's records may be used for both criminal and civil court proceedings. Various records and logs should be maintained and secured along with incident reports. Some of these records may include:

- Activity logs
- Exposure records
- Hot zone entry and exit logs
- Personal protective equipment logs
- Any local, state, and federal regulatory reports

All reports should be completed and filed in a timely manner. While each local agency will have its unique reporting requirements, federal regulatory requirements may be more stringent.

Incident reports and all associated supplemental reports must be kept for a definitive length of time. These requirements should be reviewed with department and municipal administrators. Because employee medical records may be affected by these reports, OSHA requires that hazmat incident reports be secured and maintained for the length of an employee's tenure, plus thirty years.

NOTE: Records, logs, reports, photos, videos, and other documents must be complete and accurate. These documents may be used in legal proceedings, training, after action reviews, and other investigations.

Chapter Review

1. What are the benefits of a well-planned and careful demobilization?
2. What kinds of documents and records should be filed during the termination phase?

14-1
Participate in a postincident critique. [NFPA 1072, 7.5.1, 7.6.1]

WARNING: If this skill involves the use of actual hazardous material samples, hazardous materials can cause serious injury or fatality. Appropriate personal protective equipment (PPE) must be worn and safety precautions must be followed. The following skill sheet demonstrates general steps; specific hazmat incidents may differ in procedure. Always follow the AHJ's procedures for specific incidents.

Step 1: Provide assistance in scheduled incident debriefings and critiques.

Step 2: Communicate operational observations at debriefings and critiques.

Step 3: Report and document incident operations.

Step 4: Complete, forward, and file required reports, records, and supporting documents.

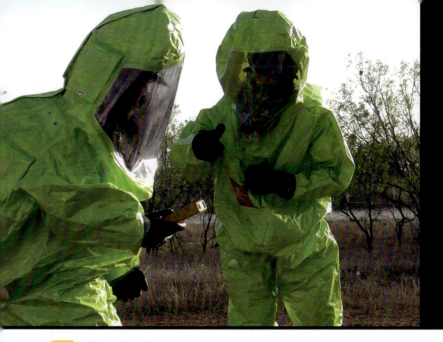

Appendices

Contents

Appendix A
Chapter and Page Correlation to NFPA 1072 Requirements 560

Appendix B
Protective Clothing Materials and Manufacturing Processes 561

Appendix A

Chapter and Page Correlation to NFPA 1072 Requirements

NFPA 1072 Competencies	Chapter References	Page References
7.2.1	2, 4, 5, 6, 8	29-45, 83-130, 135-156, 161-212, 249-319
7.2.2	7	217-245
7.2.3	8	249-319
7.2.4	2, 3, 4, 13	29-45, 49-79, 83-130, 481-544
7.2.5	5, 6, 13	135-156, 161-212, 481-544
7.3.1	9	323-348
7.3.2	11	375-434
7.3.3	12	439-470
7.3.4	5, 9	135-156, 323-348
7.4.1	10	353-370
7.4.2	11	375-434
7.4.3.1	11, 13	375-434, 481-544
7.4.3.2	11, 13	375-434, 481-544
7.4.3.3	11, 13	375-434, 481-544
7.4.3.4	11, 13, 14	375-434, 481-544, 549-557
7.4.4.1	12	439-470
7.4.4.2	11, 12	375-434, 439-470
7.5.1	14	549-557
7.6.1	14	549-557

Appendix B

Protective Clothing Materials and Manufacturing Processes

In order to protect the user, CPC is made of different materials that can withstand the threats posed by the challenge chemicals to which the suit has been tested. Different materials will work best in certain situations and may have limitations based on the hazard at hand. While no one suit material will stand up to the myriad of chemicals to which it may be exposed, the user must have an understanding of what materials are available and what protection they will offer.

DuPont®, Dow®, Lakeland®, and others manufacture materials used in the construction of CPC. Many of these manufacturers also have their own line of CPC products. More information on manufacturer materials and products can be found in product catalogs and on their websites.

A chemical protective garment can protect the wearer from harm. While the suit material is vitally important, the way it is manufactured or assembled is equally important. Seam construction can vary with the style of clothing and can also have a bearing on the suit's certification levels. The seam of a garment is where two sections are joined together. The following is an explanation of the different types of seams that are used in the manufacturing process of CPC:

- **Surged/sewn seam** — Seam where three threads are interlocked around the raw edge of two piles of material (**Figure B.1**).

- **Bound seam** — A clean finished binding that encapsulates the raw edges of two piles of fabric. All layers are sewn through with a chain stitch (**Figure B.2**).

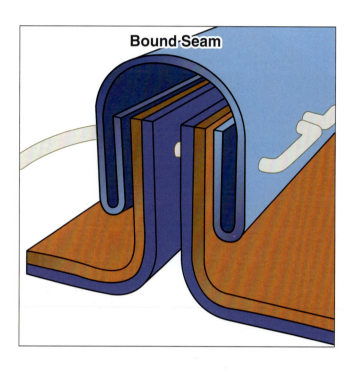

- **Ultrasonic joined seam** — Two sections of materials are lapped and acoustically welded. This allows for a clean seam with no needle holes (**Figure B.3**).

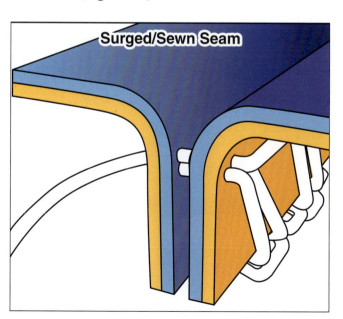

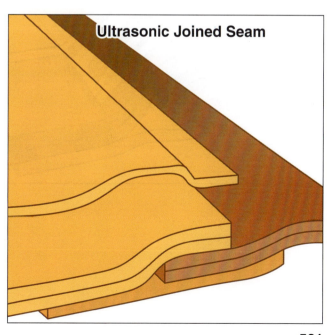

- **Taped seam** — Sewn seam is covered with a layer of compatible material and then heat sealed **(Figure B.4)**.

- **Glued and bonded seams** — Glued seam sealed with a thermal bonded overlayer **(Figure B.6)**.

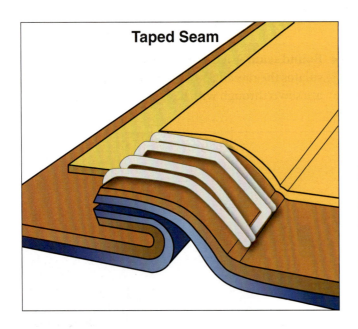

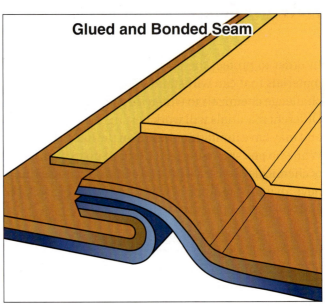

- **Double-taped seam** — Similar to the taped seam but both sides of the seam are taped and then heat sealed **(Figure B.5)**.

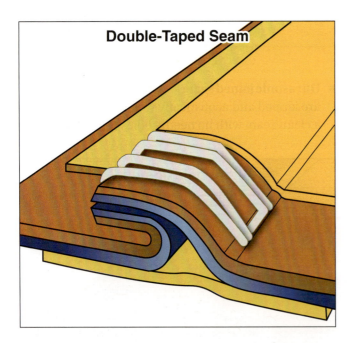

Courtesy of Barry Lindley.

Glossary

Glossary

A

Absorption — Penetration of one substance into the structure of another, such as the process of picking up a liquid contaminant with an absorbent.

Acetone Peroxide (TATP) — Triacetone triperoxide (TATP) is typically a white crystalline powder with a distinctive acrid (bleach) smell and can range in color from a yellowish to white color. *Similar to* Hexamethylene triperoxide diamine (HMTD).

Acid — Compound containing hydrogen that reacts with water to produce hydrogen ions; a proton donor; a liquid compound with a pH less than 7. Acidic chemicals are corrosive.

Action Options — Specific operations performed in a specific order to accomplish the goals of the response objective.

Activation Energy — Minimum energy that starts a chemical reaction when added to an atomic or molecular system.

Activity — Rate of decay of the isotope in terms of decaying atoms per second. Measured in becquerels (Bq) for small quantities of radiation, and curies (Ci) for large quantities of radiation.

Acute — Characterized by sharpness or severity; having rapid onset and a relatively short duration.

Acute Health Effects — Health effects that occur or develop rapidly after exposure to a hazardous substance.

Adsorb — To collect a liquid or gas on the surface of a solid in a thin layer.

Adsorption — Adherence of a substance in a liquid or gas to a solid. This process occurs on the surface of the adsorbent material.

After-Action Report (AAR) — A concise report that details and analyzes incident operations, provides lessons learned from the incident, and makes recommendations for improvement in future responses.

Agroterrorism — Terrorist attack directed against agriculture, such as food supplies or livestock.

Air-Aspirating Foam Nozzle — Foam nozzle designed to provide the aeration required to make the highest quality foam possible; most effective appliance for the generation of low-expansion foam.

Air Bill — Shipping document prepared from a bill of lading that accompanies each piece or each lot of air cargo. *Similar to* Bill of Lading *and* Waybill.

Air-Purifying Respirator (APR) — Respirator that removes contaminants by passing ambient air through a filter, cartridge, or canister; may have a full or partial facepiece.

Air-Reactive Material — Substance that reacts or ignites when exposed to air at normal temperatures. *Also known as* Pyrophoric.

Alkane — A saturated hydrocarbon, with hydrogen in every possible location. All bonds are single bonds. *Also known as* Paraffin.

Alkene — An unsaturated hydrocarbon with at least one double bond between carbon atoms. *Also known as* Olefin.

Alkyne — An unsaturated hydrocarbon with at least one triple bond. *Also known as* Acetylene.

Allergen — Material that can cause an allergic reaction of the skin or respiratory system. *Also known as* Sensitizer.

Alloy — Substance or mixture composed of two or more metals (or a metal and nonmetallic elements) fused together and dissolved into each other to enhance the properties or usefulness of the base metal.

Ambulatory — People, often responders, who are able to understand directions, talk, and walk unassisted.

American Society of Mechanical Engineers (ASME) — Voluntary standards-setting organization concerned with the development of technical standards, such as those for respiratory protection cylinders.

Ammonium Nitrate and Fuel Oil (ANFO) — High explosive blasting agent made of common fertilizer mixed with diesel fuel or oil; requires a booster to initiate detonation.

Anhydrous — Material containing no water.

Anion — Atom or group of atoms carrying a negative charge.

Antibiotic — Antimicrobial agent made from a mold or a bacterium that kills or slows the growth of bacteria; examples include penicillin and streptomycin. Antibiotics are ineffective against viruses.

Antibody — Specialized protein produced by a body's immune system when it detects antigens (harmful substances). Antibodies can only neutralize or remove the effects of their analogous antigens.

Antidote — Substance that counteracts the effects of a poison or toxin.

Antigen — Toxin or other foreign substance that triggers an immune response in a body.

Aqueous Film Forming Foam (AFFF) — Synthetic foam concentrate that, when combined with water, can form a complete vapor barrier over fuel spills and fires and is a highly effective extinguishing and blanketing agent on hydrocarbon fuels.

Aromatic Hydrocarbon — A hydrocarbon with bonds that form rings. *Also known as* Aromatics, *or* Arene.

Asphyxiant — Any substance that prevents oxygen from combining in sufficient quantities with the blood or from being used by body tissues.

Atom — The smallest complete building block of ordinary matter in any state.

Atomic Number — Number of protons in an atom.

Atomic Stability — Condition where an atom has a filled outer shell and is not seeking electrons. Stable atoms also have the same number of protons and electrons.

Atomic Weight — Physical characteristic relating to the mass of molecules and atoms. A relative scale for atomic weights has been adopted, in which the atomic weight of carbon has been set at 12, although its true atomic weight is 12.01115.

Authority Having Jurisdiction (AHJ) — An organization, office, or individual responsible for enforcing the requirements of a code or standard, or approving equipment, materials, an installation, or a procedure.

Autoclave — A device that uses high-pressure steam to sterilize objects.

Autoignition Temperature — The lowest temperature at which a combustible material ignites in air without a spark or flame. (NFPA 921)

Autoinjector — Spring-loaded syringe filled with a single dose of a lifesaving drug.

Autorefrigeration — Rapid chilling of a liquefied compressed gas as it transitions from a liquid state to a vapor state.

Awareness Level — Lowest level of training established by the National Fire Protection Association for personnel at hazardous materials incidents.

B

Bacteria — Microscopic, single-celled organisms.

Baffle — Partition placed in vehicular or aircraft water tanks to reduce shifting of the water load when starting, stopping, or turning.

Bank-Down Application Method — Method of foam application that may be employed on an ignited or unignited Class B fuel spill. The foam stream is directed at a vertical surface or object that is next to or within the spill area; foam deflects off the surface or object and flows down onto the surface of the spill to form a foam blanket. *Also known as* Deflection.

Bar — Unit of pressure measurement; not part of the SI. Equals 100 000 Pa.

Base — Any alkaline or caustic substance; corrosive water-soluble compound or substance containing group-forming hydroxide ions in water solution that reacts with an acid to form a salt.

Base Metal — In hazardous materials containers, the structural material of a containment vessel itself, independent of welding materials and external supports.

Basic Solution — Solution that has a pH between 7 and 14.

Beam — Structural member subjected to loads, usually vertical loads, perpendicular to its length.

Becquerel (Bq) — International System unit of measurement for radioactivity, indicating the number of nuclear decays/disintegrations a radioactive material undergoes in a certain period of time.

Berm — Temporary or permanent barrier intended to control the flow of water. *Similar to* Dike *and* Dam.

Bill of Lading — Shipping paper used by the trucking industry (and others) indicating origin, destination, route, and product; placed in the cab of every truck tractor. This document establishes the terms of a contract between a shipper and a carrier. It serves as a document of title, contract of carriage, and receipt for goods. *Similar to* Air Bill *and* Waybill.

Binary Explosive — A type of explosive device or material with two components that are explosive when combined but not separately.

Bioassay — Scientific experiment in which live plant or animal tissue or cells are used to determine the biological activity of a substance. *Also known as* Biological Assessment *or* Biological Assay.

Biodegradable — Capable of being broken down into innocuous products by the actions of living things, such as microorganisms.

Biological Agent — Viruses, bacteria, or their toxins which are harmful to people, animals, or crops. When used deliberately to cause harm, may be referred to as a Biological Weapon.

Biological Toxin — Poison produced by living organisms.

Blister Agent — Chemical warfare agent that burns and blisters the skin or any other part of the body it contacts. *Also known as* Vesicant *and* Mustard Agent.

Boiling Liquid Expanding Vapor Explosion (BLEVE) — Rapid vaporization of a liquid stored under pressure upon release to the atmosphere following major failure of its containing vessel. Failure is the result of over-pressurization caused by an external heat source, which causes the vessel to explode into two or more pieces when the temperature of the liquid is well above its boiling point at normal atmospheric pressure.

Boiling Point — Temperature of a substance when the vapor pressure equals atmospheric pressure. At this temperature, the rate of evaporation exceeds the rate of condensation. At this point, more liquid is turning into gas than gas is turning back into a liquid.

Bond Energy — The amount of energy needed to break covalent bonds.

Bonding — Connection of two objects with a metal chain or strap in order to neutralize the static electrical charge between the two. *Similar to* Grounding.

Breach — To make an opening in a structural obstacle (such as a masonry wall) without compromising the overall integrity of the wall to allow access into or out of a structure for rescue, hoseline operations, ventilation, or to perform other functions.

Bung — Cork, plug, or other type of stopper used in a barrel, cask, drum, or keg.

C

Calibrate — Operations to standardize or adjust a measuring instrument.

Calibration — Set of operations used to standardize or adjust the values of quantities indicated by a measuring instrument.

Calibration Test — Set of operations used to make sure that an instrument's alerts all work at the recommended levels of hazard detected. *Also known as* Bump Test *and* Field Test.

Canadian Transportation Emergency Centre (CANUTEC) — Canadian center that provides fire and emergency responders with 24-hour information for incidents involving hazardous materials; operated by Transport Canada, a department of the Canadian government.

Capacity Stencil — Number stenciled on the exterior of a tank car to indicate the volume of the tank. *Also known as* Load Limit Marking.

Carbon Dioxide (CO_2) — Colorless, odorless, heavier than air gas that neither supports combustion nor burns; used in portable fire extinguishers as an extinguishing agent to extinguish Class B or C fires by smothering or displacing the oxygen. CO_2 is a waste product of aerobic metabolism.

Carbon Monoxide (CO) — Colorless, odorless, dangerous gas (both toxic and flammable) formed by the incomplete combustion of carbon. It combines with hemoglobin more than 200 times faster than oxygen does, decreasing the blood's ability to carry oxygen.

Carcinogen — Cancer-producing substance.

CAS® Number — Number assigned by the American Chemical Society's Chemical Abstract Service that uniquely identifies a specific compound.

Case Identifier — Alphabetic and/or numeric characters used to identify a case.

Catalyst — Substance that modifies (usually increases) the rate of a chemical reaction without being consumed in the process.

Cation — Atom or group of atoms carrying a positive charge.

CBRNE — Abbreviation for Chemical, Biological, Radiological, Nuclear, and Explosive. These categories are often used to describe WMDs and other hazardous materials characteristics.

Cell Lysis Equipment — Machinery used to break down the membrane of a cell.

Celsius Scale — International temperature scale on which the freezing point is 0°C (32°F) and the boiling point is 100°C (212°F) at normal atmospheric pressure at sea level. *Also known as* Centigrade Scale.

Chain of Custody — Continuous changes of possession of physical evidence that must be established in court to admit such material into evidence. In order for physical evidence to be admissible in court, there must be an evidence log of accountability that documents each change of possession from the evidence's discovery until it is presented in court.

Chemical Agent — Chemical substance that is intended for use in warfare or terrorist activities to kill, seriously injure, or incapacitate people through its physiological effects. *Also known as* Chemical Warfare Agents.

Chemical Asphyxiant — Substance that reacts to prevent the body from being able to use oxygen. *Also known as* Blood Agent.

Chemical Assessment — An organized approach at quantifying the risks associated with the potential exposure to the chemical.

Chemical Attack — Deliberate release of a toxic gas, liquid, or solid that can poison people and the environment.

Chemical Burn — Injury caused by contact with acids, lye, and vesicants such as tear gas, mustard gas, and phosphorus.

Chemical Degradation — Process that occurs when the characteristics of a material are altered through contact with chemical substances.

Chemical Energy — Potential energy stored in the internal structure of a material that may be released during a chemical reaction or transformation.

Chemical Inventory List (CIL) — Formal tracking document showing details of stored chemicals including location, manufacturer, volume, container type, and health hazards.

Chemical Properties — Relating to the way a substance is able to change into other substances. Chemical properties reflect the ability to burn, react, explode, or produce toxic substances hazardous to people or the environment.

Chemical Protective Clothing (CPC) — Clothing designed to shield or isolate individuals from the chemical, physical, and biological hazards that may be encountered during operations involving hazardous materials.

Chemical Transportation Emergency Center (CHEMTREC®) — Center established by the American Chemistry Council that supplies 24-hour information for incidents involving hazardous materials.

Chemical Warfare Agent — Chemical substance intended for use in warfare or terrorist activity; designed to kill, seriously injure, or seriously incapacitate people through its physiological effects.

Chime — Reinforcement ring at the top (head) of a barrel or drum.

CHLOREP — Program administered and coordinated by The Chlorine Institute to provide an organized and effective system for responding to chlorine emergencies in the United States and Canada, operating 24 hours a day/7 days a week with established phone contacts.

Choking Agent — Chemical warfare agent that attacks the lungs, causing tissue damage.

Chronic — Marked by long duration; recurring over a period of time.

Chronic Health Effect — Long-term health effect resulting from exposure to a hazardous substance.

Class B Foam Concentrate — Foam fire-suppression agent designed for use on ignited or unignited Class B flammable or combustible liquids. *Also known as* Class B Foam.

Class D Fire — Fires of combustible metals such as magnesium, sodium, and titanium.

Cloud — Ball-shaped pattern of an airborne hazardous material where the material has collectively risen above the ground or water at a hazardous materials incident.

Code of Federal Regulations (CFR) — Rules and regulations published by executive agencies of the U.S. federal government. These administrative laws are just as enforceable as statutory laws (known collectively as federal law), which must be passed by Congress.

Coffer Dam — Narrow, empty space (void) between compartments or tanks of a vessel that prevents leakage between them; used to isolate compartments or tanks.

Cold Tapping — Fastening a nozzle or outlet onto a container's tank or piping to assist with the removal or transfer of a product.

Cold Zone — Safe area outside of the warm zone where equipment and personnel are not expected to become contaminated and special protective clothing is not required; the Incident Command Post and other support functions are typically located in this zone. *Also known as* Support Zone.

Colorimetric Indicator Tube — Small tube filled with a chemical reagent that changes color in a predictable manner when a controlled volume of contaminated air is drawn through it. *Also known as* Detector Tube.

Combustible Gas Detector — Device that detects the presence and/or concentration of predefined combustible gases in a defined area. May require additional features to indicate the results to an operator.

Combustible Gas Indicator (CGI) — Electronic device that indicates the presence and explosive levels of combustible gases, as relayed from a combustible gas detector.

Combustible Liquid — Liquid having a flash point at or above 100°F (37.8°C) and below 200°F (93.3°C), per NFPA.

Compatibility Group Letter — Indication on an explosives placard expressed as a letter that categorizes different types of explosive substances and articles for purposes of stowage and segregation.

Compound — Substance consisting of two or more elements that have been united chemically.

Compressed Gas — Gas that, at normal temperature, exists solely as a gas when pressurized in a container, as opposed to a gas that becomes a liquid when stored under pressure.

Computer-Aided Management of Emergency Operations (CAMEO) — A system of software applications that assists emergency responders in the development of safe response plans. It can be used to access, store, and evaluate information critical in emergency response.

Concentration — (1) Percentage (mass or volume) of a material dissolved in water (or other solvent). (2) Quantity of a chemical material inhaled for purposes of measuring toxicity. (3) Quantity of a material in relation to a larger volume of gas or liquid.

Condensation — Process of a gas turning into a liquid state.

Cone — Triangular-shaped pattern of an airborne hazardous material release with a point source at the breach and a wide base downrange.

Confined Space — Space or enclosed area not intended for continuous occupation, having limited (restricted access) openings for entry or exit, providing unfavorable natural ventilation and the potential to have a toxic, explosive, or oxygen-deficient atmosphere.

Confinement — The process of controlling the flow of a spill and capturing it at some specified location.

Consist — Rail shipping paper that contains a list of cars in the train by order; indicates the cars that contain hazardous materials. Some railroads include information on emergency operations for the hazardous materials on board with the consist. *Also known as* Train Consist.

Contagious — Capable of transmission from one person to another through contact or close proximity.

Container — (1) Article of transport equipment that is: (a) of a permanent character and strong enough for repeated use; (b) specifically designed to facilitate the carriage of goods by one or more modes of transport without intermediate reloading; and (c) fitted with devices permitting its ready handling, particularly its transfer from one mode to another. The term "container" does not include vehicles. *Also known as* Cargo Container or Freight Container. (2) Box of standardized size used to transport cargo by truck or railcar when transported over land or by cargo vessels at sea; sizes are usually 8 by 8 by 20 feet or 8 by 8 by 40 feet (2.5 m by 2.5 m by 6 m or 2.5 m by 2.5 m by 12 m).

Containment — The act of stopping the further release of a material from its container.

Contaminant — Foreign substance that compromises the purity of a given substance.

Contamination — Impurity resulting from mixture or contact with a foreign substance.

Continuous Underframe Tank Car — Construction of a rail tank car that includes full support of the tank car. The underframe rests on the truck assembly during transport. *Also known as* Full Sill.

Control — To contain, confine, neutralize, or extinguish a hazardous material or its vapor.

Convulsant — Poison that causes convulsions.

Correction Factor — Manufacturer-provided number that can be used to convert a specific device's read-out to be applicable to another function. *Also known as* Conversion Factor, Multiplier *and* Response Curve.

Corrosive — Capable of causing damage by gradually eroding, rusting, or destroying a material.

Counts per Minute (CPM) — Measure of ionizing radiation in which a detection device registers the rate of returns over time. Primarily used to detect particles, not rays.

Covalent Bond — Chemical bond formed between two or more nonmetals. This chemical bond results in a nonsalt.

Critical Point — The end point of an equilibrium curve. In liquid and vapor response, the conditions under which liquid and its vapor can coexist.

Cross Contamination — Contamination of people, equipment, or the environment outside the hot zone without contacting the primary source of contamination. *Also known as* Secondary Contamination.

Cryogen — Gas that is converted into liquid by being cooled below -130°F (-90°C). *Also known as* Refrigerated Liquid *and* Cryogenic Liquid.

Cryogenic Liquid Storage Tank — Heavily insulated, vacuum-jacketed tanks used to store cryogenic liquids; equipped with safety-relief valves and rupture disks.

Curie (Ci) — English System unit of measurement for radioactivity, indicating the number of nuclear decays/disintegrations a radioactive material undergoes in a certain period of time.

Cyber Terrorism — Premeditated, politically motivated attack against information, computer systems, computer programs, and data which result in violence against noncombatant targets by subnational groups or clandestine agents.

Cylinder — Enclosed container with a circular cross-section used to hold a range of materials. Uses include compressed breathing air, poisons, or radioactive materials. *Also known as* Tank *or* Bottle.

D

Dam — Actions to prevent or limit the flow of a liquid or sludge past a certain area.

Dangerous Goods — (1) Any product, substance, or organism included by its nature or by regulation in any of the nine United Nations classifications of hazardous materials. (2) Alternate term used in Canada and other countries for hazardous materials. (3) Term used in the U.S. and Canada for hazardous materials aboard aircraft.

Datasheet — Document that includes important information regarding a specific utility or resource in a standardized format.

Debriefing — A gathering of information from all personnel that were involved in incident operations.

Decomposition — Chemical change in which a substance breaks down into two or more simpler substances. Result of oxygen acting on a material that results in a change in the material's composition; oxidation occurs slowly, sometimes resulting in the rusting of metals.

Decontamination — Process of removing a hazardous foreign substance from a person, clothing, or area. *Also known as* Decon.

Dedicated Tank Car — Rail tank car that is specked to meet particular parameters unique to the product including pressure relief device, linings, valves, fittings, and attachments. This type of car is often used for a single specified purpose for the life of the car, and may be marked to indicate that exact purpose.

Defending in Place — Taking offensive action to protect persons in immediate danger at hazmat incidents.

Defensive Operations — Operations in which responders seek to confine the emergency to a given area without directly contacting the hazardous materials involved.

Deflagrate — To explode (burn quickly) at a rate of speed slower than the speed of sound.

Demobilization — The process of identifying assets on the scene that are no longer needed and returning them to service.

Density — Mass per unit of volume of a substance; obtained by dividing the mass by the volume.

Detection Limit — The smallest quantity of a material that is identifiable within a stated confidence level.

Detonate — To explode or cause to explode. The level of explosive capability will directly affect the speed of the combustion reaction.

Detonation — Explosion with an energy front that travels faster than the speed of sound.

Detonator — Device used to trigger less sensitive explosives, usually composed of a primary explosive; for example, a blasting cap. Detonators may be initiated mechanically, electrically, or chemically.

Dewar — All-metal container designed for the movement of small quantities of cryogenic liquids within a facility; not designed or intended to meet Department of Transportation (DOT) requirements for the transportation of cryogenic materials.

Diatomic Molecules — Molecules composed of only two atoms that may or may not be the same element.

Dike — Actions using raised embankments or other barriers to prevent movement of liquids or sludges to another area.

Dilution — Application of water to a water-soluble material to reduce the hazard.

Direct-Reading Instrument — A tool that indicates its reading on the tool itself, without requiring additional resources. Each instrument is designed for a specific monitoring purpose.

Discovery — Means by which the plaintiff (one party) obtains information from the opposing party (defendant) to prove its allegation.

Disinfection — Any process that eliminates most biological agents; disinfection techniques may target specific entities. Often uses chemicals.

Dispersion — Act or process of being spread widely.

Dissociation (Chemical) — Process of splitting a molecule or ionic compounds into smaller particles, especially if the process is reversible. *Opposite of* Recombination.

Divert — Actions to direct and control movement of a liquid or sludge to an area that will produce less harm.

Division Number — Subset of a class within an explosives placard that assigns the product's level of explosion hazard.

Dose — Quantity of a chemical material ingested or absorbed through skin contact for purposes of measuring toxicity.

Dose-Response Relationship — Comparison of changes within an organism per amount, intensity, or duration of exposure to a stressor over time. This information is used to determine action levels for materials such as drugs, pollutants, and toxins.

Dosimeter — Detection device used to measure an individual's exposure to an environmental hazard such as radiation or sound.

Drainage Time — Amount of time it takes foam to break down or dissolve. *Also known as* Drainage, Drainage Dropout Rate, *or* Drainage Rate.

Dry Powder — Extinguishing agent suitable for use on combustible metal fires.

Duet Rule — Atoms with only one shell will attempt to maintain two electrons to fill the outer shell at all times, whether by gaining or losing electrons. A complete outer electron shell makes elements very stable.

Dust Explosion — Rapid burning (deflagration), with explosive force, of any combustible dust. Dust explosions generally consist of two explosions: a small explosion or shock wave creates additional dust in an atmosphere, causing the second and larger explosion.

E

Eduction — Process used to mix foam concentrate with water in a nozzle or proportioner; concentrate is drawn into the water stream by the Venturi method. *Also known as* Induction.

Electricity — Form of energy resulting from the presence and flow of charged particles.

Electrochemical Gas Sensor — Device used to measure the concentration of a target gas by oxidizing or reducing the target gas and then measuring the current.

Electron — Subatomic particle with a physical mass and a negative electric charge.

Elevated Temperature Material — Material that when offered for transportation or transported in bulk packaging is (a) in a liquid phase and at temperatures at or above 212°F (100°C), (b) intentionally heated at or above its liquid phase flash points of 100°F (38°C), or (c) in a solid phase and at a temperature at or above 464°F (240°C).

Emergency Breathing Support System (EBSS) — Escape-only respirator that provides sufficient self-contained breathing air to permit the wearer to safely exit the hazardous area; usually integrated into an airline supplied-air respirator system.

Emergency Decontamination — The physical process of immediately reducing contamination of individuals in potentially life-threatening situations, with or without the formal establishment of a decontamination corridor.

Emergency Response Guidebook (ERG) — Manual that aids emergency response and inspection personnel in identifying hazardous materials placards and labels; also gives guidelines for initial actions to be taken at hazardous materials incidents. Developed jointly by Transport Canada (TC), U.S. Department of Transportation (DOT), the Secretariat of Transport and Communications of Mexico (SCT), and with the collaboration of CIQUIME (Centro de Información Química para Emergencias).

Emissivity — Measure of an object's ability to radiate thermal energy.

Encapsulating — Completely enclosed or surrounded, as in a capsule.

End-of-Service-Time Indicator (ESTI) — Warning device that alerts the user that the respiratory protection equipment is about to reach its limit and that it is time to exit the contaminated atmosphere; its alarm may be audible, tactile, visual, or any combination thereof.

Endothermic Reaction — Chemical reaction in which a substance absorbs heat energy.

Engulfment — Dispersion of material as defined in the General Emergency Behavior Model (GEBMO); an engulfing event occurs when matter and/or energy disperses and forms a danger zone.

Glossary 569

Evacuation — Controlled process of leaving or being removed from a potentially hazardous location, typically involving relocating people from an area of danger or potential risk to a safer place.

Evaporation — Process of a solid or liquid turning into gas.

Evaporation Rate — Speed at which some material changes from a liquid to a vapor. Materials that change readily to gases are considered volatile.

Evidence — Information collected and analyzed by an investigator.

Excepted Packaging — Container used for transportation of materials that have very limited radioactivity.

Exothermic Reaction — Chemical reaction between two or more materials that changes the materials and produces heat.

Expansion Ratio — 1) Volume of a substance in liquid form compared to the volume of the same number of molecules of that substance in gaseous form. 2) Ratio of the finished foam volume to the volume of the original foam solution. *Also known as* Expansion.

Explosive — Any material or mixture that will undergo an extremely fast self-propagation reaction when subjected to some form of energy.

Explosive Ordnance Disposal (EOD) — Emergency responders specially trained and equipped to handle and dispose of explosive devices. *Also called* Hazardous Devices Units *or* Bomb Squad.

Exposure — (1) Contact with a hazardous material, causing biological damage, typically by swallowing, breathing, or touching (skin or eyes). Exposure may be short-term (acute exposure), of intermediate duration, or long-term (chronic exposure). (2) People, property, systems, or natural features that are or may be exposed to the harmful effects of a hazardous materials emergency.

Exposure Limit — Maximum length of time an individual can be exposed to an airborne substance before injury, illness, or death occurs.

Extinguish — To put out a fire completely.

F

Fahrenheit Scale — Temperature scale on which the freezing point is 32°F (0°C) and the boiling point at sea level is 212°F (100°C) at normal atmospheric pressure.

Ferrous Metal — Metal in which iron is the main constituent element; carbon and other elements are added to the iron to create a variety of metals with various magnetic properties and tensile strengths; varieties include cast and wrought iron, steel and steel alloys; stainless steel, and high-carbon steel.

Fire Point — Temperature at which a liquid fuel produces sufficient vapors to support combustion once the fuel is ignited. Fire point must exceed five seconds of burning duration during the test. The fire point is usually a few degrees above the flash point.

Flame Ionization Detector (FID) — Gas detector that oxidizes all oxidizable materials in a gas stream, and then measures the concentration of the ionized material.

Flame-Resistant (FR) — Material that does not support combustion and is self-extinguishing after removal of an external source of ignition.

Flammability — Fuel's susceptibility to ignition.

Flammable Liquid — Any liquid having a flash point below 100°F (37.8°C) and a vapor pressure not exceeding 40 psi absolute (276 kPa) {2.76 bar}, per NFPA.

Flammable Range — Range between the upper flammable limit and lower flammable limit in which a substance can be ignited. *Also known as* Explosive Range.

Flaring — Controlled release and disposal of flammable gases or liquids through a burning process.

Flash Point — Minimum temperature at which a liquid gives off enough vapors to form an ignitable mixture with air near the surface of the liquid.

Fluorimeter — Device used to detect the fluorescence of a material, especially as pertains to the fluorescent qualities of DNA and RNA.

Forensic Evidence — Evidence obtained by scientific methods that is usable in court, for example finger prints, blood testing, or ballistics.

Fourier Transform Infrared (FT-IR) Spectroscopy — Device that uses a mathematical process to convert detection data onto the infrared spectrum.

Frameless Tank Car — Direct attachment of a rail tank car to the truck assembly. This type of construction transfers all of the stresses of transport from the railcar to the stub sill assembly and the tank itself. *Known as* Stub Sill.

Freezing Point — Temperature at which a liquid becomes a solid at normal atmospheric pressure.

Frostbite — Local tissue damage caused by prolonged exposure to extreme cold.

Fusible Plug — Safety device in pressurized vessels that consists of a threaded metal cylinder with a tapered hole drilled completely through its length; the hole is filled with a metal that has a low, predetermined melting point.

G

Gamma-Ray Spectrometer — Apparatus used to measure the intensity of gamma radiation as compared to the energy of each photon.

Gas — Compressible substance, with no specific volume, that tends to assume the shape of a container. Molecules move about most rapidly in this state.

Gas Chromatograph (GC) — Apparatus used to detect and separate small quantities of volatile liquids or gases via instrument analysis. *Also known as* Gas-Liquid Partition Chromatography (GLPC).

Geiger-Mueller (GM) Detector — Detection device that uses GM tubes to measure ionizing radiation. *Also known as a* Geiger Counter.

Geiger-Mueller (GM) Tube — Sensor tube used to detect ionizing radiation. This tube is one element of a Geiger-Mueller detector.

General Emergency Behavior Model (GEBMO) — Model used to describe how hazardous materials are accidentally released from their containers and how they behave after the release.

Geographic Information Systems (GIS) — Computer software application that relates physical features on the earth to a database to be used for mapping and analysis. The system captures, stores, analyzes, manages, and presents data that refers to or is linked to a location.

Globally Harmonized System of Classification and Labeling of Chemicals (GHS) — International classification and labeling system for chemicals and other hazard communication information, such as safety data sheets.

Glovebox — Sealed container equipped with long-cuff gloves on one facet to allow handling of materials within the container. Commonly used in laboratories and incubators where a vacuum or sterile environment is needed.

Gray (Gy) — SI unit of ionizing radiation dose, defined as the absorption of one joule of radiation energy per one kilogram of matter.

Grounding — Reducing the difference in electrical potential between an object and the ground by the use of various conductors; similar to *Bonding*.

G-Series Agents — Nonpersistent nerve agents initially synthesized by German scientists.

H

Half-Life — The time required for a radioactive material to reduce to half of its initial value.

Halogenated Agent — Chemical compounds (halogenated hydrocarbons) that contain carbon plus one or more elements from the halogen series. Halon 1301 and Halon 1211 are most commonly used as extinguishing agents for Class B and Class C fires. *Also known as* Halogenated Hydrocarbons.

Hazard — Condition, substance, or device that can directly cause injury or loss; the source of a risk.

Hazard and Risk Assessment — Formal review of the hazards and risks that may be encountered by firefighters or emergency responders; used to determine the appropriate level and type of personal and respiratory protection that must be worn. *Also known as* Hazard Assessment.

Hazard Class — Group of materials designated by the Department of Transportation (DOT) that shares a major hazardous property.

Hazard-Control Zones — System of barriers surrounding designated areas at emergency scenes, intended to limit the number of persons exposed to a hazard and to facilitate its mitigation. A major incident has three zones: Restricted (Hot) Zone, Limited Access (Warm) Zone, and Support (Cold) Zone. EPA/OSHA term: Site Work Zones. *Also known as* Control Zones *and* Scene Control Zones.

Hazardous Material — Any substance or material that poses an unreasonable risk to health, safety, property, and/or the environment if it is not properly controlled during handling, storage, manufacture, processing, packaging, use, disposal, or transportation.

Hazardous Materials Profile — A chemical size-up based upon the suspected identity, or not, of a chemical hazard. This is validated with monitoring and detection equipment upon performing a reconnaissance entry. Profiling allows the hazmat technician to predict hazards and validate the actual entry conditions even if the product is not positively identified. *Also known as* Hazard Profile.

Hazardous Materials Technician — Individual trained to use specialized protective clothing and control equipment to control the release of a hazardous material.

Hazardous Waste Operations and Emergency Response (HAZWOPER) — U.S. regulations in Title 29 (Labor) *CFR* 1910.120 for cleanup operations involving hazardous substances and emergency response operations for releases of hazardous substances.

Head Pressure — Pressure exerted by a stationary column of water, directly proportional to the height of the column.

Head Shield — Layer of puncture protection added to the head of tanks. Head shields may or may not be visible, depending on the construction of the tank and the type of protection provided.

Heat — Form of energy associated with the motion of atoms or molecules in solids or liquids that is transferred from one body to another as a result of a temperature difference between the bodies, such as from the sun to the earth. To signify its intensity, it is measured in degrees of temperature.

Heat Cramps — Heat illness resulting from prolonged exposure to high temperatures; characterized by excessive sweating, muscle cramps in the abdomen and legs, faintness, dizziness, and exhaustion.

Heat Exhaustion — Heat illness caused by exposure to excessive heat; symptoms include weakness, cold and clammy skin, heavy perspiration, rapid and shallow breathing, weak pulse, dizziness, and sometimes unconsciousness.

Heat Induced Tear — Rupture of a container caused by overpressure, often along the seam. This type of failure primarily occurs in low-pressure containers transporting flammable/combustible liquids.

Heat Rash — Condition that develops from continuous exposure to heat and humid air; aggravated by clothing that rubs the skin. Reduces the individual's tolerance to heat.

Heat Sink — In thermodynamics, any material or environment that absorbs heat without changing its physical state or appreciably changing temperature.

Heat Stroke — Heat illness in which the body's heat regulating mechanism fails; symptoms include (a) high fever of 105° to 106° F (40.5° to 41.1° C), (b) dry, red, hot skin, (c) rapid, strong pulse, and (d) deep breaths or convulsions. May result in coma or even death. *Also known as* Sunstroke.

Hemispheric Release — Semicircular or dome-shaped pattern of airborne hazardous material that is still partially in contact with the ground or water.

Hexamethylene Triperoxide Diamine (HMTD) — Peroxide-based white powder high explosive organic compound that can be manufactured using nonspecialized equipment. Sensitive to shock and friction during manufacture and handling. *Similar to* acetone peroxide (TATP).

High Explosive — Explosive that decomposes extremely rapidly (almost instantaneously) and has a detonation velocity faster than the speed of sound.

High-Hazard Flammable Trains (HHFT) — Trains that have a continuous block of twenty or more tank cars loaded with a flammable liquid or thirty-five or more cars loaded with a flammable liquid dispersed through a train.

Homemade Explosive (HME) — Explosive material constructed using common household chemicals. The finished product is usually highly unstable.

Hot Tapping — Using welding or cutting to attach a nozzle or outlet onto a container's tank or piping to assist with the removal or transfer of a product.

Hot Zone — Potentially hazardous area immediately surrounding the incident site; requires appropriate protective clothing and equipment and other safety precautions for entry. Typically limited to technician-level personnel. *Also known as* Exclusion Zone.

Hydrocarbon — Organic compound containing only hydrogen and carbon and found primarily in petroleum products and coal.

Hydrogen Cyanide (HCN) — Colorless, toxic, and flammable liquid until it reaches 79° F (26° C). Above that temperature, it becomes a gas with a faint odor similar to bitter almonds; produced by the combustion of nitrogen-bearing substances.

Hydronium — Water molecule with an extra hydrogen ion (H_3O^+). Substances/solutions that have more hydronium ions than hydroxide ions have an acidic pH.

Hydrophilic — Material that is attracted to water. This material may also dissolve or mix in water.

Hydrophobic — Material that is incapable of mixing with water.

Hydrostatic Test — Testing method that uses water under pressure to check the integrity of pressure vessels.

Hydroxide — Water molecule missing a hydrogen ion (HO^-). Substances/solutions that have more hydroxide ions than hydronium ions have a basic (alkaline) pH.

Hypergolic — Substance that ignites when exposed to another substance.

Hypothermia — Abnormally low body temperature.

I

Ignition Temperature — Minimum temperature to which a fuel (other than a liquid) in air must be heated in order to start self-sustained combustion independent of the heating source.

Immediately Dangerous to Life and Health (IDLH) — Description of any atmosphere that poses an immediate hazard to life or produces immediate irreversible, debilitating effects on health; represents concentrations above which respiratory protection should be required. Expressed in parts per million (ppm) or milligrams per cubic meter (mg/m^3); companion measurement to the permissible exposure limit (PEL).

Immiscible — Incapable of being mixed or blended with another substance.

Immunoassay (IA) — Test to measure the concentration of an analyte (material of interest) within a solution.

Improvised Explosive Device (IED) — Any explosive device constructed and deployed in a manner inconsistent with conventional military action.

Incendiary Device — (1) Contrivance designed and used to start a fire. (2) Any mechanical, electrical, or chemical device used intentionally to initiate combustion and start a fire. *Also known as* Explosive Device.

Incident Commander (IC) — Person in charge of the incident command system and responsible for the management of all incident operations during an emergency.

Incident Management System (IMS) — System described in NFPA 1561, *Standard on Emergency Services Incident Management System and Command Safety*, that defines the roles, responsibilities, and standard operating procedures used to manage emergency operations. Such systems may also be referred to as Incident Command Systems (ICS).

Incidental Release — Spill or release of a hazardous material where the substance can be absorbed, neutralized, or otherwise controlled at the time of release by employees in the immediate release area, or by maintenance personnel who are not considered to be emergency responders.

Industrial Packaging — Container used to ship radioactive materials that present limited hazard to the public and the environment, such as smoke detectors.

Inert Gas — Gas that does *not* normally react chemically with another substance or material; any one of six gases: helium, neon, argon, krypton, xenon, and radon.

Infectious — Transmittable; able to infect people.

Infectious Substance — Substance that is known, or reasonably expected, to contain pathogens.

Infrared — Invisible electromagnetic radiant energy at a wavelength in the visible light spectrum greater than the red end but lower than microwaves.

Infrared Thermometer — Non-contact measuring device that detects the infrared energy emitted by materials and converts the energy factor into a temperature reading. *Also known as* Temperature Gun.

Inhalation Hazard — Any material that may cause harm via inhalation.

Inhibitor — Material that is added to products that easily polymerize in order to control or prevent an undesired reaction. *Also known as* Stabilizer.

Initial Isolation Distance — Distance within which all persons are considered for evacuation in all directions from a hazardous materials incident.

Initial Isolation Zone — Circular zone, with a radius equivalent to the initial isolation distance, within which persons may be exposed to dangerous concentrations upwind of the source and may be exposed to life-threatening concentrations downwind of the source.

Instrument Response Time — Elapsed time between the movement (drawing in) of an air sample into a monitoring/detection device and the reading (analysis) provided to the user. *Also known as* Instrument Reaction Time.

Intermediate Bulk Container (IBC) — Rigid (RIBC) or flexible (FIBC) portable packaging, other than a cylinder or portable tank, that is designed for mechanical handling with a maximum capacity of not more than 3 cubic meters (3,000 L, 793 gal, or 106 ft^3) (49CFR178.700).

Intermodal Container — Freight containers designed and constructed to be used interchangeably in two or more modes of transport. *Also known as* Intermodal Tank, Intermodal Tank Container, *and* Intermodal Freight Container.

International System of Units (SI) — Modern form of the metric system of measurement that standardizes mathematical quantification.

Intrinsically Safe — Describes equipment that is approved for use in flammable atmospheres; must be incapable of releasing enough electrical energy to ignite the flammable atmosphere.

Inverse Square Law — Physical law that states that the amount of radiation present is inversely proportional to the square of the distance from the source of radiation.

Ion — Atom that has lost or gained an electron, thus giving it a positive or negative charge.

Ion Mobility Spectrometry (IMS) — Technique used to separate and identify ionized molecules. The ionize molecules are impeded in travel via a buffer gas chosen for the type of detection intended. Larger ions are slowed more than smaller ions; this difference provides an indication of the ions' size and identity.

Ionic Bond — Chemical bond formed by the transfer of electrons from a metal element to a nonmetal element. This chemical bond results in two oppositely charged ions.

Ionization — Process in which an atom or molecule loses electrons.

Ionization Potential — Energy required to free an electron from its atom or molecule.

Ionize — Process in which an atom or molecule gains a negative or positive charge by gaining or losing electrons.

Ionizing Radiation — Radiation that causes a chemical change in atoms by removing their electrons.

Irritant — Liquid or solid that, upon contact with fire or exposure to air, gives off dangerous or intensely irritating fumes. *Also known as* Irritating Material.

Isolation Perimeter — Outer boundary of an incident that is controlled to prevent entrance by the public or unauthorized persons.

Isotope — Atoms of a chemical element with the usual number of protons in the nucleus, but an unusual number of neutrons; has the same atomic number but a different atomic mass from normal chemical elements.

J

Jubilee Pipe Patch — Modification of a commercial hose clamp; consists of a cylindrical sheet of metal with flanges at each end. The sheet metal can be wrapped over packing around a pipe leak. The flanges can be attached to one another using screws and nuts to form a seal.

L

Label — Four-inch-square diamond-shaped marker required by federal regulations on individual shipping containers that contain hazardous materials, and are smaller than 640 cubic feet (18 m^3).

Lethal Concentration (LC) — Concentration of an inhaled substance that results in the death of the entire test population. Expressed in parts per million (ppm), milligrams per liter (mg/liter), or milligrams per cubic meter (mg/m^3); the lower the value, the more toxic the substance.

Lethal Dose (LD) — Concentration of an ingested or injected substance that results in the death of the entire test population. Expressed in milligrams per kilogram (mg/kg); the lower the value, the more toxic the substance.

Level A PPE — Highest level of skin, respiratory, and eye protection that can be given by personal protective equipment (PPE), as specified by the U.S. Environmental Protection Agency (EPA); consists of positive-pressure self-contained breathing apparatus, totally encapsulating chemical-protective suit, inner and outer gloves, and chemical-resistant boots.

Level B PPE — Personal protective equipment that affords the highest level of respiratory protection, but a lesser level of skin protection; consists of positive-pressure self-contained breathing apparatus, hooded chemical-protective suit, inner and outer gloves, and chemical-resistant boots.

Level C PPE — Personal protective equipment that affords a lesser level of respiratory and skin protection than levels A or B; consists of full-face or half-mask APR, hooded chemical-resistant suit, inner and outer gloves, and chemical-resistant boots.

Level D PPE — Personal protective equipment that affords the lowest level of respiratory and skin protection; consists of coveralls, gloves, and chemical-resistant boots or shoes.

Limits of Recovery — A container's design strength or ability to hold contents at pressure.

Line-of-Sight — Unobstructed, imaginary line between an observer and the object being viewed.

Liquefied Gas — Confined gas that at normal temperatures exists in both liquid and gaseous states.

Liquefied Natural Gas (LNG) — Natural gas stored under pressure as a liquid.

Liquefied Petroleum Gas (LPG) — Any of several petroleum products, such as propane or butane, stored under pressure as a liquid.

Liquid — Incompressible substance with a constant volume that assumes the shape of its container; molecules flow freely, but substantial cohesion prevents them from expanding as a gas would.

Liquid Splash-Protective Clothing — Chemical-protective clothing designed to protect against liquid splashes per the requirements of NFPA 1992, *Standard on Liquid Splash-Protective Suits for Hazardous Chemical Emergencies*; part of an EPA Level B ensemble.

Local Emergency Planning Committee (LEPC) — Community organization responsible for local emergency response planning. Required by SARA Title III, LEPCs are composed of local officials, citizens, and industry representatives with the task of designing, reviewing, and updating a comprehensive emergency plan for an emergency planning district; plans may address hazardous materials inventories, hazardous material response training, and assessment of local response capabilities.

Local Emergency Response Plan (LERP) — Plan detailing how local emergency response agencies will respond to community emergencies; required by U.S. Environmental Protection Agency (EPA) and prepared by the Local Emergency Planning Committee (LEPC).

Low Explosive — Explosive material that deflagrates, producing a reaction slower than the speed of sound.

Lower Flammable (Explosive) Limit (LFL) — Lower limit at which a flammable gas or vapor will ignite and support combustion; below this limit the gas or vapor is too *lean* or *thin* to burn (too much oxygen and not enough gas, so lacks the proper quantity of fuel). *Also known as* Lower Explosive Limit (LEL).

Low-Pressure Storage Tank — Class of fixed-facility storage tanks that are designed to have an operating pressure ranging from 0.5 to 15 psi (3.45 kPa to 103 kPa) {0.03 bar to 1.03 bar}.

M

Manway — Opening that is large enough to admit a person into a tank trailer or dry bulk trailer. This opening is usually equipped with a removable, lockable cover. *Also known as* Manhole.

Mass Casualty Incident — Incident that results in a large number of casualties within a short time frame, as a result of an attack, natural disaster, aircraft crash, or other cause that is beyond the capabilities of local logistical support.

Mass Decontamination — Process of decontaminating large numbers of people in the fastest possible time to reduce surface contamination to a safe level. It is typically a gross decon process utilizing water or soap and water solutions to reduce the level of contamination, with or without a formal decontamination corridor or line.

Mass Spectrometer — Apparatus used to ionize a chemical and then measure the masses within the sample.

Maximum Allowable Working Pressure (MAWP) — A percentage of a container's test pressure. Can be calculated as the pressure that the weakest component of a vessel or container can safely maintain.

Maximum Safe Storage Temperature (MSST) — Temperature below which the product can be stored safely. This is usually 20-30 degrees cooler than the SADT temperature, but may be much cooler depending on the material.

Mechanical Energy — Energy possessed by objects due to their position or motion, the sum of potential and kinetic energy.

Median Lethal Concentration, 50 Percent Kill (LC_{50}) — Concentration of an inhaled substance that results in the death of 50 percent of the test population. LC_{50} is an inhalation exposure expressed in parts per million (ppm), milligrams per liter (mg/liter), or milligrams per cubic meter (mg/m³); the lower the value, the more toxic the substance.

Median Lethal Dose, 50 Percent Kill (LD_{50}) — Concentration of an ingested or injected substance that results in the death of 50 percent of the test population. LD_{50} is an oral or dermal exposure expressed in milligrams per kilogram (mg/kg); the lower the value, the more toxic the substance.

Melting Point — Temperature at which a solid substance changes to a liquid state at normal atmospheric pressure.

Memorandum of Understanding (MOU) — Form of written agreement created by a coalition to make sure that each member is aware of the importance of his or her participation and cooperation.

Mercaptan — A sulfur-containing organic compound often added to natural gas as an odorant. Natural gas is odorless; natural gas treated with mercaptan has a strong odor. *Also known as a* Thiol.

Metadata — Information that provides background and detail about other types of information.

Meth Cook — 1) Person who generates methamphetamine in an illicit lab. 2) Area with evidence of production of methamphetamine.

Methamphetamine (Meth) — Central nervous system stimulant drug that can be produced in small labs. Low dosage medical uses include controlling weight, narcolepsy, and attention deficit hyperactivity disorder. Recreational uses include euphoriant and aphrodisiac qualities. At all dosages, misuse of this drug presents a high risk for personal and social harm.

Micron — Unit of length equal to one-millionth of a meter.

Mild Steel — Class of steel in which a low-level of carbon is the primary alloying agent; available in a variety of formable grades. *Also called* Carbon Steel.

Millimeters of Mercury (mmHg) — Unit of pressure measurement; not part of the SI. Currently defined as a rate rounded to 133 Pascals. Rough equivalent to 1 torr.

Millirem (mrem) — One thousandth of one Roentgen Equivalent in Man (rem).

Miscibility — Two or more liquids' capability to mix together.

Miscible — Materials that are capable of being mixed in all proportions.

Mitigate — (1) To cause to become less harsh or hostile; to make less severe, intense or painful; to alleviate. (2) Third of three steps (locate, isolate, mitigate) in one method of sizing up an emergency situation.

Mixture — Substance containing two or more materials not chemically united.

Mobile Data Terminal (MDT) — Mobile computer that communicates with other computers on a radio system.

Molecular Weight (MW) — Average mass of one molecule. This can be calculated as the sum of the atomic masses of the component atoms.

Monomer — A molecule that may bind chemically to other molecules to form a polymer.

Multiuse Detectors — Device with several types of equipment in one handheld device. Used to detect specific types of materials in an atmosphere.

Munitions — Military reserves of weapons, equipment, and ammunition.

N

National Fire Protection Association (NFPA) — U.S. nonprofit educational and technical association devoted to protecting life and property from fire by developing fire protection standards and educating the public. Located in Quincy, Massachusetts.

National Incident Management System - Incident Command System (NIMS-ICS) — The U.S. mandated incident management system that defines the roles, responsibilities, and standard operating procedures used to manage emergency operations; creates a unified incident response structure for federal, state, and local governments.

National Pipe Thread Taper (NPT) — U.S. standard for pipe threading developed with the intent to create a fluid-tight seal.

Nerve Agent — A class of toxic chemical that works by disrupting the way nerves transfer messages to organs.

Neutralization — Chemical reaction in water in which an acid and base react quantitatively with each other until there are no excess hydrogen or hydroxide ions remaining in the solution.

Neutron — Component of the nucleus of an atom that has a neutral electrical charge yet produces highly penetrating radiation; ultrahigh energy particle that has a physical mass but no electrical charge.

Nondispersive Infrared (NDIR) Sensor — Simple spectroscope that can be used as a gas detector.

Nonflammable — Incapable of combustion under normal circumstances; normally used when referring to liquids or gases.

Nonintervention Operations — Operations in which responders take no direct actions on the actual problem.

Nonionizing Radiation — Series of energy waves composed of oscillating electric and magnetic fields traveling at the speed of light. Examples include ultraviolet radiation, visible light, infrared radiation, microwaves, radio waves, and extremely low frequency radiation.

Nonpersistent Chemical Agent — Chemical agent that generally vaporizes and disperses quickly, usually in less than 10 minutes.

Nucleus — The positively charged central part of an atom, consisting of protons and neutrons.

O

Occupancy — (1) General fire and emergency services term for a building, structure, or residency. (2) Building code classification based on the use to which owners or tenants put buildings or portions of buildings. Regulated by the various building and fire codes. *Also known as* Occupancy Classification.

Octet Rule — Atoms with two or more shells will attempt to maintain eight electrons to fill the outermost shell at all times, whether by gaining or losing electrons. A complete outer electron shell makes elements very stable.

Offensive Operations — Operations in which responders take aggressive, direct action on the material, container, or process equipment involved in an incident.

Olfactory Fatigue — Gradual inability of a person to detect odors after initial exposure; can be extremely rapid with some toxins, such as hydrogen sulfide.

Operations Level — Level of training established by the National Fire Protection Association allowing first responders to take defensive actions at hazardous materials incidents.

Operations Mission-Specific Level — Level of training established by the National Fire Protection Association allowing first responders to take additional defensive tasks and limited offensive actions at hazardous materials incidents.

Organic Peroxide — Any of several organic derivatives of the inorganic compound hydrogen peroxide.

Organophosphate Pesticides — Chemicals that kill insects by disrupting their central nervous systems; these chemicals inactivate acetylcholinesterase, an enzyme which is essential to nerve function in insects, humans, and many other animals.

Other Regulated Material (ORM) — Material, such as a consumer commodity, that does not meet the definition of a hazardous material and is not included in any other hazard class but possesses enough hazardous characteristics that it requires some regulation; presents limited hazard during transportation because of its form, quantity, and packaging.

Overpack — (1) To enclose or secure a container by placing it in a larger container. (2) An outer container designed to enclose or secure an inner container.

Oxidation — Chemical process that occurs when a substance combines with an oxidizer such as oxygen in the air; a common example is the formation of rust on metal.

Oxidation Number — A theoretical number assigned to individual atoms and ions to track whether an oxidation-reduction reaction has taken place. *Also known as* Oxidation Level.

Oxidation-Reduction (Redox) Reaction — Chemical reaction that results in a molecule, ion, or atom gaining or losing an electron. *Also known as* Redox Reaction.

Oxidizer — Any material that readily yields oxygen or other oxidizing gas, or that readily reacts to promote or initiate combustion of combustible materials. (Reproduced with permission from NFPA 400-2010, *Hazardous Materials Code*, Copyright©2010, National Fire Protection Association)

Oxidizing Agent — Substance that oxidizes another substance; can cause other materials to combust more readily or make fires burn more strongly. *Also known as* Oxidizer.

P

Packaging — Shipping containers and their markings, labels, and/or placards.

Pandemic — Epidemic occurring over a very wide area (several countries or continents), usually affecting a large proportion of the population.

Parts Per Billion (ppb) — Method of expressing the concentration of very dilute solutions of one substance in another, normally a liquid or gas, based on volume; expressed as a ratio of the volume of contaminants (parts) compared to the volume of air (billion parts).

Parts Per Million (ppm) — Method of expressing the concentration of very dilute solutions of one substance in another, normally a liquid or gas, based on volume; expressed as a ratio of the volume of contaminants (parts) compared to the volume of air (million parts). The common unit of measure is equivalent to 1 microgram [1 µg] per liter of water or kg of solid, or a micro liter [1 µL] volume of gas in one liter of air.

Pascals (Pa) — SI unit of measure used to indicate internal pressure and stress on a container.

Pathogen — Biological agent that causes disease or illness.

Penetration — Process in which a hazardous material enters an opening or puncture in a protective material.

Periodic Table of Elements — Organizational chart showing chemical elements arranged in order by atomic number, electron configuration, and chemical properties.

Permeation — Process in which a chemical passes through a protective material on a molecular level.

Permissible Exposure Limit (PEL) — Maximum time-weighted concentration at which 95 percent of exposed, healthy adults suffer no adverse effects over a 40-hour work week; an 8-hour time-weighted average unless otherwise noted. PELs are expressed in either parts per million (ppm) or milligrams per cubic meter (mg/m³). They are commonly used by OSHA and are found in the NIOSH *Pocket Guide to Chemical Hazards*.

Persistence — Length of time a chemical agent remains effective without dispersing.

Persistent Chemical Agent — Chemical agent that remains effective in the open (at the point of dispersion) for a considerable period of time, usually more than 10 minutes.

Person-Borne Improvised Explosives Device (PBIED) — Improvised explosive device carried by a person. This type of IED is often employed by suicide bombers, but may be carried by individuals coerced into carrying the bomb.

pH — Measure of the acidity or alkalinity of a solution.

pH Indicator — Chemical detector for hydronium ions (H_3O^+) or hydrogen ions (H^+). Indicator equipment includes impregnated papers and meters.

Phase — Distinguishable part in a course, development, or cycle; aspect or part under consideration. In chemistry, a change of phase is marked by a shift in the physical state of a substance caused by a change in heat.

Phosphine — Colorless, flammable, and toxic gas with an odor of garlic or decaying fish; ignites spontaneously on contact with air. Phosphine is a respiratory tract irritant that attacks the cardiovascular and respiratory systems, causing pulmonary edema, peripheral vascular collapse, and cardiac arrest and failure.

Photoionization Detector (PID) — Gas detector that measures volatile compounds in concentrations of parts per million and parts per billion.

Photon — Weightless packet of electromagnetic energy, such as X-rays or visible light.

Physical Properties — Properties that do not involve a change in the chemical identity of the substance, but affect the physical behavior of the material inside and outside the container, which involves the change of the state of the material. Examples include boiling point, specific gravity, vapor density, and water solubility.

Pipeline and Hazardous Materials Safety Administration (PHMSA) — Branch of the U.S. Department of Transportation (DOT) that focuses on pipeline safety and related environmental concerns.

Placard — Diamond-shaped sign that is affixed to each side of a structure or a vehicle transporting hazardous materials to inform responders of fire hazards, life hazards, special hazards, and reactivity potential. The placard indicates the primary class of the material and, in some cases, the exact material being transported; required on containers that are 640 cubic feet (18 m^3) or larger.

Plume — Irregularly shaped pattern of an airborne hazardous material where wind and/or topography influence the downrange course from the point of release.

Poison — Any material, excluding gases, that when taken into the body is injurious to health.

Polar Solvent — 1) A material in which the positive and negative charges are permanently separated, resulting in their ability to ionize in solution and create electrical conductivity. Examples include water, alcohol, esters, ketones, amines, and sulfuric acid. 2) Flammable liquids with an attraction for water.

Polarity — Property of some molecules to have discrete areas with negative and positive charges.

Polymer — Large molecule composed of repeating structural units (monomers).

Polymerase Chain Reaction (PCR) — Technique in which DNA is copied to amplify a segment of DNA to diagnose and monitor a disease or to forensically identify an individual.

Polymerization — Chemical reactions in which two or more molecules chemically combine to form larger molecules; this reaction can often be violent.

Positive-Pressure Ventilation (PPV) — Method of ventilating a room or structure by mechanically blowing fresh air through an inlet opening into the space in sufficient volume to create a slight positive pressure within and thereby forcing the contaminated atmosphere out the exit opening.

Postincident Analysis (PIA) — Overview and critique of an incident including feedback from members of all responding agencies. Typically takes place within two weeks of the incident. In the training environment it may be used to evaluate student and instructor performance during a training evolution.

Postincident Critique — Discussion of the incident during the Termination phase of response. Discussion includes responders, stakeholders, and command staff, to determine facets of the response that were successful and areas that can be improved upon.

Powered Air-Purifying Respirator (PAPR) — Motorized respirator that uses a filter to clean surrounding air, then delivers it to the wearer to breathe; typically includes a headpiece, breathing tube, and a blower/battery box that is worn on the belt.

Preincident Survey — Assessment of a facility or location made before an emergency occurs, in order to prepare for an appropriate emergency response. *Also known as* Preplan.

Pressure — Force per unit area exerted by a liquid or gas measured in pounds per square inch (psi) or kilopascals (kPa).

Pressure Relief Device (PRD) — An engineered valve or other device used to control or limit the pressure in a system or vessel, often by venting excess pressure.

Pressure Relief Valve — Pressure control device designed to eliminate hazardous conditions resulting from excessive pressures by allowing this pressure to release in manageable quantities.

Pressure Storage Tank — Class of fixed facility storage tanks divided into two categories: low-pressure storage tanks and pressure vessels.

Pressure Vessel — Fixed-facility storage tanks with operating pressures above 15 psi (103 kPa) {1.03 bar}.

Primary Explosive — High explosive that is easily initiated and highly sensitive to heat; often used as a detonator. *Also known as* Initiation Device.

Protective Action Distance — Downwind distance from a hazardous materials incident within which protective actions should be implemented.

Proton — Subatomic particle with a physical mass and a positive electric charge.

Public Safety Sample — Hazardous materials collected at an incident and used to help inform response and mitigation options.

Public Safety Sampling — Techniques used to collect materials found at a Hazmat/WMD incident that result in a forensically usable and legally defensible sample. Samples are often used when determining response and mitigation options.

Purge — To expel an inert gas through a device's hosing and/or intake system to remove any residual contaminants.

R

Radiation — Energy from a radioactive source emitted in the form of waves or particles, as a result of the decay of an atomic nucleus; process known as *radioactivity*. *Also called* Nuclear Radiation.

Radiation Absorbed Dose (rad) — English System unit used to measure the amount of radiation energy absorbed by a material; its International System equivalent is gray (Gy).

Radiation-Exposure Device (RED) — Powerful gamma-emitting radiation source used as a weapon.

Radioactive Decay — Process in which an unstable radioactive atom loses energy by emitting ionizing radiation and conversion electrons.

Radioactive Material (RAM) — Material with an atomic nucleus that spontaneously decays or disintegrates, emitting radiation as particles or electromagnetic waves at a rate of greater than 0.002 microcuries per gram (Ci/g).

Radioisotope — Unstable atom that releases nuclear energy.

Radiological Dispersal Device (RDD) — Conventional high explosives wrapped with radioactive materials; designed to spread radioactive contamination over a wide area. *Also known as* Dirty Bomb.

Radiological Dispersal Weapons (RDW) — Devices that spread radioactive contamination without using explosives; instead, radioactive contamination is spread using pressurized containers, building ventilation systems, fans, and mechanical devices.

Railcar Initials and Numbers — Combination of letters and numbers stenciled on rail tank cars that may be used to get information about the car's contents from the railroad's computer or the shipper. *Also known as* Reporting Marks.

Rain-Down Application Method — Foam application method that directs the stream into the air above the unignited or ignited spill or fire, allowing the foam to float gently down onto the surface of the fuel.

Raman Spectrometer — Apparatus used to observe the absorption, scattering, and shifts in light when sent through a material. The results are unique to the molecule.

Rapid Intervention Crew or Team (RIC/RIT) — Two or more firefighters designated to perform firefighter rescue; they are stationed outside the hazard and must be standing by throughout the incident. *Previously known as* Rapid Intervention Team (RIT).

Reactive Material — Substance capable of chemically reacting with other substances; for example, material that reacts violently when combined with air or water.

Reactivity — Ability of a substance to chemically react with other materials, and the speed with which that reaction takes place.

Reagent — Chemical that is known to react to another chemical or compound in a specific way, often used to detect or synthesize another chemical.

Recovery — Situation where the victim is determined or presumed to be dead, and the goal of the operation is to recover the body.

Reducing Agent — Fuel that is being oxidized or burned during combustion. *Also known as* Reducer.

Refrigerated Intermodal Container — Cargo container having its own refrigeration unit. *Also known as* Reefer.

Remediation — Fixing or correcting a fault, error, or deficiency.

Resonant Bond — Type of chemical bond in which electrons move freely between the compound atoms. *Also known as* Delocalized Bond.

Response Model — Framework for resolving problems or conflicts using logic, research, and analysis.

Response Objective — Statement based on realistic expectations of what can be accomplished when all allocated resources have been effectively deployed that provide guidance and direction for selecting appropriate strategies and the tactical direction of resources.

Response Option — Specific operations performed in a specific order to accomplish the goals of the response objective.

Retain — Actions to contain a liquid or sludge in an area where it can be absorbed, neutralized, or removed. Often used as a longer-term solution than other similar product control methods.

Rickettsia — Specialized bacteria that live and multiply in the gastrointestinal tract of arthropod carriers, such as ticks and fleas.

Ring Stiffener — Circumferential tank shell stiffener that helps to maintain the tank cross section.

Riot Control Agent — Chemical compound that temporarily makes people unable to function, by causing immediate irritation to the eyes, mouth, throat, lungs, and skin.

Risk-Based Response — Method using hazard and risk assessment to determine an appropriate mitigation effort based on the circumstances of the incident.

Roentgen (R) — English System unit used to measure radiation exposure, applied only to gamma and X-ray radiation; the unit used on most U.S. dosimeters.

Roentgen Equivalent in Man (rem) — English System unit used to express the radiation absorbed dose (rad) equivalence as pertaining to a human body; used to set radiation dose limits for emergency responders. Applied to all types of radiation.

Roll-On Application Method — Method of foam application in which the foam stream is directed at the ground at the front edge of the unignited or ignited liquid fuel spill; foam then spreads across the surface of the liquid. *Also known as* Bounce.

Route of Entry — Pathway via which hazardous materials get into (or affect) the human body.

S

Safety Data Sheet (SDS) — Reference material that provides information on chemicals that are used, produced, or stored at a facility. Form is provided by chemical manufacturers and blenders; contains information about chemical composition, physical and chemical properties, health and safety hazards, emergency response procedures, and waste disposal procedures. *Also known as* Material Safety Data Sheet (MSDS) *or* Product Safety Data Sheet (PSDS).

Safety Officer — Member of the IMS command staff responsible to the Incident Commander for monitoring and assessing hazardous and unsafe conditions and developing measures for assessing personnel safety on an incident. *Also known as* Incident Safety Officer.

Safety Relief Device — Device on cargo tanks with an operating part held in place by a spring; the valve opens at preset pressures to relieve excess pressure and prevent failure of the vessel.

Saponification — Reaction between an alkali and a fatty acid that produces soap.

Saturation — The concentration at which the addition of more solute does not increase the levels of dissolved solute.

Scintillator — Material that glows (luminesces) when exposed to ionizing radiation.

Secondary Device — Bomb or other weapon placed at the scene of an ongoing emergency response that is intended to cause casualties among responders; secondary explosive devices are designed to explode after a primary explosion or other major emergency response event has attracted large numbers of responders to the scene.

Secondary Explosive — High explosive that is designed to detonate only under specific circumstances, including activation from the detonation of a primary explosive. *Also known as* Main Charge Explosive.

Self-Accelerating Decomposition Temperature (SADT) — Lowest temperature at which product in a typical package will undergo a self-accelerating decomposition. The reaction can be violent, usually rupturing the package, dispersing original material, liquid and/or gaseous decomposition products considerable distances.

Self-Contained Breathing Apparatus (SCBA) — Respirator worn by the user that supplies a breathable atmosphere that is either carried in or generated by the apparatus and is independent of the ambient atmosphere. Respiratory protection is worn in all atmospheres that are considered to be Immediately Dangerous to Life and Health (IDLH). *Also known as* Air Mask *or* Air Pack.

Self-Reading Dosimeter (SRD) — Detection device that displays the cumulative reading without requiring additional processing. *Also known as* Direct-Reading Dosimeters (DRDs) *and* Pencil Dosimeters.

Shell — Layer of electrons that orbit the nucleus of an atom. The innermost shell can hold up to two electrons, and each subsequent shell can hold eight. *Also known as* Orbit, Orbital, *and* Ring.

Sheltering in Place — Having occupants remain in a structure or vehicle in order to provide protection from a rapidly approaching hazard, such as a fire or hazardous gas cloud. *Opposite of* evacuation. *Also known as* Protection-in-Place, Sheltering, *and* Taking Refuge.

Short-Term Exposure Limit (STEL) — Fifteen-minute time-weighted average that should not be exceeded at any time during a workday; exposures should not last longer than 15 minutes and should not be repeated more than four times per day with at least 60 minutes between exposures.

Sievert (Sv) — SI unit of measurement for low levels of ionizing radiation and their health effect in humans.

Site Characterization — Size-up and evaluation of hazards, problems, and potential solutions of a site.

Situational Awareness — Perception of the surrounding environment and the ability to anticipate future events.

Size-Up — Ongoing evaluation of influential factors at the scene of an incident.

SKED® — Lightweight, compact device for patient packaging; shaped to accommodate a long backboard; may be used with a rope mechanical advantage system.

Slurry — Suspension formed by a quantity of granulated or powdered solid material that is not completely soluble mixed into a liquid.

Soft Patch — Combination of gasket material and either wooden plugs or wedges inserted into a hole to patch a leak.

Solid — Substance that has a definite shape and size; the molecules of a solid generally have very little mobility.

Solubility — Degree to which a solid, liquid, or gas dissolves in a solvent (usually water).

Soluble — Capable of being dissolved in a liquid (usually water).

Solution — Uniform mixture composed of two or more substances.

Solvent — A substance that dissolves another substance (solute), resulting in a third substance (solution).

Specific Gravity — Mass (weight) of a substance compared to the weight of an equal volume of water at a given temperature. A specific gravity less than one indicates a substance lighter than water; a specific gravity greater than one indicates a substance heavier than water.

Specification Marking — Stencil on the exterior of a tank car indicating the standards to which the tank car was built; may also be found on intermodal containers and cargo tank trucks.

Spectrometer — Apparatus used to measure the intensity of a given sample based on a predefined spectrum such as wavelength or mass.

Spectrophotometer — Apparatus used to measure the intensity of light as an aspect of its color.

Spectroscopy — Study of the results when a material is dispersed into its component spectrum. *Also known as* Spectrography.

Staging Area — Prearranged, temporary strategic location, away from the emergency scene, where units assemble and wait until they are assigned a position on the emergency scene; these resources (personnel, apparatus, tools, and equipment) must then be able to respond within three minutes of being assigned. Staging Area Managers report to the Incident Commander or Operations Section Chief, if one has been established.

Standard Operating Procedure (SOP) — Standard methods or rules in which an organization or fire department operates to carry out a routine function. Usually these procedures are written in a policies and procedures handbook and all firefighters should be well versed in their content.

Standard Transportation Commodity Code (STCC) — Numerical code on the waybill used by the rail industry to identify the commodity. *Also known as* STCC Number.

Static Electricity — Accumulation of electrical charges on opposing surfaces, created by the separation of unlike materials or by the movement of surfaces.

Sterilization — Any process that destroys biological agents and other life forms. Often uses heat.

Stokes Basket — Wire or plastic basket-type litter suitable for transporting patients from locations where a standard litter would not be easily secured, such as a pile of rubble, a structural collapse, or the upper floor of a building; may be used with a harness for lifting.

Street Clothes — Clothing that is anything other than chemical protective clothing or structural firefighters' protective clothing, including work uniforms and ordinary civilian clothing.

Strong Oxidizer — Substance that readily gives off large quantities of oxygen, thereby stimulating combustion; produces a strong reaction by readily accepting electrons from a reducing agent (fuel).

Structural Firefighters' Protective Clothing — General term for the equipment worn by fire and emergency services responders; includes helmets, coats, pants, boots, eye protection, gloves, protective hoods, self-contained breathing apparatus (SCBA), and personal alert safety system (PASS) devices.

Sublimation — Vaporization of a material from the solid to vapor state without passing through the liquid state.

Supervisory Control and Data Acquisition (SCADA) — System that monitors and controls coded signals from preset locations within an infrastructure (pipeline system), industry (manufacturing system), or facility (building system).

Supplied Air Respirator (SAR) — Atmosphere-supplying respirator for which the source of breathing air is not designed to be carried by the user; not certified for fire fighting operations. *Also known as* Airline Respirator System.

Surface Acoustic Wave (SAW) Sensor — Device that senses a physical phenomenon. Electrical signals are transduced to mechanical waves, and then back to electrical signals for analysis.

Surfactant — Chemical that lowers the surface tension of a liquid; allows water to spread more rapidly over the surface of Class A fuels and penetrate organic fuels.

Synergistic Effect — Phenomenon in which the combined properties of substances have an effect greater than their simple arithmetical sum of effects.

Systemic Effect — Damage spread through an entire system; opposite of a local effect, which is limited to a single location.

T

Tapping — Process to attach a nozzle or outlet onto a container's tank or piping to assist with the removal or transfer of a product.

T-Code — Portable tank instruction code used to identify intermodal containers used to transport hazardous materials. This set of codes replace the IMO type listings.

Technical Decontamination — Using chemical or physical methods to thoroughly remove contaminants from responders (primarily entry team personnel) and their equipment; usually conducted within a formal decontamination line or corridor following gross decontamination. *Also known as* Formal Decontamination.

Termination — The phase of an incident in which emergency operations are completed and the scene is turned over to the property owner or other party for recovery operations.

Tertiary Explosive — High explosive that require initiation from a secondary explosive. Tertiary explosives are often categorized with secondary explosives. *Also known as* Blasting Agents.

Thermal Burn — Injury caused by contact with flames, hot objects, and hot fluids; examples include scalds and steam burns.

Thermal Imager — Electronic device that forms images using infrared radiation. *Also known as* Thermal Imaging Camera.

Thermal Insulation — Materials added to decrease heat transfer between objects in proximity to each other.

Threshold Limit Value (TLV®) — Maximum concentration of a given material in parts per million (ppm) that may be tolerated for an 8-hour exposure during a regular workweek without ill effects.

Threshold Limit Value/Ceiling (TLV®/C) — Maximum concentration of a given material in parts per million (ppm) that should not be exceeded, even instantaneously.

Torr — Unit of pressure measurement; not part of the SI. Measured as 1/760 of a standard atmosphere.

Toxic — Poisonous.

Toxic Industrial Material (TIM) — Industrial chemical that is toxic at a certain concentration and is produced in quantities exceeding 30 tons (30 tonnes) per year at any one production facility; readily available and could be used by terrorists to deliberately kill, injure, or incapacitate people. *Also known as* Toxic Industrial Chemical (TIC).

Toxic Inhalation Hazard (TIH) — Volatile liquid or gas known to be a severe hazard to human health during transportation.

Toxicity — Degree to which a substance (toxin or poison) can harm humans or animals. Ability of a substance to do harm within the body.

Toxicology — Study of the adverse effects of chemicals on living organisms.

Toxin — Substance that has the property of being poisonous.

Transient Evidence — Material that will lose its evidentiary value if it is unpreserved or unprotected; for example, blood in the rain.

Transmutation — Conversion of one element or isotope into another form or state.

Transport Index (TI) — Number placed on the label of a package expressing the maximum allowable radiation level in millirem per hour at 1 meter (3.3 feet) from the external surface of the package.

Transportation Mode — Technologies used to move people and/or goods in different environments; for example, rail, motor vehicles, aviation, vessels, and pipelines.

Trench Foot — Foot condition resulting from prolonged exposure to damp conditions or immersion in water; symptoms include tingling and/or itching, pain, swelling, cold and blotchy skin, numbness, and a prickly or heavy feeling in the foot. In severe cases, blisters can form, after which skin and tissue die and fall off.

Triage — System used for sorting and classifying accident casualties to determine the priority for medical treatment and transportation.

Type A Packaging — Container used to ship radioactive materials with relatively high radiation levels.

Type B Packaging — Container used to ship radioactive materials that exceed the limits allowed by Type A packaging, such as materials that would present a radiation hazard to the public or the environment if there were a major release.

Type C Packaging — Container used to ship highly reactive radioactive materials intended for transport via aircraft.

U

UN/NA Number — Four-digit number assigned by the United Nations to identify a specific hazardous chemical. North America (DOT) numbers are identical to UN numbers, unless the UN number is unassigned.

Unified Command (UC) — In the Incident Command System, a shared command role in which all agencies with geographical or functional responsibility establish a common set of incident objectives and strategies. In unified command there is a single incident command post and a single operations chief at any given time.

Unstable Material — Materials that are capable of undergoing chemical changes or that can violently decompose with little or no outside stimulus.

Upper Flammable Limit (UFL) — Upper limit at which a flammable gas or vapor will ignite; above this limit the gas or vapor is too *rich* to burn (lacks the proper quantity of oxygen). *Also known as* Upper Explosive Limit (UEL).

V

Vacuum Relief Valve — Pressure control device designed to introduce outside air into a container during offloading operations.

Valve — Mechanical device with a passageway that controls the flow of a liquid or gas.

Vapor Density — Weight of pure vapor or gas compared to the weight of an equal volume of dry air at the same temperature and pressure. A vapor density less than one indicates a vapor lighter than air; a vapor density greater than one indicates a vapor heavier than air.

Vapor Dispersion — Action taken to direct or influence the course of airborne hazardous materials.

Vapor Explosion — Occurrence when a hot liquid fuel transfers heat energy to a colder, more volatile liquid fuel. As the colder fuel vaporizes, pressure builds in a container and can create shockwaves of kinetic energy.

Vapor Pressure — The pressure at which a vapor is in equilibrium with its liquid phase for a given temperature; liquids that have a greater tendency to evaporate have higher vapor pressures for a given temperature.

Vapor Suppression — Action taken to reduce the emission of vapors at a hazardous materials spill.

Vapor-Protective Clothing — Gas-tight chemical-protective clothing designed to meet NFPA 1991, *Standard on Vapor-Protective Ensembles for Hazardous Materials Emergencies*; part of an EPA Level A ensemble.

Vector — An animate intermediary in the indirect transmission of an agent that carries the agent from a reservoir to a susceptible host.

Vehicle-Borne Improvised Explosives Device (VBIED) — An improvised explosive device placed in a car, truck, or other vehicle. This type of IED typically creates a large explosion.

Ventilation — Systematic removal of heated air, smoke, gases or other airborne contaminants from a structure and replacing them with cooler and/or fresher air to reduce damage and facilitate fire fighting operations.

Virus — Simplest type of microorganism that can only replicate itself in the living cells of its hosts. Viruses are unaffected by antibiotics.

Viscosity — Measure of a liquid's internal friction at a given temperature. This concept is informally expressed as thickness, stickiness, and ability to flow.

Volatility — Ability of a substance to vaporize easily at a relatively low temperature.

W

Warm Zone — Area between the hot and cold zones that usually contains the decontamination corridor; typically requires a lesser degree of personal protective equipment than the Hot Zone. *Also known as* Contamination Reduction Zone *or* Contamination Reduction Corridor.

Water Solubility — Ability of a liquid or solid to mix with or dissolve in water.

Water-Reactive Material — Substance, generally a flammable solid, that reacts when mixed with water or exposed to humid air.

Waybill — Shipping paper used by a railroad to indicate origin, destination, route, and product; a waybill for each car is carried by the conductor. *Similar to* Air Bill *and* Bill of Lading.

Weapon of Mass Destruction (WMD) — Any weapon or device that is intended or has the capability to cause death or serious bodily injury to a significant number of people through the release, dissemination, or impact of toxic or poisonous chemicals or their precursors, a disease organism, or radiation or radioactivity; may include chemical, biological, radiological, nuclear, or explosive (CBRNE) type weapons.

Wet Chemistry — Branch of analysis with a focus on chemicals in their liquid phase.

Wireless Information System for Emergency Responders (WISER) — This electronic resource brings a wide range of information to the hazmat responder such as chemical identification support, characteristics of chemicals and compounds, health hazard information, and containment advice.

Witness — Person called upon to provide factual testimony before a judge or jury.

Z

Zeroing — Resetting an instrument to read at normal (baseline) levels in fresh air.

Index

Index

A

AAR. *See* After Action Report (AAR); American Association of Railroads (AAR)
AAR 2014, cryogenic railway tank cars, 295
AAR Specifications for Tank Cars, 293, 294
Absorption decontamination method
    advantages and disadvantages, 446
    defined, 442
    hazardous materials, 449
    procedures, 522
    resources for using, 449
    uses for, 448
Absorption route of entry, 162
ACADA (automatic chemical agent detector alarm), 192
Access entryways in nonpressure cargo tanks, 267
Access to Employee Exposure and Medical Records (Title 29 CFR 1910.1020), 367
Accessibility considerations for decontamination operations, 457
Accuracy of monitoring and detection equipment, 138
Acetone peroxide (TATP), 120
Acetylene, 84
ACGIH (American Conference of Governmental Industrial Hygienists®), 164–168, 231
Acid
    defined, 94
    fuming sulfuric acid, 334
    pH paper/strip detection of, 180
Acronyms
    ALARA, 115, 185
    ALERT, 127
    APIE. *See* APIE
    B-NICE, 100
    COBRA, 100
    DUMBELS, 102–103
    GEDAPER, 327
    ISHP, 232
    NBC, 100
    READY, 415
    SLUDGEM, 102
    SMART, 335
Action levels, 143–144
Action plan. *See* Incident Action Plan (IAP)
Activation energy, 76
Activity logs, 364
Activity of radioactivity, 99
Acute dose, 114
Acute exposure, 163
Acute Exposure Guideline Level-1 (AEGL-1), 166
Acute Exposure Guideline Level-2 (AEGL-2), 166
Acute Exposure Guideline Level-3 (AEGL-3), 166
Adamsite, 106
Adsorb, defined, 198
Adsorption decontamination method
    advantages and disadvantages, 446
    defined, 442
    procedures, 522
    uses for, 449
AEGL-1 (Acute Exposure Guideline Level-1), 166
AEGL-2 (Acute Exposure Guideline Level-2), 166
AEGL-3 (Acute Exposure Guideline Level-3), 166
Aerosol, mobile like a gas, 31
AFFF (Aqueous Film Forming Foam), 517
After Action Report (AAR)
    defined, 554
    information included, 340
    purpose of, 554–555
Agricultural chemicals, 110–111
AHJ. *See* Authority having jurisdiction (AHJ)
AIHA (American Industrial Hygiene Association), 166–167
Air bill, 306
Air cooling, 412
Air cylinder buffers, 421
Air cylinder capacities, 421
Air disinfectants, 451
Air freight cargo, 305–306
Air management, 417, 422
Air mask, 378
Air monitoring during product removal and transfer, 506
Air pack, 378
Air supply
    mission work duration considerations, 422
    PPE limitations, 384
Airline respiratory system, 380, 397
Air-purifying respirator (APR)
    combination particulate- and vapor-and-gas removing, 376
    combined respirators, 383–384
    defined, 380
    facepieces, 381
    filters, 376, 380–381
    limitations, 382
    NIOSH standards, 377, 378
    particulate-removing, 376, 382–383
    powered air-purifying respirator (PAPR)
        defined, 376
        mechanism, 383
        responder use of, 377
        types of, 383
        uses for, 381, 383
    supplied-air hoods, 384
    uses for, 381
    vapor-and-gas removing, 376, 383
Air-reactive materials, 72, 88–89
A-Kit, 499, 500
ALARA (As Low As Reasonably Achievable), 115, 185
Alcohol disinfectant, 451
Aldehyde disinfectant, 451
ALERT acronym, 127
Alkali metals, 60, 61
Alkaline earths, 60, 61–62
Alkalis, 94
Alkane, 84
Alkene, 84
Alkyne, 84
Alloy
    defined, 56
    fires, 91–92
    high strength low alloy steel, 488
    metal, 56
    production process, 56
    reactivity of, 57
    solutions, 55
ALOHA, 236
Alpha (α) particles, 96, 112
Aluminum chloride, 90
Aluminum containers, 487, 504
Aluminum phosphide, 90
Ambulatory, defined, 445
Ambulatory victim decontamination, 445, 469–470, 472–473

American Association of Railroads (AAR)
    AAR 2014, cryogenic railway tank cars, 295
    AAR Specifications for Tank Cars, 293, 294
    Emergency Handling of Hazardous Materials in Surface Transportation, 227
    railway tank car class numbers and approving authority, 282
    railway tank car specification markings, 283–284
American Chemical Society, 224
American Chemistry Council, 142
American Conference of Governmental Industrial Hygienists® (ACGIH), 164–168, 231
American Industrial Hygiene Association (AIHA), 166–167
American National Standards Institute (ANSI), 231, 379
American Society of Mechanical Engineers (ASME)
    certification plates, 263
    defined, 263
    pressure vessels, 294
Ammonia gas, 90
Ammonia refrigeration systems, 508
Amplification of monitoring and detection equipment, 137
Analysis reaction, 75
Analyze the incident (Step 1 of APIE)
    Awareness Level responders, 12–13
    generally, 11
    hazmat technician, 18–20
    Operations Level responders, 14–15
Anhydrous, 34
Animal decontamination, 464
Anion, defined, 68
ANSI (American National Standards Institute), 231, 379
Anthrax incidents in the U.S., 468
Antibody, 187
Antigen, 187
APIE
    book organization, 1–2
    response model, 326
    Step 1: Analyze the incident
        Awareness Level responders, 12–13
        generally, 11
        hazmat technician, 18–20
        Operations Level responders, 14–15
    Step 2: Plan the initial response, 20–22
        action plan development, 22
        Awareness Level responders, 13
        decontamination procedures, 21
        generally, 11
        Operations Level responders, 15
        personal protective equipment selection, 21
        response objectives and options, 21
    Step 3: Implement the response
        Awareness Level responders, 13
        control functions, 23
        decontamination, 23
        generally, 11
        ICS/IMS duties, 22–23
        Operations Level responders, 15
        personal protective equipment, 23
    Step 4: Evaluate progress
        Awareness Level responders, 14
        generally, 11
        hazmat technician responsibilities, 23–24
        Operations Level responders, 15
Apparatus, decontamination of, 456
Appearance of materials, 38
Apps, 237–238
APR. *See* Air-purifying respirator (APR)
Aqueous Film Forming Foam (AFFF), 517
Area monitoring, 145–146
Arene, 85

Aromatic hydrocarbons, 68–69, 84–85
As Low As Reasonably Achievable (ALARA), 115, 185
ASME. *See* American Society of Mechanical Engineers (ASME)
Assay, 187
Atmospheric pressure barometer, 30
Atmospheric storage tanks, 310–311
Atmospheric underground storage tank, 311
Atom, 49–50
Atomic number, 50, 58
Atomic stability, 65
Atomic theory, 49–56
    atoms, 49–50
    compounds
        atomic stability, 65–66
        defined, 50
        nonsalts, 54–55
        salts, 53–54
        sodium chloride, 51
    elements, 50–52
        atomic stability, 65–66
        compounds, 50, 51
        diatomic molecules, 50, 51
        metalloids, 51
        metals, 52
        nonmetals, 52
        overview, 50
    mixture, 55–56
        alloy, 56, 57
        defined, 55
        slurry, 56
        solutions, 55–56
Atomic weight, 50, 58
Austenitic stainless containers, 488
Authority having jurisdiction (AHJ)
    action levels, 143, 185
    air management, 417, 422
    Awareness Level actions, 2–3
    calibration of monitoring instruments, 139
    chemical-protective clothing, 394, 395
    cooling vests, 412
    decontamination corridor entry, 463
    decontamination selection, 341
    detection devices, 172
    emergency procedures, 345
    evidence chain of custody, 152
    evidence collection, 151, 152
    forms and logs, 364
    hazardous materials profile, 221
    hazmat technician training standards, 17, 18
    Incident Action Plan development, 329
    medical surveillance plan, 343, 344
    Mission Specific level, 16
    notification procedures, 336, 337
    personal protective equipment selection and use, 341
    pH paper/strip use, 180
    PPE selection, 407–408
    pre-entry evaluation, 344
    sampling plans, 148, 149
    vapor dispersion, 517
    wet chemistry, 199
Autoignition temperature, 44
Automatic chemical agent detector alarm (ACADA), 192
Automatic pumps on monitoring equipment, 139
Autorefrigeration, 41, 508
Awareness Level responders
    actions allowed, 2–3
    APIE responsibilities, 11
    qualifications, 12–16
        analyzing the incident, 12–13

evaluating progress, 14
implementing the response, 13
planning the response, 13

B

Backpack bomb, 124
Backup personnel, 345
Bacteria, 106
Badge dosimeter, 168, 184–185, 210
Baffle, 266
Bags
    bulk bags, 257
    non-bulk containers, 250–251
Ballistic protection, 404
Bar, defined, 32
Barges, 305
Base
    corrosives, 94
    defined, 94
    pH paper/strip detection of, 180
Base cover, 462
Base metal, 490
Batch plant, 314
Beam, 298
Beam-type intermodal tank, 298
Bear claw, 180
Becqueral (Bq), 99, 168
Behavior of products, predicting, 20, 217–218
BEIs® (Biological Exposure Indices), 166
Benner, Ludwig, 327
Benzene, 68–69, 85
Berm, 459
Beta (β) particles, 96, 113
Betts valves for product removal and transfer, 506
Big bags, 257
Bill of lading, 266, 292
Binary salts, 53
Bioassay, 109
Biodegradable materials, 450
Biological, nuclear, incendiary, chemical, explosive (B-NICE), 100
Biological agents and toxins, 106–111
    bacteria, 106
    defined, 106
    dose-response relationships, 108
    exposures, 169
    federal assistance for incidents, 107
    incubation period after exposure, 169
    infectious dose, 169
    measuring and expressing dose and concentration, 108–109
    pesticides and agricultural chemicals, 110–111
    rickettsia, 107
    toxicity, 109
    toxicology, 107
    toxins, 107
    viruses, 106
Biological degradation, 446, 449–450
Biological Exposure Indices (BEIs®), 166
Biological immunoassay indicators, 187
Biphenyl, 85
B-Kit, 499, 500
Blanketing for product control, 484, 517–518
BLEVE. *See* Boiling liquid expanding vapor explosion (BLEVE)
Blister agents
    defined, 103
    M256A1 Detector Kit, 182–183
Blood agents
    generally, 104
    M256A1 Detector Kit, 182–183
B-NICE (biological, nuclear, incendiary, chemical, explosive), 100

Bobtail tank, 275
Body armor, 386, 404
BoE (Bureau of Explosives), 227
Boiling liquid expanding vapor explosion (BLEVE)
    container fire control, 504
    defined, 293
    hazards of gases, 34
    pressure containers, 293
    strategic goals and tactical objectives, 338
    temperature and vapor pressure in containers, 41
Boiling point, 42–43, 45
Boiling-point elevation, 43
Bombs
    backpack, 124
    bomb robot, 128
    bomb squad, 116, 129–130, 243
    bomb suits, 386, 404
    box bomb, 124
    briefcase, 124
    duffle bag, 124
    knapsack, 124
    letter bomb, 125–126
    mail bomb, 116, 125–126
    package, 125–126
    pipe bomb, 116, 123–124
    plastic bottle bomb, 124
    Public Safety Bomb Disposal (PSBD), 243
    satchel, 124
    suicide bomb, 126–128
    tennis ball bomb, 125
Bonding, 65–70
    atomic stability, 65–66
    bond energy, 70
    covalent bonds, 66–67
    defined, 507
    diatomic molecules, 66
    Duet Rule, 65
    hazmat response, 70
    ionic bonds, 67–68, 70
    Octet Rule, 65
    procedures for use, 537
    resonant bonds, 68–69
    unstable materials, 66
Booming for product control, 484
Boot liners, 398
Bottles, 251–252, 253
Bottom outlet valves, 291
Box bomb, 124
Box patch, 497
Box truck, 260
Boxcars, 296
Boxes, 256
Box-type intermodal tank, 298
Breaches in cargo tanks, 503–504. *See also* Leak control
Break bulk carrier, 304
Breakdown reaction, 75
Breakthrough time for permeation, 409
Breathing air cylinder capacities, 421
Briefcase bomb, 124
Briefing, safety, 416
Brushes for decontamination, 462
Buddy systems, 345, 416, 419
Bulk bags, 257
Bulk sacks, 257
Bump test, 140
Bung leaks, 513
Bureau of Explosives (BoE), 227
Burns, 94–95
Butterfly valve, 509

C

Calibration of monitoring and detection instruments, 139-143
  bump test, 140
  calibration, defined, 139
  CGI gases, 173
  correction factors, 140-142
  cross-sensitivities, 142-143
  interference, 142-143
  photoionization detectors (PIDs), 186-187
  relative response, 140-142
CAM (chemical agent monitor), 192
CAMEO (Computer-Aided Management of Emergency Operations), 234-236
Canada
  Canadian National Energy Board, 306
  continuity (chain of custody), 150
  first responder offensive tasks, 14
  liquefied flammable gas carrier regulations, 304
  pipelines, 306-307
  Transport Canada (TC)
    CANUTEC. See Canadian Transportation Emergency Centre (CANUTEC)
    Emergency Handling of Hazardous Materials in Surface Transportation, 227
    Emergency Response Guidebook (ERG), 226
    railway tank car specification markings, 283-284
Canadian Transportation Emergency Centre (CANUTEC)
  defined, 142
  instrument correction factor, 142
  purpose of, 240-241
Canine decontamination, 464
CANUTEC. See Canadian Transportation Emergency Centre (CANUTEC)
Capacity stencil, 284-285
Carbon dioxide (CO_2)
  carbonated beverages, 55, 56, 255
  combustion, 74
  dry ice, 43
  expansion ratio, 32
  grenades, 125
  sublimation, 43
Carbon monoxide (CO)
  combustion, 74
  electrochemical cell sensors, 175
  exposure, 162
Carbon steel, 487
Carbonated beverages, 55, 56, 255
Carboys, 251-252
Carcinogen, 189
Cargo containers, 146. See also Containers
Cargo tank product control, 501-507
  bonding and grounding, 506-507, 537
  fire control, 504
  leak control, 502-504
  product removal and transfer, 505-507, 536
Cargo tank truck, 260
Cargo vessels, 304
Cas® number, 224
Catalyst, 74-75, 76
Cation, defined, 68
Caustics, 94
CBRN (Chemical, Biological, Radiological and Nuclear) Standard for Open-Circuit Self-Contained Breathing Apparatus (SCBA), 377-378
CBRNE (chemical, biological, radiological, nuclear, and explosive), 99, 244
CCD (Hawley's Condensed Chemical Dictionary), 229
CDC. See Centers for Disease Control and Prevention (CDC)
CECOM (National Coordination of Civil Protection), 241
Celsius scale, 38
Centers for Disease Control and Prevention (CDC)
  hazmat technician training, 18
  infectious agent decontamination, 467
Centigrade scale, 38
Centro de Información Química para Emergencias (CIQUIME), 226
Certified Equipment List, 379
CFR. See Code of Federal Regulations (CFR)
CGA. See Compressed Gas Association (CGA)
CGI. See Combustible gas indicator (CGI)
Chain of custody, 148, 150, 152
Change of state/phase, 41
Chemical, Biological, Radiological and Nuclear (CBRN) Standard for Open-Circuit Self-Contained Breathing Apparatus (SCBA), 377-378
Chemical, biological, radiological, nuclear, and explosive (CBRNE), 99, 244
Chemical, ordinance, biological, radiological agents (COBRA), 100
Chemical Abstract Service (CAS®), 224
Chemical agent detection papers, 182-183
Chemical agent monitor (CAM), 192
Chemical Agent Permeation and Penetration Resistance Against Distilled Sulfur Mustard (HD [military designation]) and Sarin (GB [military designation]), 379
Chemical asphyxiant, 104
Chemical assessment for decision-making, 224-225
Chemical bonding. See Bonding
Chemical burn, 94-95
Chemical carriers, 303
Chemical datasheet, 234
Chemical decontamination, 440, 450, 454
Chemical degradation, 391
Chemical dictionary, 229
Chemical Hazards Emergency Medical Management (CHEMM), 237
Chemical Hazards Response Information System (CHRIS), 227
Chemical Inventory List (CIL), 233
Chemical protective clothing (CPC). See also Personal protective equipment (PPE)
  decontamination procedures, 463
  defined, 390
  degradation, 391, 410
  documentation and written PPE program, 426, 427-428, 434
  flash protection, 408
  inspection, 425-426, 434
  liquid splash-protective clothing, 391-392, 432
  maintenance and storage, 426-427, 434
  mission specific operations, 394, 395
  penetration, 391, 410
  permeation, 391
  psychological issues during use, 414
  purpose of, 390
  selection factors
    degradation, 410
    generally, 405-407
    hazmat technician selection of CPC, 21
    lower explosive limit, 407-408
    penetration, 410
    permeation, 409-410
    resistance/compatibility, 408-410
  service life, 395
  for specific hazards, 218
  standards, 386-387
  suit integrity, 420
  testing, 426, 434
  vapor-protective clothing, 393-394
  written management programs, 394
Chemical Transportation Emergency Center (CHEMTREC®)
  chlorine emergency plan (CHLOREP), 243
  defined, 142
  instrument correction factor, 142
  pesticide decontamination, 467
  purpose of, 240

radiation decontamination, 467
Chemical warfare agents, 100–106
    blister agents/vesicants, 103
    blood agents, 104
    choking agents, 104
    nerve agents, 100–103
    riot control agents/irritants, 105–106
Chemical/biological degradation, 446, 449–450
Chemical-specific APR, 384
CHEMM (Chemical Hazards Emergency Medical Management), 237
CHEMTREC®. *See* Chemical Transportation Emergency Center (CHEMTREC®)
Chime leaks, 513
Chip measurement system (CMS), 179
Chloramine, 75
Chlorate-based explosives, 121
Chlorates, 88
CHLOREP (chlorine emergency plan), 242–243
Chlorinated hydrocarbons, 86
Chlorine
    as choking agent, 104
    compounds, 51
    container valves, 510
    diatomic molecules, 50
    emergency kits, 259–260
    high-pressure cargo tanks, 274
    intermodal tanks for, 301
    ionic bonds, 67–68
    molecular weight, 37
    pressure railway tank cars for, 293, 294
    toxicity of, 63
    vapor density, 36
Chlorine emergency plan (CHLOREP), 242–243
Chlorine Institute
    chlorine emergency kits, 259–260
    Chlorine Emergency Plan (CHLOREP), 242–243
    Emergency Kit "A," 499
    Emergency Kit "B," 499
    Emergency Kit "C," 500
    tank inlet and discharge outlet standards, 301
Choking agents, 104
CHRIS *(Chemical Hazards Response Information System)*, 227
Chronic dose, 114
Chronic exposure, 163
CIL (Chemical Inventory List), 233
Cinnabar (HgS), 53
CIQUIME (Centro de Información Química para Emergencias), 226
C-Kit, 500
Class 3 ensembles, 392
Class 7 radioactive materials, 114
Class B foam, 517
Class D fire
    combustible metals and alloys, 91–92
    defined, 90
    dry powder agents, 91
    extinguishers, 91–92
    test fires, 92
Cleaning solution analysis, 456
Closed-circuit SCBA
    air supply limitations, 384
    high-pressure, 379
    low-pressure, 379
    positive-pressure, 379
    pressure-demand, 379
    purifying exhaled air, 375
Clothing, 385–395
    body armor, 386, 404
    bomb suits, 386, 404
    chemical protective. *See* Chemical protective clothing (CPC)
    flame-resistant clothing, 390
    high temperature clothing, 389–390
    overview, 385–386
    standards at hazmat/WMD incidents, 386–387
    structural firefighters' clothing, 388, 403, 430
    vapor-protective clothing, 23, 386, 433
CMS (chip measurement system), 179
Coast Guard, 227
COBRA (chemical, ordinance, biological, radiological agents), 100
Code of Federal Regulations (CFR)
    division numbers, 117–118
    Hazard Communication regulation (29 *CFR* 1910.1200), 218
    Hazardous Materials Table (49 *CFR* 172.101), 234, 299
    NIOSH Regulation 42 *CFR* Part 84, *Approval of Respiratory Protective Devices*, 378, 379
    OSHA 29 *CFR* 1910.134, *Respiratory Protection*, 377, 378, 419
    Process Safety Management of Highly Hazardous Chemicals regulation (29 *CFR* 1910.119), 218
    Regulation 29 *CFR* 1910.132, *Personal Protective Equipment*, 387
    Regulation 29 *CFR* 1910.156, *Fire Brigades*, 387
    Regulation 40 CFR Part 311, *Worker Protection*, 387
    Title 29 CFR 1910.120. *See* Hazardous Waste Operations and Emergency Response (HAZWOPER, Title 29 CFR 1910.120)
    Title 29 CFR 1910.1000, Toxic and Hazardous Substances, 344
    Title 29 CFR 1910.1020, Access to Employee Exposure and Medical Records, 367
    Title 49 *CFR 178*, intermodal tanks, 299
    Title 49 *CFR* 178.3, drums, 252
    Title 49 *CFR* 178.37, Y cylinders, 256
    Title 49 *CFR*, bulk and non-bulk packaging, 250
    Title 49 *CFR Part 179*, railway tank cars, 293, 294
Coffer dam, 303
Cold emergencies
    decontamination operations, 464–465
    PPE-related stresses, 413–414
Cold tapping, 519
Colorimetric detection methods, 176–183
    colorimetric indicator tube/chip, 177–179, 204–205
    M8, M9, M256 chemical agent detection papers, 182–183
    pH paper/strips
        generally, 179–181
        hazardous materials profile, 221
        procedures for use, 206
    purpose of, 176–177
    reagent papers
        generally, 181–182
        hazardous materials profile, 221
        procedures for use, 207
Combustible gas indicator (CGI)
    calibration, 173
    correction factors, 173–175
    hazardous materials profile, 222
    mechanism, 172–173
    types, 172
    uses for, 171
    Wheatstone Bridge, 172
Combustible liquid, 38
Combustible metals, 40, 91–92
Combustible range, 38
Combustion, 73–74
Commodity name stencils, 286
Communication. *See also* Documentation; Reports
    facepiece limitations, 384
    hand signals, 418–419, 444
    hazard communication briefing, 24
    hazardous materials automated cargo communications for efficient and safe shipments (HM-ACCESS), 233

Incident Radio Communications (ICS Form 205), 329, 342
radio
    Incident Radio Communications (ICS Form 205), 329, 342
    Radio Requirement Work Sheet (ICS Form 216), 342
    use with encapsulating suit, 399
use with encapsulating suit, 399

Compound
    atomic stability, 65–66
    defined, 50
    nonsalts, 54–55
    salts, 53–54
    sodium chloride, 51

Compressed Gas Association (CGA)
    CGA 341, cryogenic liquid tank, 278
    cylinder identification, 254
    vapor density, 36

Compressed gas tube trailer, 260

Compressed gases
    chlorine Emergency Kit "A," 259
    defined, 33
    hazards, 33–34
    non-refrigerated gas intermodal tanks, 300–301
    tube trailers, 280

Computer-Aided Management of Emergency Operations (CAMEO), 234–236

Concentration, 77
Condensation, 42–43
Cone roof tank, 310
Confined space, 146
Confinement of hazardous material, 338–339, 482
Consist, 283
Container vessel, 304

Containers
    air freight cargo, 305–306
    air monitoring, 146
    bulk vs. non-bulk, 250
    condition, evaluating, 319
    damage assessment, 19, 486–492
    for decontaminated items, 462
    fixed facility, 309–313
        cryogenic tanks, 313
        nonpressure tanks, 309–311
        pressure tanks, 312–313
    highway cargo containers, 260–281
        corrosive liquid tanks, 270–273
        cryogenic tanks, 276–279
        dry bulk carriers, 281
        high-pressure cargo tanks, 273–276
        low-pressure cargo tanks, 267–270
        nonpressure cargo tanks, 263–267
        specification plates, 261–263
        tank markings, 261
        tube trailers, 279–280
    illicit use, 314
    improvised explosive devices (IEDs), 121, 123
    intermediate bulk container (IBC), 256–258
    intermodal containers, 297–302
        beam type, 298
        box type, 298
        chlorine Emergency Kit "C," 260
        defined, 297
        description, 299
        liquids and solid hazardous materials, 300
        non-refrigerated liquefied compressed gases, 300–301
        refrigerated liquefied gases, 301–302
        T-code, 298–299
    marine tank vessels, 302–305
        barges, 305
        cargo vessels, 304
        tankers, 302–304
    mismarked, 223
    multiple states of matter in the same container, 30
    non-bulk, 249–256
        bags, 250–251
        bottles, 251–252
        boxes, 256
        bulk vs., 250
        carboys, 251–252
        cylinders, 253–256
        drums, 252–253
        multicell packaging, 256
        pails, 252–253
    non-regulated use, 314
    other regulated material (ORM), 250
    pressurized. *See* Pressurized containers
    product control
        behavior prediction, 492–494, 530–531
        damage assessment, 486–492
    railway tank cars. *See* Railway tank cars
    shape as clue to presence of hazardous materials, 12, 13
    ton container (pressure drum), 258–260

Containment of hazardous material. *See also* Product control
    decontamination corridor layout, 461
    defined, 339, 482
    patching. *See* Patching for materials containment
    plugging. *See* Plugging for materials containment
    selection of technique, 532

Containment pools, 462
Contaminant, 111

Contamination
    avoidance, 417–418
    defined, 111, 417
    exposure vs., 162–163, 417
    radioactive material, 112
    technician considerations for, 112

Continuity for sampling, 150
Continuous underframe tank car, 286
Control of product release. *See* Product control
Conversion factor, 169, 174–175. *See also* Correction factor
Cooling vests, 412

Correction factor
    combustible gas indicators (CGIs), 173–175
    defined, 169
    limitations, 174–175
    monitoring and detection equipment, 140–142
    photoionization detectors (PIDs), 173–175

Corridors for decontamination
    ambulatory victims, 472–473
    berm, 459
    containment, 461
    equipment, 461–463
    layout, 459–463
    marking, 444
    nonambulatory victims, 474–475
    personnel, 461
    procedures, 463
    sample layout, 460
    segregation of the decon line, 461

Corrosives
    acids, 94
    bases, 94
    Emergency Response Guidebook numbers, 226
    field testing for, 147
    hazards, 94–95
    liquid tank release, 485
    liquid tanks, 270–273
        anatomy, 271
        characteristics, 272

construction features, 273
  identification, 270
 pH paper/strip detection of, 181
 water-reactive material, 90
Cost as PPE selection factor, 407
Counts per minute (cpm), 169
Covalent bonds, 66–67
Covered top floating roof tank, 310
Covered top floating roof tank with geodesic dome, 311
Covering for product control, 484, 517–518
CPC. See Chemical protective clothing (CPC)
Cracks in containers, 490
Criminal incidents, evidence collection for, 146–147, 150–151
Criminal suspect decontamination, 464
Critical point, 44
Critical pressure, 44
Critical temperature, 44
Critique. See Postincident critique
Cross-sensitivities of monitoring and detection equipment, 142–143
Cryogen, defined, 276
Cryogenic cylinders, 254–255
Cryogenic intermodal tanks, 301
Cryogenic liquid
  cold emergencies from, 413
  cylinders, 254–255
  defined, 33, 276
  gas changed into, 30
  hazards, 33–34
  heat transfer, 231
  helium, 64
  railway tank cars, 295–296
  release hazards, 255
  storage tank, 313
  tank cars, 282
Cryogenic tanks, 276–279
  characteristics, 278
  construction features, 277–278
  identification, 277
  liquid storage tank, 313
  uses for, 276
Curie (Ci), 99, 168
Cyanogen chloride (CK), 104
Cyclohexyl sarin (GF), 101
Cylinder
  boxes, 256
  construction features, 254
  cryogenic cylinders, 254–255
  defined, 253
  DOT-3AA cylinder, 256
  DOT-3AAX cylinder, 256
  identification, 253–254
  multicell packaging, 256
  walls, 511
  Y cylinders, 256

D

Damage assessment
  container information, 487–488
    aluminum, 487
    austenitic stainless, 488
    high strength low alloy steel, 488
    steel, 487–488
  estimating container damage, 19
  potential container stress, 488
  primary concerns, 486
  size-up, 487
  types of container damage, 488–492
    cracks, 490

  dents, 490
  heat-affected areas, 491–492
  scores and gouges, 490–491
Damage Assessment of Railroad Tank Cars Involved in Accidents, 489
Damage tolerance analysis (DTA), 490
Dams used for product control, 338–339, 459, 523
Dangerous goods. See Hazardous materials
Datasheet, 234
Debriefing
  After Action Report (AAR), 554–555
  defined, 550
  documentation, 555–556
  effectiveness, 551
  objectives, 550
  postincident critique, 552–554, 557
  on-scene, 340
  terminating the incident, 24
  timing for, 552
D.E.C.I.D.E. response model, 327
Decomposition, 75
Decomposition temperature, 94
Decontamination, 439–476
  ambulatory victims, 445, 469–470, 472–473
  apparatus, 456
  chemical protective clothing (CPC), 463
  corridors. See Corridors for decontamination
  Decontamination Leader, 358
  doffing after decon, 424, 425
  effectiveness assessment, 455–456
  equipment, 461–463
  evidence, 152, 465–466
  hazmat technician responsibilities, 23
  hazmat/WMD incidents, 468, 476
  implementation, 457–466
    animals, 464
    cold weather, 464–465
    corridor layout, 459–463
    criminal suspects, 464
    evidence collection, 465–466
    factors, 458–459
    law enforcement and military personnel, 463
    medical monitoring, 461, 465
    rescue device, 462–463
    security, 463, 465
    site selection, 457–458
    termination, 466
  infectious agents, 467
  methods, 440–455
    absorption, 442, 446, 448–449, 522
    adsorption, 442, 446, 449, 522
    advantages and disadvantages, 446–447
    chemical decon, 440, 450, 454
    chemical/biological degradation, 446, 449–450
    dilution, 446, 450, 527
    disinfection, 446, 451
    dry decon, 440, 441
    emergency decon, 443
    evaporation, 447, 451–452
    gross decon, 442
    isolation and disposal, 440, 441, 447, 452
    mass decon, 443–445, 468, 469–471
    neutralization, 447, 452–453
    physical decon, 440, 442, 450
    selection of, 341, 439
    solidification, 447, 453
    sterilization, 447, 453–454
    technical decon. See Technical decontamination
    vacuuming, 447, 454
    washing, 447, 455

wet decon, 440, 441
mission work duration, 422
nonambulatory victims, 445, 471, 474–475
pesticides, 467
plan outline, 343
PPE selection factors, 407
radiation, 467
samples, 152
showers, 461, 462, 465
Technician Level selection of, 21
terrorist events, 468
Defensive operations
    actions taken, 334
    defined, 333
    product control, 482, 483–485, 522–529
    strategies, 331
    uses for, 334
Deflagrate, 117
Degradation
    biological, 446, 449–450
    chemical, 391
    chemical protective clothing (CPC), 410
    decontamination, 449–450
    gloves, 409
Delocalized bond, 68
Demobilization, 549–550
Density of materials, 35
Dents in containers, 490
Department of Energy (DOE)
    exposure limits terminology, 167
    radiation decontamination, 467
Department of Homeland Security (DHS)
    hazmat technician training, 18
    protective clothing and equipment standards, 386
    respiratory protection standards, 377
Department of Transportation (DOT)
    air freight cargo, 305–306
    bulk and non-bulk packaging, 249
    cryogenic cylinders, 255
    cryogenic liquids, 64
    cylinders, 253, 255
    damage tolerance analysis (DTA) for metals, 490
    DOT-3AA cylinder, 256
    DOT-3AAX cylinder, 256
    DOT/TC cargo container standards, 260
    Emergency Handling of Hazardous Materials in Surface Transportation, 227
    Emergency Response Guidebook (ERG), 226
    hazard class, 325
    Hazardous Materials Table (49 *CFR* 172.101), 299
    intermediate bulk container (IBC), 256
    multi-unit tank car tank (DOT 110 and DOT 106), 258
    nameplates, 262
    Pipeline and Hazardous Materials Safety Administration (PHMSA), 232, 233, 306
    pipelines, 306, 309
    railway tank car class numbers and approving authority, 282
    railway tank car specification markings, 283–284
    specification plates, 263
    tank markings, 261
    UN/NA number, 224, 227
Detection devices
    action levels, 143–144
    calibration, 139–143
        bump test, 140
        calibration, defined, 139
        CGI gases, 173
        correction factors, 140–142
        cross-sensitivities, 142–143
        interference, 142–143
        photoionization detectors (PIDs), 186–187
        relative response, 140–142
    clues to presence of hazardous materials, 13
    decontamination effectiveness, 456
    direct-reading instrument, 138
    features, 135–136
    flame ionization detector (FID), 147, 190
    fluorimeter, 199
    Fourier Transform Infrared (FT-IR) spectroscopy, 194
    gamma-ray spectrometer, 193
    gas chromatograph, 190–191
    halogenated hydrocarbon meters, 189–190
    hazmat technician training, 19
    instrument response time, 138
    ion mobility spectrometry (IMS), 192
    maintenance, 152–153, 156
    mass spectroscopy, 191
    mercury detection, 197–198
    photoionization detector. *See* Photoionization detector (PID)
    polymerase chain reaction, 199
    potential causes of damage, 153
    purpose of, 135
    radiation, 183–185
        dosimeters and badges, 184–185, 210
        Geiger-Mueller (GM) detector, 183–184
        hazardous materials profile, 221
        procedures for use, 208–209
        scintillation detectors, 184
    Raman spectrometer, 195–196
    sensor-based instruments. *See* Sensor-based detection instruments
    spectrophotometer, 194–195
    surface acoustic wave, 192–193
    technology, 136–138, 170
    tic tracer leak detector, 189
    wet chemistry, 198–199
Detection limit, 137
Detector tube, 177
Detonate, 118
Detonator, 118
Dewar flask, 255
DHS. *See* Department of Homeland Security (DHS)
Diatomic gases, 66
Diatomic molecules, 50, 51, 66
Dictionary, chemical, 229
Diking for product control, 459, 524
Dilution
    decontamination, 446, 450
    procedures, 527
Direct-reading instrument, 138
Disinfection, 446, 451
Dispersion characteristics for chemical assessment, 225
Dissociation (chemical), 94
Distance, to limit radiation exposure, 115
Distilled sulfur mustard (HD), 379
Diversion for product control, 525
Division assignment list (ICS Form 204), 329
Division number, 117–118
DNA fluoroscopy, 199
Documentation
    chemical protective clothing (CPC), 426, 427–428, 434
    critique records, 555
    debriefing records, 555
    entry/exit logs, 365
    exposure records, 365
    facility documents, 233
    National Fire Incident Reporting System (NFIRS), 555
    post-exposure monitoring, 367
    records management, 555–556

reimbursement logs, 555
research, 220
terminating the incident, 24, 555–556
DOE. *See* Department of Energy (DOE)
Doffing of PPE, 424–425, 430–433
Dome cover
    leaks, 502–503
    nonpressure cargo tanks, 267
Dome roof tank, 312
Donning of PPE, 423–424, 430–433
Dose, defined, 107
Dose-response relationship, 108
Dosimeter, 168, 184–185, 210
DOT. *See* Department of Transportation (DOT)
DOT 51, portable tanks, 500
DOT 105, pressure railway tank cars, 293, 294
DOT 105J500W, chlorine tank car, 500
DOT 106, multi-unit tank car tank, 258
DOT 110, multi-unit tank car tank, 258
DOT 111, tank cars, 292
DOT 112, pressure railway tank cars, 293
DOT 113, cryogenic railway tank cars, 295
DOT 114, pressure railway tank cars, 293
DOT 406, nonpressure cargo tanks
    anatomy, 264
    characteristics, 265
    construction features, 266–267
    fire control, 504
    highway cargo container standards, 260
    leak control, 502
    product control, 501
    product removal and transfer, 505
    products shipped, 263
DOT 407, low-pressure cargo tanks
    construction features, 270
    identification, 267–270
    leak control, 501, 502
    product removal and transfer, 505, 506
DOT 412, corrosive liquid tanks
    anatomy, 271
    identification, 270, 272
    leak control, 501, 502
    product removal and transfer, 505
DOT MC 331, cargo tank, 500
Drains as decontamination operation consideration, 457–458
Drums
    bung leaks, 513
    chime leaks, 513
    configurations, 252
    construction materials, 252
    hazmat considerations, 253
    lifting method of overpacking, 516
    liners, 514
    overpacking, 514–516
    sidewall punctures, 513–514
    upender, 515
    uses for, 252
Dry bulk carriers, 281, 304
Dry bulk container, 260
Dry decontamination, 440, 441
Dry ice, 43
Dry powder, 91
Dry van truck, 260
DTA (damage tolerance analysis), 490
Duet Rule, 65
Duffle bag bomb, 124
DUMBELS indicators, 102–103
Durene, 85
Dust, mobile like a gas, 31
Dust mask, 383
Dynamite activation energy, 76

E

EBSS (emergency breathing support system), 380
e-HASP, 341
Eight Step Incident Management Process©, 327
Elbow patch, 498
Electrical supply as decontamination operation consideration, 457
Electricity, static, 506
Electrochemical cell sensor, 175
Electromagnetic spectrum, radiation, 97
Electron
    defined, 50
    oxidizing agents gain electrons, 73
    reducing agents lose electrons, 73
    structure, 49
Electronic technical resources, 234–240
    Chemical Hazards Emergency Medical Management (CHEMM), 237
    Computer-Aided Management of Emergency Operations (CAMEO), 234–236
    E-Plan for First Responders, 237
    Internet resources, 240
    mobile applications (apps), 237–238
    mobile devices, 234
    National Pipeline Mapping System (NPMS), 238–239
    Palmtop Emergency Action for Chemicals (PEAC®)-WMD, 237
    plume modeling, 239–240
    smart phones, 234
    weather data, 239
    Wireless Information System for Emergency Responders (WISER), 236–237
Elements, 50–52. *See also* Periodic Table of Elements
    atomic stability, 65–66
    compounds, 50, 51
    diatomic molecules, 50, 51
    metalloids, 51
    metals, 52
    nonmetals, 52
    overview, 50
EMA (ethyl methacrylate), 223
Emergency breathing support system (EBSS), 380
Emergency decontamination, 443
Emergency Handling of Hazardous Materials in Surface Transportation, 227
Emergency Kit "A," 259
Emergency Kit "B," 259
Emergency Kit "C," 260
Emergency Management Institute, 18
Emergency Operations Center (EOC), 337
Emergency procedures, 345
Emergency response centers, 240–242
Emergency Response Guidebook (ERG)
    Awareness Level responder use of, 13
    chemical groups, 226
    defined, 226
    initial isolation distances, 226–227
    purpose of, 226, 328
    safety precautions, 328–329
Emergency Response Planning Guideline, Level 1 (ERPG-1), 166
Emergency Response Planning Guideline, Level 2 (ERPG-2), 167
Emergency Response Planning Guideline, Level 3 (ERPG-3), 167
Emergency Response Team (ERT), 243
Emergency Transportation System for the Chemical Industry (SETIQ), 241, 242
Emissivity of an object, 189
EMS ensembles, 404
Encapsulating suit
    boot liners, 398

gloves, 399
liquid splash-protective clothing, 391–392
End-of-service-time indicator (ESTI), 379
Endothermic reaction, 71, 72
Energy expenditures during sports and fire fighting activities, 422
English System of measurement
    counts per minute (CPM), 169
    curie (Ci), 99, 168
    radiation absorbed dose (rad), 169
    roentgen (R), 169
    roentgen equivalent in man (rem), 169
Entry log, 365
Entry Team Leader, 357–358, 365
Environment. *See also* Weather
    air consumption and, 422
    cold emergencies, 413–414
    heat emergencies, 411–413
    interference of monitoring and detection equipment, 143
    PPE selection factors, 405
    PPE selection factors for unknown environments, 407–408
    protection as incident priority, 331
Environmental Protection Agency (EPA)
    Emergency Response Team (ERT), 243
    exposure limits terminology, 166
    Incident Command requirements, 356
    Incident Management System for hazmat incidents, 353
    Level A ensemble, 393, 396
    Level B ensemble, 391, 392
    local emergency response plan, 332
    medical surveillance plan, 344
    mercury incident, 197–198
    Oil Bond® for solidification decontamination, 453
    plume modeling software, 239
    Regulation 40 CFR Part 311, *Worker Protection*, 387
    Uniform Hazardous Waste Manifest, 232
EOC (Emergency Operations Center), 337
EOD (Explosive Ordnance Disposal) team, 116, 243
EPA. *See* Environmental Protection Agency (EPA)
E-Plan for First Responders, 237
Equipment
    Certified Equipment List, 379
    decontamination operations, 461–463
    detection. *See* Detection devices
    monitoring. *See* Monitoring devices
    PPE. *See* Personal protective equipment (PPE)
    specialized plugging equipment, 499–501
ERG. *See* Emergency Response Guidebook (ERG)
ERPG-1 (Emergency Response Planning Guideline, Level 1), 166
ERPG-2 (Emergency Response Planning Guideline, Level 2), 167
ERPG-3 (Emergency Response Planning Guideline, Level 3), 167
ERT (Emergency Response Team), 243
ESTI (end-of-service-time indicator), 379
Ethyl methacrylate (EMA), 223
Ethylene, cryogenic tank cars for transporting, 295
Ethylene polymerization reaction, 74
Evacuation protocols, 419
Evaluate progress (Step 4 of APIE)
    Awareness Level responders, 14
    generally, 11
    hazmat technician responsibilities, 23–24
    Operations Level responders, 15
Evaporation
    decontamination method, 447, 451–452
    defined, 43
    rate, 43
Evidence
    chain of custody, 148, 150, 152
    collection and preservation, 152
    collection techniques, 150–152
    criminal incidents, 146–147, 150–151
    decontamination of, 152, 465–466
    defined, 147
    investigator duties, 151
    sampling procedures, 147
Excepted packaging, 314
Exit log, 365
Exothermic reaction
    defined, 71
    fire or explosion hazards, 72
    from water-reactive material, 90
Expansion, 32
Expansion ratio, 32
Explosive Ordnance Disposal (EOD) team, 116, 243
Explosives and incendiaries, 116–118
    categories, 118
    divisions, 117
    Emergency Response Guidebook numbers, 226
    examples, 116
    Explosive Ordnance Disposal (EOD), 116, 243
    explosive range, 38, 39
    field testing for, 147
    high explosives, 118
    improvised explosive devices, 116
    low explosives, 117–118
    primary explosives, 118
    secondary explosives, 118
    understanding the danger, 116–117
Exposure
    acute, 163
    biological, 169
    chronic, 163
    contamination vs., 162–163, 417
    defined, 111, 417
    delayed symptoms, 551
    duration, as PPE selection factor, 405
    immediately dangerous to life and health (IDLH), 164
    limits, 164–168
    post-exposure monitoring, 367
    predicting likely behavior, 494
    radiation, 114–115
    radiological, 168–169
    records, 365
    reporting, 366–367, 466
    routes of entry, 161–162
        absorption, 162
        defined, 161
        ingestion, 161–162
        inhalation, 161, 224
        injection, 162
    toxic gas, 162
Extinguishing a fire
    Class D fires and extinguishers, 91–92
    dry powder agents, 91
    water-reactive material, 90–91

F

Facepiece
    air-purifying respirator (APR), 381
    limitations, 384
Facility documents, 233
Fahrenheit scale, 38
Families of elements
    Group I – Alkali metals, 60, 61
    Group II – Alkaline earths, 60, 61–62
    Group VII – Halogens, 60, 62–63
    Group VIII – Noble gases, 60, 64
    in the Periodic Table of Elements, 59

FBI WMD Coordinator (WMDC), 243–244
Federal Bureau of Investigation, Weapons of Mass Destruction Coordinator (WMDC), 243–244
Federal Emergency Management Agency (FEMA)
    Emergency Management Institute, 18
    hazmat technician training, 18
Federal Railroad Administration (FRA), 489
FEMA. *See* Federal Emergency Management Agency (FEMA)
Ferrous metal, 487–488
FIBC (flexible intermediate bulk container), 257
Fiberboard boxes, 256
Fiberboard drums, 252
FID (flame ionization detector), 147, 190.
Field screening samples, 147–150
    chain of custody, 148, 150
    minimum testing for, 147
    plans, 147–149
Field test, 140
Filters
    air-purifying respirator (APR), 376, 380–381
    high-efficiency particulate air (HEPA), 382, 454
Finance/Administration Section, demobilization roles, 550
Fire Brigades (OSHA Regulation 29 CFR 1910.156), 387
Fire control
    cargo tanks, 504
    defined, 338
    goals and objectives, 338
Fire extinguishment, 90–92
Fire fighting, energy expenditures during, 422
Fire point, 38
Fire Protection Guide to Hazardous Materials, 228–229
Fire service ensembles, 403
Fire-entry suits, 389–390
Fires, Class D, 90, 91–92
Fireworks as IEDs, 125
First responder, OSHA definition, 2
Fittings, pressurized container, 508–513, 538–540
Fixed facility containers, 309–313
    cryogenic tanks, 313
    nonpressure tanks, 309–311
    pressure tanks, 312–313
Flame ionization detector (FID), 147, 190
Flame-resistant (FR) protective clothing, 390
Flame-retardant protective clothing, 390
Flammability
    defined, 39
    field testing for, 147
    flash point, 38
    hazards, 224
Flammable liquid, 38
Flammable range, 38–40
Flaring, 519–520
Flash point, 38, 39, 45
Flexible intermediate bulk container (FIBC), 257
Floating drugstore, 303
Fluid consumption for heat exposure prevention, 411–412
Fluorimeter, 199
Fluorine, 63, 68
Fluorine ionic bond, 67
Foam
    Aqueous Film Forming Foam (AFFF), 517
    Class B foam, 517
Fog, mobile like a gas, 31
Forms and logs. *See specific ICS Form*
Fourier Transform Infrared (FT-IR) spectroscopy, 194
FR (flame-resistant) protective clothing, 390
FRA (U.S. Federal Railroad Administration), 489
Frameless tank car, 285
Freezing point, 42

Freon®, 297
Friction-sensitive reactive materials, 92–93
Frostbite, 413
FT-IR (Fourier Transform Infrared) spectroscopy, 194
Fuel vapor-to-air mixture, 39
Full sill tank car, 286
Fuming sulfuric acid, 334
Fusible plug threads, 511
Fusible plugs, 509–511

G

Gamma (γ) rays, 97, 113
Gamma-ray spectrometer, 193
Gas
    ammonia, 90
    change into cryogenic liquids, 30
    characteristics, 31–32
    combustible gas indicators (CGIs). *See* Combustible gas indicator (CGI)
    compressed
        chlorine Emergency Kit "A," 259
        defined, 33
        hazards, 33–34
        non-refrigerated gas intermodal tanks, 300–301
        tube trailers, 280
    cryogenic liquids, 33–34
    defined, 29
    diatomic, 66
    flammable gas generation from water-reactive materials, 89
    inert, 61
    initial isolation distance, 227
    items acting like gas, 30–31
    liquefied, 33–34
    liquefied natural gas (LNG), 276, 303–304
    liquefied petroleum gas (LPG)
        defined, 293
        marine carriers for transport, 303–304
        pressure railway tank cars for, 293
        storage, 33
    noble gases, 60, 64
    phosphine, 90
    physical states, 29–31
    toxic gases from water-reactive materials, 90
    vapor pressure, 32
Gas chromatograph (GC), 186, 190–191
Gas main valve, 509
Gas masks, 377
Gas-liquid partition chromatography (GLPC), 190
Gasoline
    flash point and molecular weight, 39
    specific gravity, 36
    vapor pressure and vapor content, 45
Gauging devices, 291
GC (gas chromatograph), 186, 190–191
GEDAPER response model, 327
Geiger Counter, 183
Geiger-Mueller (GM) detector, 183–184
Geiger-Mueller (GM) tube, 184
General service (nonpressure) railway tank cars, 291–293
    construction features, 292–293
    identification, 292
    uses for, 291–292
Geographic information system (GIS), 238, 240
GHS (Globally Harmonized System) of classification and labeling of chemicals, 224, 231–232
GIS (geographic information system), 238, 240
Globally Harmonized System (GHS) of classification and labeling of chemicals, 224, 231–232
Gloves

Index 593

construction materials, 388
    encapsulating suit, 399
    permeation/degradation resistance guide, 409
GLPC (gas-liquid partition chromatography), 190
GM (Geiger-Mueller) detector, 183–184
GM (Geiger-Mueller) tube, 184
Gold, atomic number and atomic weight, 58
Golf tees used for plugging, 495
Gouges in containers, 490–491
Gross decontamination, 442
Gross rinse, 461
Gross tool drop, 461
Grounding, 506–507, 537
Group I – Alkali metals, 60, 61
Group II – Alkaline earths, 60, 61–62
Group level critique, 554
Group Supervisor, 356–357
Group VII – Halogens, 60, 62–63
Group VIII – Noble gases, 60, 64
G-series agents, 101–102
Guide to Fire Hazard Properties of Flammable Liquids, Gases, and Volatile Solids (NFPA 325), 229
Guide to Hazardous Chemical Reactions (NFPA 491), 229

H

Half-life, 98–99
Halogenated agent, 86
Halogenated hydrocarbon meters, 189–190
Halogenated hydrocarbons, 86
Halogens, 60, 62–63
Hand signals, 418–419, 444
Hawley's Condensed Chemical Dictionary (CCD), 229
Hazard and risk assessment, 323–324
Hazard assessment, 323
Hazard class, 325
Hazard communication briefing, 24
Hazard Communication regulation (29 *CFR* 1910.1200), 218
Hazard Communication Standard (HCS), 218, 233
Hazard identification manual, 231
Hazard profile
    defined, 221
    tactic to change the profile, 56
Hazardous Chemicals Data (NFPA 49), 229
Hazardous devices units, 116
Hazardous materials
    clues to presence of, 12–13
    corrosives, 94–95
    data sheet, 342
    decontamination method selection, 476
    Fire Protection Guide to Hazardous Materials, 228–229
    hazardous liquid release incident, 485
    inorganic compounds, 83
    organic compounds, 83–87
        halogenated agents, 86
        hydrocarbon derivatives, 86–87
        hydrocarbons, 83–85
    oxidizing agents, 87–88
        chlorates and perchlorates, 88
        defined, 72
        disinfectants, 451
        electrons, gaining, 73
        inorganic peroxides, 87
        organic peroxides, 87–88
    predicting behavior, 531
    product control, 532
    profile, 221–223
    protective clothing, 386–387. *See also* Chemical protective clothing (CPC); Personal protective equipment (PPE)
    radiation. *See* Radiation
    reactive materials, 88–94
        air-reactive, 88–89
        alkali metals, 61
        defined, 53, 88
        hazards, 88
        light-sensitive, 93
        magnesium, 62
        shock- and friction-sensitive, 92–93
        temperature-sensitive, 93–94
        water-reactive, 89–92
    skin contact problems, 385–386
    WMDs. *See* Weapons of mass destruction (WMDs)
Hazardous materials automated cargo communications for efficient and safe shipments (HM-ACCESS), 233
Hazardous Materials Code (NFPA 400-2010), 39
Hazardous Materials Table (49 *CFR* 172.101), 234, 299
Hazardous materials technician, 16–25
    analyzing the incident, 18–20
        container damage, 19
        detecting, monitoring, sampling hazardous materials, 19
        predicting behavior, 20
        product identification, 19
    defined, 16
    evaluating progress, 23–24
    implementing the planned response, 22–23
        control functions, 23
        decontamination, 23
        ICS/IMS duties, 22–23
        personal protective equipment, 23
    planning the response, 20–22
        action plan development, 22
        decontamination procedures, 21
        personal protective equipment selection, 21
        response objectives and options, 21
    terminating the incident, 24–25
    terminology, 16
    training, 17–18
        NFPA standards, 17–18
        OSHA standards, 17
        sources, 18
Hazardous Substance Data Bank, 236
Hazardous Waste Operations and Emergency Response (HAZWOPER, Title 29 CFR 1910.120)
    buddy system, 419
    defined, 17
    immediately dangerous to life and health (IDLH) atmosphere, 164
    Incident Command requirements, 356
    medical monitoring, 366
    medical surveillance plan, 343, 344
    postincident critique, 24
    protective clothing standards, 387
    site safety plan, 341
    terminating the incident, 550
    terminology, 2
    worker safety, 218
Hazmat. *See* Hazardous materials
Hazmat Branch Director/Group Supervisor, 356–357
Hazmat Logistics Officer, 359
HazMatIQ©, 327
HAZWOPER. See Hazardous Waste Operations and Emergency Response (HAZWOPER, Title 29 CFR 1910.120)
HCS (Hazard Communication Standard), 218, 233
Head shield, 287
Heat and air consumption needs, 422
Heat cramp, 411
Heat exhaustion, 411
Heat exposure prevention, 411–413
Heat from water-reactive material, 90

Heat index, 413
Heat induced tear, 293, 325
Heat rash, 411
Heat sink, 504
Heat stroke, 411
Heat-affected areas, container damage assessment, 491–492
Heating coils on railway tank cars, 288
Helium
    atom, 49
    atomic number and atomic weight, 58
    characteristics, 64
    transported via tube trailers, 279
HEPA (high-efficiency particulate air) filter, 382, 454
Hexamethylene triperoxide diamine (HMTD), 120
HF (hydrogen fluoride) vapor density, 37
High explosive, 118
High strength low alloy containers, 488
High temperature-protective clothing, 389–390
High-angle technical rescue, 403
High-efficiency particulate air (HEPA) filter, 382, 454
High-pressure bobtail tank, 275
High-pressure cargo tanks, 273–276
    anatomy, 273
    characteristics, 275
    construction features, 274, 276
    identification, 274
    uses for, 273–274
High-pressure SCBA, 379
Highway cargo containers, 260–281
    corrosive liquid tanks, 270–273
    cryogenic tanks, 276–279
    dry bulk carriers, 281
    high-pressure cargo tanks, 273–276
    low-pressure cargo tanks, 267–270
    nonpressure cargo tanks, 263–267
    specification plates, 261–263
    tank markings, 261
    tube trailers, 279–280
Hildebrand, Michael S., 327
HM-ACCESS (hazardous materials automated cargo communications for efficient and safe shipments), 233
HME. *See* Homemade explosives (HMEs)
HMTD (hexamethylene triperoxide diamine), 120
Homeland Security. *See* Department of Homeland Security (DHS)
Homemade explosives (HMEs), 118–121. *See also* Improvised explosive devices (IEDs)
    chlorate-based explosives, 121
    components, 119
    nitrate-based explosives, 121
    peroxide-based explosives, 120
Hook bolts and patch, 497–498
Hopper car, 296, 297
Horizontal pressure vessel, 312
Horizontal tank, 310
Hot tapping, 518
Hot wash, 550
Hydrocarbons
    alkanes, 84
    alkenes, 84
    alkynes, 84
    aromatics, 84–85
    characteristics, 84
    defined, 83
    derivatives, 86–87
    halogenated hydrocarbon meters, 189–190
    methane molecule, 83
    molecular weight, 41
Hydrogen
    Duet Rule of atoms, 65
    flammable gas generation from water-reactive materials, 89
    gas, 33
    hydrogen chloride, 90
    hydrogen cyanide (AC), 104
    hydrogen fluoride (HF) vapor density, 37
    hydrogen peroxide, 87
    molecular structure, 50
    in the Periodic Table of Elements, 59
    protons, 50
Hydrophilic, 34
Hydrophobic, 34
Hydrostatic test, 253
Hydroxide, 90
Hypergolic materials, 72
Hypothermia, 413

I

IA (immunoassay), 187
IAP. *See* Incident Action Plan (IAP)
IBC (intermediate bulk container), 256–258
IC. *See* Incident Commander (IC)
ICAD (improved chemical agent detector), 192
Ice cooling, 412
ICS (Incident Command System), defined, 353
ICS Form 201 (Site Security, Control, and Communications), 329, 342
ICS Form 202 (Incident Objectives), 329, 342
ICS Form 203 (Roles, Responsibilities, and Authority), 329, 342
ICS Form 204 (division assignment list), 329
ICS Form 205 (Incident Radio Communications), 329, 342
ICS Form 206 (Medical Plan), 330, 342
ICS Form 208 HM (site safety plan), 330, 341
ICS Form 214 (Unit/Activity Log), 356
ICS Form 216 (Radio Requirement Work Sheet), 342
ICS Form numbers and titles table, 364
ICS/IMS (Incident Command System/Incident Management System), technician performing assigned duties, 22–23
IDLH. *See* Immediately dangerous to life and health (IDLH)
IED. *See* Improvised explosive devices (IEDs)
IFR (inherently flame resistant) protective clothing, 390
IFSTA Principles of Risk Management, 326
Illicit container use, 314
Illicit labs, lab packs, 516
IM 101 portable tank, 299, 300
IM 102 portable tank, 299, 300
Immediately dangerous to life and health (IDLH)
    buddy system, 416, 419
    defined, 164
    exposure limits terminology, 165
    Level B ensemble for use in, 400
    Level D ensemble for use in, 402
    respiratory protection
        air-purifying respirator (APR), 382
        limitations, 384
        self-contained breathing apparatus (SCBA), 379
        standards, 377, 378
    SAR with escape (egress), 380
Immiscible materials, 79
Immunoassay (IA), 187
IMO Type 5 intermodal tank, 300
IMO Type 7 container, 301
Implement the response (Step 3 of APIE)
    Awareness Level responders, 13
    generally, 11
    hazmat technician, 22–23
        control functions, 23
        decontamination, 23
        ICS/IMS duties, 22–23
        personal protective equipment, 23

Operations Level responders, 15
Improved chemical agent detector (ICAD), 192
Improvised explosive devices (IEDs), 121-130. *See also* Homemade explosives (HMEs)
    carbon dioxide grenades, 125
    containers, 121, 123
    defined, 121
    fireworks, 125
    identification, 122-123
    letter bombs, 125-126
    mail bombs, 116, 125-126
    M-devices, 125
    munitions, 122
    package bombs, 125-126
    Person-Borne Improvised Explosive Devices (PBIEDs), 126-128
    pipe bombs, 116, 123-124
    plastic bottle bombs, 124
    response to incidents, 129-130
    satchel, backpack, knapsack, duffle bag, briefcase, box bombs, 124
    targeted attacks, 122
    tennis ball bombs, 125
    Vehicle-Borne Improvised Explosive Devices (VBIEDs), 128-129
IMS. *See* Incident Management System (IMS)
IMS (ion mobility spectrometry), 192
Incendiaries. *See* Explosives and incendiaries
Incident Action Plan (IAP)
    development guidance
        ICS Form 201 (incident briefing including site sketch), 329, 342
        ICS Form 202 (incident objectives), 329, 342
        ICS Form 203 (organizational assignment list), 329
        ICS Form 204 (division assignment list), 329
        ICS Form 205 (Incident Radio Communications), 329
        ICS Form 206 (medical plan), 330, 342
        ICS Form 208 HM (site safety plan), 330, 341
    development of, 329-341
        decontamination selection, 341
        by hazmat technicians, 22
        incident priorities, 330-331
        PPE selection, 341
        preincident surveys, 330
        response modes, 331-334
        strategic goals and tactical objectives, 335-340
    information included, 354
    Unified Command use of, 355-356
Incident briefing (ICS Form 201), 329, 342
Incident command, 353-355
Incident Command Post, 355
Incident Command System (ICS), defined, 353
Incident Command System/Incident Management System (ICS/IMS), technician performing assigned duties, 22-23
Incident Commander (IC)
    area monitoring, 145
    command structure, 354
    demobilization, 549, 550
    evaluating progress, 23
    general responsibilities, 354
    specific responsibilities, 355
    technical research, 217
Incident levels, 361-363
Incident Management System (IMS), 353-356
    assigned function procedures, 369
    defined, 353
    forms and logs, 364-365
    hazmat incident levels and evaluation, 361-364, 370
    incident command, 353-355
        command structure, 354
        general responsibilities, 354
        specific responsibilities, 355
    medical monitoring and exposure reporting, 366-367

    Section Leaders and Supervisors, 356-361
    Unified Command, 355-356
Incident Objectives (ICS Form 202), 329, 342
Incident priorities, 330-331
Incident Radio Communications (ICS Form 205), 329, 342
Incident Safety Officer, 358
Incubation period for biological exposures, 169
Indian Springs Manufacturing, 259
Individual monitoring, 144-145
Industrial materials properties, 230-231
Industrial packaging, 314
Inert gas, 61
Infectious agent decontamination, 467
Infectious dose (ID) for biological exposures, 169
Information and Research Officer, 360-361
Information centers. *See* Technical information centers and specialists
Infrared radiation, 188
Infrared spectrophotometer, 194-195
Infrared thermometer, 188-189, 212
Ingestion route of entry, 161-162
Inhalation route of entry, 161, 224
Inherently flame resistant (IFR) protective clothing, 390
Inhibitor, 75
Initial response. *See* Plan the initial response (Step 2 of APIE)
Initiation device, 118
Injection route of entry, 162
Inorganic
    compounds, 83
    nonsalts, 54
    peroxides, 87
Inspection
    chemical protective clothing (CPC), 425-426, 434
    decontamination effectiveness, 456
    pre-entry, 415
Instability of elements and compounds, 66
Instrument reaction time, 138
Instrument response time, 137, 138
Insulation
    railway tank car safety feature, 288
    thermal, 288
Interagency Radiological Assistance Plan (IRAP), 243
Interference of monitoring and detection equipment, 137, 142-143
Interferometer, 194
Intermediate bulk container (IBC), 256-258
Intermodal container, 297-302
    beam type, 298
    box type, 298
    chlorine Emergency Kit "C," 260
    defined, 297
    description, 299
    freight container, 297
    liquids and solid hazardous materials, 300
    non-refrigerated liquefied compressed gases, 300-301
    refrigerated liquefied gases, 301-302
    tank container, 297
    T-code, 298-299
International Maritime Organization, 227
International Standards Organization (ISO), respiratory protection standards, 377
International System of Units (SI)
    becquerel (Bq), 99, 168
    defined, 32
    radiation absorbed dose (rad), 112
    sievert (Sv), 169
Internet resources, 240
Interoperability of equipment as selection factor, 407
Interpreting hazard and response information, 245
Inverse square law of radiation, 115
Inverted overpacking method, 516

Iodine sublimation, 43
Ion, defined, 68
Ion mobility spectrometry (IMS), 192
Ionic bonds, 67–68, 70
Ionization detectors, flame, 147, 190
Ionization potential (IP), 186
Ionize, 97
Ionizing radiation, 96–98, 112–113
IP (ionization potential), 186
IRAP (Interagency Radiological Assistance Plan), 243
Irritants, 105–106
ISHP acronym, 232
ISO (International Standards Organization), respiratory protection standards, 377
Isolation and disposal decontamination, 440, 441, 447, 452
Isolation perimeter, 335–336
Isolation steps, 462

J

Jane's CBRN Response Handbook, 229
Jane's Chem-Bio Handbook, 229
Japan, sarin attacks, 468
Jargon, 2
JCAD (joint chemical agent detector), 192
Jerry can, 251
Joint chemical agent detector (JCAD), 192
Jubilee pipe patch, 498

K

Kelvin scale, 38
Key terms, 3
Knapsack bomb, 124
Krypton, 65

L

Labels as clues to presence of hazardous materials, 12, 13
Laboratories, 313
Laboratory packs, 516
Laboratory Respirator Protection Level (LRPL), 379
Lamp strength, 186
Laser hazards, 196
Law enforcement ensembles, 404, 463
Leak control
    cargo tanks, 502–504
    defined, 482
    patching
        box patch, 497
        elbow patch, 498
        hook bolts and patch, 497–498
        jubilee pipe patch, 498
        large holes, 520
        pipe patches, 498, 535
        procedures, 534, 535
        saddle patch, 498
        soft patch, 498
        uses for, 496
    plugging
        A-Kit, 499, 500
        B-Kit, 499, 500
        C-Kit, 500
        equipment for, 499–501
        large holes, 520
        methods, 495
        Midland Emergency Response Kit, 500–501
        procedures, 495–496, 533
        propane A- and B-Kits, 500
        recovery vessels, 501
    wooden wedges and golf tees, 495
LEL (lower explosive limit), 39, 407
LEL meter (combustible gas detector), 222
LEPC (Local Emergency Planning Committee), 233, 332
LERP (local emergency response plan), 332, 336
Lesak, David, 327
Lethal concentration 50 (LC_{50}), 109
Lethal dose 50 (LD_{50}), 109
Letter bomb, 125–126
Level A ensemble
    components, 396, 397–399
    defined, 396
    pressure test, 426
    protection provided, 399
    uses for, 399–400
    vapor-protective clothing, 393
Level B ensemble
    components, 396, 400
    defined, 396
    liquid splash-protective clothing, 391, 392
    protection provided, 400
    uses for, 400–401
Level C ensemble
    components, 396, 401
    defined, 396
    donning and doffing, 431
    protection provided, 401
    uses for, 401–402
Level D ensemble
    components, 396, 402
    defined, 396
    protection provided, 402
    uses for, 402
Level I incident, 361, 362
Level II incident, 361, 362
Level III incident, 362
Levels 5/4/3/2/1 hazardous materials incidents (NIMS Incident Typing System), 362–363
Levels of Concern (LOC), 165
Lewisite, 103
LFL (lower flammable limit), 38–39
Life safety as incident priority, 330–331
Life Safety Initiatives, 343
Lifter roof tank, 311
Light bar test of CPC, 426
Lighting, decontamination considerations, 457
Light-sensitive reactive materials, 93
Liquefied flammable gas carrier, 303–304
Liquefied gas, 33–34
Liquefied natural gas (LNG), 276, 303–304
Liquefied petroleum gas (LPG)
    defined, 293
    marine carriers for transport, 303–304
    pressure railway tank cars for, 293
    storage, 33
Liquid
    bulk carrier, 304
    characteristics, 31
    defined, 29
    initial isolation distance, 227
    intermodal tanks for hazardous materials, 300
    mobility, 30
    oxygen, 34
    physical states, 29–31
    splash-protective clothing, 391–392, 432
Lithium, 57, 92
Lithium nitride, 90
LNG (liquefied natural gas), 276, 303–304
Load limit marking, 284–285

LOC (Levels of Concern), 165
Local Emergency Planning Committee (LEPC), 233, 332
Local emergency response plan (LERP), 332, 336
Location as clue to presence of hazardous materials, 12
Logistics Officer, 359
Logistics Section, demobilization roles, 550
Logs
    activity logs, 364
    entry/exit, 365
    ICS Form 214 (Unit/Activity Log), 356
Low explosive, 117, 118
Lower explosive limit (LEL), 39, 407
Lower flammable limit (LFL), 38–39
Low-pressure cargo tanks, 267–270
    anatomy, 268
    characteristics, 269
    construction features, 270
    identification, 267–270
Low-pressure SCBA, 379
Low-pressure storage tank, 312
LPG. *See* Liquefied petroleum gas (LPG)
LRPL (Laboratory Respirator Protection Level), 379
Lucite, 223

M

M8, M9, M256 chemical agent detection papers, 182–183
Magnesium
    extinguishing fires, 90
    locations for, 92
    reactive material, 62
Mail bomb, 116, 125–126
Main charge explosive, 118
Maintenance
    chemical protective clothing (CPC), 426–427, 434
    monitoring and detection equipment, 152–153, 156
    open-circuit self-contained breathing apparatus, 378
    personal protective equipment, 426–427, 434
    PPE selection factors, 407
    sampling equipment, 152–153, 156
Manhole, 267, 289
Manuals. *See* Technical references
Manways, 289
Mapping Applications for Response, Planning and Local Operational Tasks (MARPLOT), 235–236
Marine tank vessels, 302–305
    barges, 305
    cargo vessels, 304
    chemical carriers, 303
    liquefied flammable gas carriers, 303–304
    petroleum carriers, 302
    tankers, 302–304
Markings
    cargo tank, 261
    clues to presence of hazardous materials, 12, 13
    pipelines, 307
    railway tank cars, 282–285, 286
    specification markings, 283–284
MARPLOT (Mapping Applications for Response, Planning and Local Operational Tasks), 235–236
Mask, particle, 383
Mass decontamination, 443–445, 468, 469–471
Mass spectrometer, 191
Material Safety Data Sheet (MSDS), 38, 218, 231
Matter
    molecular weight. *See* Molecular weight (MW)
    multiple states in the same container, 30
    phase changes and related properties
        phase, defined, 41

        relationships of physical properties, 44–45
        temperature changes, 42–44
    physical properties, 34–38
        appearance, 38
        density, 35
        odor, 37
        specific gravity, 35–36
        vapor density, 36–37
        viscosity, 37
    physical states, 29–34
        chemical assessment, 224
        gas, 29–34
        liquid, 29–31, 33–34
        solid, 29–31
    polarity, 41
    pressure, 40–41
        vapor expansion, 41
        vapor pressure, 40–41
    temperature, 38–40
        flammable range, 38–40
        flash points and fire points, 38, 39
MAWP (maximum allowable working pressure), 260
Maximum allowable working pressure (MAWP), 260
Maximum safe storage temperature (MSST), 44, 93
MC (motor carrier) standards, 260
MC 306, nonpressure cargo tanks
    anatomy, 264
    characteristics, 265
    construction features, 266–267
    fire control, 504
    highway cargo container standards, 260
    leak control, 502
    product control, 501
    product removal and transfer, 505
    products shipped, 263
MC 307, low-pressure cargo tanks
    construction features, 270
    identification, 267–270
    leak control, 501, 502
    product removal and transfer, 505, 506
MC 312, corrosive liquid tanks
    anatomy, 271
    identification, 270–272
    leak control, 501, 502
    product release incident, 485
    product removal and transfer, 505
MC 331, high-pressure cargo tanks, 273–276, 500
MC 338, cryogenic tanks, 276–279
M-devices, 125
MDT (mobile data terminal), 129
Measurement conversion, 4–6
Median lethal concentration 50 (LC_{50}), 109
Median lethal dose 50 (LD_{50}), 109
Medical monitoring, 366–367
    decontamination, 461, 465
    immediate care, 367
    initial and annual hazmat team physical, 366–367
    planned procedures, 366
    post-exposure monitoring, 367
Medical Officer, 360
Medical paperwork requirements, 365
Medical Plan (ICS Form 206), 330, 342
Medical surveillance plan, 343–344
Megger kit, 507
Melting point, 42
Memorandum of Understanding, SCBA certification, 379
Merck Index, 230
Merck Index Online, 230
Mercury detection, 197–198

Mercury incident, 197–198
Metadata, 238
Metalloids
    difficulty identifying, 51
    in the Periodic Table of Elements, 59
Metals
    alkali metals, 60, 61
    alloy, 56
    aluminum, 487, 504
    base metal, 490
    characteristics, 52
    combustible, 40, 91–92
    damage tolerance analysis, 490
    drums, 252
    ferrous metal, 487–488
    metal oxide sensors, 175
    in the Periodic Table of Elements, 59
    rust, 73, 75
    steel
        austenitic stainless, 488
        carbon steel, 487
        composition, 56
        container dents, 490
        containers, 487–488
        high strength low alloy, 488
        mild steel, 487
Methane
    chemical combustion, 73
    molecular structure, 66, 83, 86, 87
    vapor density, 36
Methanol, 86, 87
Methyl methacrylate (MMA), 223
Methylene chloride specific gravity, 36
Met-L-X, 91
Metric conversions, 4–6
Mexico
    Emergency Transportation System for the Chemical Industry (SETIQ), 241, 242
    National Coordination of Civil Protection (CECOM), 241
    Secretariat of Transport and Communications of Mexico (SCT), Emergency Response Guidebook (ERG), 226
Midland Emergency Response Kit, 500–501
Mild steel, 487
Military personnel decontamination, 463
Military standards for using PPE, 396
Milligrams per cubic meter (mg/m^3), 109
Milligrams per kilogram (mg/kg), 109
Milligrams per square centimeter (mg/cm^2), 109
Milligrams per square meter (mg/m^2), 109
Millimeters of mercury (mmHg), 32
Mine Safety and Health Administration (MSHA), 379
Miscibility, 79
Miscible material, 78
Mission Specific level, 15–16
Mission specific operations using chemical-protective clothing, 394
Mission statement, 422
Mission work duration, 420–422
Mission-oriented protective posture (MOPP) ensemble, 405
Mist, mobile like a gas, 31
Mixed load container, 260
Mixing materials, 77–79
    concentration, 77
    hazards, 77
    miscibility, 79
    polarity, 79
    solubility, 78
Mixture
    alloy, 56, 57
    defined, 55

    slurry, 56
    solutions, 55–56
MMA (methyl methacrylate), 223
mmHg (millimeters of mercury), 32
Mobile applications (apps), 237–238
Mobile data terminal (MDT), 129
Mobile devices, 234
Molasses viscosity, 37
Molecular weight (MW)
    defined, 37
    flash point and, 38, 39
    states of matter and, 41
    vapor density and, 36–37
Molecule
    diatomic, 50, 51, 66
    ethyl methacrylate, 223
    methane, 66, 83
    methyl methacrylate, 223
    styrene, 69
    toulene, 69
    water, 50
Monitoring, medical. *See* Medical monitoring
Monitoring devices
    action levels, 143–144
    area monitoring, 145–146
    automatic pumps, 139
    calibration, 139–143
        bump test, 140
        correction factors, 140–142
        cross-sensitivities, 142–143
        defined, 139
        interference, 142–143
        relative response, 140–142
    cargo containers, 146
    clues to presence of hazardous materials, 13
    confined spaces, 146
    decontamination effectiveness, 456
    direct-reading instrument, 138
    features, 135–136
    hazmat technician training, 19
    individual monitoring, 144–145
    instrument response time, 138
    maintenance, 152–153, 156
    passive instruments, 138
    potential causes of damage, 153
    purpose of, 135
    technology, 136–138
Monomer, 74
MOPP (mission-oriented protective posture) ensemble, 405
Motor carrier (MC) standards, 260
Motor oil
    flash point and molecular weight, 39
    vapor pressure and vapor content, 45
MSA SAW MiniCAD®, 192
MSDS (Material Safety Data Sheet), 38, 218, 231
MSHA (Mine Safety and Health Administration), 379
MSST (maximum safe storage temperature), 44, 93
Multicell packaging, 256
Multi-gas meter, 176, 202–203
Multiplier, 169
Multi-use detector, 176, 202–203
Munitions, 122
Mustard agents, 103
Mustard gas, 95
MW. *See* Molecular weight (MW)
Mylar®, 184

N

Nameplate, 262
National Association of Chemical Industries, 241
National Coordination of Civil Protection (CECOM), 241
National Fallen Firefighters Foundation (NFFF), 343
National Fire Academy
    GEDAPER response model, 327
    hazmat technician training, 18
    SMART modeling for strategy development, 335
National Fire Incident Reporting System (NFIRS), 555
National Fire Protection Association (NFPA). *See also specific NFPA*
    buddy systems and backup personnel, 345
    combustible liquid, 38
    CPC maintenance, testing, inspection, and record keeping, 425
    Fire Protection Guide to Hazardous Materials, 228-229
    flammable liquid, 38
    hazmat technician training standards, 17-18
    medical monitoring, 366
    pressure tanks, 312
    respiratory protection standards, 377-378
    SCBA certification, 379
    terminology, 2-3
National Incident Management System (NIMS)
    development of action plans, 22
    forms and logs, 364
    hazmat technician training, 18
    incident management systems, 353
    Incident Typing System, 362-363
    NIMS-ICS forms for health and safety plan, 341-342
National Incident Management System-Incident Command System (NIMS-ICS), 353
National Institute for Occupational Safety and Health (NIOSH)
    Certified Equipment List, 379
    Chemical, Biological, Radiological and Nuclear (CBRN) Standard for Open-Circuit Self-Contained Breathing Apparatus (SCBA), 377
    Chemical Agent Permeation and Penetration Resistance Against Distilled Sulfur Mustard (HD [military designation]) and Sarin (GB [military designation]), 379
    combustible metal reaction, 40
    exposure limits terminology, 165, 166, 168
    immediately dangerous to life and health (IDLH) definition, 164
    ionization potential numbers, 174
    Laboratory Respirator Protection Level (LRPL), 379
    Level B ensemble, 400
    NIOSH Pocket Guide to Chemical Hazards (NPG), 228
    Pocket Guide, 174
    PPE selection factors, 405
    Regulation 42 *CFR* Part 84, *Approval of Respiratory Protective Devices*, 378, 379
    respiratory protection standards, 377-378
    RgasD (relative gas density), 36
    SARs as Type C respirator, 380
    SCBA certification, 379
    solubility information, 78
    Standard for Chemical, Biological, Radiological, and Nuclear (CBRN) Air-Purifying Escape Respirator and CBRN Self-Contained Escape Respirator, 378
    Standard for Chemical, Biological, Radiological, and Nuclear (CBRN) Full Facepiece Air-Purifying Respirator (APR), 377
National Institute of Standards and Technology (NIST), SCBA certification, 379
National Oceanic and Atmospheric Administration (NOAA), 234, 413
National Pesticide Information Center (NPIC), 467
National Pipeline Mapping System (NPMS), 238-239
National Response Team, 361-362
National Pipe Thread Taper (NPT), 506
National Weather Service Heat Index, 413
NBC (nuclear, biological, chemical), 100

NDIR (nondispersive infrared) sensor, 176
Negative catalyst, 75
Negative pressure intake, 454
Nerve agents
    common agents and characteristics, 101
    defined, 100
    G-series agents, 101-102
    M256A1 Detector Kit, 182-183
Nerve gas, 101
Neutralization
    decontamination, 447, 452-453
    product control, 484
Neutron
    defined, 50
    no charge, 49
    radiation, 96, 113
NFFF (National Fallen Firefighters Foundation), 343
NFIRS (National Fire Incident Reporting System), 555
NFPA. *See* National Fire Protection Association (NFPA)
NFPA 49, *Hazardous Chemicals Data*, 229
NFPA 325, *Guide to Fire Hazard Properties of Flammable Liquids, Gases, and Volatile Solids*, 229
NFPA 400-2010, *Hazardous Materials Code*, 39
NFPA 472, *Standard for Professional Competence of Responders to Hazardous Materials/Weapons of Mass Destruction Incidents (2008)*
    Hazmat Branch Director/Group Supervisor certification, 356
    hazmat technician requirements, 17
    Safety Officer training, 358
    specialty competencies, 18
NFPA 473, *Standard for Competencies for EMS Personnel Responding to Hazardous Materials/Weapons of Mass Destruction Incidents (2008)*, 18
NFPA 491, *Guide to Hazardous Chemical Reactions*, 229
NFPA 497, *Recommended Practice for the Classification of Flammable Liquids, Gases, or Vapors or of Hazardous (Classified) Locations for Electrical Installations in Chemical Process Areas*, 229
NFPA 499, *Recommended Practice for the Classification of Combustible Dusts and of Hazardous (Classified) Locations for Electrical Installations in Chemical Process Areas*, 229
NFPA 704, *Standard System for the Identification of the Hazards of Materials for Emergency Response*
    hazmat markings and colors, 12, 13
    purpose of, 229
NFPA 1072, *Standard for Hazardous Materials/Weapons of Mass Destruction Emergency Response Personnel Professional Qualifications, 2016 Edition*
    best practices, 11
    Hazmat Branch Director/Group Supervisor, 356
    hazmat technician competencies, 16
    hazmat technician training standards, 17, 18
    railway tank cars, 282
    Technician Level certification requirements, 1
NFPA 1500, *Standard on Fire Department Occupational Safety and Health Program*, 344, 358
NFPA 1561, *Standard on Emergency Services Incident Management System and Command Safety*, 353
NFPA 1582, *Standard on Comprehensive Occupational Medical Program for Fire Department*, 344
NFPA 1584, *Standard on the Rehabilitation Process for Members during Emergency Operations and Training Exercises*, 344, 412
NFPA 1852, *Standard on Selection, Care, and Maintenance of Open-Circuit Self-Contained Breathing Apparatus (SCBA)*, 378
NFPA 1981, *Standard on Open-Circuit Self-Contained Breathing Apparatus (SCBA) for Emergency Services*, 378, 379
NFPA 1991, *Standard on Vapor-Protective Ensembles for Hazardous Materials Emergencies*
    Level A ensemble, 398-399
    purpose of, 386
    requirements, 393

NFPA 1992, *Standard on Liquid Splash-Protective Ensembles and Clothing for Hazardous Materials Emergencies*
    Level B ensemble, 400
    minimum design criteria, 391
    regulations, 387
NFPA 1994, *Standard on Protective Ensembles for First Responders to CBRN Terrorism Incidents*
    ensemble components, 392
    PPE selection factors, 405
    regulations, 387
Nickel-63, 192
NIMS. *See* National Incident Management System (NIMS)
NIMS-ICS (National Incident Management System-Incident Command System), 353
NIMS-ICS forms for health and safety plan, 341–342
9/11 terrorism attacks, 468
NIOSH. *See* National Institute for Occupational Safety and Health (NIOSH)
NIOSH Pocket Guide to Chemical Hazards (NPG), 228
NIST (National Institute of Standards and Technology), SCBA certification, 379
Nitrate-based explosives, 121
Nitrogen
    diatomic molecules, 50
    lithium reactivity with, 57
NOAA (National Oceanic and Atmospheric Administration), 234
NOAA National Weather Service Heat Index, 413
Noble gases, 60, 64
Noded spheroid tank, 312
Noll, Gregory G., 327
Nonambulatory victim decontamination, 445, 471, 474–475
Non-bulk containers, 249–256
    bags, 250–251
    bottles, 251–252
    boxes, 256
    bulk vs., 250
    carboys, 251–252
    cylinders, 253–256
    drums, 252–253
    multicell packaging, 256
    pails, 252–253
Nondispersive infrared (NDIR) sensor, 176
Nonencapsulating liquid splash-protective clothing, 391–392
Nonintervention operations
    product control, 483
    response models, 332–333
    response modes, 331
Nonionizing radiation, 96–98
Nonmetals
    characteristics, 52
    in the Periodic Table of Elements, 59
Nonpressure cargo tanks, 263–267
    anatomy, 264
    characteristics, 265
    construction features, 266–267
    fixed facility containers, 309–311
    identification, 265–266
    intermodal tanks, 300
    liquid tanks, 265
    railway tank cars, 291–293
    storage tanks, 310–311
    tank cars, 282
    uses for, 263
Non-refrigerated liquefied compressed gas intermodal tanks, 300–301
Non-regulated container use, 314
Nonsalts, 54–55
Nonspec tank, 260
Notification of emergency response plans, 336–337

NPG (NIOSH Pocket Guide to Chemical Hazards), 228
NPIC (National Pesticide Information Center), 467
NPMS (National Pipeline Mapping System), 238–239
NPT (National Pipe Thread Taper), 506
Nuclear, biological, chemical (NBC), 100
Nuclear radiation, 314. *See also* Radiation
Nucleus, defined, 49

O

Objective 13, evaluating the condition of a hazardous materials container, 319
Objectives. *See also* Strategic goals and tactical objectives
    debriefing, 550
    fire control, 338
    hazmat technician response, 21
    ICS Form 202 (Incident Objectives), 329, 342
    Incident Action Plan, 335–340
    initial response, 21
    Objective 13, evaluating the condition of a hazardous materials container, 319
    personal protective equipment, 337
    planning the response, 347
    response, 15
Occupancy as clue to presence of hazardous materials, 12, 13
Occupational Safety and Health Administration (OSHA)
    buddy systems and backup personnel, 345, 419
    CPC maintenance, testing, inspection, and record keeping, 425
    debriefing process, 340
    decontamination effectiveness, 455
    exposure limits terminology, 165, 168
    first responder offensive tasks, 14
    hazard communication briefing, 24
    Hazard Communication Standard (HCS), 218, 233
    Hazmat Branch Director/Group Supervisor, 357
    hazmat technician training standards, 17
    immediately dangerous to life and health (IDLH) definition, 164
    Incident Management System for hazmat incidents, 353
    medical paperwork requirements, 365
    OSHA 29 *CFR* 1910.134, *Respiratory Protection*, 377, 378, 419
    oxygen concentration action levels, 170
    records management, 556
    Regulation 29 *CFR* 1910.120. *See* Hazardous Waste Operations and Emergency Response (HAZWOPER, Title 29 CFR 1910.120)
    Regulation 29 *CFR* 1910.132, *Personal Protective Equipment*, 387
    Regulation 29 *CFR* 1910.156, *Fire Brigades*, 387
    SCBA certification, 379
    site safety plan, 341
    terminology, 2–3
    Title 29 CFR 1910.1000, Toxic and Hazardous Substances, 344
    Title 29 CFR 1910.1020, Access to Employee Exposure and Medical Records, 367
    Title 29 CFR 1910.120. *See* Hazardous Waste Operations and Emergency Response (HAZWOPER, Title 29 CFR 1910.120)
    transportation placards, labels, and markings, 12, 13
Octet Rule, 65
Odor of chemicals, 37
Offensive operations
    defined, 334
    first responders, 14
    hazmat technician responsibilities, 22
    product control, 482, 485–486
    strategies, 331
Office of Pipeline Safety (OPS), 238
Oil. *See* Motor oil
Oil Bond® for solidification decontamination, 453
Olefin, 84
Oleum, 334
Open top floating roof tank, 310

Open-circuit SCBA
  air supply limitations, 384
  high-pressure, 379
  limitations, 377
  low-pressure, 379
  maintenance, 378
  mechanics, 376
  NFPA standards, 378, 379
  NIOSH standards, 377
  positive-pressure, 379
  pressure-demand, 379
Operating range of monitoring and detection equipment, 137
Operations Level responders
  actions allowed, 2–3
  analyzing the incident, 14–15
  APIE responsibilities, 11
  decontamination procedures, 21
  evaluating progress, 15
  implementing the response, 15
  Mission Specific competencies, 16
  personal protective equipment, 21
  planning the response, 15
  qualifications
    analyzing the incident, 14–15
    evaluating progress, 15
    implementing the response, 15
    planning the response, 15
Operations Mission Specific level, 15–16
Operations Section Chief, 356
Operations Section, demobilization roles, 550
OPS (U.S. Office of Pipeline Safety), 238
Orbitals, 49, 50
Orbits, 49, 50
Organic compounds, 83–87
  halogenated agents, 86
  hydrocarbon derivatives, 86–87
  hydrocarbons, 83–85
Organic nonsalts, 54
Organic peroxides, 87–88
Organization of this book, 1–2
Organizational assignment list (ICS Form 203), 329
Organometallic compound, 66
Organophosphate pesticides, 110
ORM (other regulated material), 250
OSHA. See Occupational Safety and Health Administration (OSHA)
Other regulated material (ORM), 250
Overpack
  defined, 23, 514
  drums, 514–516
  laboratory packs, 516
  nonbulk container/radioactive materials package, 541–544
  product control, 514–516
Oxidation, 72–74
  defined, 73
  level, 61
  number, 61
  oxidation-reduction (redox) reaction, 72
Oxidizer
  defined, 39, 72
  field testing for, 147
Oxidizing agent, 87–88
  chlorates and perchlorates, 88
  defined, 72
  disinfectants, 451
  electrons, gaining, 73
  inorganic peroxides, 87
  organic peroxides, 87–88
Oxygen
  detection devices, 170–172
  diatomic molecules, 50
  displacement, 170
  liquid oxygen, 34
  protons, 50

P

Pa (Pascals), 32
Package bomb, 125–126
Packing nut, 512
Pails, 252–253
Palmtop Emergency Action for Chemicals (PEAC®)-WMD, 237
Paperless hazard communication program, 233
PAPR. See Powered air-purifying respirator (PAPR)
Paraffin, 84
Paratech®, 520
Participant level critique, 554
Particle mask, 383
Particulate size and chemical reactions, 76–77
Particulate-removing air-purifying respirator (APR), 376, 382–383
Parts-per-billion (ppb), 109
Parts-per-million (ppm), 108–109
Pascals (Pa), 32
Patching for materials containment
  box patch, 497
  elbow patch, 498
  hook bolts and patch, 497–498
  jubilee pipe patch, 498
  large holes, 520
  pipe patches, 498, 535
  procedures for use, 534, 535
  saddle patch, 498
  soft patch, 498
  uses for, 496
PBIEDs (Person-Borne Improvised Explosive Devices), 126–128
PCR (polymerase chain reaction), 199
PEAC®-WMD (Palmtop Emergency Action for Chemicals), 237
PEL (Permissible Exposure Limit), 165, 168
PEL Ceiling Limit (PEL-C), 165, 168
PEL-C (PEL Ceiling Limit), 165, 168
Penetration of chemical protective clothing, 391, 410
Perchlorates, 88
Periodic Table of Elements, 56–59
  atomic number, 58
  defined, 56
  families, 59
  illustration, 57
  material types, 59
  reading the table, 58
Permeation
  breakthrough time, 409
  chemical protective clothing (CPC), 391
  defined, 391, 409
  indicators, 410
  PPE selection factors, 409–410
  structural firefighters' clothing, 388
Permissible Exposure Limit (PEL), 165, 168
Peroxide-based explosives, 120
Peroxides, 87–88
Personal protective equipment (PPE), 375–434
  clothing, 385–395
    body armor, 386, 404
    bomb suits, 386, 404
    chemical protective. See Chemical protective clothing (CPC)
    flame-resistant clothing, 390
    high temperature clothing, 389–390
    overview, 385–386
    standards at hazmat/WMD incidents, 386–387
    structural firefighters' clothing, 388, 403, 430

vapor-protective clothing, 23, 386, 433
design, 405–406
documentation and written PPE program, 427–428, 434
donning and doffing, 423–425, 430–433
EMS ensembles, 404
fire service ensembles, 403
inspection, 425–426, 434
law enforcement ensembles, 404
Level A ensemble
    components, 396, 397–399
    defined, 396
    pressure test, 426
    protection provided, 399
    uses for, 399–400
    vapor-protective clothing, 393
Level B ensemble
    components, 396, 400
    defined, 396
    liquid splash-protective clothing, 391, 392
    protection provided, 400
    uses for, 400–401
Level C ensemble
    components, 396, 401
    defined, 396
    donning and doffing, 431
    protection provided, 401
    uses for, 401–402
Level D ensemble
    components, 396, 402
    defined, 396
    protection provided, 402
    uses for, 402
maintenance and storage, 426–427, 434
PPE-related stresses, 410–414
    cold emergencies, 413–414
    heat emergencies, 411
    heat-exposure prevention, 411–413
    psychological issues, 384, 414
procedures for use, 414–425
    air management, 417
    buddy system, 419
    communication, 418–419
    contamination avoidance, 417–418
    donning and doffing of PPE, 423–425, 430–433
    mission work duration, 420–422
    pre-entry inspection, 415
    safety briefing, 416
    suit integrity, 420
respiratory protection. *See* Respiratory protection
selection factors, 404–410
    chemical resistance/compatibility, 408–410
    data interpretation, 218–220
    general factors, 405–407
    hazardous materials profile, 221
    incident response plan, 341
    procedures, 429
    responder mission, 402
    thermal protective clothing, 408
    unknown environments, 407–408
selection sheet, 343
strategic goals and tactical objectives, 337
Technical Level selection of, 21
technician use of, 23
testing, 426, 434
using monitoring and detection equipment while wearing, 136
vapor-protective clothing, 23
Person-Borne Improvised Explosive Devices (PBIEDs), 126–128
Pesticides
    categories and functions, 110

decontamination, 467
defined, 110
National Pesticide Information Center, 467
organophosphate, 110
transporting, 111
types and target organisms, 110
Petroleum carriers, 302
pH indicator, 176
pH meters, 176, 201
pH of water, 179
pH paper/strips
    generally, 179–181
    hazardous materials profile, 221
    procedures for use, 206
Phase, 41–45
    defined, 41
    relationships of physical properties, 44–45
    temperature changes, 42–44
        boiling and condensation, 42–43
        critical points, 44
        freezing, 42
        melting, 42
        sublimation, 43
Phenolic disinfectant, 451
PHMSA (Pipeline and Hazardous Materials Safety Administration), 232, 233, 306
Phosgene, 104
Phosgene oxime, 103
Phosphine, 88
Phosphine gas, 90
Phosphorus
    air-reactive materials, 89
    chemical burns from, 95
    toxic gases from, 90
Photoionization detector (PID)
    calibration of monitoring instruments, 186–187
    colorimetric indicator tube, 177
    correction factors, 173–175
    defined, 94
    hazardous materials profile, 222
    ionization potential and lamp strength, 186
    limitations, 186
    mechanism, 185–186
    procedures for use, 211
    purpose of, 185
Photomultiplier tube, 184
Photon, 113
Physical decontamination, 440, 442, 450
Physical examinations for personnel, 366–367
Physical fitness for heat exposure prevention, 412
Physical properties of materials, 34–38
    appearance, 38
    density, 35
    odor, 37
    specific gravity, 35–36
    vapor density, 36–37
    viscosity, 37
Physical states of matter, 29–34
    chemical assessment, 224
    gas, 29–34
    liquid, 29–31, 33–34
    solid, 29–31
PIA (postincident analysis), 340, 552
PID. *See* Photoionization detector (PID)
Pipe bomb, 116, 123–124
Pipe patches, 498, 535
Pipelines, 306–309
    construction features, 307–309
    identification, 306–307

markers, 307
Pipeline and Hazardous Materials Safety Administration (PHMSA), 232, 233, 306
principles of operation, 306
valves, 308
PIV (post indicator valve), 509
Placards as clue to presence of hazardous materials, 12, 13
Plan annex, 343
Plan the initial response (Step 2 of APIE)
Awareness Level responders, 13
generally, 11
hazmat technician, 20–22
action plan development, 22
decontamination procedures, 21
personal protective equipment selection, 21
response objectives and options, 21
Operations Level responders, 15
Planning Section, demobilization roles, 550
Planning the response, 323–348
Incident Action Plan development, 329–341
decontamination selection, 341
by hazmat technicians, 22
incident priorities, 330–331
PPE selection, 341
preincident surveys, 330
response modes, 331–334
strategic goals and tactical objectives, 335–340
response models, 326–329
APIE, 326
D.E.C.I.D.E., 327
defined, 326
Eight Step Incident Management Process©, 327
Emergency Response Guidebook (ERG), 328–329
GEDAPER, 327
HazMatIQ©, 327
response objectives and options, 347
risk vs. harm, 326
risk-based response, 323–325
site safety plan, 341–345
backup and rapid intervention, 345
components, 342–343
emergency procedures, 345
ICS Form 208 HM, 330, 341
medical surveillance plan, 343–344
post-entry evaluation, 344–345
pre-entry evaluation, 344
procedures for development, 348
Plastic
bottle bomb, 124
drums, 252
Plexiglas, 223
Plugging for materials containment
A-Kit, 499, 500
B-Kit, 499, 500
C-Kit, 500
equipment for, 499–501
large holes, 520
methods, 495
Midland Emergency Response Kit, 500–501
procedures, 495–496, 533
propane A- and B-Kits, 500
recovery vessels, 501
wooden wedges and golf tees, 495
Plume modeling, 239–240
Pneumatically unloaded hopper cars, 297
Poison control centers, 242
Polar solvent, 78
Polarity, 41, 79
Polymer, 74

Polymerase chain reaction (PCR) devices, 199
Polymerization, 74–75
catalysts, 74–75, 76
defined, 44, 74
inhibitors, 75
Portable pumps for product removal and transfer, 505
Portable tanks. *See* Intermodal container
Positive-pressure SCBA, 379
Positive-pressure ventilation (PPV), 452
Post indicator valve (PIV), 509
Post-entry evaluation, 344–345
Post-exposure monitoring, 367
Postincident analysis (PIA), 340, 552
Postincident critique
components, 553
defined, 340, 552
group level critique, 554
participant level critique, 554
participation, 553–554, 557
questions to answer, 553
records, 555
section level critique, 554
terminating the incident, 24–25
Potassium, 61
Potassium hydroxide, 90
Powered air-purifying respirator (PAPR)
defined, 376
mechanism, 383
responder use of, 377
types of, 383
uses for, 381, 383
PPE. *See* Personal protective equipment (PPE)
PPV (positive-pressure ventilation), 452
Predicting behavior of products
amount of product, 494
container features, 493
exposures and response conditions, 494
internal pressure of the tank, 493–494
likely effect of hazardous materials incidents, 531
procedures, 530
product identification, 492–493
product properties, 493
temperature of the product, 494
Pre-entry evaluation, 344
Pre-entry inspection, 415
Preincident survey
clues to presence of hazardous materials, 12, 13
defined, 323
Incident Action Plan development, 330
Preplan, 323
Pressure
atmospheric pressure barometer, 30
critical pressure, 44
predicting likely behavior, 493–494
states of matter and, 30, 40
temperature, relationship to, 44
vapor expansion, 41
vapor pressure, 40–41
Pressure differential pumps for product removal and transfer, 505–506
Pressure drum, 258–260
Pressure intermodal tank, 300
Pressure railway tank cars, 293–294
Pressure relief devices
on cylinders, 254
pressure relief valve, 290, 291
on railway tank cars, 290, 291
Pressure storage tanks, 312–313
Pressure tank cars, 282
Pressure test of CPC, 426

Pressure vessel, 312
Pressure-demand SCBA, 379
Pressurized containers, 507–513
    chlorine container valves, 510
    cylinder wall, 511
    fittings, 508–513, 538–540
    fusible plug threads, 511
    fusible plugs, 509–511
    shutoff device procedures, 508
    shutoff valve, 512
    valve blowout, 511–512
    valve gland, 512
    valve inlet thread, 512
    valve seat, 512
    valve stem blowout, 512–513
Primary explosive, 118
Process Safety Management of Highly Hazardous Chemicals regulation (29 *CFR* 1910.119), 218
Product control, 481–544
    blanketing/covering, 484, 517–518
    cargo tanks, 501–507
        bonding and grounding, 506–507, 537
        fire control, 504
        leak control, 502–504
        product removal and transfer, 505–507, 519, 536
    container behavior prediction, 492–494, 530–531
    container damage assessment, 486–492
    defensive operations, 482, 483–485, 522–529
    drums, 513–514
    hazardous liquid release incident, 485
    hazmat technician responsibilities, 23
    nonintervention, 483
    offensive operations, 482, 485–486
    overpacking, 23, 514–516, 541–544
    pressurized containers, 507–513
        chlorine container valves, 510
        cylinder wall, 511
        fittings, 508–513, 538–540
        fusible plug threads, 511
        fusible plugs, 509–511
        shutoff device procedures, 508
        shutoff valve, 512
        valve blowout, 511–512
        valve gland, 512
        valve inlet thread, 512
        valve seat, 512
        valve stem blowout, 512–513
    product containment
        patching. *See* Patching for materials containment
        plugging. *See* Plugging for materials containment
        selection of technique, 532
    purpose of, 481
    specialized techniques, 518–520
        cold tapping, 519
        flaring, 519–520
        hot tapping, 518
        product transfer, 519
        vent and burn, 520
        venting, 520
    vapor dispersion, 483, 517, 527
    vapor suppression, 517, 527
Product identification
    chemical assessment, 224
    interpreting information, 19
    sample collection for, 147, 155
    sources for clues, 336
Product information and behavior prediction, 217–218
Product removal and transfer, 505–507, 536
Product Safety Data Sheet (PSDS), 38

Propane A- and B-Kits, 500
Property protection as incident priority, 331
Protection
    property and the environment as incident priority, 331
    responder and public safety, 337–338
Proton, 49
Proximity suits, 389
PSBD (Public Safety Bomb Disposal), 243
PSDS (Product Safety Data Sheet), 38
Psychological stress of PPE, 384, 414
Public Safety Bomb Disposal (PSBD), 243
Purge, 136
Purpose of this manual, 1
Pyrene G-1, 91
Pyrophoric materials, 72, 88–89

Q
Q fever, 107

R
Radiation, 95–99
    activity, 99
    chemical assessment, 225
    decontamination, 467
    defined, 314
    detection instruments, 183–185
        dosimeters and badges, 184–185, 210
        Geiger-Mueller (GM) detector, 183–184
        hazardous materials profile, 221
        procedures for use, 208–209
        scintillation detectors, 184
    electromagnetic spectrum, 97
    Emergency Response Guidebook numbers, 226
    exposures, 168–169
    half-life, 98–99
    health hazards, 114
    infrared, 188
    inverse square law of radiation, 115
    ionizing, 96–98, 112–113
    neutron, 96, 113
    nonionizing, 96–98
    nuclear, 314
    penetrating powers of, 96
    protection from exposure, 114–115
    radioactive decay, 98
    thermal radiation thermometer, 189
Radiation absorbed dose (rad), 112, 169
Radiation survey meters, 168
Radiation thermometer, 189
Radio
    Incident Radio Communications (ICS Form 205), 329, 342
    Radio Requirement Work Sheet (ICS Form 216), 342
    use with encapsulating suit, 399
Radioactive and nuclear WMDs, 111–115
    exposure and contamination, 111–112
    health hazards, 114
    ionizing radiation, 112–113
    protection from exposure, 114–115
    radiation absorbed dose (rad), 112
Radioactive material (RAM)
    defined, 95
    packaging, 314–317
        excepted, 314
        industrial, 314
        Type A, 315
        Type B, 315
        Type C, 316–317
    radioactive isotope half-life, 98–99

uses for, 95
Radioactivity
    defined, 314
    field testing for, 147
Radioisotope, 98
Rail burn, container damage assessment, 491–492
Railcar initials and numbers, 282–283
Railway tank cars, 282–297
    boxcars, 296
    capacity stencil, 284
    cryogenic tank cars, 295–296
    Damage Assessment of Railroad Tank Cars Involved in Accidents, 489
    fittings, 289–291
        bottom outlet valves, 291
        gauging devices, 291
        manways, 289
        safety relief devices, 290, 291
        sample lines, 290
        thermometer wells, 291
        valves and venting devices, 289–290
    general service (nonpressure) tank cars, 291–293
    hopper car, 296, 297
    markings, 282–285, 286
    pneumatically unloaded hopper cars, 297
    pressure tank cars, 293–294
    product transfer, 519
    railcar initials and numbers, 282–283
    refrigerated cars, 297
    safety features, 287–291
        head shields, 287
        heating coils, 288
        insulation, 288
        skid protection, 289
        thermal protection, 288
        top and bottom shelf couplers, 288
    structure, 285–286
RAM. *See* Radioactive material (RAM)
Raman spectrometer, 195–196
Range of concentration of monitoring and detection equipment, 137
Rankine scale, 38
Rapid intervention crew or team (RIC/RIT), 345
Reactions, 70–77
    combustion, 73–74
    decomposition, 75
    emergency response considerations, 71
    endothermic, 71, 72
    exothermic, 71, 72
    fundamentals, 75–77
    mixing incompatible chemicals, 76
    oxidation, 72–74
    oxidation-reduction (redox) reaction, 72
    particulate size, 76–77
    polymerization, 74–75
        catalysts, 74–75, 76
        defined, 44, 74
        inhibitors, 75
    synergistic, 75
Reactive elements and compounds, 66
Reactive material, 88–94
    air-reactive, 88–89
    alkali metals, 61
    defined, 53, 88
    hazards, 88
    light-sensitive, 93
    magnesium, 62
    shock- and friction-sensitive, 92–93
    temperature-sensitive, 93–94
    water-reactive, 89–92

Reactivity
    chemical assessment, 225
    defined, 88
    product and container, 503
READY acronym, 415
Reagent papers
    generally, 181–182
    hazardous materials profile, 221
    procedures for use, 207
Recombination, 94
Recommended Exposure Limit (REL), 166, 168
Recommended Practice for the Classification of Combustible Dusts and of Hazardous (Classified) Locations for Electrical Installations in Chemical Process Areas (NFPA 499), 229
Recommended Practice for the Classification of Flammable Liquids, Gases, or Vapors and of Hazardous (Classified) Locations for Electrical Installations in Chemical Process Areas (NFPA 497), 229
Recon assessment, 221
Records. *See* Documentation
Recovery, 339–340
Recovery vessels, 501
Reducer, 72
Reducing agent, 72–73
Reference manuals. *See* Technical references
Refrigeration
    autorefrigeration, 41, 508
    liquid gas intermodal tanks, 301–302
    liquid tanks, 277
    liquids, 276, 277
    refrigerated cars, 297
Regulation 29 *CFR* 1910.132, *Personal Protective Equipment*, 387
Regulation 29 *CFR* 1910.156, *Fire Brigades*, 387
Regulation 40 *CFR* Part 311, *Worker Protection*, 387
Rehabilitation
    heat exposure prevention, 412
    NFPA 1584 standards, 344, 412
Reimbursement logs, 555
REL (Recommended Exposure Limit), 166, 168
Relative density, 35
Relative gas density (RgasD), 36
Relative response of monitoring and detection equipment, 140–142
Reliability of monitoring and detection equipment, 138
Removal of product, 505–507
Reporting marks, 282–283
Reports
    AAR. *See* After Action Report (AAR)
    exposure, 366–367, 466
    National Fire Incident Reporting System (NFIRS), 555
    terminating the incident, 24
Rescue, 338
Rescue device decontamination operations, 462–463
Research, 217–245
    chemical assessment, 224–225
    data interpretation for PPE selection, 218–220
    documentation, 220
    electronic technical resources, 234–240
        Chemical Hazards Emergency Medical Management (CHEMM), 237
        Computer-Aided Management of Emergency Operations (CAMEO), 234–236
        E-Plan for First Responders, 237
        Internet resources, 240
        mobile applications (apps), 237–238
        mobile devices, 234
        National Pipeline Mapping System (NPMS), 238–239
        Palmtop Emergency Action for Chemicals (PEAC®)-WMD, 237
        plume modeling, 239–240
        smart phones, 234
        weather data, 239

Wireless Information System for Emergency Responders (WISER), 236-237
facility documents, 233
hazardous materials profile, 221-223
Information and Research Officer, 360-361
interpreting hazard and response information, 245
product information and behavior prediction, 217-218
risk-based response, 217, 222
safety data sheet (SDS), 231-232
shipping papers, 232-233
specialist roles, 217
technical information, 220
technical information centers and specialists, 240-244
    chlorine emergency plan (CHLOREP), 242-243
    emergency response centers, 240-242
    EPA Emergency Response Team (ERT), 243
    Explosive Ordnance Disposal (EOD), 116, 243
    FBI WMD Coordinator (WMDC), 243-244
    federal resources, 244
    Interagency Radiological Assistance Plan (IRAP), 243
    poison control centers, 242
    WMD-Civil Support Teams, 243
technical references, 225-233
    Chemical Hazards Response Information System (CHRIS), 227
    cryogenic heat transfer, 231
    Emergency Handling of Hazardous Materials in Surface Transportation, 227
    Emergency Response Guidebook (ERG), 226-227. See also *Emergency Response Guidebook (ERG)*
    Fire Protection Guide to Hazardous Materials, 228-229
    Hawley's Condensed Chemical Dictionary (CCD), 229
    Jane's CBRN Response Handbook, 229
    Merck Index, 230
    NIOSH Pocket Guide to Chemical Hazards (NPG), 228
    reference manuals, 225-231, 236-237
    Sax's Dangerous Properties of Industrial Materials, 230-231
    Symbol Seeker: Hazard Identification Manual, 231
    Threshold Limit Values and Biological Exposure Indices (TLVs® & BEIs®), 231
Research Officer, 217
Resonant bonds, 68-69
Resources
  federal resources for hazardous materials incidents, 244
  references. See Technical references
  technical decontamination, 448
  technical information, 220
Respiratory protection, 375-385
  airline respiratory system, 380, 397
  APR. See Air-purifying respirator (APR)
  emergency breathing support system (EBSS), 380
  equipment limitations, 384-385
  PAPR. See Powered air-purifying respirator (PAPR)
  powered air-purifying respirator (PAPR), 383
  Respiratory Protection Management Program, 394
  SCBA. See Self-contained breathing apparatus (SCBA)
  standards at hazmat/WMD incidents, 377-378
  supplied air respirator (SAR), 376, 380
  Type C respirator, 380
Responders, OSHA definition, 2
Response curve, 169
Response models, 326-329
  APIE, 326
  D.E.C.I.D.E., 327
  defined, 326
  Eight Step Incident Management Process©, 327
  Emergency Response Guidebook (ERG), 328-329
  GEDAPER, 327
  HazMatIQ©, 327
Response modes, 331-334
  defensive, 331, 333-334
  nonintervention, 331, 332-333
  offensive, 331, 334
  strategic mode selection factors, 331-332
    size/complexity, 332
    time, 332
    value, 332
Response objectives and options, 15, 21
Retention for product control, 526
RgasD (relative gas density), 36
RIBC (rigid intermediate bulk container), 257, 258
Rickettsia, 106, 107
RIC/RIT (rapid intervention crew or team), 345
Rigid intermediate bulk container (RIBC), 257, 258
Rings, 49, 50
Riot control agents/irritants, 105-106
Risk vs. harm, 326
Risk-based response
  decision chart, 217
  defined, 217
  hazardous materials incidents, 323-325
  hazardous materials profile, 222
  for hazmat incidents, 21
  offensive operations, 485
  technical research, 217
RNA fluoroscopy, 199
Road burn, container damage assessment, 491-492
Roentgen (R), 169
Roentgen equivalent in man (rem), 169
Roles, Responsibilities, and Authority (ICS Form 203), 329, 342
Roll on/roll off vessel, 304
Rolling slide-in overpacking method, 515-516
Routes of entry, 161-162
  absorption, 162
  defined, 161
  ingestion, 161-162
  inhalation, 161, 224
  injection, 162
Royal Society of Chemistry, 230
Rust on metals, 73, 75

S

Saddle patch, 498
SADT (self-accelerating decomposition temperature), 44, 93-94
Safe Refuge Area Manager, 359-360
Safety briefing, 416
Safety buffer, 420-421
Safety data sheet (SDS)
  chemical reaction information, 76
  decomposition temperature, 94
  defined, 38
  information dissemination from, 218
  information included, 38, 231
  sections, 232
  solubility information, 78
Safety Officer, 358-359
Safety relief devices, 290, 291
Salt water
  concentration, 77
  mixtures, 55-56
Salts
  bases and water to form, 94
  binary, 53
  characteristics, 53-54
  examples, 68
  hazards, 54
Sampling
  chain of custody, 148, 150

collection for hazard identification, 147, 155
collection procedures, 154
decontamination of, 152
evidence, 146-147
field screening, 147-150
hazmat technician training, 19
maintenance of equipment, 152-153, 156
minimum testing for, 147
plans, 147-149
sample lines, 290
team qualifications, 149
wipe sampling, 456
SAR. *See* Supplied air respirator (SAR)
SARA Title III, 233, 234
Saranex/Tyvek®, 395
Sarin (GB), 100, 101, 379, 468
Satchel bomb, 124
Saturation, 137
SAW (surface acoustic wave) sensor, 192-193
Sax's Dangerous Properties of Industrial Materials, 230-231
SBCCOM (U.S. Army Soldier and Biological Chemical Command), 379
SCADA (Supervisory Control and Data Acquisition), 308
SCBA. *See* Self-contained breathing apparatus (SCBA)
Science Officer, 217
Scintillation detectors, 184
Scintillator, 113
Scope of this manual, 1
Scores in containers, 490-491
SCT (Secretariat of Transport and Communications of Mexico), Emergency Response Guidebook (ERG), 226
SCT-306, nonpressure cargo tanks, 265
SCT-307, low-pressure cargo tanks, 269
SCT-331, high-pressure cargo tanks, 275
SCT-338, cryogenic liquid tank, 278
SCT-412, corrosive liquid tank, 272
SDS. *See* Safety data sheet (SDS)
Secondary explosive, 118
Secretariat of Transport and Communications of Mexico (SCT), Emergency Response Guidebook (ERG), 226
Section Leaders and Supervisors, 356-361
    Decontamination Leader, 358
    Entry Team Leader, 357-358
    Hazmat Branch Director/Group Supervisor, 356-357
    Hazmat Logistics Officer, 359
    Information and Research Officer, 360-361
    Medical Officer, 360
    Safe Refuge Area Manager, 359-360
    Safety Officer, 358-359
    Site Access Control Leader, 358
    structure, 356
Section level critique, 554
Security
    decontamination operations, 463, 465
    Department of Homeland Security (DHS)
        hazmat technician training, 18
        protective clothing and equipment standards, 386
        respiratory protection standards, 377
    ICS Form 201 (Site Security, Control, and Communications), 329, 342
Segregation of the decontamination line, 461
Selectivity of monitoring and detection equipment, 137
Self-accelerating decomposition temperature (SADT), 44, 93-94
Self-contained breathing apparatus (SCBA)
    air management, 417
    breathing air cylinder capacities, 421
    certification, 379
    closed-circuit
        air supply limitations, 384
        high-pressure, 379
        low-pressure, 379
        positive-pressure, 379
        pressure-demand, 379
        purifying exhaled air, 375
    components, 378-379
    defined, 378
    IDLH certification, 379
    open-circuit
        air supply limitations, 384
        high-pressure, 379
        limitations, 377
        low-pressure, 379
        maintenance, 378
        mechanics, 376
        NFPA standards, 378, 379
        NIOSH standards, 377
        positive-pressure, 379
        pressure-demand, 379
    worn with liquid splash-protective clothing, 392
    worn with vapor-protective ensembles, 393
Self-reading dosimeter (SRD), 184-185
Semiconductors in the Periodic Table of Elements, 59
Senses and clues to presence of hazardous materials, 12
Sensitivity of monitoring and detection equipment, 137
Sensor-based detection instruments, 170-189
    biological immunoassay indicators, 187
    colorimetric methods, 176-183
    combustible gas indicators (CGIs), 172-173, 222
    correction factors, 173-175
    electrochemical cell sensors, 175
    infrared thermometers, 188-189, 212
    metal oxide sensors, 175
    multi sensor instruments, 176, 202-203
    oxygen indicators, 170-172
    pH meters, 176, 201
    photoionization detector (PID)
        calibration of monitoring instruments, 186-187
        colorimetric indicator tube, 177
        correction factors, 173-175
        defined, 94
        hazardous materials profile, 222
        ionization potential and lamp strength, 186
        limitations, 186
        mechanism, 185-186
        procedures for use, 211
        purpose of, 185
    radiation, 183-185, 221
    thermal imagers, 187-188
SETIQ (Emergency Transportation System for the Chemical Industry), 241, 242
Shelf couplers on railway tank cars, 288
Shell, 49, 50
Shielding, to limit radiation exposure, 115
Shipping papers, 232-233
Shock-sensitive reactive materials, 92-93
Short-Term Exposure Limit (STEL), 165
Shower for decontamination, 461, 462, 465
Shutoff devices, pressurized containers, 508, 512
SI. *See* International System of Units (SI)
Sievert (Sv), 169
Signals, hand, used for communication, 418-419, 444
Site access control, 358-359
Site Access Control Leader, 358
Site safety plan, 341-345
    backup and rapid intervention, 345
    components, 342-343
    emergency procedures, 345
    ICS Form 208 HM, 330, 341
    medical surveillance plan, 343-344
    post-entry evaluation, 344-345

pre-entry evaluation, 344
Site Security, Control, and Communications (ICS Form 201), 329, 342
Site selection for decontamination operations, 457–458
Site sketch (ICS Form 201), 329, 342
16 Life Safety Initiatives, 343
Size/complexity factor in selection of response mode, 332
Size-up, damage assessment, 487
SKED®, 462
Skid protection, 289
Slide-in overpacking method, 515
Slip-over overpacking method, 516
SLUDGEM indicators, 102
Slurry, 56
SMART modeling, 335
Smart phones, 234
Smell of chemicals, 37
Soap bubble test of CPC, 426
Soap polarity, 79
Societal restoration as incident priority, 331
Sodium
    characteristics, 61
    compounds, 51
    ionic bonds, 67–68
    molecular structure, 50
    unstable atoms, 65
Sodium chloride
    binary salts, 53
    compounds, 51
    oxidation-reduction reaction, 72
    uses for, 68
Sodium fluoride, 68
Sodium hydroxide, 90
Soft patch, 498
Soldier and Biological Chemical Command (SBCCOM), 379
Solid
    characteristics, 31
    defined, 29
    initial isolation distance, 227
    intermodal tanks for hazardous materials, 300
    mobility, 30
    physical states, 29–31
    solubility, 79
Solidification decontamination, 447, 453
Solubility, 78
Soluble materials
    decontamination, 459
    defined, 78
Solution containers, 462
Solutions, 55–56
Solvent, polar, 78
Soman (GD), 100, 101
SOP/G. *See* Standard operating procedures/guidelines (SOP/Gs)
Spec. 51 portable tank, 299, 300
Specialists. *See* Technical information centers and specialists
Specific gravity, 35–36
Specification markings, 283–284
Specification plates, 261–263
Specificity of monitoring and detection equipment, 137
Spectrography, 191
Spectrometer, 191
Spectrophotometer, 194–195
Spectroscopy, 191
Spherical pressure vessel, 312
Spheroid tank, 312
Spill control, 482
Sports, energy expenditures during, 422
SRD (self-reading dosimeter), 184–185
Stability of elements and compounds, 65–66
Stabilizer, 75

Stabilizing the container for product removal and transfer, 506
Stabilizing the incident as incident priority, 331
Stainless containers, 488
Standard deviation, 177
Standard for Chemical, Biological, Radiological, and Nuclear (CBRN) Air-Purifying Escape Respirator and CBRN Self-Contained Escape Respirator, 378
Standard for Chemical, Biological, Radiological, and Nuclear (CBRN) Full Facepiece Air-Purifying Respirator (APR), 377
Standard for Competencies for EMS Personnel Responding to Hazardous Materials/Weapons of Mass Destruction Incidents (2008) (NFPA 473), 18
Standard for Hazardous Materials/Weapons of Mass Destruction Emergency Response Personnel Professional Qualifications, 2016 Edition. See NFPA 1072, *Standard for Hazardous Materials/Weapons of Mass Destruction Emergency Response Personnel Professional Qualifications, 2016 Edition*
Standard for Professional Competence of Responders to Hazardous Materials/Weapons of Mass Destruction Incidents (2008). See NFPA 472, *Standard for Professional Competence of Responders to Hazardous Materials/Weapons of Mass Destruction Incidents (2008)*
Standard on Comprehensive Occupational Medical Program for Fire Department (NFPA 1582), 344
Standard on Emergency Services Incident Management System and Command Safety (NFPA 1561), 353
Standard on Fire Department Occupational Safety and Health Program (NFPA 1500), 344, 358
Standard on Liquid Splash-Protective Ensembles and Clothing for Hazardous Materials Emergencies. See NFPA 1992, *Standard on Liquid Splash-Protective Ensembles and Clothing for Hazardous Materials Emergencies*
Standard on Open-Circuit Self-Contained Breathing Apparatus (SCBA) for Emergency Services (NFPA 1981), 378, 379
Standard on Protective Ensembles for First Responders to CBRN Terrorism Incidents. See NFPA 1994, *Standard on Protective Ensembles for First Responders to CBRN Terrorism Incidents*
Standard on Selection, Care, and Maintenance of Open-Circuit Self-Contained Breathing Apparatus (SCBA) (NFPA 1852), 378
Standard on the Rehabilitation Process for Members during Emergency Operations and Training Exercises (NFPA 1584), 344, 412
Standard on Vapor-Protective Ensembles for Hazardous Materials Emergencies. See NFPA 1991, *Standard on Vapor-Protective Ensembles for Hazardous Materials Emergencies*
Standard operating procedures/guidelines (SOP/Gs)
    air management calculations, 417
    chemical-protective clothing, 394, 395
    emergency response plans, 336
    military guidelines for using PPE, 396
    planning the response, 13
Standard System for the Identification of the Hazards of Materials for Emergency Response. See NFPA 704, *Standard System for the Identification of the Hazards of Materials for Emergency Response*
Standard Transportation Commodity Code (STCC)
    defined, 225
    emergency handling of hazardous materials, 227
    product identification, 224
    shipping papers, 232
    STCC Number, 225, 232
Stannic chloride, 90
States of matter. *See* Physical states of matter
Static electricity, 506
STCC. *See* Standard Transportation Commodity Code (STCC)
Steel
    austenitic stainless, 488
    carbon steel, 487
    composition, 56
    container dents, 490
    containers, 487–488
    high strength low alloy, 488
    mild steel, 487
STEL (Short-Term Exposure Limit), 165

Stenciled commodity names, 286
Sterilization decontamination, 447, 453–454
Stokes basket, 462
Storage of chemical protective clothing, 426–427, 434
Strategic goals and tactical objectives, 335–340
    confinement, 338–339
    containment, 339
    fire control, 338
    identification, 336
    isolation, 335–336
    notification, 336–337
    protection, 337–338
    recovery and termination, 339–340
    rescue, 338
Strategic mode selection factors, 331–332
Stress, 410–414
    container physical damage, 488
    mechanical stress on containers, 492
    PPE-related
        cold emergencies, 413–414
        heat emergencies, 411
        heat-exposure prevention, 411–413
        psychological issues, 384, 414
Structural firefighters' protective clothing, 388, 403, 430
Stub sill tank car, 285
Styrene, 69, 85
Sublimation, 43
Suicide bomb, 126–128
Suicide vest, 127
Sulfur, 50
Sunstroke, 411
Supersacks, 257
Supervisors. *See* Section Leaders and Supervisors
Supervisory Control and Data Acquisition (SCADA), 308
Supplied air respirator (SAR)
    air compressors, 376
    components, 380
    defined, 380
    limitations, 380
    SAR with escape (egress), 380
    Type C respirator, 380
Supplied-air hoods, 384
Surface acoustic wave (SAW) sensor, 192–193
Surfactant, 455
Survey. *See* Preincident survey
Symbol Seeker: Hazard Identification Manual, 231
Synergistic effect, 75

T

T-50 intermodal tank, 300
T-75 tank, 301
Tabun (GA), 100, 101
Tactile inspection of CPC, 426
Tailboard review, 550
Tank, 253
Tank markings, 261
Tank motor vehicle, 260
Tank truck, 260
Tankers
    chemical carriers, 303
    liquefied flammable gas carriers, 303–304
    petroleum carriers, 302
Tapping for product control, 518–519
Targeted attacks, 122
Tarps for equipment drop, 462
TATP (acetone peroxide), 120
TC. *See* Transport Canada (TC)
TC 306, nonpressure cargo tanks, 265
TC 307, low-pressure cargo tanks, 269
TC 312, corrosive liquid tank, 272
TC 331, high-pressure cargo tanks, 275
TC 338, cryogenic liquid tank, 278
TC 341, cryogenic liquid tank, 278
TC 406, nonpressure cargo tanks, 260, 265
TC 407, low-pressure cargo tanks, 269
TC 412, corrosive liquid tank, 272
T-Code, 298–299
Tear gas, 95
Technical decontamination
    advantages and disadvantages of methods, 446–447
    defined, 446
    procedures, 442, 473–475
    selection of correct procedures, 448
    uses for, 468
Technical information centers and specialists, 240–244
    chlorine emergency plan (CHLOREP), 242–243
    emergency response centers, 240–242
    EPA Emergency Response Team (ERT), 243
    Explosive Ordnance Disposal (EOD), 116, 243
    FBI WMD Coordinator (WMDC), 243–244
    federal resources, 244
    Interagency Radiological Assistance Plan (IRAP), 243
    poison control centers, 242
    WMD-Civil Support Teams, 243
Technical markings as clue to presence of hazardous materials, 12, 13
Technical references, 225–233
    Chemical Hazards Response Information System (CHRIS), 227
    cryogenic heat transfer, 231
    Emergency Handling of Hazardous Materials in Surface Transportation, 227
    Emergency Response Guidebook (ERG), 226–227. *See also Emergency Response Guidebook (ERG)*
    Fire Protection Guide to Hazardous Materials, 228–229
    Hawley's Condensed Chemical Dictionary (CCD), 229
    Jane's CBRN Response Handbook, 229
    Merck Index, 230
    NIOSH Pocket Guide to Chemical Hazards (NPG), 228
    reference manuals, 225–231, 236–237
    Sax's Dangerous Properties of Industrial Materials, 230–231
    Symbol Seeker: Hazard Identification Manual, 231
    Threshold Limit Values and Biological Exposure Indices (TLVs® & BEIs®), 231
Technical rescue, high-angle, 403
Technical research. *See* Research
Technical Specialist, 217
Technician Level responders. *See* Hazardous materials technician
Technology
    electronic technical resources, 234–240
    monitoring and detection equipment, 136–138, 170
    paperless hazard communication program, 233
TECP (totally encapsulating chemical-protective) clothing, 393
TEEL-0 (Temporary Emergency Exposure Limits, Level 0), 167
TEEL-1 (Temporary Emergency Exposure Limits, Level 1), 167
TEEL-2 (Temporary Emergency Exposure Limits, Level 2), 167
TEEL-3 (Temporary Emergency Exposure Limits, Level 3), 167
Temperature
    autoignition temperature, 44
    Celsius scale, 38
    critical temperature, 44
    decomposition temperature, 94
    effect on colorimetric tubes/chips, 178
    Fahrenheit scale, 38
    flammable range, 38–40
    flash points and fire points, 38, 39
    high temperature-protective clothing, 389–390
    maximum safe storage temperature (MSST), 44, 93
    phase changes, 42–44

boiling and condensation, 42–43
    critical points, 44
    freezing, 42
    melting, 42
    sublimation, 43
  predicting likely behavior, 494
  pressure, relationship to, 44
  self-accelerating decomposition temperature (SADT), 44, 93–94
  states of matter and, 30, 31
  temperature-sensitive reactive materials, 93–94
  thermal detector, 222
  vapor pressure relationship with, 41
Temperature gun, 188–189, 212
Temporary Emergency Exposure Limits, Level 0 (TEEL-0), 167
Temporary Emergency Exposure Limits, Level 1 (TEEL-1), 167
Temporary Emergency Exposure Limits, Level 2 (TEEL-2), 167
Temporary Emergency Exposure Limits, Level 3 (TEEL-3), 167
Tennis ball bombs, 125
Terminating the incident
  After Action Report (AAR), 340, 554–555
  after decon operations, 466
  debriefing, 550–552
  defined, 550
  delayed exposure symptoms, 551
  documentation, 555–556
  hazmat technician responsibilities, 24–25
  postincident critique, 552–554, 557
  strategic goals, 339–340
  termination, defined, 339
Terminology, 2–3
Terrain considerations for decontamination operations, 457
Terrorism
  anthrax incidents in the U.S., 468
  decontamination during, 468
  federal assistance for incidents, 107
  law enforcement ensembles, 404
  9/11, 468
  protective clothing standards, 387
  sarin attacks in Japan, 468
  WMDs. *See* Weapons of mass destruction (WMDs)
Testing
  chemical protective clothing (CPC), 425, 426, 434
  corrosives, 147
  explosives and incendiaries, 147
  field screening samples, 147
  flammability, 147
  oxidizer, 147
  personal protective equipment, 426, 434
  radioactivity, 147
  sampling equipment, 147
Thermal burn, 95
Thermal detector, 222
Thermal imager (TI), 187–188
Thermal imaging camera (TIC), 187–188
Thermal insulation, 288
Thermal protection
  defined, 288
  PPE selection factors, 408
  purpose of, 288
Thermal radiation thermometer, 189
Thermometer, infrared, 188–189
Thermometer wells, 291
Threshold Limit Value® (TLV®), 165, 168
Threshold Limit Value®-Ceiling (TLV®-C), 166, 168
Threshold Limit Value®-Short-Term Exposure Limit (TLV®-STEL), 165, 168
Threshold Limit Value®-Time-Weighted Average (TLV®-TWA), 165, 168
Threshold Limit Values and Biological Exposure Indices (TLVs® & BEIs®), 231

TI (thermal imager), 187–188
TI (transport index), 317
TIC (thermal imaging camera), 187–188
Tic tracer leak detector, 189
TIH. *See* Toxic inhalation hazard (TIH)
Time, to limit radiation exposure, 115
Time factor in selection of response mode, 332
Title 29 *CFR* 1910.119, Process Safety Management of Highly Hazardous Chemicals, 218
Title 29 CFR 1910.120. *See* Hazardous Waste Operations and Emergency Response (HAZWOPER, Title 29 CFR 1910.120)
Title 29 *CFR* 1910.1000, Toxic and Hazardous Substances, 344
Title 29 *CFR* 1910.1020, Access to Employee Exposure and Medical Records, 367
Title 29 *CFR* 1910.1200, Hazard Communication regulation, 218
Title 49 *CFR 178*, intermodal tanks, 299
Title 49 *CFR* 178.3, drums, 252
Title 49 *CFR* 178.37, Y cylinders, 256
Title 49 *CFR*, bulk and non-bulk packaging, 250
Title 49 *CFR Part 179*, railway tank cars, 293, 294
TLV® (Threshold Limit Value®), 165, 168
TLV®-C (Threshold Limit Value-Ceiling), 166, 168
TLV®-STEL (Threshold Limit Value®-Short-Term Exposure Limit), 165, 168
TLV®-TWA (Threshold Limit Value®-Time-Weighted Average), 165, 168
TLVs® & BEIs® (Threshold Limit Value® and Biological Exposure Indices), 231
TNT, 116–117
Toluene, 85, 116
Ton container (pressure drum), 258–260
Top and bottom shelf couplers, 288
Torr, defined, 32
Totally encapsulating chemical-protective (TECP) clothing, 393
Tote bags, 257
Totes, 256, 257
Toulene, 69
Toxic and Hazardous Substances (OSHA 29 CFR 1910.1000), 344
Toxic gases, 90
Toxic inhalation hazard (TIH)
  chemical assessment, 224
  defined, 252
  inhalation route of entry, 161
Toxicity
  decontamination factor, 459
  defined, 108
  hazards, 224
  measuring, 109
Toxicology, 107
Toxin, 106, 107
Trailers
  compressed gas tube trailer, 260, 280
  dry bulk, 281
  helium transported via tube trailers, 279
  highway tube trailers, 279–280
Train
  cars. *See* Railway tank cars
  consist, 283
  derailment in Cherry Valley, IL, 333
  derailment in Farragut, TN, 334
  hazard and risk assessment, 323
Training
  hazmat technician, 17–18
    NFPA standards, 17–18
    OSHA standards, 17
    sources, 18
  monitoring and detection devices, 19
  personal protective equipment, 423
  Safety Officer, 358
Transfer of product, 505–507, 519, 536
Transmutation, 98

Transport Canada (TC)
    CANUTEC. *See* Canadian Transportation Emergency Centre (CANUTEC)
    Emergency Handling of Hazardous Materials in Surface Transportation, 227
    Emergency Response Guidebook (ERG), 226
    railway tank car specification markings, 283–284
Transport index (TI), 317
Transportation, *Emergency Handling of Hazardous Materials in Surface Transportation*, 227
Transportation of Dangerous Goods Directorate, 240
Travel time, 421
Trench foot, 413
Tube trailers, 279–280
Type A packaging, 315
Type B packaging, 315
Type C packaging, 316–317
Type C respirator, 380
Tyvek®, 395

U

U238 half-life, 98–99
UC. *See* Unified Command (UC)
UEL (upper explosive limit), 39
UFL (upper flammable limit), 39
UN Recommendations, 256–257
Underground storage tank, 311
Underwriters Laboratories certification of monitoring and detection equipment, 136
Undressing area, 461
Unified Command (UC)
    defined, 355
    Incident Action Plan, 356
    structure, 355
Uniform Hazardous Waste Manifest, 232
Unit/Activity Log (ICS Form 214), 356
United Nations Recommendations on the Transportation of Dangerous Goods, 256–257
United States
    anthrax incidents, 468
    Army Soldier and Biological Chemical Command (SBCCOM), 379
    Coast Guard, 227
    Department of Health and Human Services, 237
    DOE. *See* Department of Energy (DOE)
    DOT. *See* Department of Transportation (DOT)
    Federal Bureau of Investigation, Weapons of Mass Destruction Coordinator (WMDC), 243–244
    Federal Railroad Administration (FRA), 489
    liquefied flammable gas carrier regulations, 304
    National Response Team, 361–362
    Office of Pipeline Safety (OPS), 238
    pipelines, 306–307
    U.S. Fire Administration, 555
    Weapon of Mass Destruction-Civil Support Team (WMD-CST), 243
UN/NA datasheet, 234
UN/NA number, 224, 227
Unstable material, 66
Upper explosive limit (UEL), 39
Upper flammable limit (UFL), 39

V

V agent (VX), 100, 101–102
Vacuum relief valve, 290, 291
Vacuum trucks for product removal and transfer, 506
Vacuuming decontamination, 447, 454
Value factor in selection of response mode, 332
Valve
    airline respirators, 397
    Betts, 506
    blowout, 511–512
    bottom outlet valves, 291
    butterfly, 509
    chlorine container valves, 510
    defined, 289
    gas main, 509
    gland, 512
    inlet thread, 512
    pipelines, 308
    post indicator, 509
    pressure relief valve, 254, 290, 291
    on railway tank cars, 289–290
    remote shutoff, 508–509, 538–539
    seat, 512
    shutoff, 512
    stem blowout, 512–513
    vacuum relief valve, 290, 291
Vapor, mobile like a gas, 31
Vapor content, boiling point relationship to flash point, 45
Vapor density, 36–37
Vapor dispersion, 483, 517, 527
Vapor expansion, 41
Vapor pressure, 40–41, 45
Vapor suppression, 517, 527
Vapor-and-gas-removing air-purifying respirator (APR), 376, 383
Vapordome roof tank, 311
Vapor-protective clothing
    defined, 393
    donning and doffing, 433
    limitations, 393–394
    NFPA 1991 standards, 386, 393
    safety procedures, 23
    using with SCBA/SAR, 393
VBIEDs (Vehicle-Borne Improvised Explosive Devices), 128–129
Vehicle-Borne Improvised Explosive Devices (VBIEDs), 128–129
Vent and burn of hazardous materials, 520
Venting devices on railway tank cars, 289–290
Venting of hazardous materials, 520
Vesicants, 103
Vetter®, 520
Virus, 106
Viscosity, 37
Visual clues to presence of hazardous materials, 12
Visual inspection
    chemical protective clothing (CPC), 426
    decontamination effectiveness, 456
Visual Plumes, 239
V-method of overpacking, 515–516
Volatility, 43, 147

W

Wash station, 461
Washing decontamination, 447, 455
Wastewater detector strip, 182
Water
    boiling point, 42–43
    compounds, 53
    decontamination with, 444
    electrolysis, 75
    heat exposure prevention, 412
    molecular structure, 50
    pH concentration, 179
    reactive materials. *See* Water-reactive materials
    salt water
        concentration, 77
        mixtures, 55–56
    solubility, 78, 79

specific gravity, 36
supply as decontamination operation consideration, 458
transport for decontamination operations, 463
viscosity, 37
Water-reactive materials, 57, 89–92
    Class D fires and extinguishers, 91–92
    corrosive solutions, 90
    decontamination, 459
    defined, 72
    fire extinguishment, 90–91
    flammable gas generation, 89
    heat generation, 90
    response to fires with, 90
    toxic gases, 90
Waterways as decontamination operation consideration, 457–458
Waybill, 266, 292, 306
Weapons of mass destruction (WMDs)
    biological agents and toxins, 106–111
        bacteria, 106
        defined, 106
        dose-response relationships, 108
        exposures, 169
        federal assistance for incidents, 107
        incubation period after exposure, 169
        infectious dose, 169
        measuring and expressing dose and concentration, 108–109
        pesticides and agricultural chemicals, 110–111
        rickettsia, 107
        toxicity, 109
        toxicology, 107
        toxins, 107
        viruses, 106
    B-NICE, 100
    CBRNE, 99, 244
    chemical warfare agents, 100–106
        blister agents/vesicants, 103
        blood agents, 104
        choking agents, 104
        nerve agents, 100–103
        riot control agents/irritants, 105–106
    COBRA, 100
    decontamination method selection, 476
    explosives and incendiaries, 116–118
        categories, 118
        divisions, 117
        examples, 116
        high explosives, 118
        improvised explosive devices, 116
        low explosives, 117–118
        primary explosives, 118
        secondary explosives, 118
        understanding the danger, 116–117
    hazardous materials, 3. *See also* Hazardous materials
    hazards, 99
    homemade explosives/improvised explosive devices, 118–121
        chlorate-based explosives, 121
        components, 119
        nitrate-based explosives, 121
        peroxide-based explosives, 120
    improvised explosive devices (IEDs), 121–130
        carbon dioxide grenades, 125
        containers, 121, 123
        defined, 121
        fireworks, 125
        identification, 122–123
        letter bombs, 125–126
        mail bombs, 116, 125–126
        M-devices, 125
        munitions, 122
        package bombs, 125–126
        Person-Borne Improvised Explosive Devices (PBIEDs), 126–128
        pipe bombs, 116, 123–124
        plastic bottle bombs, 124
        response to incidents, 129–130
        satchel, backpack, knapsack, duffle bag, briefcase, box bombs, 124
        targeted attacks, 122
        tennis ball bombs, 125
        Vehicle-Borne Improvised Explosive Devices (VBIEDs), 128–129
    NBC (nuclear, biological, chemical), 100
    Palmtop Emergency Action for Chemicals (PEAC®)-WMD, 237
    product control, 532
    protective clothing, 386–387
    radioactive and nuclear, 111–115
    ranking of threats, 100
    responder competence, 17–18. *See also* NFPA 472, *Standard for Professional Competence of Responders to Hazardous Materials/Weapons of Mass Destruction Incidents (2008)*
    threat spectrum, 99–100
    Weapon of Mass Destruction-Civil Support Team (WMD-CST), 243
    Weapons of Mass Destruction Coordinator (WMDC), 243–244
Weather. *See also* Environment
    cold emergencies, 413–414
    cold weather decontamination, 464–465
    decontamination considerations, 457
    heat emergencies, 411–413
    impact on properties of materials, 31
    technical resources, 239
    weather stations for data, 239
    wind direction considerations for decontamination, 457
Welds, container damage assessment, 491–492
Wet chemistry, 187, 198–199
Wet decontamination, 440, 441
Wheel burn, container damage assessment, 491–492
WHO (World Health Organization), ionizing radiation as carcinogenic, 114
Wind chill, 413, 414
Wind direction considerations for decontamination operations, 457
Wipe sampling, 456
Wireless Information System for Emergency Responders (WISER), 236–237
WISER (Wireless Information System for Emergency Responders), 236–237
WMDC (Weapons of Mass Destruction Coordinator), 243–244
WMD-CST (Weapon of Mass Destruction-Civil Support Team), 243
WMDs. *See* Weapons of mass destruction (WMDs)
Wood boxes, 256
Wooden wedges used for plugging, 495
Work mission statement, 422
Work rotation for heat exposure prevention, 412
Work time duration, 420–422
Worker Protection (EPA Regulation 40 CFR Part 311), 387
Workload and air supply duration, 422
World Health Organization (WHO), ionizing radiation as carcinogenic, 114
Written PPE plan, 427–428
Written resources as clue to presence of hazardous materials, 12, 13

X

X-rays
    chronic exposure to, 163
    defined, 97
    technician response, 113
XT boxcar, 296

Y

Y cylinders, 256
Yvorra, James G., 327